Series on Chemistry, Energy and the Environment

Topics in Enantioselective Catalysis
Recent Achievements and Future Challenges

Series on Chemistry, Energy and the Environment

ISSN: 2529-7716

Series Editors: Karl M. Kadish *(University of Houston, USA)*
Roger Guilard *(University of Bourgogne, France)*

The aim of this series on "Chemistry, Energy and the Environment" is to bring together authoritative contributions where the multidisciplinary character of Chemistry and its relationship to energy and the environment can be illustrated. In each volume, the latest advances in bio-, energy- and environmentally-related fields are described in a way that unambiguously shows the major contributions from chemists at the top of their field who describe new and improved technologies for the future.

Published

Vol. 9   *Topics in Enantioselective Catalysis: Recent Achievements and Future Challenges*
edited by Angela Marinetti

Vol. 8   *Progress in Layered Double Hydroxides: From Synthesis to New Applications*
edited by Morena Nocchetti and Umberto Costantino

Vol. 7   *Supramolecular Catalysts: Design, Fabrication and Applications*
edited by Leyong Wang and Cheng-Yong Su

Vol. 6   *Advanced Green Chemistry*
Part 2: From Catalysis to Chemistry Frontiers
edited by István T. Horváth and Max Malacria

Vol. 5   *Bioinspired Chemistry: From Enzymes to Synthetic Models*
edited by Marius Réglier

Vol. 4   *Prospects for Li-ion Batteries and Emerging Energy Electrochemical Systems*
edited by Laure Monconduit and Laurence Croguennec

For the complete list of volumes in this series, please visit
www.worldscientific.com/series/scee

∾ Volume 9 ∽

Series on Chemistry, Energy and the Environment

Topics in Enantioselective Catalysis
Recent Achievements and Future Challenges

Edited by

Angela Marinetti
CNRS-ICSN, France & Paris-Saclay University, France

Series Editors
Karl M. Kadish
University of Houston, USA

Roger Guilard
Université de Bourgogne, France

World Scientific

NEW JERSEY · LONDON · SINGAPORE · BEIJING · SHANGHAI · HONG KONG · TAIPEI · CHENNAI · TOKYO

Published by

World Scientific Publishing Co. Pte. Ltd.
5 Toh Tuck Link, Singapore 596224
USA office: 27 Warren Street, Suite 401-402, Hackensack, NJ 07601
UK office: 57 Shelton Street, Covent Garden, London WC2H 9HE

British Library Cataloguing-in-Publication Data
A catalogue record for this book is available from the British Library.

Series on Chemistry, Energy and the Environment — Vol. 9
TOPICS IN ENANTIOSELECTIVE CATALYSIS
Recent Achievements and Future Challenges

Copyright © 2022 by World Scientific Publishing Co. Pte. Ltd.

All rights reserved. This book, or parts thereof, may not be reproduced in any form or by any means, electronic or mechanical, including photocopying, recording or any information storage and retrieval system now known or to be invented, without written permission from the publisher.

For photocopying of material in this volume, please pay a copying fee through the Copyright Clearance Center, Inc., 222 Rosewood Drive, Danvers, MA 01923, USA. In this case permission to photocopy is not required from the publisher.

ISBN 978-981-124-842-9 (hardcover)
ISBN 978-981-124-843-6 (ebook for institutions)
ISBN 978-981-124-844-3 (ebook for individuals)

For any available supplementary material, please visit
https://www.worldscientific.com/worldscibooks/10.1142/12589#t=suppl

Typeset by Stallion Press
Email: enquiries@stallionpress.com

Printed in Singapore

Preface

The stereochemical control of chiral units is undoubtedly the most advanced and challenging level of selectivity to be achieved when complex organic molecules are targeted. Toward this well-recognized goal, enantioselective organo- and organometallic catalysis represent privileged tools as far as they make use of sub-stoechiometric amounts of chiral auxiliaries, while possibly attaining extremely high levels of enantioselectivity. Enantioselective catalytic methods notably entail the same practical, economical and environmentally benign features of other catalytic processes that have been especially valued from the very beginning by industrial stakeholders, and even more with the emergence of sustainable development concerns. Decades of active research have led some branches of enantioselective catalysis to reach full maturity, including practical academic and industrial applications, as typified notably by enantioselective metal-promoted hydrogenation reactions. Nevertheless, the continuous opening of new frontiers in synthesis endlessly creates new needs for chiral catalysts, new concepts and methods. Overall, this state of things still motivates extensive fundamental research.

In this context, books or series dedicated to Enantioselective Catalysis can only give a glimpse, a few snapshots of this extremely wide and diverse field. Therefore, to be as far as possible representative of the area, of its state of art and current challenges, this book covers rather disparate topics by assembling short contributions from many authors in their respective fields of expertise. The book is organized into three parts that are devoted to ligands or catalysts design, to recent catalytic approaches and to selected classes of reactions, respectively.

In the first part of this book (Chapters 1–7) that relates to catalyst design and catalytic methods, mainstream topics in enantioselective catalysis have been voluntarily discarded so as to illustrate exploratory fields and shed light on some emerging strategies, without prejudging of their future developments and achievements.

Thus, modern catalyst design is typified first by bio-hybrid DNA-based metal complexes. These catalysts take advantage from the helical chirality of DNA strands, and their covalent or supramolecular bonding to the active metal center, to create a suitable chiral environment for the control of catalytic processes (Chapter 1). The role of supramolecular interactions in catalyst design is illustrated also by properly functionalized metal- or organocatalysts whose catalytic activity can be regulated by weakly bonding additives in an enzyme-like bio-inspired approach (Chapter 3). On a practical point of view, this approach provides a suitable tool to access catalysts libraries for the easy optimization of selected catalytic processes. Moreover, it has the potential of enlightening subtle effects of reaction conditions on the catalytic behaviors.

Chapter 2 provides an additional example of supramolecular effect, the spontaneous self-assembly of ligands in the coordination sphere of the metal, by focusing on chiral secondary phosphine oxides and their metal complexes. In this case, the phosphorus function itself enables secondary bonding interactions leading to good stereochemical control in selected catalytic reactions.

Innovative catalytic methods are typified in Part 2 by some methods and techniques that are currently emerging but have been barely considered so far for applications in enantioselective processes. Thus, chapters in this section include a general description of the method or appropriate general references showing its relevance in organic synthesis, and moves then to the few known examples of enantioselective variants. For instance, Chapter 6 focuses on the prominent Wittig, Mitsunobu and related phosphine-promoted reactions. It shows how catalytic variants can be implemented via *in situ* reduction of the resulting phosphine oxides (P^{III}/P^{V} redox cycling) and summarizes then a couple of recent achievements toward enantioselective processes that have been attained using chiral phosphines. Owing to the major synthetic implications of these classes of reactions, it makes no doubt that the field will attract more and more attention in the future.

Along the same line, the use of frustrated Lewis pairs has emerged as an alternative activation mode that may replace or complement the classical metal activation in catalytic processes. So far frustrated pairs involving either chiral acids or chiral bases components remain extremely rare, even though some of them, notably those displaying boranes or phosphenium ions as Lewis acid components, have demonstrated high efficiency and selectivity in enantioselective hydrogenations and hydrosilylation reactions. These pioneering studies are summarized in Chapter 5. This overview clearly highlights the severe current limitations and reveals the predictable potential of these methods.

In a different framework, it can be seen also that modern technologies may not have been implemented yet extensively in enantioselective catalysis. As a representative example, Chapter 7 recalls the input of the 3D-printing technology in organic synthesis processes, with special focus on flow-reaction devices. It illustrates how the method can be implemented to modulate and optimize stereoselective processes and clearly shows the lack of extensive investigations in this field.

Other innovative catalytic methods are simply based on alternative chemical approaches and concepts. An unquestionable example is given in Chapter 4 that shows how the classical paradigm of using ligands as chiral auxiliaries in organometallic catalysis has been reversed with the use of cationic achiral metal complexes combined to chiral counteranions. While highly efficient, the method remains largely underdeveloped so far.

Modern challenges in enantioselective catalysis may also relate to special classes of advanced reactions whose enantioselective variants are currently tackled by following rather diverse strategies. This issue is illustrated in Chapter 8 by the amination of $C(sp^3)$-H bonds, a highly convenient method for the synthesis of chiral amines from easily available, non-functionalized substrates. Although some advanced have been achieved during the last twenty years or so, enantioselective reactions are still eagerly sought through the appropriate design of metal catalysts (Ru, Co, Fe, Rh, Ir) that operate via a variety of mechanisms.

Finally, this book also encompasses recent advances related to well-established enantioselective reactions, i.e. metal-promoted hydrogenations (Chapter 9) and iridium-promoted allylic substitution reactions (Chapter 11). The scope of these reactions being extremely large, the two chapters focus on selected catalyst classes (bifunctional metal complexes) and selected substrates (enolate surrogates) respectively. Moreover, since enantioselective organocatalysis can't be overlooked in this book, Chapter 10 typifies enantioselective acid promoted reactions by focusing on reactions involving N-acyliminium ions as key intermediates and phosphoric acids, thioureas and others as chiral catalysts. Overall, the last three chapters are intended to complement the other sections on exploratory and rather unachieved approaches, by typifying efficient, wide-scope enantioselective catalytic processes as versatile synthetic tools of academic and possibly industrial relevance.

Altogether, this book combines mainstream topics and pioneering studies in order to emphasize at once the established synthetic potential of the field, its endlessly renewed challenges, its increasingly ambitious goals and innovative research directions. The most frustrating feature of the book is to be limited to a number of items incommensurably small with respect to the myriad of new and creative strategies in enantioselective catalysis that emerge in the recent literature. It is equally frustrating to showcase more mature areas and applications by focused examples only, although these fields are often exhaustively reviewed elsewhere. Obviously, a myriad of other choices might have been done, with the very same legitimacy. Further issues in the series might complement this modest contribution in the future.

We hope that the non-conventional choice to focus mainly on promising but largely unachieved research topics will make this book inspiring and stimulating for the scientific community involved in enantioselective catalysis.

Contents

Preface v

Chapter 1 DNA-Based Asymmetric Catalysis: Past,
Present and Future 1
*Sidonie Aubert, Nicolas Duchemin, Jin-Lei Zhang,
Michael Smietana and Stellios Arseniyadis*

List of Abbreviations	2
I. Introduction	3
II. Supramolecular Assembly	3
A. Intercalators and Groove Binders	5
B. Increasing the Applicability of DNA-Based Catalysis	10
III. The Covalent Approach	13
A. Copper Catalysis	14
B. Palladium- and Iridium-Catalyzed Reactions	16
IV. Expanding the Applicability of Biohybrid Catalysis	17
A. G-Quadruplex, an Alternative Architecture	17
B. Tuning the Selectivity Outcome Using Mirror-Image DNAs	22
C. Improving Scalability and Sustainability Using Solid-Supported DNA	23
D. Bringing DNA into the Realm of Asymmetric Photocatalysis	23
V. Conclusion	26
VI. Acknowledgments	27
VII. References	27

Chapter 2 Chiral Secondary Phosphine Oxides as
           Preligands in Enantioselective Catalysis   31
*Romain Membrat, Didier Nuel, Laurent Giordano
and Alexandre Martinez*

List of Abbreviations   32
  I. Introduction   33
  II. Synthesis, Structure, and Properties of SPOs
    and the Corresponding Coordination Complexes   35
  III. P-Chirogenic SPOs in Enantioselective Catalysis   38
  IV. Other Chiral SPOs in Enantioselective Catalysis   40
    A. SPOs with Backbone Located Chirality   40
    B. SPOs with Combined P-located and Backbone
      Located Chirality   44
  V. Chiral Mixed Ligand Strategy   47
  VI. Conclusion   53
  VII. References   54

Chapter 3 Supramolecular Regulation in
           Enantioselective Catalysis   59
Matthieu Raynal and Anton Vidal-Ferran

List of Abbreviations   60
  I. Introduction   61
  II. Supramolecular Regulation of Bidentate
    Ligands in Metal Complexes   63
    A. Supramolecular Regulation with Aza-Crown-Ether
      Motifs as the Regulation Sites   64
    B. Supramolecular Regulation with Linear
      Polyether Motifs as Regulation Sites 65
  III. Induction of Chirality by Chiral Regulating Agents 70
    A. Bidentate Ligands with a Remote Chiral
      Regulating Agent   71
    B. Ion-paired Chiral Ligands   76
    C. Supramolecular Helical Catalysts   80
  IV. Supramolecular Regulation in Organocatalysis   85

V. Conclusions and Outlook	90
VI. Acknowledgments	91
VII. References	92

Chapter 4 Chiral Counterions in Enantioselective Organometallic Catalysis — 95

Louis Fensterbank, Cyril Ollivier, Antoine Roblin and Marion Barbazanges

List of Abbreviations	96
I. Introduction	97
II. Reactions Involving Formation of C–H Bonds	102
III. Reactions Involving Formation of C–Heteroatom Bonds	103
A. Heterocyclizations *via* Hydroaminations or Hydroalkoxylations	103
B. Diamination of Alkenes	109
C. Overman Rearrangement	110
IV. Reactions Involving Formation of C–C Bonds	111
A. Cycloisomerization Reactions	111
B. Tandem Cycloisomerization/Addition Reactions	114
C. [2+2+2] Cycloaddition Reactions	116
D. Hydrovinylation of Olefins	119
V. Conclusion	120
VI. Acknowledgments	120
VII. References	121

Chapter 5 Enantioselective Catalysis by Nonmetal Frustrated Lewis Pairs — 123

Armen Panossian, Julien Bortoluzzi and Frédéric R. Leroux

List of Abbreviations	124
I. Introduction	124
II. Mechanistic Aspects of Enantioselective Catalysis by FLPs	126
III. FLP Catalysts Featuring Chiral Lewis Acid Moieties	128
A. Boron-based Chiral Lewis Acids	128

	B. Phosphorus-based Chiral Lewis Acids	141
IV.	FLP Catalysts Featuring Chiral Lewis Base Moieties	142
V.	Chiral Intramolecular FLP Catalysts	148
	A. Phosphorus/Boron-based Chiral Intramolecular FLPs	149
	B. Nitrogen/Boron-based Chiral Intramolecular FLPs	150
	C. Oxygen/Boron-based Chiral Intramolecular FLPs	151
VI.	Rationalization of Enantioselectivity	151
VII.	Conclusion	155
VIII.	Acknowledgments	156
IX.	References	156

Chapter 6 Organocatalytic and Asymmetric Processes via P^{III}/P^{V} Redox Cycling   161

Charlotte Lorton and Arnaud Voituriez

List of Abbreviations		162
I.	Introduction	162
II.	Toward the Development of Enantioselective CWRs	163
	A. Catalytic Wittig Reactions	163
	1. Initial studies and prerequisites for the development of phosphine-catalyzed P^{III}/P^{V} redox reactions	163
	2. Further optimization and development of CWR	168
	B. Asymmetric Wittig Reactions	172
	1. Asymmetric wittig–type olefination reactions using stoichiometric amounts of phosphorus reagents	172
	2. Enantioselective CWRs	174
	3. Application in total synthesis	175
III.	Tandem Reactions Including Wittig and aza-Wittig Reactions	176
	A. Tandem Michael Addition/Wittig Olefination	177
	1. A catalytic tandem reaction	177
	2. Catalytic and asymmetric michael addition/wittig reactions	180

B. Tandem Staudinger/aza-Wittig Reaction 183
  1. Catalytic tandem processes 183
  2. Catalytic and asymmetric staudinger/aza-wittig reaction 184
IV. The Catalytic Mitsunobu Reaction 186
  A. Introduction to the Mitsunobu Reaction 186
  B. Catalytic Mitsunobu Reactions *via* P^{III}/P^V Processes 188
  C. Catalytic Mitsunobu Reactions *via* P^V/P^V Processes 189
V. Conclusion and Outlook 191
VI. Acknowledgments 192
VII. References 192

Chapter 7 Recent Advances in the Use of Three-dimensional-printed Devices in Organic Synthesis Including Enantioselective Catalysis 197

Sergio Rossi, Alessandra Puglisi, Laura Maria Raimondi and Maurizio Benaglia

List of Abbreviations 198
I. Introduction 199
II. 3D-Printed Devices in Nonstereoselective Reactions 207
  A. Catalytic 3D-printed Devices in Batch Reactions 209
  B. 3D-Printed Devices for In-flow Reactions 223
    1. Reactions under heterogeneous conditions 224
    2. Reactions under homogeneous conditions 226
  C. 3DP for Drugs and Pharmaceuticals Applications 236
III. 3D-Printed Devices in Stereoselective Reactions 240
IV. Conclusion 244
V. Acknowledgments 245
VI. References 245

Chapter 8 Catalytic Strategies for the Enantioselective Amination of $C(sp^3)$–H Bonds 255

Philippe Dauban and Tanguy Saget

List of Abbreviations 256
I. Introduction 256

II. Catalytic Asymmetric C(sp^3)–H Amination with Azides	259
A. Metal-catalyzed Concerted Processes	259
B. Metal-catalyzed Radical Processes	261
C. Biocatalytic Amination with Engineered Enzymes	263
III. Catalytic Asymmetric C(sp^3)–H Amination with Dioxazolones	266
A. Introduction	266
B. Design of Chiral Hydrogen-bond-donor Catalysts	267
C. Design of Catalysts with Chiral Hydrophobic Pockets	268
D. Design of an Achiral Cp*M(III)/Chiral Carboxylic Acid Hybrid Catalytic System	270
IV. Catalytic Asymmetric C(sp^3)–H Amination with Iodine(III) Oxidants	271
A. Introduction	271
B. Design of a Chiral Rhodium(II)–Carboxamidate Complex	272
C. Design of a Chiral Rhodium(II)–Carboxylate Complex	273
D. Design of a Bisoxazoline Ligand with Quaternary Stereocenters	274
E. Design of Supramolecular Metal Complexes	275
V. Conclusion	278
VI. Acknowledgments	278
VII. References	278

Chapter 9 Bifunctional Homogeneous Catalysts Based on Ruthenium, Rhodium and Iridium in Asymmetric Hydrogenation   281

Christophe Michon and Francine Agbossou-Niedercorn

List of Abbreviations	283
I. Introduction	284
II. ATHs Using Bifunctional Catalysts	287
A. Ruthenium-Catalyzed Transfer Hydrogenations	287
1. Noyori-type catalysts	287
2. Shvo-type catalysts	305
B. Rhodium-Catalyzed Transfer Hydrogenations	307

  1. Diamine-tethered η^5-cyclopentadienyl
   rhodium complexes with active NH functions   307
  2. Amino-acid-derived amides and thioamides
   with active NH functions   309
 C. Iridium-Catalyzed Transfer Hydrogenations   312
  1. Amino- and amido iridium(III) catalysts with
   active NH functions   312
  2. Octahedral chiral-at-metal iridium(III)
   catalysts with π-stacking and H-bonding ligands   316
 D. Osmium-Catalyzed Transfer Hydrogenations   318
III. AHs Using Bifunctional Catalysts   319
 A. Ruthenium-Catalyzed Hydrogenations   319
  1. Shvo-type catalysts   319
  2. Noyori-type catalysts   320
 B. Rhodium-Catalyzed Hydrogenations   332
  1. Catalysts implying hydrogen bondings between
   ligands and either substrates or halide counteranions   332
  2. Catalysts implying ion pairs as noncovalent
   interactions between ligands and substrates   342
 C. Iridium-Catalyzed Hydrogenations   345
  1. Iridium catalysts involving active NH functions
   and related cases   346
  2. Catalysts implying ion pair or hydrogen
   bonding interactions   349
  3. Iridium phosphates and triflylphosphoramides
   as bifunctional catalysts   354
IV. Conclusion   356
V. Acknowledgments   357
VI. References   357

Chapter 10   Organocatalytic Enantioselective Reactions
      Involving Cyclic *N*-Acyliminium Ions   369

Milane Saidah and Laurent Commeiras

List of Abbreviations   370
 I. Introduction   371
 II. Enantioselective Intramolecular Additions to NAI Intermediates   373

A. Ion-pairing Catalysis with Chiral Phosphoric Acids	374
1. Nucleophilic additions of indole derivatives	375
2. Nucleophilic additions of electron-rich arenes	378
B. Ion-paring Catalysis with Chiral Thioureas-based Catalysts	380
1. Nucleophilic additions of indole and pyrrole derivatives	380
2. Nucleophilic additions of olefins	383
C. Enamine Catalysis	384
III. Enantioselective Intermolecular Additions to NAI Intermediates	386
A. Ion-pairing Catalysis with CPAs and Analogs	387
1. Nucleophilic additions of indole derivatives	387
2. Nucleophilic additions of hydroxystyrenes and hydrazones	395
3. Nucleophilic additions of sulfur, phosphorus, and hydride nucleophiles	396
B. Ion-paring Catalysis with Chiral Thioureas-based Catalysts	400
C. Enamine Catalysis	401
IV. Conclusion	403
V. Acknowledgments	404
VI. References	404

Chapter 11 Enolate Surrogates and Unusual Nucleophiles in Stereoselective Iridium-catalyzed Allylic Substitutions   407

Pierre Bouillac, Manuel Barday, Thierry Constantieux and Muriel Amatore

List of Abbreviations	409
I. Introduction	411
II. Iridium-catalyzed Allylations of Silyl Enolates	415
A. Enantioselective α-Allylation of Silyl Enolates	416
1. α-allylation of silyl enol ethers as ketone surrogates	416
2. α-allylation of silyl ketene acetals as ester surrogates	417

	3. α-allylation of α,β-unsaturated silyl enol ethers	419
	4. α-allylation of silylated enols of vinylogous esters and amides	421
	B. Enantioselective γ-Allylation of Silyl Dienolates Derived from Dioxinones	422
	C. Conclusion	424
III.	Iridium-catalyzed Stereoselective and Stereodivergent Allylations of Other Enolate Surrogates	424
	A. Stereoselective α-Allylations	425
	1. α-allylation of enol ethers and enamines as aldehydes surrogates	425
	2. Enantioselective α-allylation of enamines, enamides, and other ketone surrogates	427
	3. Enantioselective α-allylation of orthoesters as ester surrogates	429
	4. Enantioselective α-allylation of ketene aminals as amide surrogates	430
	B. Stereodivergent α-Allylations Under Dual Catalysis	432
	1. Stereodivergent α-allylation of aldehydes under dual iridium and enamine catalysis	433
	2. Stereodivergent γ-allylation of α,β-unsaturated aldehydes under dual iridium and enamine catalysis	438
	3. Stereodivergent α-allylations under dual iridium and lewis base catalysis	438
	C. Conclusion	441
IV.	Iridium-catalyzed Enantioselective Allylation of Other Carbon Nucleophiles	443
	A. Enantioselective Allylation of Organometallic Reagents	443
	1. Allylation of organozinc reagents	443
	2. Allylation of boron, silicon, and silver-based reagents	447
	B. Stereoselective Allylation of Hydrazones and Imines	450
	C. Conclusion	452

V. Iridium-catalyzed Enantioselective Allylation
of Electron-rich Unsaturated Substrates and
Other Weak Nucleophiles 452
   A. Enantioselective Allylation of C=C Double Bonds 453
      1. Enantioselective allylation of olefins 453
      2. Stereoselective allylations of indoles and pyrroles 454
      3. Enantioselective C-allylation of anilines 463
      4. Enantioselective C-allylation of phenols 464
      5. Stereoselective allylations involving
cation-π cyclizations of polyenes 468
   B. Enantioselective Allylations of Benzylic
Nucleophiles and Allylic Aromatization of
Methylene-substituted Heterocycles 469
   C. Conclusion 471
VI. Conclusion 471
VII. Acknowledgments 472
VIII. References 472

Index   479

© 2022 World Scientific Publishing Company
https://doi.org/10.1142/9789811248436_0001

1 DNA-Based Asymmetric Catalysis: Past, Present and Future

Sidonie Aubert,* Nicolas Duchemin,* Jin-Lei Zhang,* Michael Smietana[†,‡] and Stellios Arseniyadis[*,§]

*Queen Mary University of London, Department of Chemistry, Mile End Road, London, E1 4NS, UK
[†]Université de Montpellier, Institut des Biomolécules Max Mousseron, CNRS, ENSCM, Place Eugène Bataillon, 34095 Montpellier, France
[‡]michael.smietana@umontpellier.fr
[§]s.arseniyadis@qmul.ac.uk

Table of Contents

List of Abbreviations	2
I. Introduction	3
II. Supramolecular Assembly	3
A. Intercalators and Groove Binders	5
B. Increasing the Applicability of DNA-Based Catalysis	10
III. The Covalent Approach	13
A. Copper Catalysis	14
B. Palladium- and Iridium-Catalyzed Reactions	16

IV. Expanding the Applicability of Biohybrid Catalysis	17
A. G-Quadruplex, an Alternative Architecture	17
B. Tuning the Selectivity Outcome Using Mirror-Image DNAs	22
C. Improving Scalability and Sustainability Using Solid-Supported DNA	23
D. Bringing DNA into the Realm of Asymmetric Photocatalysis	23
V. Conclusion	26
VI. Acknowledgments	27
VII. References	27

List of Abbreviations

A	adenosine
bp	base pair
bipy	bipyridine
C	cytidine
CD	circular dichroism
ct-DNA	calf thymus-deoxyribonucleic acid
d	day(s)
dmbpy	4,4′-Dimethyl-2,2′-dipyridyl
DNA	deoxyribonucleic acid
ds-DNA	double stranded deoxyribonucleic acid
ee	enantiomeric excess
G	guanosine
H33258	Hoechst 33258
h	hour(s)
IR	infrared
min	minute(s)
ON	oligonucleotide
RNA	ribonucleic acid
rt	room temperature
st-DNA	salmon testes deoxyribonucleic acid
T	thymidine
T_m	melting temperature

| U | uridine |
| UV–Vis | ultraviolet–visible |

I. Introduction

The widespread demand for chiral compounds has risen drastically in recent years, mainly driven by the demands of the pharmaceutical industry. As a result, the development of new enantioselective catalytic tools has become a central focus within the synthetic organic chemistry community and beyond. The use of transition metals coupled with chiral ligands to catalyse specific transformations has become particularly popular, but often requires the use of expensive, sometimes toxic, and perhaps sensitive catalytic systems. Biohybrid catalysis has recently emerged as a valuable alternative to standard catalysis as it often offers enhanced catalytic activity, high levels of selectivity, and also environment-friendly conditions.[1–7] Many efforts have been devoted to the use of proteins as chiral scaffolds, where key interactions provided by the second coordination sphere of the metalloenzymes allow to control both the rate of the reaction and the stereoinduction.[8–13] Inspired by this concept, DNA-based asymmetric catalysis was introduced by Roelfes and Feringa taking advantage of the inherent chirality of the right-handed double helix of DNA.[14] Since then, the concept has been successfully applied to many enantioselective carbon–carbon and carbon–heteroatom bond-forming reactions,[15–21] which will all be reviewed here.

II. Supramolecular Assembly

DNA-based catalysis presents several interesting features. Indeed, it does not require the use of a chiral ligand bound to a given metal, and the DNA from natural sources is an abundant and inexpensive source of chirality, making it attractive for large-scale applications. Moreover, DNA is biodegradable and reactions are run in aqueous media, thus limiting the pollution and safety hazards.

DNA-based biohybrid catalysts are composed of a nonchiral catalyst, most of the time a transition metal complex, which is incorporated within the chiral supramolecular framework constituted by the DNA double

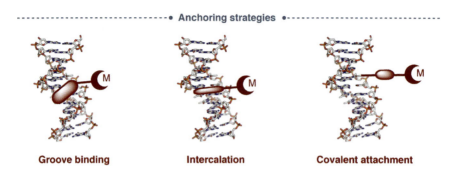

Figure 1. Anchoring strategies — Left: supramolecular approach through groove binding; middle: supramolecular approach through base intercalation; and right: covalent approach.

helix. The transition metal, the ligand, and DNA have synergistic functions: the transition metal usually activates the substrate enabling the catalytic process, the ligand finely tunes the interactions with DNA, which ultimately provides a well-organized chiral environment and enough space for the substrates.

Performing DNA-based catalysis is challenging as it aims to place the catalytically active transition metal complex in the DNA helix. Indeed, due to the second coordination sphere provided by DNA, a chirality transfer occurs, resulting in enantioselectivity and possible additional rate acceleration. The incorporation of the transition metal complex within the DNA scaffold can be achieved using either specific supramolecular interactions or by covalently attaching the ligand to the DNA duplex, both strategies having their own pros and cons (Figure 1).[22]

A number of DNA-based catalysts have been developed and evaluated on various synthetic transformations over the years proving their synthetic utility. Noncanonical DNA structures have also been considered and evaluated in the context of biohybrid catalysis. An overview of these strategies will be detailed in the following sections.

As mentioned earlier, the supramolecular approach is based on specific interactions between the bioscaffold (e.g. st-DNA [salmon testes-DNA], ct-DNA, or any defined sequences) and the metal-binding moiety;

these include hydrophobic, π-stacking, electrostatic, and hydrogen-bonding interactions. Although intercalation requires conformational changes to accommodate the stacking of a planar aromatic system between two base pairs, the groove-binding mode involves direct interaction of the bound molecule with the edges of the base pairs in either of the grooves and thus does not disturb the helix. Both strategies are easy to set up and enable rapid optimization; however, the ligands usually do not exhibit a strong sequence or base-pairing affinity, resulting in a heterogeneous mixture of catalysts.

A. Intercalators and Groove Binders

In 2005, Roelfes and Feringa reported the first example of a direct transfer of chirality using a nonchiral ligand bound to a DNA duplex in a noncovalent fashion.[14] The first generation of ligands **L1–L4** was introduced based on a well-known DNA intercalator, 9-amino acridine,[23] linked to a Cu^{II} complex via a bidentate N-ligand and a suitable spacer. The biohybrid catalyst was used to promote a Diels–Alder cycloaddition between azachalcone **1** and cyclopentadiene **2**, a transformation known to benefit from the aqueous environment due to hydrophobic effects (Figure 2).[24] The reaction yielded the corresponding cycloadduct **3** with an almost complete *endo* selectivity (up to 98:2 *endo/exo* ratio) and up to 53%

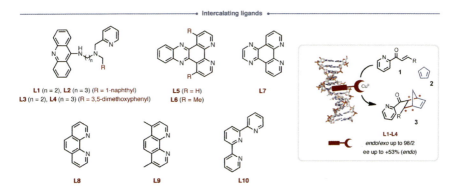

Figure 2. First- and second-generation DNA catalysts. Initial application to the Diels–Alder reaction between azachalcone **1** and cyclopentadiene **2** using **L1–L4**.[14]

enantiomeric excess (ee) for the *endo* isomer (for R = *p*-MeOPh). Cu^{II} acted as a Lewis acid by activating the dienophile. The results obtained with these first-generation ligands (**L1–L4**) showed a correlation between the ees, the nature of the side chain of the amine ligand (CH_2R), and the length of the spacer. As a general trend, the side chain must contain an aromatic ring, which suggests a π–π interaction with the dienophile. Interestingly, when using ligands **L1** and **L3**, which incorporate the shortest spacer ($n = 2$), the opposite enantiomers were obtained with respect to ligands with $n = 3$ or 4, thus showcasing the importance of the design of the ligand. Moreover, elongation of the spacer appeared to be detrimental for the selectivity, indicating that a proximity to the DNA is critical for an efficient chirality transfer.

Kinetic and structural studies revealed that the DNA strand acted solely as the chiral auxiliary and no rate acceleration could be observed.[25] As a matter of fact, the reaction rate slightly decreased in the presence of DNA and the first-generation ligands. The reaction proceeded at the edge of the DNA groove as the DNA intercalator, 9-amino acridine, is linked to the Cu^{II} complex *via* a spacer. As a result, the DNA-catalyzed reaction was not much different from the reaction occurring in solution in the absence of DNA. Regarding the DNA itself, the presence of AT-rich sequences resulted in a drop in selectivity whereas GC-rich sequences led to higher ees. It is also worth mentioning that the selectivity outcome, in other words the choice of the major enantiomer, is highly ligand dependent.

Taking into account these observations, a second generation of ligands **L5–L12**, in which the DNA- and metal-binding moieties are both integrated, was introduced (Figure 2). In general, these ligands are planar, symmetrical, and achiral, and due to the absence of spacer the metal center is very close to the DNA helix. The best selectivities were obtained with 4,4′-Dimethyl-2,2′-dipyridyl (dmbpy) **L12**.[26] The two methyl groups seem to alter the geometry of the substrate/catalyst complex, thus leading to a better fit within the DNA helix. The resulting more compact complex increases the probability for the diene to approach selectively from one of the faces of the prochiral Cu^{II}-bound dienophile (the *Si* face), thus resulting in the preferential formation of one of the enantiomers. The ees > 97% were obtained in the Diels–Alder reaction for R = *p*-MeOPh, *p*-NO_2Ph, and *t*-Bu.

The presence of DNA does not only contribute to the selectivity, it also increases the kinetics as demonstrated by the 58-fold rate enhancement, which was not observed with the first-generation ligands.[27] The reaction is likely to occur in the vicinity of the DNA duplex, which provides the best environment to stabilize the transition state. The influence of the DNA sequence was also investigated. The study revealed that the presence of three consecutive G bases, called G-tracts, in a dodecamer duplex gives rise to the highest levels of enantioselectivity. Interestingly, the DNA sequences that induced the highest ees also induced the highest rate enhancements.

The mode of binding of these second-generation ligands was fully investigated in order to elucidate the mechanism by which the chirality transfer occurs.[28] Hence, by taking the Diels–Alder reaction as a model reaction, a set of analytical studies including electron paramagnetic resonance (EPR), Raman spectroscopy, ultraviolet–visible (UV–Vis) absorption, and circular dichroism spectroscopy revealed that ligands **L5–L10** were intercalators, whereas dmbpy **L12**, which is the catalyst that provides the highest ees and acceleration rates, mainly binds to the minor groove of the DNA and is more dynamic and flexible (Figure 2). However, no correlation could be made between the binding mode and the ees obtained, as other parameters like the binding affinity of the complexes also come to play.

It was also demonstrated that the dienophile (*cf* the azachalcone **1**), has a certain affinity for DNA through an intercalative binding mode, thus increasing the effective molarity of the substrate.[29] Hence, Cu^{II}-dmbpy **L12** benefits from both its groove-binding properties (it does not compete for the same binding site as the substrate) and the increased local concentration. Another important parameter that was evaluated was the dynamic binding of the metal complex allowing both partners to interact. All these parameters taken together tend to explain the rate acceleration and highlight the power of the second coordination sphere interactions.

Interestingly, Roelfes and coworkers showed that a change in the denticity of the ligand had an influence on the selectivity outcome of the reaction. Hence, the use of tridentate ligands instead of the commonly used bidentate ligands reversed the selectivity outcome.[30] One enantiomer of the Diels–Alder reaction was formed in >99% ee when using dmbpy

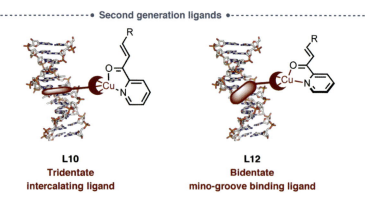

Figure 3. Binding of azachalcone to Cu^{II}–L10 and Cu^{II}–L12 complexes in a DNA-based catalysis.[30]

L12, while the opposite enantiomer was obtained in −92% ee with the intercalator L10 under otherwise identical conditions. Based on computational studies, the geometry around the Cu^{II} appeared to be crucial for the enantiomeric preference. With L12, two coordination sites remain available for the substrate in the equatorial plane of Cu(II) (Figure 3, right), whereas with L10 only one site remains available in that plane, thus forcing the substrate to bind to the axial site (Figure 3, left). Hence, depending on the denticity of the ligand, the interaction with DNA differs, which forces the nucleophile to approach from one prochiral face of the electrophile rather than the other.

Following these results, a new class of dienophiles was used. First introduced by Evans and coworkers, α,β-unsaturated 2-acyl imidazoles 4 appeared as particularly appealing substrates[31,32] as the imidazole moiety can be easily transformed through methylation and subsequent addition of nucleophiles.[33] In particular, Roelfes and coworkers reported the first DNA-based asymmetric Diels–Alder reactions with these substrates in the presence of Cu^{II}-dmbpy L12 affording the major *endo* product 5 with up to 98% ee (Figure 4).[34]

Another kind of minor groove binders was explored, based on Hoechst 33258 (H33258), which exhibits a strong affinity for AT-rich regions of B-DNA duplexes (H33258, Figure 5).[35,36] H33258 is characterized by various key features that are essential for the interactions with DNA. These include a positive charge at physiological pH enabling an

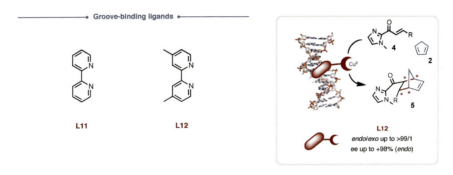

Figure 4. DNA-catalyzed Diels–Alder reaction between α,β-unsaturated 2-acyl imidazole **4** and cyclopentadiene **2**.[34]

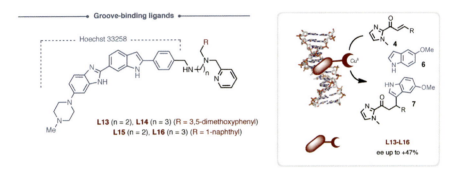

Figure 5. DNA-catalyzed Friedel–Crafts alkylation using H33258-derived ligands.[37]

electrostatic interaction with the anionic phosphate backbone of DNA and a crescent shape fitting the narrow minor groove. Taking advantage of the high sequence-specificity of H33258, Arseniyadis, Smietana, and coworkers synthesized a library of bifunctional DNA ligands **L13–L16** bearing both a minor groove–binding moiety and a chelating site to coordinate a metal center (Figure 4).[37] The two units were linked *via* a spacer and evaluated for their catalytic activity in the presence of DNA. As a general trend, flexible ligands showed higher affinity toward ct-DNA (calf thymus-DNA) and sequences rich in AT base pairs compared to rigid ones. The Cu(II) complexes were tested in the Michael-type addition of indoles to α,β-unsaturated 2-acyl imidazoles **4** (Figure 5) and afforded complete conversion of albeit moderate selectivities (up to 47% ee).

B. Increasing the Applicability of DNA-Based Catalysis

Since the pioneering work of Roelfes and Feringa on the Diels–Alder reaction, DNA-based asymmetric catalysis using Cu^{II}-dmbpy **L12** has also been applied to Friedel–Crafts alkylations of indoles **6** and pyrroles **8**, affording up to 93% and 81% ee, respectively (Figure 6).[38] This was later extended to RNA-based catalysts; however, the selectivities were only moderate (up to 54% ee).[39] Interestingly, an intramolecular Friedel–Crafts alkylation was developed by Sugiyama and coworkers, but it required the use of an intercalative ligand instead of the groove-binder

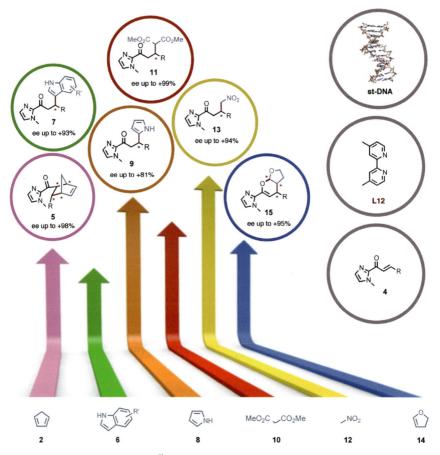

Figure 6. st-DNA/Cu^{II}–**L12**–catalyzed C–C bond-forming reactions.

L12 to reach ees as high as 71%, demonstrating the importance of a well-designed catalytic pocket.[40,41]

The minor groove–binding strategy has also been extended to other C–C bond-forming reactions including Michael additions using various nucleophiles such as dimethylmalonate **10**, nitromethane **12**,[42] malononitrile, or cyanoacetate,[43] with ees ranging from 84% to 99%. More recently, the first example of an asymmetric DNA-catalyzed inverse electron-demand hetero–Diels–Alder reaction was reported. The reaction allows the formation of fused bicyclic O,O-acetals in high yields and excellent diastereo- and enantioselectivities (up to >99:1 dr, up to 95% ee).[44] Finally, several other reactions such as enantioselective fluorinations using Selectfluor^{TM} and epoxide hydrolyses have been successfully reported with ees up to 74% and 63%, respectively.[45,46]

As a general trend, all of these transformations showcased the superiority of the groove-binding strategy over the intercalative approach.

One potential limitation in DNA-based catalysis is the need of an aqueous medium to perform the reaction. Indeed, this raises some solubility issues, which can however be circumvented by adding water-miscible cosolvents such as alcohols, but also DMSO, DMF, or even acetonitrile. All these solvents either prevent the DNA from precipitating or prevent B-type helix destabilization. Moreover, some interesting solvent effects have been observed in both Friedel–Crafts alkylations and Michael additions, most probably due to the faster dissociation of the product from the Cu^{II} complex in polar media.[47] Moreover, the presence of a cosolvent allows to work at lower temperatures and thus potentially increase the selectivity. Ionic liquids and glymes (glycol diethers) have also been shown to be compatible with DNA catalysis.[48]

In addition to C–C bond-forming reactions, DNA catalysis has also been used to form C–O bonds in a highly stereoselective fashion. Roelfes and Feringa reported the diastereoselective and enantioselective syn-hydration of α,β-unsaturated 2-acyl imidazoles **4**, affording the corresponding β-hydroxy product **15** in up to 72% ee (Figure 7).[49] Interestingly, in this case, the minor groove–binder **L12** led to an unusual poor selectivity, whereas the first-generation intercalating ligands **L1–L4** induced the highest ees. Moreover, due to the reversibility of the process, the selectivity was highly dependent on the conversion rate, as

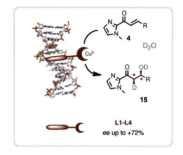

Figure 7. st-DNA/Cu^{II}–catalyzed *syn*-hydration of α,β-unsaturated 2-acyl imidazoles.[49]

epimerization was observed after 24 h. Although the reaction proceeded only with alkyl-substituted substrates, this represented a great breakthrough in the field of asymmetric catalysis, as this was the first example reported so far, thus confirming the importance of developing the field of biohybrid catalysis.

A systematic study was conducted later to correlate the observed selectivity with the structure of the ligands.[50] In the case of the hydration, the first coordination sphere plays a crucial role as it appears to be highly organized, thanks to the second coordination sphere. The intercalating ligands allow an ideal positioning of the complex in the first coordination sphere, compared to the minor-groove binders. This was also observed in the *oxa*-Michael additions involving alcohols instead of water (up to 86% ee).[51]

In all the abovementioned examples, Cu^{II} behaves as a Lewis acid. Roelfes and coworkers extended the field of biohybrid catalysis to organometallic transformations, by reporting the first DNA-catalyzed asymmetric intramolecular cyclopropanation (Figure 8).[52] Starting from α-diazo-β-keto sulfones **16**, this reaction afforded the corresponding cyclopropanes **17** in up to 84% ee. The reaction proceeds through the formation of a Cu^I–carbene complex, which requires strong intercalative ligands to limit the access of H$_2$O and thus prevent any side reaction.

More recently, Roelfes and coworkers have developed a carbene-transfer reaction based on the use of Fe^{III} and a porphyrin-based ligand **L17**, known to strongly bind to DNA (Figure 9).[53] Although the selectivities obtained were only moderate (up to 53% ee), this showed that DNA

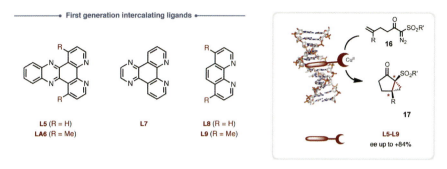

Figure 8. st-DNA/Cu^{II}–catalyzed intramolecular cyclopropanation.[52]

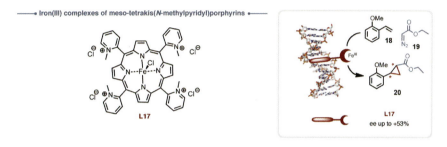

Figure 9. st-DNA/Fe^{III}-catalyzed intermolecular cyclopropanation.[53]

catalysis is not limited to Cu-catalyzed transformations and could find synthetic applications in other fields.

III. The Covalent Approach

Although the supramolecular approach enables an easy access to the catalyst, it does not allow an exact positioning of the cofactor within the DNA framework and thus prevents any rational around the sequence selectivity. In contrast, covalent anchoring ensures the precise positioning of the catalytic entity within a given DNA strand, thus resulting in a homogeneous mixture of catalysts. This precise positioning of the ligand is warranted in the covalent approach as the ligand is directly attached onto a specific nucleobase, prior to the hybridization of the two complementary single-stranded oligonucleotides (ON). Moreover, this also allows to fine-tune the base opposite to the cofactor. However, as this strategy involves synthetic modifications of ON, the design, synthesis, and catalytic evaluation of new sequences can sometimes be time consuming.[54]

A. Copper Catalysis

Roelfes and coworkers developed a covalent strategy based on a modular assembly of three different sequences: an oligonucleotide **ON1** functionalized with a bipyridine (bipy) moiety, a nonfunctionalized oligonucleotide **ON2** and a counter-strand **ON3** complementary to both **ON1** and **ON2** (Figure 10).[55] After hybridization and introduction of Cu^{II}, the catalytically active complex ideally positioned at the center of the duplex **ON4** was evaluated in the model Diels–Alder reaction. The results showed not only that the selectivity was strongly sequence-dependent but also that the length of the linker in **ON1** was crucial. Fine-tuning of the system allowed to achieve up to 93% ee (for R = Ph), which compared favorably with the selectivities obtained with the same sequence but through a supramolecular approach.

Later, the same group exploited the ability of cisplatin to strongly coordinate DNA to develop a new covalent anchorage strategy.[56] To do so, a cisplatin-like unit [cis-Cl_2Pt(diamine)] was attached to a modified bipy and combined then with st-DNA, resulting in the biohybrid catalyst **ON5** (Figure 11). The latter was evaluated in the asymmetric Friedel–Crafts alkylations of indoles **6** with enones **4** and in Diels–Alder cycloadditions. These experiments gave both good yields and good enantioselectivities (up to 80% ee), albeit inferior to the ones obtained through the supramolecular approach.

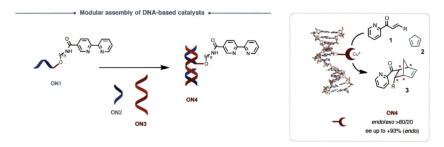

Figure 10. DNA-based Diels–Alder reactions using a covalent anchorage strategy.[55]

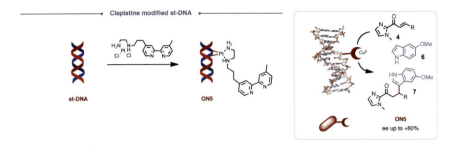

Figure 11. Covalent anchorage strategy using cisplatin as a DNA binder.[56]

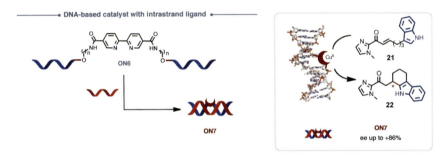

Figure 12. Covalent bonding strategy using a bipy-bridged ON.[57]

More recently, Park and Sugiyama reported yet another covalent strategy featuring a DNA-hybrid catalyst bearing a bipy motif attached to the ON in an intrastrand fashion (**ON6**, Figure 12).[57] Once hybridized with the proper counter-strand, the newly synthesized catalysts **ON7** were evaluated in a copper-promoted asymmetric intramolecular Friedel–Crafts alkylation. Interestingly, the counter-base facing the bipy ligand (up to 86% ee obtained with cytosine), the size of the binding pocket (the best selectivities were obtained with the propandiyl linker, $n = 3$), the nature of the bases located next to the bipy (GC pairs appeared more suitable), and the central position of the ligand in the strand, all had an influence on the enantioselectivity of the catalytic reactions.

Taking into account the benefits of having a cytosine as a counter-base, Park and Sugiyama also explored a ligandoside approach (i.e. the use of metal coordinating base pair), driven by the high affinity that some divalent cations exhibit for particular nucleotides, thus eliminating the need of having an external ligand.[58] Triethylene glycol or alkyl linkers were therefore incorporated into DNA backbones, across from the cytosine, in the same way as for the bipy derivative in Figure 12. The aim was to design an appropriately sized catalytic pocket. Interestingly, by simply varying the neighboring bases around the central cytosine and by using the triethylene glycol linker, up to −97% ee could be obtained in the asymmetric Diels–Alder cycloaddition. Interestingly, these catalysts led to the opposite enantiomer with respect to the previously reported Cu(II)-dmbpy/st-DNA catalytic systems, thus proving that an adapted design of the active site enables a fine-tuning of the selectivity.

Concomitantly, Carell and coworkers developed yet another ligand-free methodology and synthesized multiple Cu^{II}-containing DNA strands by incorporating modified nucleobases able to chelate metal ions.[59] This strategy afforded some promising rate accelerations in the Diels–Alder reaction; however, only moderate enantioselectivities were obtained (up to 39% ee).

B. Palladium- and Iridium-Catalyzed Reactions

In 2007, Kamer and coworkers introduced a diphenylphosphine moiety onto a modified uridine, the 5-iodo-2′-deoxyuridine **ON8**, through a Pd^{II}-catalyzed cross-coupling reaction (Figure 13).[60] This ON-based phosphine ligand **ON9** was evaluated on a Pd^{II}-catalyzed allylic amination affording the desired allylic amine **25** in quantitative yield and up to 80% ee. Interestingly, the enantiopreference of the reaction depended on the solvent used and no conversion was observed when running the reaction in water. Jäschke and coworkers reported the synthesis of several

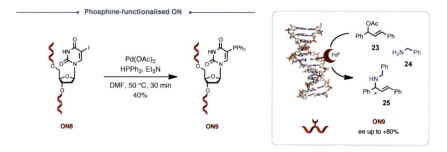

Figure 13. Pd-catalyzed allylic amination using a phosphine-modified ON.[60]

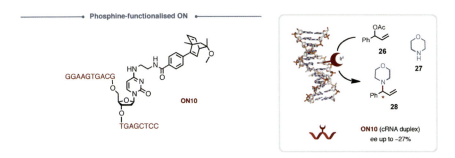

Figure 14. Ir-catalyzed allylic amination using a chiral diene–modified ON.[62]

other DNA--based phosphine-containing ligands;[61] however, their catalytic activity has not been reported so far.

Jäschke and coworkers also reported an allylic amination using a DNA-diene-Ir^{I} catalyst prepared by incorporating a chiral diene onto a deoxyuridine nucleoside in a 19-mer DNA sequence (ON10).[62] The latter afforded up to 92% yield of the desired amination product **28**, however, the selectivity remained modest (up to 27% ee, Figure 14).

IV. Expanding the Applicability of Biohybrid Catalysis

Despite all these promising results, DNA-based asymmetric catalysis still remains in its infancy. For the past several years, efforts have been dedicated to broaden the scope and address relevant synthetic challenges. To achieve this goal, other noncanonical structures of DNA, such as G-quadruplexes (G4DNA), mirror-image L-DNA, or even solid-supported DNA, have been considered. DNA has also been used to template-specific reactions such as light-induced cycloadditions.

A. G-Quadruplex, an Alternative Architecture

The structure of DNA is not limited to duplexes. Indeed, other high-order architectures are known, such as the **G4DNA** supramolecular assembly.[63] **G4DNA** consists in the assembly of four guanidines through Hoogsteen base pairings formed by one or more DNA molecules, generating a tetramer.[64] Other tetramers present in the rest of the loop can stack above

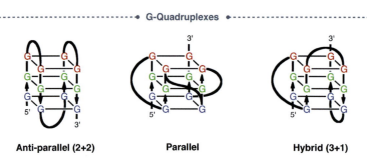

Figure 15. G4DNA topologies.

each other resulting in diverse folding topologies: parallel, antiparallel (2+2), and hybrid (3+1) (Figure 15). Their formation and stability are dependent on monovalent cations, preferably K^+ and Na^+. Additional stability can be obtained by the stacking of ligands at the external face of the assembly through π–π bonding interactions with an affinity for the terminal G-tetrads. This enabled to position a metal center in close proximity to the chiral environment provided by the DNA. Due to their complex topologies, these structures present a greater diversity that can easily be tuned. Additionally, when functionalized with lipophilic tails, **G4DNA** can self-assemble into micelles.[65]

In 2010, Moses and coworkers reported the first example of an asymmetric **G4DNA**-catalyzed transformation based on a **G4DNA**–Cu^{II} complex.[66] The selectivities in the Diels–Alder cycloaddition between cyclopentadiene **2** and enone **1** (see Figure 2), albeit modest (up to 34% ee), could be tuned by varying the sequence of the loop, the topology of the **G4DNA** (namely parallel or antiparallel architectures), and the orientation of the ligands.

Later, Li and coworkers studied the same Diels–Alder cycloaddition, but applying a ligandoside approach. Divalent cations (Cu^{2+}, Zn^{2+} and Cd^{2+}) exhibited good affinity for the nucleic acids and good catalytic activity, reaching up to 74% ee and a sevenfold rate acceleration.[67] This ligand-free strategy was highly tunable, thanks to different additives. Indeed, the presence of Na^+ cations stabilized the antiparallel conformation of **G4DNA** whereas when using PEG200, **G4DNA** switched from antiparallel to parallel topology, also resulting in a reversal of the enantiopreference, leading to −71% ee (Figure 16). The same group also evaluated higher-order **G4DNA**, consisting in multiple consecutive units connected through TTA loops. Each unit worked in synergy with the others resulting in an enhanced selectivity (up to 92% ee). The use of K^+, which has a larger ionic radius than Na^+, induced a rearrangement of the **G4DNA** into a stable hybrid conformation, leading again to a reversal of the enantioselectivity.[68] Besides the selectivity, the presence of K^+ allowed a 68-fold enhancement of the reaction rate.

An unexpected switch of selectivity was also observed when using NH_4^+ instead of K^+ (ees up to 90%).[69] Although both cations are roughly

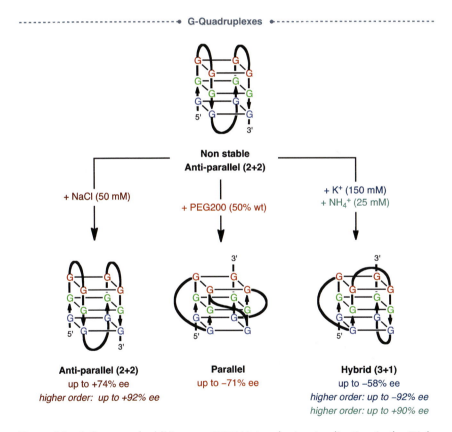

Figure 16. Influence of additives on **G4DNA** topologies. Implication in the Diels–Alder reaction between cyclopentadiene **2** and enone **1**.[67–69]

the same size and bind similarly to **G4DNA** in its hybrid conformation, the assembly appears to be more stable in the presence of K$^+$ as shown by UV melting experiments, which could explain the different selectivity observed.

The use of **G4DNA** was also successfully applied to Cu^{II}-catalyzed Friedel–Crafts reactions reaching up to 75% ee.[70]

The ligandoside approach proved to be efficient in terms of selectivity and reactivity; however, the absence of ligands makes it difficult to rationalize the results. Consequently, ligands such as dmbpy **L12** (for **L12**, see Figure 2) were reintroduced. Li and coworkers, for example, described the **G4DNA**-catalyzed oxidation of the thioanisole derivatives **29** (R = phenyl, substituted aryls and heteroaryls), in the presence of hydrogen peroxide, using the **G4DNA-1–Cu^{II}** complex that features the dmbpy **L12** ligand (Figure 17).[71] Upon addition of K$^+$, **G4DNA** folded into a stable antiparallel conformation and provided a good catalyst allowing full conversion and up to 77% ee. Base mutation into the sequence revealed that the integrity of the quadruplex structure is mandatory in order to retain both selectivity and reactivity levels.

Other studies involving noncovalent interactions of the ligand with the chiral scaffold were reported with dmbpy **L12**,[72] but also porphyrin-derived ligands[73] and terpyridine **L10**,[74] leading to moderate-to-high enantioselectivities in Michael additions and Diels–Alder reactions.

Li and coworkers developed yet another class of DNA-based catalysts coining a **G4DNA** and an Fe-porphyrin to catalyse a highly enantioselective cyclopropanation reaction (Figure 18).[75] Interestingly, two of their

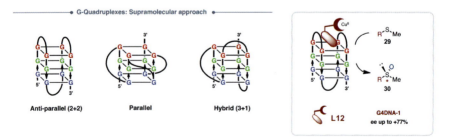

Figure 17. Sulfoxidations promoted by **G4DNA-1/L12**–Cu^{II} complexes.[71]

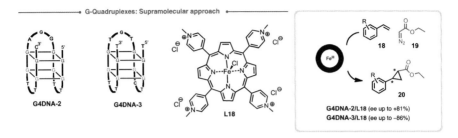

Figure 18. G4DNA2/L18 and G4DNA3/L18-catalyzed cyclopropanation.[75]

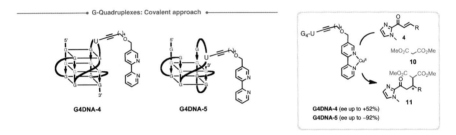

Figure 19. G4DNA-4 and G4DNA-5-catalyzed Michael additions.[76,77]

best catalysts, **G4DNA-2** (up to +81% ee) and **G4DNA-3** (up to −86% ee), obtained after several rounds of site mutation starting from the thrombin-binding aptamer **G4DNA** induced reversed selectivities.

As said before, the binding mode of several ligands, including dmbpy **L12**, remained broadly unclear. To address this issue, Jäschke and coworkers covalently attached bipy units to a **G4DNA** scaffold adopting a parallel conformation (Figure 19).[76] The bipys were introduced at two different positions of the **G4DNA** scaffold through Sonogashira couplings and the resulting **G4DNA-4** and **G4DNA-5** were evaluated in the Michael addition of dimethyl malonate to the enone shown in Figure 19. In the case of **G4DNA-4**, where the 1-hexynyl side chain of the bipy ligand is attached to a loop that stabilises the parallel topology of the **G4DNA**, ees up to 52% were obtained. Interestingly, with a longer side chain, the opposite enantiomer was formed as the major stereoisomer.[77] In the case of

G4DNA-3, where the ligand is placed in a loop favoring an antiparallel folding, the opposite enantiomer was formed in up to −92% ee. In this case, however, increasing the length of the linker led to a drop in selectivity. The divergent behavior of **G4DNA-2** and **G4DNA-3** was particularly noteworthy as the two positions that have been modified in their strands are only two nucleotides apart, thus showing that the selectivity outcome can be completely shifted with appropriate modifications in the same region of the **G4DNA**.

B. Tuning the Selectivity Outcome Using Mirror-Image DNAs

As exemplified in the previous sections, playing with the denticity of the ligand (Section II.A), its position within a sequence or the topology of a **G4DNA** (Section IV.A) can help tune and even reverse the enantioselectivity. The Arseniyadis and Smietana group developed a more general strategy to reverse the selectivity of a given reaction and thus allow a straightforward access to both enantiomers. Indeed, by simply switching to L-DNA, mirror image of the naturally occurring D-DNA,[78] one can reliably reverse the selectivity outcome (Figure 20).[79] D- and L-ON sequences were assembled on a DNA synthetizer and combined then with the Cu^{II}–**L12** complexes. The catalytic activity and enantioselectivity of the enantiomeric catalysts were compared in the Michael addition of various nucleophiles to enones **4**, leading to opposite stereochemical control. This strategy was also verified when using D- and L-RNA strands.[39]

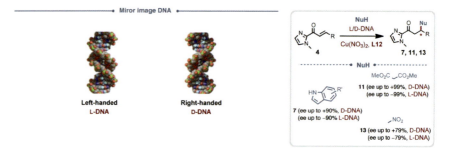

Figure 20. Reversing the stereochemical outcome of DNA-based asymmetric catalysis: DNA *vs* mirror-image DNA.[79]

C. Improving Scalability and Sustainability Using Solid-Supported DNA

One of the last advances made in the field of DNA-based asymmetric catalysis concerns the use of solid-supported DNA as a cheap, robust, and recyclable catalyst, enabling large-scale applications. DNA was first immobilized on silica through noncovalent interactions and evaluated then on the model DNA-catalyzed Diels–Alder reaction (see Figure 2) in the presence of Cu^{II}–bipy **L12**.[80] Quasi-complete conversion of the starting material could be obtained along with high enantioselectivities (up to 94% ee). Most importantly, the catalyst could be recycled up to 10 times before some erosion of the enantioselectivity could be observed. Later, Cu^{II}-containing DNA–silica mineral complexes were used in the same reaction, affording slightly improved enantioselectivities.[81]

In a similar strategy, Smietana, Arseniyadis, and coworkers used ct-DNA covalently attached to cellulose to catalyse Friedel–Crafts alkylations and Michael additions, affording results that compared with the ones obtained with st-DNA (Figure 21).[82] Interestingly, the reactions could be run under continuous flow conditions using a column loaded with supported DNA, allowing to scale up the reactions without any noticeable loss in either reactivity or selectivity.

D. Bringing DNA into the Realm of Asymmetric Photocatalysis

One of the latest developments made in the field of DNA-based asymmetric catalysis is a photocatalyzed [2+2] cycloaddition, promoted by a

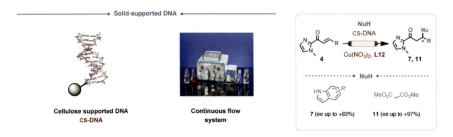

Figure 21. Cellulose-supported DNA for catalyst recycling and continuous flow process.[82]

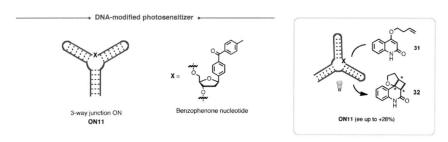

Figure 22. Three-way junction DNA-catalyzed [2+2] cycloaddition of quinolinone **29**.[83]

benzophenone-modified three-way junction DNA, reported by Wagenknecht and coworkers (Figure 22).[83] This asymmetric photosensitizing procedure has been applied to the intramolecular [2+2] cycloaddition that converts the 4-butenyloxyquinolinone **31** into the heteropolycyclic derivative **32**. The hydrophobic pocket created by the three-way junction plays a crucial role here as it is used here as a recognition site for substrate **32**. That's why the reaction failed when using the structurally different ds-DNA. Although moderate selectivities were obtained (up to 28% ee), this work opened the way to new applications of DNA-based catalysis, as far as it represents the first example of chirality transferred from a DNA scaffold to a photosensitizer.

Arseniyadis, Smietana, and coworkers have also been interested in developing a DNA-based [2+2] photocycloaddition reaction (Figure 23).[84] In contrast to the approach of Wagenknecht and coworkers, DNA was used here to template the reaction rather than to catalyse it. Interestingly, the photodimerization of **33**, which does not occur in solution in the absence of DNA, was shown to proceed effectively when DNA was added to the reaction mixture. The corresponding dimerized product **34** was obtained in high yield, although purification was quite tedious. DNA was shown to act as a template, allowing the organization of the monomers in a way suitable for the cycloaddition to proceed. Most importantly, this DNA-templated photodimerization was applied to the synthesis of dictazole B (**37**), a cyclobutane-containing marine natural product isolated form several sponges and corals from the Pacific Ocean. The condensation of the brominated (*E*)-aplysinopsin **35** and the brominated *nor*-aplysinopsin **36** led to the formation of the

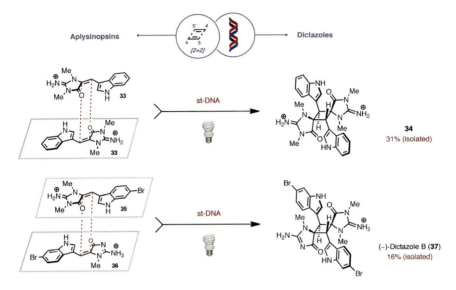

Figure 23. DNA-templated photo-induced [2+2] cycloaddition of aplysinopsins.[84]

corresponding heterodimer dictazole B (**37**)[85] in 16% yield in a single run (a 0.15-mmol scale), which compared favorably with the 3.4% yield previously obtained under solid-state conditions, after combining 28 batches, each on a 0.025-mmol scale.[86] In sharp contrast, a control experiment performed in the absence of DNA resulted in the full recovery of the starting material, which proved the role of DNA in the cycloaddition process. Moreover, this role was also confirmed by the obtention of nonracemic dictazole B (7% ee).

Interestingly, the *pseudo*-dictazole **34** could be further converted to another natural product, tubastrindole B (**38**), through a biomimetic ring-expansion featuring a trifluoroacetic acid (TFA)-mediated ionic rearrangement (Figure 24). Tubastrindole B was obtained in a good 40% isolated yield. These experiments demonstrated that this elegant and straightforward DNA-templated [2+2]/ring-opening sequence provide an entire family of aplysinospin-derived natural products.

Overall, these recent advances illustrate a promising application of DNA as a promoter in photocatalytic reactions, whose potential remains largely unexplored.

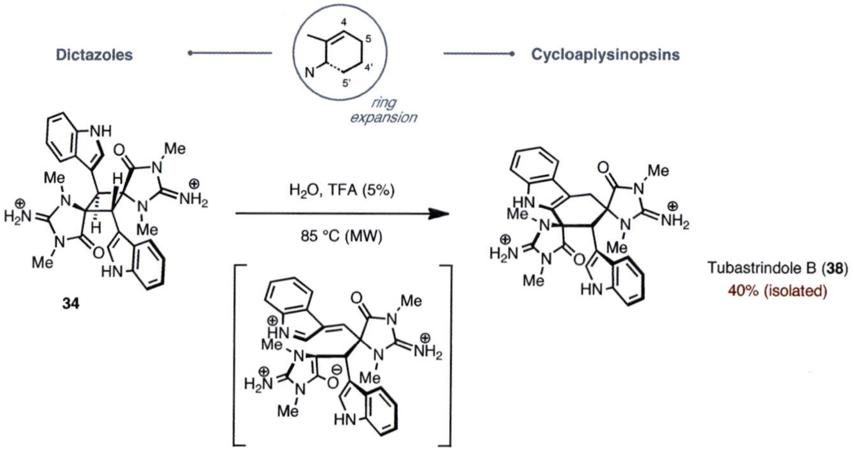

Figure 24. Ionic intramolecular rearrangement of dictazoles into cycloaplysinopsins. Synthesis of tubastindrole B.

V. Conclusion

Since the seminal work of Roelfes and Feringa, less than two decades ago, the use of DNA in the context of asymmetric catalysis has met with an ever-growing success. Since then, an impressive variety of transformations has been unveiled using various DNA-based catalytic systems. Nowadays, the field is no longer limited to Lewis acid–catalyzed transformations; it also encompasses asymmetric organometallic catalysis, through the use of Cu-, Fe-, Pd-, or Ir-based catalysts, as well as asymmetric photo-induced transformations. Most importantly, both carbon–carbon and carbon–heteroatoms bond-forming reactions are now accessible in yields and selectivities sometimes surpassing the catalytic system routinely employed in asymmetric synthesis. On a purely synthetic point of view, the field is now wide-open and innovative transformations are underway.

This search for the optimum catalytic system has also witnessed the design of new catalysts capable, through various modes of binding, to strongly influence the selectivity outcome of a reaction. In parallel, the progress made in the field of ON synthesis has allowed the design and use of efficient, covalently-anchored metallic cofactors, solid-supported ON, or even noncanonical DNA architectures. These various advances have allowed tremendous improvements in sustainability, versatility, and scalability, which ultimately resulted in the field gaining in credibility.

Nonetheless, these last two decades have also been the stage for another and potentially more important development in the field of DNA-based asymmetric catalysis. Indeed, a tremendous body of work has been dedicated to the rationalization of the chirality transfer from the DNA to any given product. The field hence merged into a rich domain, where structural biology easily rallies pure organic synthesis and computational chemistry to bear a deeper comprehension of the mechanisms at stake and decipher the various structural and experimental parameters required to reach high levels of reactivity and selectivity.

The development of efficient and rational ON-based catalysts in the context of asymmetric synthesis has been thriving and numerous research groups have joined the enterprise. However, it is not only a valuable exercise for academic research as it also holds great promise for applications in medicine and biotechnology. Additionally, the development of ON-based catalytic tools can provide a straightforward and highly selective access to molecules and scaffolds of interest for the chemical, pharmaceutical, and agrochemical industries. Moreover, the mild reaction conditions and the recyclability of the catalytic systems strongly limit the environmental impacts of DNA-based catalytic processes, making this domain perfectly fitted for the major challenges synthetic organic chemists are now facing to comply with the demanding policies of our current industrial era.

VI. Acknowledgments

We are deeply grateful to the Agence Nationale de la Recherche (ANR-2010-JCJC-715-1, ANR-2015-CE29-0021-01, ANR-20-CE07-0021-01), the China Scholarship Council (CSC), Queen Mary University of London, and the University of Montpellier for continuous support.

VII. References

1. Davies, R. R.; Distefano, M. D. *J. Am. Chem. Soc.* **1997**, *119*, 11643–11652.
2. Roy, R. S.; Imperiali, B. *Protein Eng. Des. Sel.* **1997**, *10*, 691–698.
3. Blackburn, G. M. Ed., *Nucleic Acids in Chemistry and Biology*, RSC Pub, Cambridge, UK, **2006**.

4. Letondor, C.; Humbert, N.; Ward, T. R. *Proc. Natl. Acad. Sci. USA.* **2005**, *102*, 4683–4687.
5. Mihovilovic, M. D. *J. Chem. Technol. Biotechnol.* **2007**, *82*, 1067–1071.
6. Biggs, G. S.; Klein, O. J.; Boss, S. R.; Barker, P. D. *Johnson Matthey Technol. Rev.* **2020**, *64*, 407–418.
7. Schwizer, F.; Okamoto, Y.; Heinisch, T.; Gu, Y.; Pellizzoni, M. M.; Lebrun, V.; Reuter, R.; Köhler, V.; Lewis, J. C.; Ward, T. R. *Chem. Rev.* **2018**, *118*, 142–231.
8. Steinreiber, J.; Ward, T. R. *Coord. Chem. Rev.* **2008**, *252*, 751–766.
9. Dürrenberger, M.; Ward, T. R. *Curr. Opin. Chem. Biol.* **2014**, *19*, 99–106.
10. Hoarau, M.; Hureau, C.; Gras, E.; Faller, P. *Coord. Chem. Rev.* **2016**, *308*, 445–459.
11. Wieszczycka, K.; Staszak, K. *Coord. Chem. Rev.* **2017**, *351*, 160–171.
12. Jeschek, M.; Panke, S.; Ward, T. R. *Trends Biotechnol.* **2018**, *36*, 60–72.
13. Metrano, A. J.; Chinn, A. J.; Shugrue, C. R.; Stone, E. A.; Kim, B.; Miller. S. J. *Chem. Rev.* **2020,** *120*, 11479–11615.
14. Roelfes, G.; Feringa, B. L. *Angew. Chem. Int. Ed.* **2005**, *44*, 3230–3232.
15. Boersma, A. J.; Megens, R. P.; Feringa, B. L.; Roelfes, G. *Chem. Soc. Rev.* **2010**, *39*, 2083–2092.
16. Rosati, F.; Roelfes, G. *ChemCatChem.* **2010**, *2*, 916–927.
17. Silverman, S. K. *Angew. Chem. Int. Ed.* **2010**, *49*, 7180–7201.
18. Park, S.; Sugiyama, H. *Angew. Chem. Int. Ed.* **2010**, *49*, 3870–3878.
19. Park, S.; Sugiyama, H. *Molecules.* **2012**, *17*, 12792–12803.
20. Rioz-Martínez, A.; Roelfes, G. *Curr. Opin. Chem. Biol.* **2015**, *25*, 80–87.
21. Drienovská, I.; Roelfes, G. *Isr. J. Chem.* **2015**, *55*, 21–31.
22. Duchemin, N.; Heath-Apostolopoulos, I.; Smietana, M.; Arseniyadis, S. *Org. Biomol. Chem.* **2017**, *15*, 7072–7087.
23. Lerman, L. S. *J. Mol. Biol.* **1961**, *3*, 18–30.
24. Rideout, D. C.; Breslow, R. *J. Am. Chem. Soc.* **1980**, *102*, 7816–7817.
25. Rosati, F.; Boersma, A. J.; Klijn, J. E.; Meetsma, A.; Feringa, B. L.; Roelfes, G. *Chem. Eur. J.* **2009**, *15*, 9596–9605.
26. Roelfes, G.; Boersma, A. J.; Feringa, B. L. *Chem. Commun.* **2006**, 635–637.
27. Boersma, A. J.; Klijn, J. E.; Feringa, B. L.; Roelfes, G. *J. Am. Chem. Soc.* **2008**, *130*, 11783–11790.
28. Draksharapu, A.; Boersma, A. J.; Leising, M.; Meetsma, A.; Browne, W. R.; Roelfes, G. *Dalton Trans.* **2015**, *44*, 3647–3655.
29. Draksharapu, A.; Boersma, A. J.; Browne, W. R.; Roelfes, G. *Dalton Trans.* **2015**, *44*, 3656–3663.
30. Boersma, A. J.; de Bruin, B.; Feringa, B. L.; Roelfes, G. *Chem. Commun.* **2012**, *48*, 2394–2396.
31. Mansot, J.; Vasseur, J.-J.; Arseniyadis, S.; Smietana, M. *ChemCatChem.* **2019**, *11*, 5686–5704.
32. Lauberteaux, J.; Pichon, D.; Baslé, O.; Mauduit, M.; Marcia de Figueiredo, R.; Campagne, J.-M. *ChemCatChem.* **2019**, *11*, 5705–5722.

33. Evans, D. A.; Fandrick, K. R.; Song, H.-J. *J. Am. Chem. Soc.* **2005**, *127*, 8942–8943.
34. Boersma, A. J.; Feringa, B. L.; Roelfes, G. *Org. Lett.* **2007**, *9*, 3647–3650.
35. Harshman, K. D.; Dervan, P. B. *Nucleic Acids Res.* **1985**, *13*, 4825–4835.
36. Dervan, P. *Science.* **1986**, *232*, 464–471.
37. Amirbekyan, K.; Duchemin, N.; Benedetti, E.; Joseph, R.; Colon, A.; Markarian, S. A.; Bethge, L.; Vonhoff, S.; Klussmann, S.; Cossy, J.; Vasseur, J.-J.; Arseniyadis, S.; Smietana, M. *ACS Catal.* **2016**, *6*, 3096–3105.
38. (a) Boersma, A. J.; Feringa, B. L.; Roelfes, G. *Angew. Chem. Int. Ed.* **2009**, *48*, 3346–3348. For other examples of DNA-based asymmetric Friedel-Crafts reactions, see: (b) García-Fernández, A.; Megens, R. P.; Villarino, L.; Roelfes, G. *J. Am. Chem. Soc.* **2016**, *138*, 16308–16314. (c) Mansot, J.; Aubert, S.; Duchemin, N.; Vasseur, J.-J.; Arseniyadis, S.; Smietana, M. *Chem. Sci.* **2019**, *10*, 2875–2881.
39. Duchemin, N.; Benedetti, E.; Bethge, L.; Vonhoff, S.; Klussmann, S.; Vasseur, J.-J.; Cossy, J.; Smietana, M.; Arseniyadis, S. *Chem. Commun.* **2016**, *52*, 8604–8607.
40. Park, S.; Ikehata, K.; Watabe, R.; Hidaka, Y.; Rajendran, A.; Sugiyama, H. *Chem. Commun.* **2012**, *48*, 10398–10400.
41. Petrova, G. P.; Ke, Z.; Park, S.; Sugiyama, H.; Morokuma, K. *Chem. Phys. Lett.* **2014**, *600*, 87–95.
42. Coquière, D.; Feringa, B. L.; Roelfes, G. *Angew. Chem. Int. Ed.* **2007**, *46*, 9308–9311.
43. Li, Y.; Wang, C.; Jia, G.; Lu, S.; Li, C. *Tetrahedron.* **2013**, *69*, 6585–6590.
44. Mansot, J.; Lauberteaux, J.; Lebrun, A.; Mauduit, M.; Vasseur, J.-J.; Marcia de Figueiredo, R.; Arseniyadis, S.; Campagne, J.-M.; Smietana, M. *Chem. Eur. J.* **2020**, *26*, 3519–3523.
45. Shibata, N.; Yasui, H.; Nakamura, S.; Toru, T. *Synlett.* **2007**, *2007*, 1153–1157.
46. Dijk, E. W.; Feringa, B. L.; Roelfes, G. *Tetrahedron Asymmetry.* **2008**, *19*, 2374–2377.
47. Megens, R. P.; Roelfes, G. *Org. Biomol. Chem.* **2010**, *8*, 1387–1393.
48. Zhao, H.; Shen, K. *RSC Adv.* **2014**, *4*, 54051–54059.
49. Boersma, A. J.; Coquière, D.; Geerdink, D.; Rosati, F.; Feringa, B. L.; Roelfes, G. *Nat. Chem.* **2010**, *2*, 991–995.
50. Rosati, F.; Roelfes, G. *ChemCatChem.* **2011**, *3*, 973–977.
51. Megens, R. P.; Roelfes, G. *Chem. Commun.* **2012**, *48*, 6366–6368.
52. Oelerich, J.; Roelfes, G. *Chem. Sci.* **2013**, *4*, 2013–2017.
53. Rioz-Martínez, A.; Oelerich, J.; Ségaud, N.; Roelfes, G. *Angew. Chem. Int. Ed.* **2016**, *55*, 14136–14140.
54. Singh, Y.; Murat, P.; Defrancq, E. *Chem. Soc. Rev.* **2010**, *39*, 2054–2070.
55. Oltra, N. S.; Roelfes, G. *Chem. Commun.* **2008**, 6039–6041.
56. Gjonaj, L.; Roelfes, G. *ChemCatChem.* **2013**, *5*, 1718–1721.
57. Park, S.; Zheng, L.; Kumakiri, S.; Sakashita, S.; Otomo, H.; Ikehata, K.; Sugiyama, H. *ACS Catal.* **2014**, *4*, 4070–4073.
58. Park, S.; Okamura, I.; Sakashita, S.; Yum, J. H.; Acharya, C.; Gao, L.; Sugiyama, H. *ACS Catal.* **2015**, *5*, 4708–4712.
59. Su, M.; Tomás-Gamasa, M.; Carell, T. *Chem. Sci.* **2015**, *6*, 632–638.

60. Ropartz, L.; Meeuwenoord, N. J.; van der Marel, G. A.; van Leeuwen, P. W. N. M.; Slawin, A. M. Z.; Kamer, P. C. J. *Chem. Commun.* **2007**, 1556–1558.
61. Caprioara, M.; Fiammengo, R.; Engeser, M.; Jäschke, A. *Chem. Eur. J.* **2007**, *13*, 2089–2095.
62. Fournier, P.; Fiammengo, R.; Jäschke, A. *Angew. Chem. Int. Ed.* **2009**, *48*, 4426–4429.
63. Burge, S.; Parkinson, G. N.; Hazel, P.; Todd, A. K.; Neidle, S. *Nucleic Acids Res.* **2006**, *34*, 5402–5415.
64. Nikolova, E. N.; Kim, E.; Wise, A. A.; O'Brien, P. J.; Andricioaei, I.; Al-Hashimi, H. M. *Nature.* **2011**, *470*, 498–502.
65. Cozzoli, L.; Gjonaj, L.; Stuart, M. C. A.; Poolman, B.; Roelfes, G. *Chem. Commun.* **2018**, *54*, 260–263.
66. Roe, S.; Ritson, D. J.; Garner, T.; Searle, M.; Moses, J. E. *Chem. Commun.* **2010**, *46*, 4309–4311.
67. Wang, C.; Jia, G.; Zhou, J.; Li, Y.; Liu, Y.; Lu, S.; Li, C. *Angew. Chem. Int. Ed.* **2012**, *51*, 9352–9355.
68. Wang, C.; Jia, G.; Li, Y.; Zhang, S.; Li, C. *Chem. Commun.* **2013**, *49*, 11161–11163.
69. Li, Y.; Wang, C.; Hao, J.; Cheng, M.; Jia, G.; Li, C. *Chem. Commun.* **2015**, *51*, 13174–13177.
70. Wang, C.; Li, Y.; Jia, G.; Liu, Y.; Lu, S.; Li, C. *Chem. Commun.* **2012**, *48*, 6232–6234.
71. Cheng, M.; Li, Y.; Zhou, J.; Jia, G.; Lu, S.-M.; Yang, Y.; Li, C. *Chem. Commun.* **2016**, *52*, 9644–9647.
72. Zhao, H.; Shen, K. *Biotechnol. Prog.* **2016**, *32*, 891–898.
73. Wilking, M.; Hennecke, U. *Org. Biomol. Chem.* **2013**, *11*, 6940–6945.
74. Li, Y.; Cheng, M.; Hao, J.; Wang, C.; Jia, G.; Li, C. *Chem. Sci.* **2015**, *6*, 5578–5585.
75. Hao, J.; Miao, W.; Cheng, Y.; Lu, S.; Jia, G.; Li, C. *ACS Catal.* **2020**, *10*, 6561–6567.
76. Dey, S.; Jäschke, A. *Angew. Chem. Int. Ed.* **2015**, *54*, 11279–11282.
77. Dey, S.; Rühl, C. L.; Jäschke, A. *Chem. Eur. J.* **2017**, *23*, 12162–12170.
78. Urata, H.; Shinohara, K.; Ogura, E.; Ueda, Y.; Akagi, M. *J. Am. Chem. Soc.* **1991**, *113*, 8174–8175.
79. Wang, J.; Benedetti, E.; Bethge, L.; Vonhoff, S.; Klussmann, S.; Vasseur, J.-J.; Cossy, J.; Smietana, M.; Arseniyadis, S. *Angew. Chem. Int. Ed.* **2013**, *52*, 11546–11549.
80. Park, S.; Ikehata, K.; Sugiyama, H. *Biomater. Sci.* **2013**, *1*, 1034–1036.
81. Sakashita, S.; Park, S.; Sugiyama, H. *Chem. Lett.* **2017**, *46*, 1165–1168.
82. Benedetti, E.; Duchemin, N.; Bethge, L.; Vonhoff, S.; Klussmann, S.; Vasseur, J.-J.; Cossy, J.; Smietana, M.; Arseniyadis, S. *Chem. Commun.* **2015**, *51*, 6076–6079.
83. Gaß, N.; Gebhard, J.; Wagenknecht, H.-A. *ChemPhotoChem.* **2017**, *1*, 48–50.
84. Duchemin, N.; Skiredj, A.; Mansot, J.; Leblanc, K.; Vasseur, J.-J.; Benidir, M. A.; Evanno, L.; Poupon, E.; Smietana, M.; Arseniyadis, S. *Angew. Chem. Int. Ed.* **2018**, *57*, 11786–11791.
85. Bialonska, D.; Zjawiony, J. K. *Mar. Drugs.* **2009**, *7*, 166–183.
86. Skiredj, A.; Beniddir, M. A.; Joseph, D.; Leblanc, K.; Bernadat, G.; Evanno, L.; Poupon, E. *Angew. Chem. Int. Ed.* **2014**, *53*, 6419–6424.

© 2022 World Scientific Publishing Company
https://doi.org/10.1142/9789811248436_0002

2 Chiral Secondary Phosphine Oxides as Preligands in Enantioselective Catalysis

Romain Membrat,*[,‡] Didier Nuel,[†,§] Laurent Giordano[†,¶] and Alexandre Martinez[†,∥]

*Provepharm Life Solutions, 22 rue Marc Donadile, 13013 Marseille, France
[†]Aix-Marseille University, CNRS, Centrale Marseille, iSm2, Marseille, France
[‡]romain.membrat@provepharm.com
[§]dnuel@centrale-marseille.fr
[¶]lgiordano@centrale-marseille.fr
[∥]amartinez@centrale-marseille.fr

Table of Contents

List of Abbreviations	32
I. Introduction	33
II. Synthesis, Structure, and Properties of SPOs and the Corresponding Coordination Complexes	35
III. P-Chirogenic SPOs in Enantioselective Catalysis	38
IV. Other Chiral SPOs in Enantioselective Catalysis	40

A.	SPOs with Backbone Located Chirality	40
B.	SPOs with Combined P-located and Backbone Located Chirality	44
V.	Chiral Mixed Ligand Strategy	47
VI.	Conclusion	53
VII.	References	54

List of Abbreviations

AAA	asymmetric allylic alkylation
Ac	acetyl
Ad	adamantyl
Ar	aryl
BINAP	2,2′-bis(diphenylphosphino)-1,1′-binaphthyl
BSA	bis(trimethylsilyl)acetamide
CML	"Chiral Mixed Ligand" strategy
cod	1,5-cyclooctadiene
DCM	dichloromethane
DIPAMP	ethane-1,2-diylbis[(2-methoxyphenyl)phenylphosphane]
Dnp	dinaphthyl
DFT	density functional theory
ee	enantiomeric excess
HASPO	heteroatom substituted secondary phosphine oxide
HPLC	high performance liquid chromatography
L	undefined neutral ligand
M/PA	metallic complex bearing a phosphinous acid ligand
M/PAP	metallic complex bearing a phosphinito-phosphinous acid ligand
Men	menthyl
MS3A	molecular sieves (3 Ångström)
nbd	norbornadiene
NHC	N-heterocyclic carbene
PA	Phosphinous Acid Ligand
PAP	phosphinito-phosphinous acid ligand
PPTS	pyridinium p-toluenesulfonate
Py	pyridine

NMR	magnetic nuclear resonance
SPO	secondary phosphine oxide
THF	tetrahydrofuran
TMS	trimethylsilyl

I. Introduction

Ligand design and synthesis are always crucial steps on the way to the development of efficient enantioselective catalytic processes.[1] The holy grail of this challenging task is to find ligands that combine high activity and enantioselectivity. Since the first report on the use of chiral phosphine ligands in metal-catalyzed enantioselective hydrogenation by Horner *et al.* in 1968,[2] the organic chemists community actively contrived to the conception of chiral trivalent phosphorus derivatives.[3–4] For the practical purpose, trivalent phosphines are highly valuable ligands since a broad palette of steric and electronic properties could be explored by modulation of the phosphorus substitution or replacement of carbon groups by heteroatom-containing moieties.[5–6] Trialkylphosphines are known as excellent σ-donor ligands while phosphites or phosphorus trihalides are commonly recognized as weakly π-acceptor ligands.[7] Although monophosphines can provide good ligands for enantioselective catalysis, bidentate diphosphines are commonly recognized as more efficient. Their catalytic properties have been modulated to a large extent, and a wide variety of bite angles could be surveyed by changing the length of their bridging chain.[8] It should be stressed that, contrary to most tertiary amines, where nitrogen can't be a chirogenic center because of the rapid Walden inversion at room temperature, P-chiral tertiary phosphines are configurationally stable. However, phosphorus racemization issues can arise under catalytic conditions.[9]

A substantial number of metal-catalyzed enantioselective transformations using enantiomerically pure P-chirogenic phosphines as chiral inductors have been reported, including the industrial synthesis of L-DOPA (levodopa and 1-3,4-dihydroxyphenylalanine) with the P-chirogenic DIPAMP (ethane-1,2-diylbis[(2-methoxyphenyl)phenylphosphane]).[10–13] However, phosphorus-based catalysts dramatically suffer from the high sensitivity of phosphorus to oxidation by air, which may be very prejudicial for catalyst's life time. Air sensitivity also makes

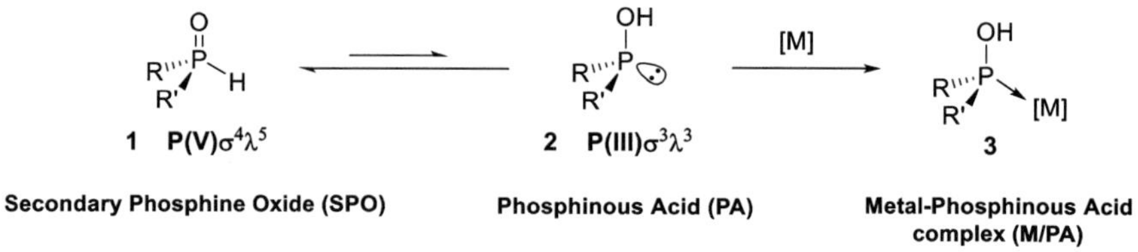

Scheme 1. Tautomeric equilibrium allowing the use of SPO as preligand for transition metals.

the synthesis and handling of phosphines very difficult. Protection of the phosphorus center by Lewis acids like BH_3 is a suitable solution to tackle these limitations, but it leads to additional synthetic steps, as long as treatment by a Lewis base is required to generate the trivalent species before complexation.[14] The synthesis and the storage of the ligands on the phosphine oxide form could be a suitable alternative, but this strategy is limited by the small number of methods reported so far for the stereospecific reduction of the P=O bonds.[15] In this context, secondary phosphine oxides (SPOs), including P-chirogenic species, are worth considering because of their peculiar properties. These compounds exist in tautomeric equilibrium between a pentavalent $P(V)\sigma^4\lambda^5$ form (Scheme 1, SPO, **1**) and a trivalent $P(III)$ $\sigma^3\lambda^3$ form (phosphinous acid form, PA, **2**) which could be trapped by a transition metal to form the corresponding complex M/PA, **3**, as demonstrated by the early works of Chatt and Heaton in 1968.[16,17]

SPOs are insensitive to oxidation, can be synthesized and stored under air atmosphere, and provide all the positive assets of tertiary phosphines through their P(III) form, including the catalytic applications of their transition metal complexes.[18] The first example of use of a PA ligand in Rh catalysis has been published by van Leeuwen and coworkers in 1986 (hydroformylation of olefins).[19] It is especially noteworthy that P(V)/P(III) equilibrium and complexation to the metal center proceed with complete retention of configuration at the phosphorus atom.[20] Thus, it is not surprising to find in the literature a substantial number of enantioselective catalytic systems based on chiral SPOs preligands, as summarized in several recent reviews.[21–25] These studies are presented briefly hereafter, by considering both P-chiral SPOs and compounds where the chirality is introduced through the ligand backbone.

II. Synthesis, Structure, and Properties of SPOs and the Corresponding Coordination Complexes

Since the very first report on an SPO published by Hamilton and Williams in 1952,[26] several methods have been disclosed for the preparation of achiral or racemic SPOs. The synthetic approaches can be categorized into two types: the nucleophilic substitutions of one or two alkoxy groups on a phosphinate or a phosphonate, and the hydrolysis of a P–OR or a P–halogen bond.[27–32] Enantiomerically pure P-chirogenic SPOs have been considered first as intermediates in the synthesis of P-chirogenic tertiary phosphines and, only recently, SPOs have been used themselves as preligands in enantioselective catalysis.[33,34] There are five general pathways to obtain SPOs in enantioenriched form, as depicted in Scheme 2, through selected examples.

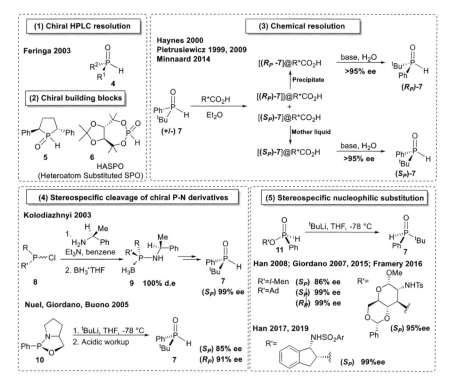

Scheme 2. General pathways for the synthesis of enantioenriched SPOs (Ref. 35–46).

In a 2003 article, Feringa, Minnard, De Vries, and coworkers reported that the enantiomers of a series of P-chirogenic SPOs can be separated easily by chiral HPLC (high performance liquid chromatography) (Scheme 2, (1)), whereas other series of SPOs as well as Heteroatom-Substituted Secondary Phosphine Oxides (HASPOs) with chiral backbones could be obtained from enantiopure substrates and precursors (Scheme 2, (2)).[35] Haynes and coworkers have developed procedures for the resolution of racemic ^{t}BuPhP(O)H, **7**, which involve the formation of diastereomeric complexes with chiral acids, such as (+)-mandelic acid. Both enantiomers were obtained in very good enantiomeric excesses (ees) (Scheme 2, (3)).[30,36–39] A fourth strategy, developed independently by Kolodiazhnyi and Buono, Giordano, Nuel, and coworkers, consists in the stereospecific hydrolysis of the P–N bonds of diastereomerically pure aminophosphine derivatives. For instance, aminophosphines **9** and oxazaphospholidines **10** have been used in these reactions. They were obtained from the corresponding chiral amines via diastereoselective reactions (Scheme 2, (4)).[40,41] A more general pathway is based on the stereospecific nucleophilic substitutions on H-phosphinates with organolithium derivatives, providing access to a broad variety of SPOs (Scheme 2, (5)). The enantiopure precursors were obtained by chiral HPLC (e.g. adamantyl phenylphosphinate[42]), preferential crystallization of diastereomeric mixtures (e.g. *l*-menthylphosphinate[43]), or by diastereoselective synthesis from chiral alcohols (e.g. glucosamide-derived phosphinates,[44] amino indenylphosphinate[45,46]).

The coordination chemistry of SPOs preligands is particularly rich because of the variety of coordination modes offered by these compounds (Scheme 3).[21] SPOs coordinate to early transition metals such as titanium, tungsten, or molybdenum by their "hard" oxygen atom (Scheme 3(a)) whereas their phosphinous acid forms, like PA, coordinate through the "soft" phosphorus atom to late transition metals such as palladium, platinum, rhodium, or ruthenium (Scheme 3, (b)).[47,48] Notably, Chatt and Heaton, Berry and coworkers, and Li provided suitable methodologies to get access to M/PA complexes from symmetrical SPOs preligands from various metal precursors.[17,49,50] A representative example, related to the synthesis of a bimetallic palladium(II) complex, is given in Scheme 3. It should be stressed also that two coordinated PA ligands can self-assemble

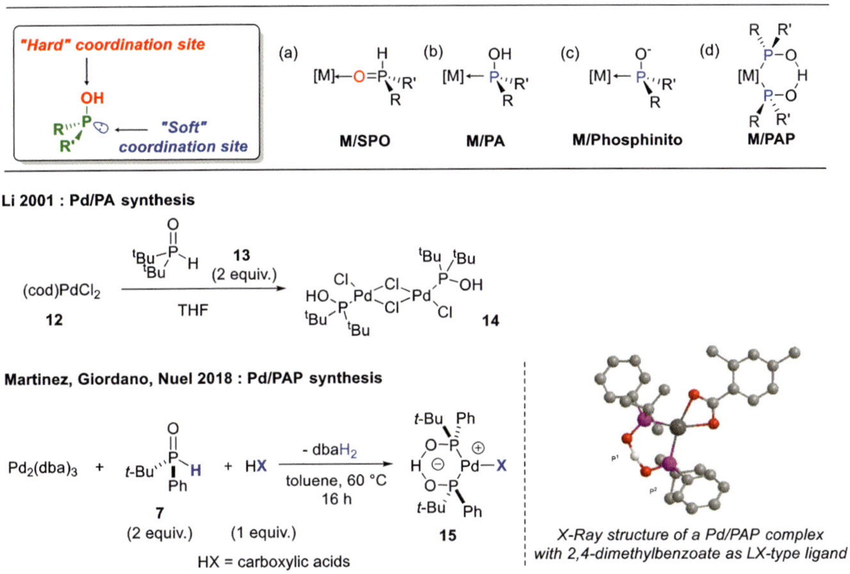

Scheme 3. Coordination modes of SPO and PA ligands. Synthesis of representative complexes (Refs. 21, 50, 56).

by hydrogen bonding, after deprotonation of one of the acid units, to mimic the structure of a bidentate diphosphine. Thus, deprotonation generates anionic phosphinito-phosphinous acid ligands in M/PAP-type complexes (Scheme 3, (d)). This original coordination mode, discovered fortuitously by Trotskaya in 1944[11] and confirmed much later independently by the Palenik and Roundhill groups,[51,52] gave new opportunities in catalysis because of the dynamic behavior of the hydrogen bond–assisted self-assembly. In 2004, Leung and coworkers provided examples of Pd/PAP complexes made from SPOs bearing a stereogeneous phosphorus atom. These reactions take place with retention of the phosphorus stereochemistry, by using NEt$_3$ as a deprotonating agent.[53] Later on, the Buono and the Hong groups developed a simpler methodology to obtain enantiopure or racemic Pd/PAP complexes from commercial metal salts, without the addition of bases.[54,55] In 2018, Nuel, Giordano, Martinez, and coworkers also reported a hydrogen transfer–based procedure to extend the scope of the self-assembly process. The palladium(II) complexes **15** are formed here from a Pd(0) starting material, Pd$_2$(dba)$_3$ (Scheme 3).[56,57]

As highlighted by Buono and coworkers in 2011, the donating ability of PA ligands can be increased further by the deprotonation of the OH function, which leads to phosphinito ligands (Scheme 3c) that are stronger donors than N-heterocyclic carbenes.[57] On a practical point of view, unprecedented properties in (enantioselective) catalysis can be expected with this class of ligands.

III. P-Chirogenic SPOs in Enantioselective Catalysis

In principle, the use of P-chirogenic ligands in organometallic enantioselective catalysis is expected to be beneficial because of the direct proximity of the metal and the P-chirogenic center. In so far, as the stereo discriminating step proceeds close to the metal center, chiral induction should be maximized in that case.[12,58–60] In 2003, Feringa, Minnaard, de Vries, and coworkers reported the first catalytic application of P-chirogenic SPOs, illustrated by the iridium-catalyzed enantioselective hydrogenation of imines (Scheme 4).[35] Among an extensive library of P-chirogenic SPOs, only (R_P)-^{t}BuPhP(O)H **7** provided appreciable ee. The joint use of **7** and pyridine as a co-ligand allowed reaching an 83% ee.

During the same year, but independently, Haynes and coworkers used the P-chirogenic SPO **7** in the asymmetric allylic alkylation (AAA) of 1,3-diphenylallyl acetate **14** (Scheme 5).[20] Good enantioselectivity was obtained by *in situ* generation of the active catalyst (72% ee), while the direct use of preformed, well-defined Pd/PA **16a** or Pd/PAP **16b** and **16c** complexes resulted in lower yields and ees (32 to 57% ee).

In 2004, Minnaard, Feringa, de Vries, and coworkers showed that Rh/SPO-**7** complexes catalyze the enantioselective hydrogenation of activated C=C double bonds. An exhaustive screening of substrates and

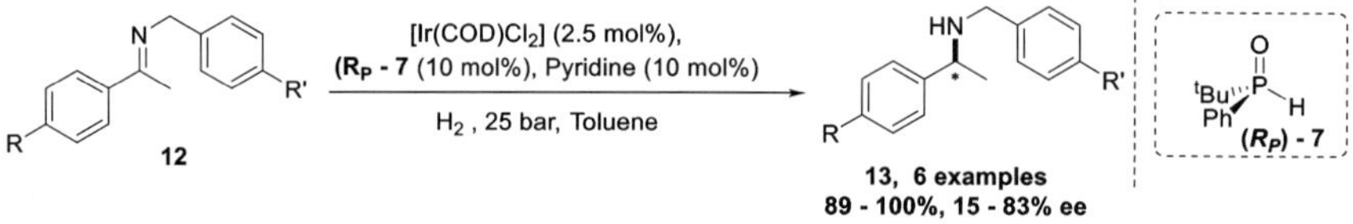

Scheme 4. First uses of P-chirogenic SPO preligand in enantioselective catalysis (Ref. 35).

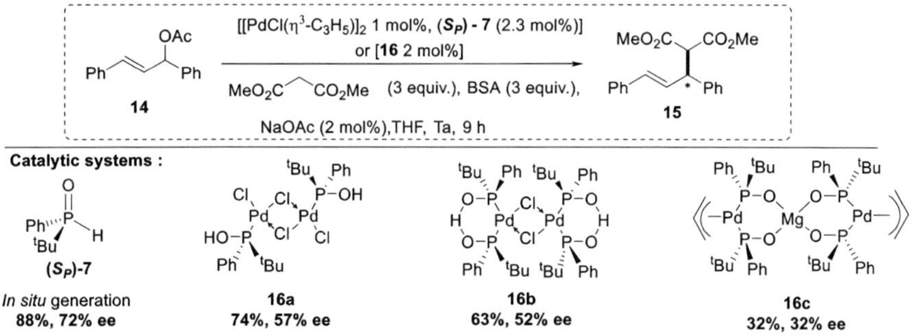

Scheme 5. AAA reactions: comparison of catalysts obtained from P-chirogenic SPO ligands (Ref. 20).

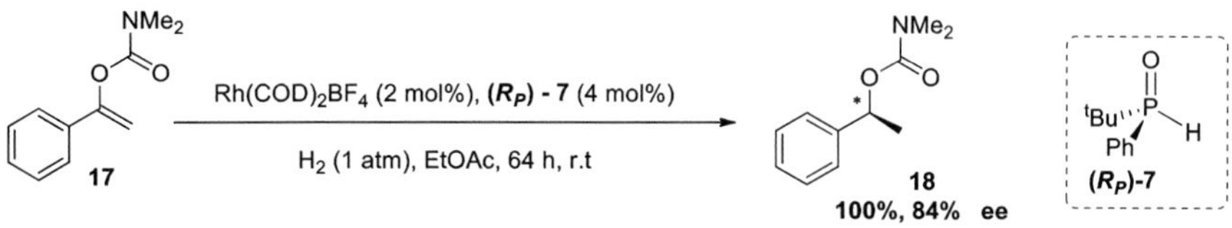

Scheme 6. Hydrogenation of an enol carbamate with Rh/P-chirogenic SPO complexes (Ref. 59).

experimental conditions led to full conversion and a maximum ee of 84% for the hydrogenation of the enol carbamate **17** (Scheme 6).[61]

The Pd/PAP and Pt/PAP complexes exhibited very suitable properties in C–C bond forming reactions through cycloaddition processes.[54,62–64] In 2005, Giordano, Buono, and coworkers demonstrated the efficiency of palladium complexes such as **21** as catalysts for the synthesis of vinyl cyclopropanes by [2+1] cycloaddition between alkynes and norbornene derivatives (Scheme 7).[54] The vinylcyclopropane **20** in Scheme 7 is chiral if the R group is different from a hydrogen atom. The P-chirogenic SPO **7** was converted into the palladium precatalyst **21** that, after activation with AgOAc, afforded the products **20a** and **20b** in moderate ee (59% and 49% ee, respectively). The ees could be enhanced, however, by using (S)-(+)-mandelic acid as an additive. Thus, for **20b**, a 95% ee could be obtained, but with a concomitant erosion of the yield (21%).[65]

This overview of the use of P-chirogenic SPOs in enantioselective catalysis shows that this family of ligands can be convenient for a variety

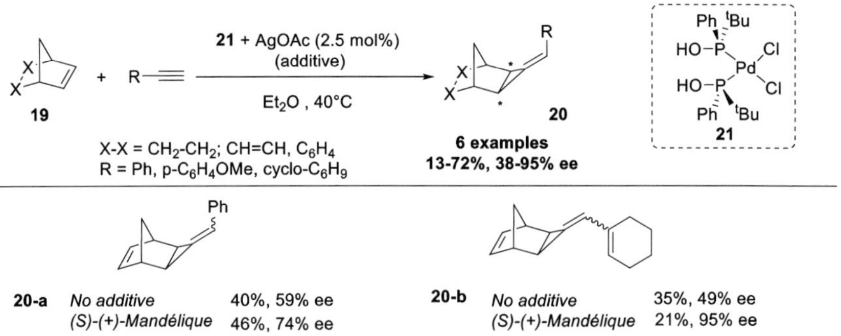

Scheme 7. Pd complexes of P-chirogenic SPOs in enantioselective [2+1] cycloadditions (Refs. 54, 65).

of transformations on model substrates. However, capping of the enantioselectivity level is quickly reached when highly functionalized substrates are targeted. Moreover, the combination of high activity and enantioselectivity is often difficult to achieve. Finally, in this field, the use of chiral additives for enantioselectivity amplification has been a successful strategy. As an alternative, the use of chiral non-P-chirogenic SPOs has been considered, as shown hereafter.

IV. Other Chiral SPOs in Enantioselective Catalysis

A. SPOs with Backbone Located Chirality

Despite a few relevant attempts, the development of P-chirogenic SPOs did not meet the intended level of performance. On another way, the use of non-P-chirogenic chiral diphosphines bearing the chirality element on the backbone proved to be very fruitful in enantioselective catalysis. The success story of (R)-BINAP (2,2′-bis(diphenylphosphino)-1,1′-binaphthyl) and analogues displaying a chiral axis and benefiting from a very high rigidity[66] encouraged the synthesis of non-P-chirogenic SPOs with axial chirality.

In 2011, van Leeuwen and coworkers used a dinuclear Rh/PAP-type complex **24a** bearing a BINOL-derived HASPO ligand **25a** for the enantioselective transfer hydrogenation of acetophenone (Scheme 8).[67] When the hydrogen-bond self-assembled bimetallic complex **24a** was used, the

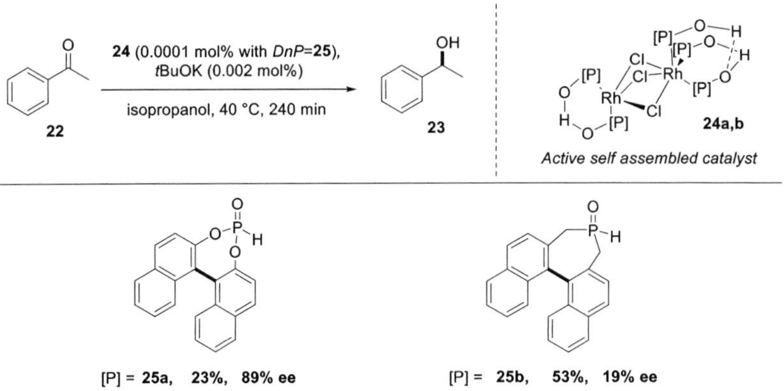

Scheme 8. Transfer hydrogenation with SPO preligands displaying a chiral axis (Ref. 67).

desired alcohol was obtained with up to 89% ee with a 1000 S/Rh ratio. The low yield (23%) was assigned to some instability of the ligand under the basic reaction conditions. To circumvent this drawback, the chiral dinaphthophosphepine derivative **25b** was prepared. The corresponding rhodium complex **25b** provided higher catalytic activity but low ee. It was postulated that these complexes operate through the generation of a metal hydride, as shown later in other series of SPO complexes.[68,69] For this hydrogenation reaction, the authors proposed a concerted, outer sphere hydrogen transfer mechanism that was supported by a computational study.

The high ee obtained with preligand **25a** in this process highlighted the good potential of HASPOs in enantioselective catalysis. The efficiency of these oxides, that feature coordination properties similar to SPOs, has been demonstrated then by a serendipitous discovery of Ding and coworkers, within their studies on the reduction of the α-substituted vinylphosphonic acid **26** (Scheme 9).[70] In this reaction, the Rh(I) complexes of phosphoramidites **28a–d** always gave poor yields and ees, except when the reaction was carried out in the presence of triethylamine. To explain this result, the authors postulated a prior hydrolysis of the phosphoramidites **28** into the corresponding HASPO **29a–d**, promoted by triethylamine. Indeed, conducting the reactions directly with the HASPO preligands **29a–d** led to similar excellent performances. Finally, in this

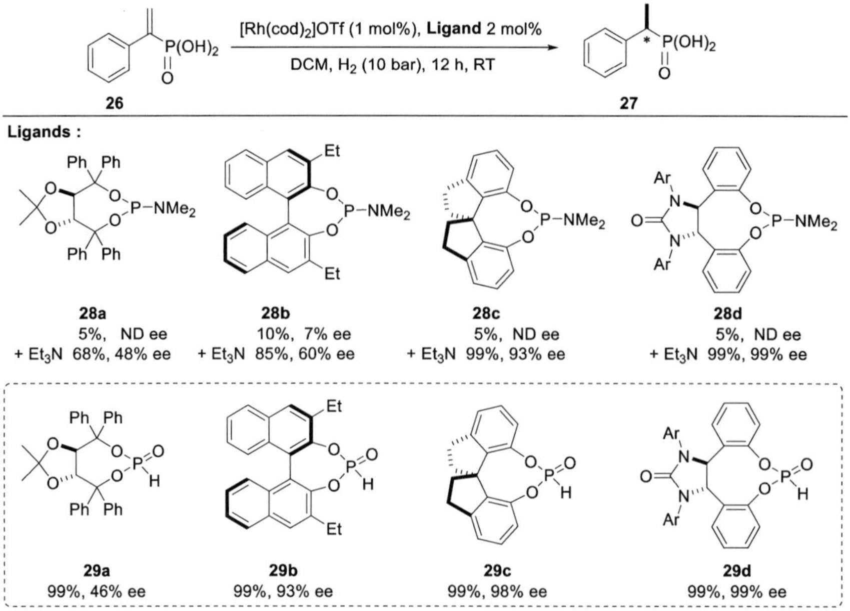

Scheme 9. HASPO preligands in the enantioselective hydrogenation of vinylphosphonic acids (Ref. 70).

study, the Ding group highlighted a library of very successful chiral HASPOs preligands, which has been extensively used then in Rh(I)-promoted hydrogenation reactions. Thus, for instance, the axially chiral preligand **29b**, combined with PPh$_3$ as the co-ligand, allowed the highly enantioselective reduction of a series of β-substituted acrylic acids: 25 examples, ee 91–99%. The method has been applied notably to the synthesis of the natural product (S)-cheryline **31** (Scheme 10).[71] Excellent enantioselectivity (up to 95% ee) was also achieved in the Rh-promoted hydrogenation of α-arylacrylic acid and β-arylbut-3-enoic acid by using HASPO **29d**, which displays central chirality.[72] This catalyst was applied to the synthesis of the anti-inflammatory drug (S)-naproxen **33** as depicted in Scheme 10.

In 2013, Cramer and coworkers reported the use of a heterobimetallic chiral catalyst for the enantioselective intramolecular hydrocarbamoylations of alkenes.[73] This nickel- and aluminum-catalyzed reaction involves activation of the C–H bond in the starting formamides **38**. It was carried

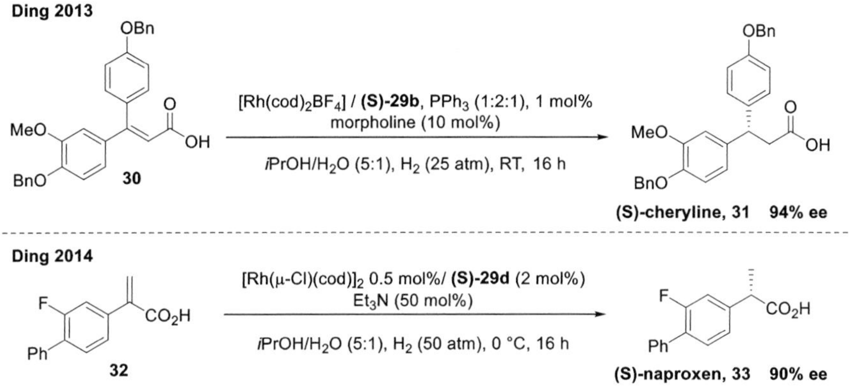

Scheme 10. Application of Rh(I)-HASPO enantioselective hydrogenation to the synthesis of natural or bioactive products (Refs. 71, 72).

out with the *C*-stereogenic HASPO **34** and provided the pyrrolidinones **42** in excellent ees (14 examples, 46–98% yields, 80–98% ee). The proposed mechanism is of particular interest because it reflects the versatility of the SPOs and HASPO coordination chemistry (Scheme 11).

The reaction pathway begins with the trapping of the P(III) form of HASPO, **35**, by AlMe$_3$, *via* coordination of the hard oxygen atom, followed by complexation of Ni(0) through the soft phosphorus atom to generate the active bimetallic chiral catalyst **37**. According to the postulated catalytic cycle, the HASPO-coordinated Ni(0) center of complex **37** is responsible for the C–H activation process, whereas coordination of the aluminum center to the substrate affords the rigidity required to enhance the enantioselectivity (up to 98% ee). A complementary experiment carried out directly using the putative intermediate **36** as the catalyst, without any additional AlMe$_3$, gave similar results, supporting the hypothesis of a bimetallic activation mechanism.

The studies summarized in this section clearly point out that SPOs or HASPOs displaying nonchirogenic phosphorus and chiral backbones can provide good catalytic activity and very high ee in enantioselective organometallic catalysis. On the other hand, the potential of P-chirogenic SPOs seems underexploited so far, to some extent. To complete this outline of the field, the next section will present ligands with combined P- and backbone located chirality.

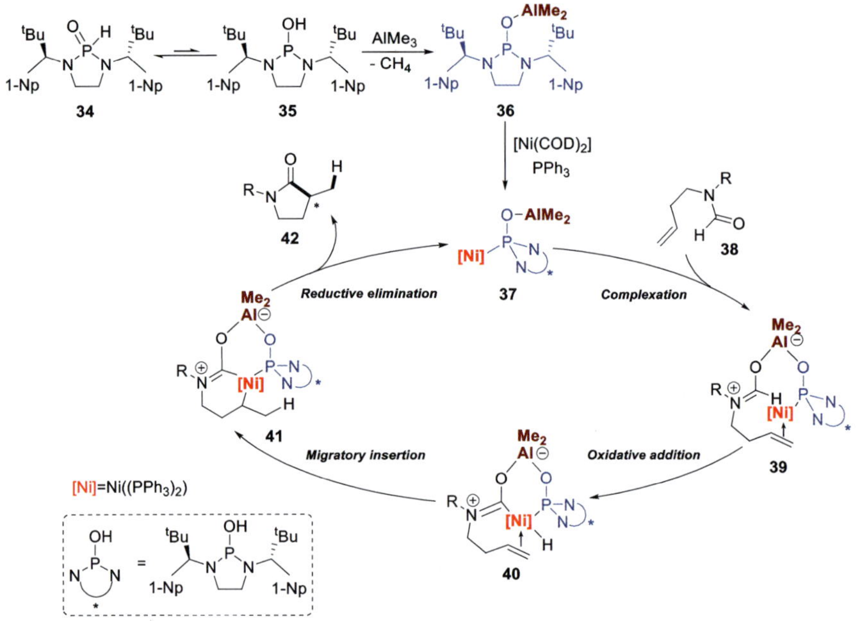

Scheme 11. HASPOs in the enantioselective hydrocarbamoylation of alkenes: coordination modes and reaction mechanism (Ref. 73).

B. SPOs with Combined P-located and Backbone Located Chirality

This section of the review will assess the performance of SPOs combining P-located chirality and backbone located chirality. Potential benefits are multiple. Such ligands generate a chiral environment around the metal center, in both the first and the second coordination spheres, which should increase enantioselection. Moreover, the combination of two different types of chirality (chirogenic phosphorus and chirality axis, for example) can result in synergistic enhancement of the enantioselectivity in catalytic processes.

As early as 2004, Hamada and coworkers proposed the diastereoselective synthesis of HASPO **43a** (diaminophosphine oxide [DIAPHOX], Scheme 12) from enantiopure aspartic acid.[74,75] Compound **43a** bears two chirogenic elements: the phosphorus atom and a carbon center with R and S configurations respectively. The use of this new preligand in the classical

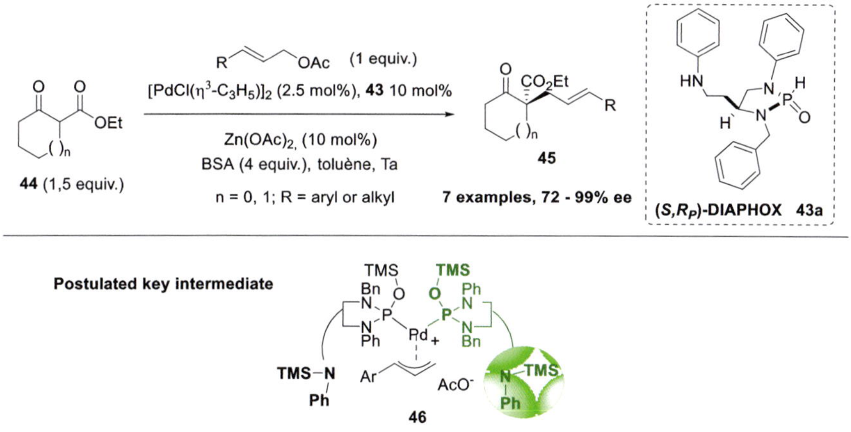

Scheme 12. First use of HASPO ligands combining P-centered chirality and backbone located chirality: AAA reactions with prochiral nucleophiles (Refs. 74, 75).

Pd-catalyzed AAA reactions in Scheme 12 resulted in very high yields and ees (72–92% ee). The reaction required the use of BSA (bis(trimethylsilyl)amide) as the base. This observation and further experiments suggested the hypothesis that the active species should contain a trivalent P-OTMS ligand with a silylated N-function, as in **46**. Thus, BSA acts as both a Brønsted base and a protective reagent for the OH group of the ligand. It was also demonstrated that the noncoordinated nitrogen atom on the sidearm of the ligand plays a crucial role in the enantiofacial discrimination of prochiral nucleophiles.

The judiciously designed preligand **43a** and some analogues have been reused by the same group in 2006 for C–N bond forming processes, that is, in the enantioselective allylic aminations of allylic acetates promoted by Ir(I) complexes (Scheme 13).[76] The influence of the substitution patterns of the aromatic groups in ligands **43** was also investigated. The best results were obtained with preligand **43h** that displays bulky *t*-Bu substituents on the aryl groups. This ligand provided excellent selectivity, yields, and ees (86% ee, 92% ee at −20°C).

DIAPHOX-type preligands **43** also proved highly efficient in the Pd-promoted enantioselective allylic amination of Morita–Baylis–Hillman adducts, giving almost quantitative conversions and enantioselectivity up to 99%. In this case also, tuning of the substitution patterns of the ligand

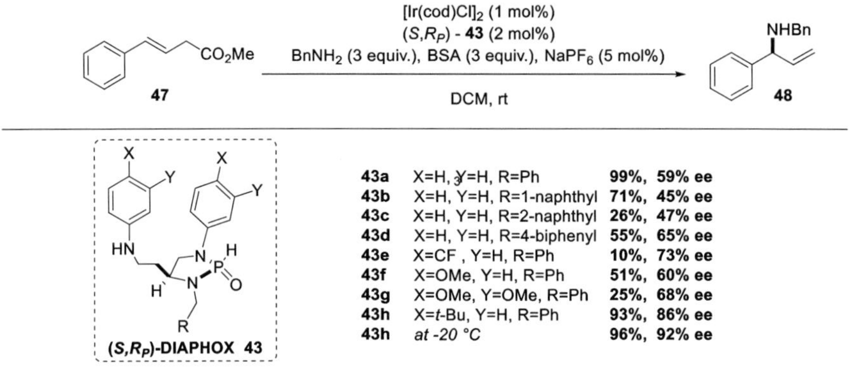
[Ir(cod)Cl]2 (1 mol%)
(S,RP) - 43 (2 mol%)
BnNH2 (3 equiv.), BSA (3 equiv.), NaPF6 (5 mol%)
DCM, rt
CO2Me
47
NHBn
48
X
Y
HN
N
H
P
O
R
(S,RP)-DIAPHOX 43
43a X=H, Y=H, R=Ph 99%, 59% ee
43b X=H, Y=H, R=1-naphthyl 71%, 45% ee
43c X=H, Y=H, R=2-naphthyl 26%, 47% ee
43d X=H, Y=H, R=4-biphenyl 55%, 65% ee
43e X=CF , Y=H, R=Ph 10%, 73% ee
43f X=OMe, Y=H, R=Ph 51%, 60% ee
43g X=OMe, Y=OMe, R=Ph 25%, 68% ee
43h X=t-Bu, Y=H, R=Ph 93%, 86% ee
43h at -20 °C 96%, 92% ee

Scheme 13. Screening of DIAPHOX-type ligands in Ir-promoted allylic amination (Ref. 76).

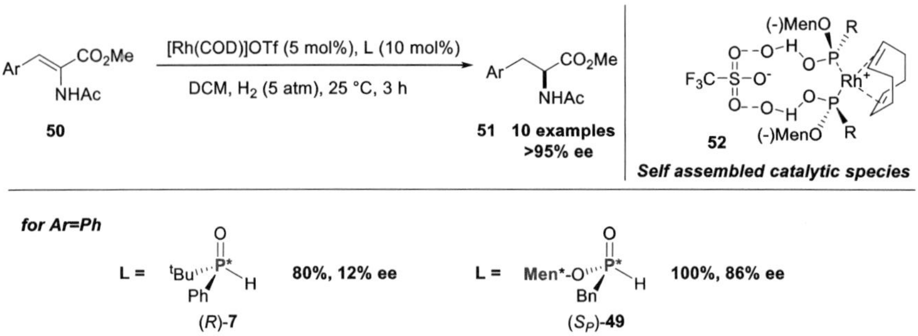
Ar
CO2Me
NHAc
50
[Rh(COD)]OTf (5 mol%), L (10 mol%)
DCM, H2 (5 atm), 25 °C, 3 h
51 10 examples
>95% ee
(-)MenO R
F3C-S-O-
Rh+
52
Self assembled catalytic species
for Ar=Ph
L = tBu P* H Ph
(R)-7
80%, 12% ee
L = Men*-O P* H Bn
(SP)-49
100%, 86% ee

Scheme 14. P-chiral SPO and (–)-menthylphosphinates in enantioselective hydrogenation (Ref. 78).

allowed optimization of the catalytic behavior, the best ligand being **43g** (X=Y=OMe).[77]

The beneficial effect of the combination of two types of chiral inductors in HASPO-type ligands has been highlighted in hydrogenation reactions also by Han and coworkers in 2014 (Scheme 14).[78] The use of the simple P-chirogenic (*S*)-^{t}BuPhP(O)H preligand **7** in the Rh-catalyzed enantioselective hydrogenation of α-acetamidocinnamates resulted in poor ee despite of a good conversion rate (10% ee on a model substrate). Moving to (–)-menthyl-(S_P)-benzylphosphinate **49**, bearing a double chiral information led to a significant enhancement of both yield and ee (86% ee on the model substrate **50** for Ar = Ph). The authors proposed a hydrogen bond–assisted self-assembly of the two HASPOs ligands involving a bridging TfO⁻ unit, thus mimicking the structure of a bidentate ligand (Scheme 14, **52**). This assumption was supported by ^{31}P magnetic nuclear resonance (NMR) spectroscopy.

In this way, the synergistic effect of multiple chirality elements has been clearly evidenced. The introduction of a backbone located chirality element on a P-chirogenic SPO or HASPO resulted in significant amplification of the enantioselectivity.

The studies summarized in this section led to a successful step toward full leveraging of chiral SPO preligands. However, the control of all the stereogeneous elements during the synthesis of the ligand could be troublesome, and reaching both high activity and enantioselectivity may be still challenging.

V. Chiral Mixed Ligand Strategy

In the early 2010s, the use of chiral SPOs in enantioselective catalysis was still in its infancy despite some specific disclosures, as described in the precedent sections. In particular, notwithstanding their high resistance to oxidation, P-chirogenic SPOs have not met so far with the same success as other P-chirogenic diphosphines such as DIPAMP. This is obviously due to the fact that SPOs have been considered only recently by the catalysis community. Moreover, a plausible explanation lies in the insufficient affinity of SPOs for metals such as rhodium, iridium, or ruthenium. Indeed, this results in ill-defined complexes, rather unstable under catalytic conditions, finally leading to moderate activity.

A solution to tackle this limitation consists in the combined use of an achiral tertiary phosphine and a chiral SPO in the same catalytic system, the so-called "Chiral Mixed Ligand" strategy (CML, Scheme 15). On the one hand, the achiral phosphine moiety offers stronger coordination to the metal center, well-defined complexes and higher catalytic activities. On the other hand, the chiral SPO preligand affords the chiral information and controls enantioselectivity. This strategy can be implemented through either bidentate SPO-phosphine ligands where the two phosphorous groups are supported by the same backbone or by combining two separate, monodentate SPO and phosphine ligands.

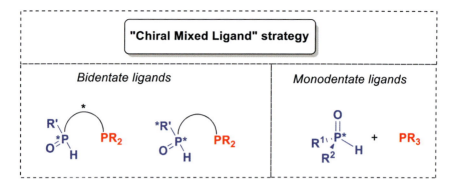

Scheme 15. The Chiral Mixed Ligand strategy.

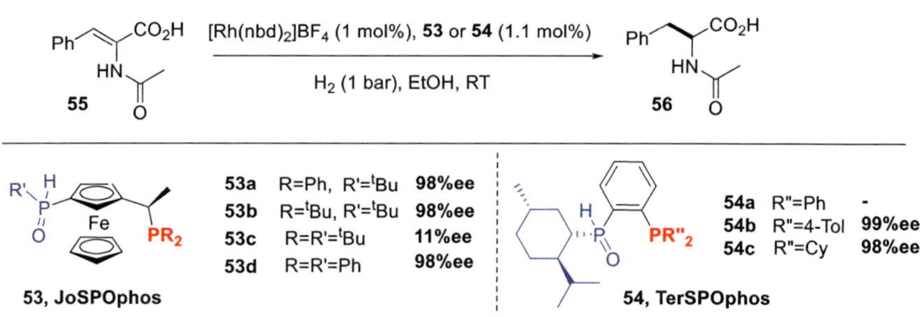

Scheme 16. Enantioselective hydrogenations with mixed P-chiral SPO-phosphine ligands (Ref. 79).

In the first approach, the tethering backbone can be either chiral or achiral (Scheme 15).

The first application of the CML strategy has been reported by Pfaltz, Pugin, and coworkers in 2010.[79] These authors illustrated the concept by synthesizing two new bidentated SPO-phosphine-type compounds, structurally related to the well-known diphosphine JosiPHOS. The first one, called JoSPOphos, combines a PR_2 and a P-chirogenic SPO units that are tethered by a planar chiral ferrocenyl backbone with an additional stereogenic carbon center (Scheme 16, **53**). The second one, named TerSPOphos, displays a PR_2 unit, a P-chirogenic SPO with a chiral menthyl substituent on phosphorus, and an achiral linker (Scheme 16, **54**).

These two families of preligands have been used for the rhodium-catalyzed hydrogenation of functionalized alkenes (dehydro-α-amino acids

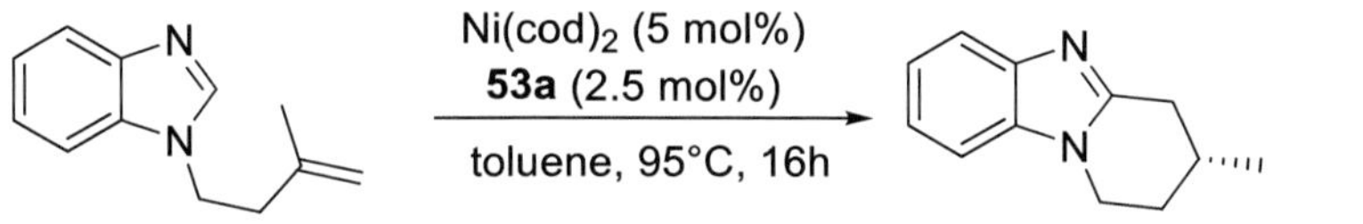

Scheme 17. Enantioselective hydroarylations through C–H activation with mixed P-chiral SPO-phosphine ligands (Ref. 81).

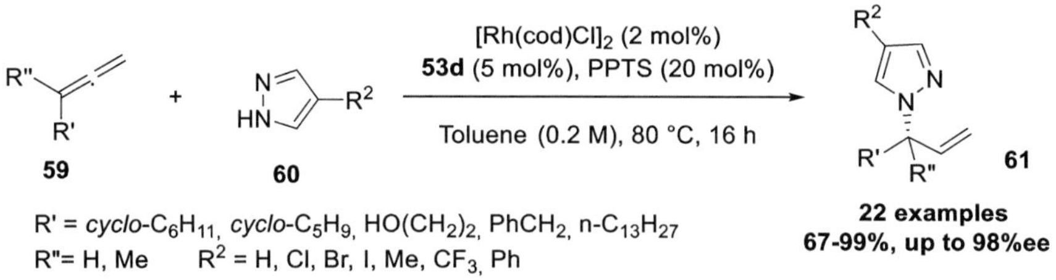

Scheme 18. Addition of pyrazoles to allenes promoted by rhodium–JoSPOphos complexes (Ref. 82).

and esters, itaconates, and β-aminoacrylates), leading to complete conversions and excellent enantioselectivity.

In 2019, inspired by the early work of Ye and coworkers on HASPO-nickel/aluminum-promoted hydroarylations,[80] Ackermann and coworkers used JoSPOphos in aluminum-free enantioselective hydroarylations (Scheme 17). The P-chirogenic-mixed ligand **53a** proved to be the most efficient ligand for these reactions that involve C–H activation of imidazoles and intramolecular hydroarylation of the pendant olefin.[81]

The same ligands, based on the CML strategy, have been subjected to further investigations by Breit and coworkers, who have made use of **53d** for the rhodium-catalyzed enantioselective N-addition of pyrazoles **60** to allenes **59**.[82] Preligand **53d** afforded the N-substituted pyrazoles **61** in excellent yields and ees (Scheme 18, 67–99% ee). When unsymmetrically disubstituted pyrazoles were used in the same reaction, N^1 addition took place with usually high regioselectivity and excellent enantioselectivity (>90% ee).

The second variant of the CML strategy has been proposed in 2013 by Ding and coworkers with the studies on the enantioselective hydrogenation

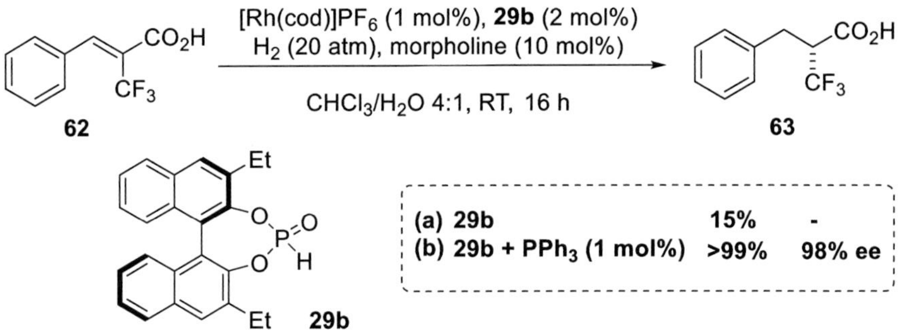

Scheme 19. Enantioselective hydrogenations with Rh(I) catalysts combining chiral and achiral ligands (Ref. 83).

of α-CF$_3$- and β-CF$_3$-substituted acrylic acids, catalyzed by Rh(I) complexes (Scheme 19).[83] The use of HASPOs alone (two equivalents) as preligands gave very poor conversion rates, revealing the low activity of the corresponding complexes. The addition of PPh$_3$ (one equivalent to Rh(I)) in combination with chiral HASPOs (two equivalents) caused a significant improvement of both conversion rates and ees in these highly challenging reactions. Thus, for instance, with the binaphthyl-based HASPO preligand **29b**, the hydrogenation of α-CF$_3$-cinnamic acid **62** led to a 99% ee, in the presence of PPh$_3$ (Scheme 19). The authors found that a biphasic mixture (CHCl$_3$/H$_2$O 4:1) was crucial to achieve almost complete conversion and high enantioselectivity. According to their hypothesis, water assists the cleavage of the [Rh^I(HASPO)$_2$]-type complex and allows PPh$_3$ to enter the coordination sphere to afford a mixed [Rh^I(HASPO)PPh$_3$] complex, which is responsible for catalysis. Under these conditions, the axially chiral HASPO **29b** allowed the hydrogenation of a large variety of α-CF$_3$ and β-CF$_3$ acrylic acids with complete conversions and excellent ees.

The mixed ligand CML strategy could be leveraged to refine the design of the catalytic species, for the better control of a given elementary step in a catalytic cycle, as demonstrated by Ding and coworkers in 2015.[84] The enantioselective hydrogenation of the β-phenyloxyacrylic acid **64**, catalyzed by Rh(I)/SPO **29b**, provided the expected product **65** in poor yield and ee, along with a large amount of by-product **66** (Scheme 20(a)). According to the postulated mechanism, the reaction pathways leading to **65** and **66,** respectively, involve intermediate **68** at their crossing point. After complexation of **64** by Rh(I) and oxidative addition of H_2, followed by a migratory insertion, intermediate **67** is formed. From **67**, reductive elimination gives the desired product **65** (Path A in Scheme 20), whereas the achiral by-product **66** is formed via the migration of the OPh group to give **70**, and subsequent by the hydrogenation of the olefinic double bond (Path B). The desired Path A could be favored by addition of the relatively electron-rich trianisylphosphine **71**, together with the electron-poor axially chiral HASPO **28b**. The mixed-ligand complex also provided substantial rate enhancement and high enantioselectivity (23 examples, ee up to 99%).

Significant breakthroughs in the development of the CML concept have been reported by Chung, Dong, Zhang, and coworkers in 2017 following a bioinspired strategy.[85] The performances of enzymes in enantioselective catalysis remain unparalleled since they benefit from multiple, cooperative, and noncovalent interactions between enzyme pockets and substrates. Among all these interactions, H-bonding and ion pairs are probably the most important ones. Inspired by these bonding properties of enzymes, the authors designed an unprecedented SPO ligand, SPO-Wudaphos, **72** (Scheme 21). The free OH group of the phosphinous acid function of **72** can be involved in hydrogen bonding interactions with substrates like carbonyls, while its nitrogen atom can be protonated to generate ionic pairs with negatively charged moieties. SPO-Wudaphos has been prepared from the chiral Ugi's amine. It combines the P-located chirality of SPO, the planar chirality of the disubstituted ferrocenyl group, and also bears a stereogenic carbon atom. It could be used as a bidentate ligand for rhodium, by coordinating both the P(III) form of the SPO moiety and the PPh_2 group. The authors demonstrated then the performance of the SPO-Wudaphos rhodium complexes in the enantioselective hydrogenation of a series of α-methylene-γ-keto carboxylic acids **73**. These experiments led to complete conversion and almost total enantioselectivity (Scheme 21, up to 95% ee). The crucial role of the multiple interactions between the ligand and the substrates has been evidenced by density functional theory (DFT) calculations.

According to the high level performance of the catalytic systems described in this last section, the CML concept constitutes so far one of the best strategies to take advantage of chiral SPO preligands in

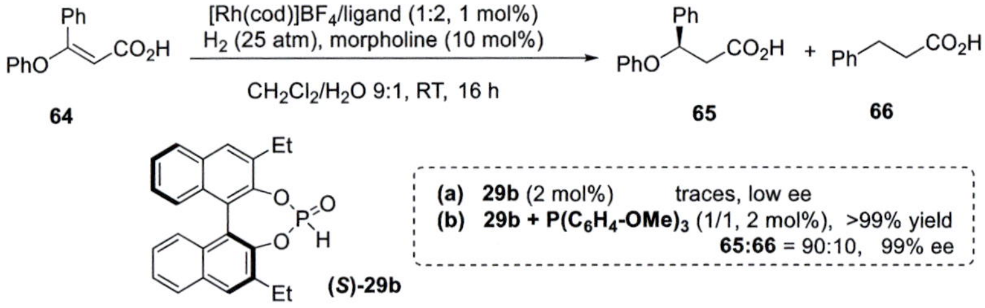

Scheme 20. Hydrogenation of β-phenoxyacrylates with mixed ligands rhodium complexes: minimizing the phenoxy elimination pathway (Ref. 84).

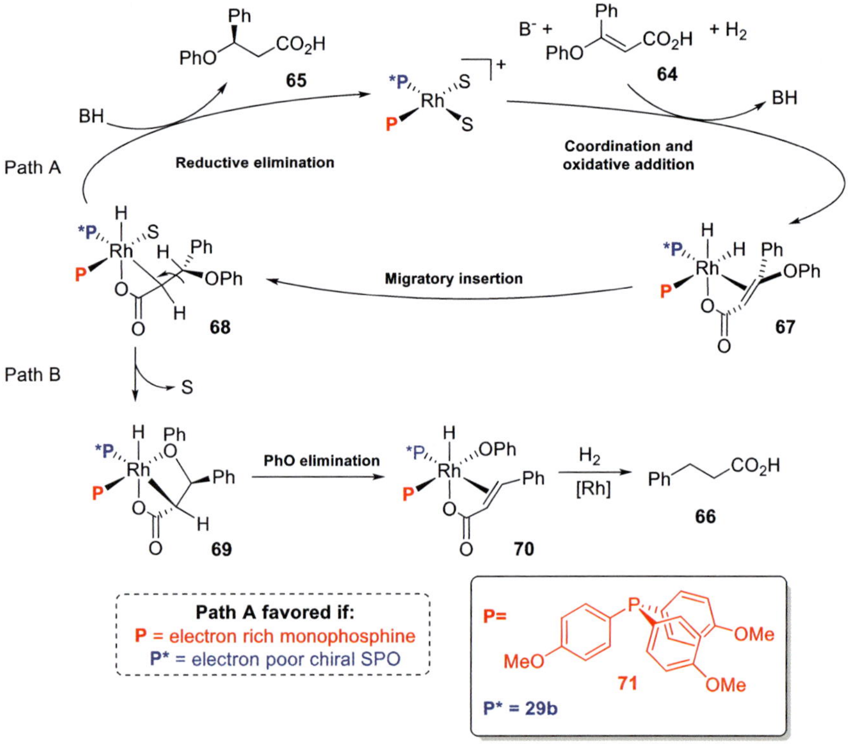

enantioselective catalysis. The synergistic effects of PR_2 and HASPO moieties and the variations of the nature and the location of the chirality on the ligand allow the achievement of optimal activity and enantioselectivity. The design of these mixed ligands requires a high level of

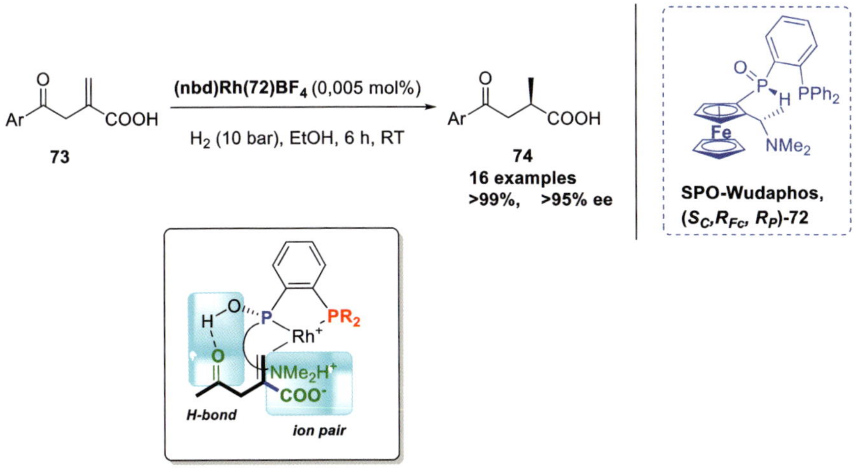

Scheme 21. Enzymes inspired design of multifunctional CMLs: enantioselective Rh-promoted hydrogenations (Ref. 85).

understanding of mechanistic issues and the identification of the key interactions involved in transition states.

VI. Conclusion

Since the beginning of this century, chiral P-chirogenic SPO were no longer reduced to the status of intermediates in the synthesis of P-chiral tertiary phosphines, and have been widely studied as chiral preligands in enantioselective catalysis. Their catalytic potential has been firmly established in recent studies. Overall, it can be stated that, in these ligands, a single stereogenic phosphorus atom is most often insufficient to achieve good enantioselectivity in catalytic processes, as illustrated by the reduced number of successful examples reported in the literature. On the other hand, SPOs displaying a chirality axis on their backbone only, or a combination of a chirality axis and a stereogenic phosphorus atom proved to be highly efficient.

The design of SPOs with chiral backbones is restricted so far mainly to axially chiral scaffolds. It is expected therefore that other types of chiral units, for example planar chiral units, will be developed further in the future. For instance, in 2018, Rowlands and coworkers reported on the diastereoselective synthesis of a planar chiral

[2.2]paracyclophane-derived SPO, but its catalytic properties haven't been investigated in depth so far.[86] Studies on the effect of supramolecular chirality would be also worthwhile.

Finally, very powerful catalytic systems based on the "CML" strategy have been developed starting from 2010. This recent approach is highly promising and, together with other successful strategies, it demonstrates that the potential of chiral SPO ligands is far from being fully exploited yet.

VII. References

1. Stradiotto, M.; Lundgren, R. J. *Ligand Design in Metal Chemistry: Reactivity and Catalysis*, John Wiley & Sons, Ltd, **2016**.
2. Knowles, W. S.; Sabacky, M. J. *Chem. Commun. (London)*. **1968**, (22), 1445–1446.
3. Horner, L.; Siegel, H.; Buthe, H. *Angew. Chem. Int. Ed.* **1968**, *7*(12), 942.
4. Dang, T. P.; Kagan, H. B. *J. Chem. Soc. Chem. Commun.* **1972**, 481.
5. Tolman, C. A. *Chem. Rev.* **1977**, *77*, 313–348.
6. Suresh, C. H.; Koga, N. *Inorg. Chem.* **2002**, *41*, 1573–1578.
7. Tolman, C. A. *J. Am. Chem. Soc.* **1970**, *92*, 2953–2956.
8. Dierkes, P.; van Leeuwen, P. W. N. M. *J. Chem. Soc. Dalton Trans.* **1999**, 1519–1529.
9. Mislow, K.; Baechler, R. D. *J. Am. Chem. Soc.* **1971**, *93*, 773–774.
10. Börner, A. *Phosphorus Ligands in Asymmetric Catlysis: Synthesis and Applications*, Wiley-VCH, Weinheim, **2008**.
11. Grinberg, A. A.; Troitskaya, A. D. *Bull. Acad. Sci. U.S.S.R., Classe Sci. Chim.* **1944**, 178.
12. Grabulosa, A. *P-stereogenic Ligands in Enantioselective Catalysis*, RSC Publishing, **2011**.
13. Huber, R.; Passera, A.; Mezzetti, A. *Chem. Commun.* **2019**, *55*, 9251–9266.
14. Ohff, M.; Holz, J. *Synthesis* **1998**, 1391–1415.
15. Hérault, D.; Nguyen, D. H.; Nuel, D.; Buono, G. *Chem. Soc. Rev.* **2015**, *44*, 2508–2528.
16. Kurscheid, B.; Wiebe, W.; Neumann, B.; Stammler, H.-G.; Hoge, B. *Eur. J. Inorg. Chem.* **2011**, 5523–5529.
17. Chatt, J.; Heaton, B. T. *J. Chem. Soc. A.* **1968**, 2745–2757.
18. Ackermann, L. *Synthesis* **2006**, 1557–1571.
19. van Leeuwen, P. W. N. M.; Roobeek, C. F.; Wife, R. L.; Frijns, J. H. J. *J. Chem. Soc. Chem. Commun.* **1986**, 31–33.
20. Dai, W.-M.; Yeung, K. K. Y.; Leung, W.-H.; Haynes, R. K. *Tetrahedron: Asymmetry* **2003**, *14*, 2821–2826.
21. Achard, T. *Chimia* **2016**, *70*, 8–19.
22. Shaikh, T. M.; Weng, C.-M.; Hong, F.-E. *Chem. Soc. Rev.* **2012**, *256*, 771–803.

23. Dubrovina, N. V.; Börner, A. *Angew. Chem. Int. Ed.* **2004**, *43*, 5883–5886.
24. Gallen, A.; Riera, A.; Verdaguer, X.; Grabulosa, A. *Catal. Sci. Technol.* **2019**, *9*, 5504–5561.
25. van Leeuwen, P. W. N. M.; Cano, I.; Freixa, Z. *ChamCatChem* **2020**, *12*, 3982–3984.
26. Williams, R. H.; Hamilton, L. A. *J. Am. Chem. Soc.* **1952**, *74*, 5418–5420.
27. Hays, H. R. *J. Org. Chem.* **1968**, *33*, 3690–3694.
28. Williams, R. H.; Hamilton, L. A. *J. Am. Chem. Soc.* **1955**, *77*, 3411–3412.
29. Wife, R. L.; van Oort, A. B.; van Doorn, J. A.; van Leeuwen, P. W. N. M. *Synthesis* **1983**, 71–73.
30. Haynes, R. K.; Au-Yeung, T.-L.; Chan, W.-K.; Lam, W.-L.; Li, Z.-Y.; Yeung, L.-L.; Chan, A. S. C.; Li, P.; Koen, M.; Mitchell, C. R.; Vonwiller, S. C. *Eur. J. Org. Chem.* **2000**, 3205–3216.
31. Quin, L. D.; Montgomery, R. E. *J. Org. Chem.* **1963**, *28*, 3315–3320.
32. Baba, G.; Pilard, J.-F.; Tantaoui, K.; Gaumont, A.-C.; Denis, J.-M. *Tetrahedron Lett.* **1995**, *36*, 4421–4424.
33. Lemouzy, S.; Nguyen, D. H.; Camy, V.; Jean, M.; Gatineau, D.; Giordano, L.; Naubron, J.-V.; Vanthuyne, N.; Hérault, D.; Buono, G. *Chem. Eur. J.* **2015**, *21*, 15607–15621.
34. Lemouzy, S.; Giordano, L.; Hérault, D.; Buono, G. *Eur. J. Org. Chem.* **2020**, 3351–3366.
35. Jiang, X.-B.; Minnaard, A. J.; Hessen, B.; Feringa, B. L.; Duchateau, A. L. L.; Adrien, J. G. O.; Boogers, J. A.; de Vries, J. G. *Org. Lett.* **2003**, *5*, 1503–1506.
36. Haynes, R. K.; Freeman, R. N.; Mitchell, C. R.; Vonwiller, S. C. *J. Org. Chem.* **1994**, *59*, 2919–2921.
37. Drabowicz, J.; Łyżwa, P.; Omelańczuk, J.; Pietrusiewicz, M.; Mikołajczyk, M. *Tetrahedron: Asymmetry* **1999**, *10*, 2757–2763.
38. Kortmann, F. A.; Chang, M.-C.; Otten, E.; Couzijn, E. P. A.; Lutz, M.; Minnaard, A. J. *Chem. Sci.* **2014**, *5*, 1322–1327.
39. Holt, J.; Maj, A. M.; Schudde, E. P.; Pietrusiewicz, M.; Sieroń, L.; Wieczorek, W.; Jerphagnon, T.; Arends, I. W. C. E.; Hanefeld, U.; Minnaard, A. J. *Synthesis* **2009**, 2061–2065.
40. Kolodiazhnyi, O. I.; Gryshkun, E. V.; Andrushko, N. V.; Freytag, M.; Jones, P. G.; Schmutzler, R. *Tetrahedron: Asymmetry* **2003**, *14*, 181–183.
41. Leyris, A.; Nuel, D.; Achard, M.; Buono, G. *Tetrahedron Lett.* **2005**, *46*, 8677–8680.
42. Gatineau, D.; Nguyen, D. H.; Vanthuyne, N.; Leclaire, J.; Giordano, L.; Buono, G. *J. Org. Chem.* **2015**, *80*, 4132–4141.
43. Leyris, A.; Bigeault, J.; Nuel, D.; Giordano, L.; Buono, G. *Tetrahedron Lett.* **2007**, *48*, 5247–5250.
44. Copey, L.; Jean-Gérard, L.; Andrioletti, B.; Framery, E. *Tetrahedron Lett.* **2016**, *57*, 543–545.
45. Han, Z. S.; Wu, H.; Xu, Y.; Zhang, Y.; Qu, B.; Li, Z.; Caldwell, D. R.; Fandrick, K. R.; Zhang, L.; Roschangar, F.; Song, J. J.; Senanayake, C. H. *Org. Lett.* **2017**, *19*, 1796–1799.

46. Li, S. G.; Yuan, M.; Topic, F.; Han, Z. S.; Senanayake, C. H.; Tsantrizos, Y. S. *J. Org. Chem.* **2019**, *84*, 7291–7302.
47. Roundhill, D. M. S.; Sperline, R. F.; Beaulied, W. B. *Coord. Chem. Rev.* **1978**, *26*, 263–279.
48. Appleby, T.; Woollins, J. D. *Coord. Chem. Rev.* **2002**, *235*, 121–140.
49. Berry, D. E.; Beveridge, K. A.; Bushnell, G. W.; Dixon, K. R. *Can. J. Chem.* **1985**, *63*, 2949–2957.
50. Li, G. Y. *Angew. Chem. Int. Ed.* **2001**, *40*, 1513–1516.
51. Naik, D. V.; Palenik, G. J.; Jacobson, S.; Carty, A. J. *J. Am. Chem. Soc.* **1974**, *96*, 2286–2288.
52. Beaulieu, W. B.; Rauchfuss, T. B.; Roundhill, D. M. *Inorg. Chem.* **1975**, *14*, 1732–1734.
53. Chan, E. Y. Y.; Zhang, Q.-F.; Sau, Y.-K.; Lo, S. M. F.; Sung, H. H. Y.; Williams, I. D.; Haynes, R. K.; Leung, W.-H. *Inorg. Chem.* **2004**, *43*, 4921–4926.
54. Bigeault, J.; Giordano, L.; Buono, G. *Angew. Chem. Int. Ed.* **2005**, *44*, 4753–4757.
55. Jung, L.-Y.; Tsai, S.-H.; Hong, F.-E. *Organometallics* **2009**, *28*, 6044–6053.
56. Vasseur, A.; Membrat, R.; Palpacelli, D.; Giorgi, M.; Nuel, D.; Giordano, L.; Martinez, A. *Chem. Commun.* **2018**, *54*, 10132–10135.
57. Martin, D.; Moraleda, D.; Achard, T.; Giordano, L.; Buono, G. *Chem. Eur. J.* **2011**, *17*, 12729–12740.
58. Lipkowitz, K. B.; D'Hue, C. A.; Sakamoto, T.; Stack, J. N. *J. Am. Chem. Soc.* **2002**, *124*, 14255–14267.
59. Crepy, K. V. L.; Imamoto, T. *Adv. Synth. Catal.* **2003**, *345*, 79–101.
60. Dutartre, M.; Bayardon, J.; Jugé, S. *Chem. Soc. Rev.* **2016**, *45*, 5771–5794.
61. Jiang, X.-B.; van den Berg, M.; Minnaard, A. J.; Feringa, B. L.; de Vries, J. G. *Tetrahedron: Asymmetry* **2004**, *15*, 2223–2229.
62. Achard, T.; Lepronier, A.; Gimbert, Y.; Clavier, H.; Giordano, L.; Tenaglia, A.; Buono, G. *Angew. Chem. Int. Ed. Engl.* **2011**, *50*, 3552–3556.
63. Lepronier, A.; Achard, T.; Giordano, L.; Tenaglia, A.; Buono, G.; Clavier, H. *Adv. Synth. Catal.* **2016**, *358*, 631–642.
64. Bigeault, J.; Giordano, L.; De Riggi, I.; Gimbert, Y.; Buono, G. *Org. Lett.* **2007**, *9*, 3567–3570.
65. Gatineau, D.; Moraleda, D.; Naubron, J.-V.; Bürgi, T.; Giordano, L.; Buono, G. *Tetrahedron: Asymmetry* **2009**, *20*, 1912–1917.
66. Berthod, M.; Mignani, G.; Woodward, G.; Lemaire, M. *Chem. Rev.* **2005**, *105*, 1801–1836.
67. Castro, P. M.; Gulyás, H.; Benet-Buchholz, J.; Bo, C.; Freixa, Z.; van Leeuwen, P. W. N. M. *Catal. Sci. Technol.* **2011**, *1*, 401–407.
68. Membrat, R.; Vasseur, A.; Martinez, A.; Giordano, L.; Nuel, D. *Eur. J. Org. Chem.* **2018**, 5427–5434.
69. Vasseur, A.; Membrat, R.; Gatineau, D.; Tenaglia, A.; Nuel, D.; Giordano, L. *ChemCatChem* **2017**, *9*, 728–732.

70. Dong, K.; Wang, Z.; Ding, K. *J. Am. Chem. Soc.* **2012**, *134*, 12474–12477.
71. Li, Y.; Dong, K.; Wang, Z.; Ding, K. *Angew. Chem. Int. Ed.* **2013**, *52*, 6748–6752.
72. Dong, K.; Li, Y.; Wang, Z.; Ding, K. *Org. Chem. Front.* **2014**, *1*, 155–160.
73. Donets, P. A.; Cramer, N. *J. Am. Chem. Soc.* **2013**, *135*, 11772–11775.
74. Nemoto, T.; Matsumoto, T.; Masuda, T.; Hitomi, T.; Katano, K.; Hamada, Y. *J. Am. Chem. Soc.* **2004**, *126*, 3690–3691.
75. Nemoto, T.; Masuda, T.; Matsumoto, T.; Hamada, Y. *J. Org. Chem.* **2005**, *70*, 7172–7178.
76. Nemoto, T.; Sakamoto, T.; Matsumoto, T.; Hamada, Y. *Tetrahedron Lett.* **2006**, *47*, 8737–8740.
77. Nemoto, T.; Fukuyama, T.; Yamamoto, E.; Tamura, S.; Fukuda, T.; Matsumoto, T.; Akimoto, Y.; Hamada, Y. *Org. Lett.* **2007**, *9*, 927–930.
78. Wang, X.-B.; Goto, M.; Han, L.-B. *Chem. Eur. J.* **2014**, *20*, 3631–3635.
79. Landert, H.; Spindler, F.; Wyss, A.; Blaser, H.-U.; Pugin, B.; Ribourdouille, Y.; Gschwend, B.; Ramalingam, B.; Pfaltz, A. *Angew. Chem. Int. Ed.* **2010**, *49*, 6873–6876.
80. Wang, Y.-X.; Qi, S.-L.; Luan, Y.-X.; Han, X.-W.; Wang, S.; Chen, H.; Ye, M. *J. Am. Chem. Soc.* **2018**, *140*, 5360–5364.
81. Loup, J.; Müller, V.; Ghorai, D.; Ackermann, L. *Angew. Chem. Int. Ed.* **2019**, *58*, 1749–1753.
82. Haydl, A. M.; Xu, K.; Breit, B. *Angew. Chem. Int. Ed.* **2015**, *54*, 7149–7153.
83. Dong, K.; Li, Y.; Wang, Z.; Ding, K. *Angew. Chem. Int. Ed.* **2013**, *52*, 14191–14195.
84. Li, Y.; Wang, Z.; Ding, K. *Chem. Eur. J.* **2015**, *21*, 16387–16390.
85. Chen, C.; Zhang, Z.; Jin, S.; Fan, X.; Geng, M.; Zhou, Y.; Wen, S.; Wang, X.; Chung, L. W.; Dong, X.-Q.; Zhang, X. *Angew. Chem. Int. Ed.* **2017**, *56*, 6808–6812.
86. Mungalpara, M. N.; Wang, J.; Coles, M. P.; Plieger, P. G.; Rowlands, G. J. *Tetrahedron* **2018**, *74*, 5519–5527.

© 2022 World Scientific Publishing Company
https://doi.org/10.1142/9789811248436_0003

3 Supramolecular Regulation in Enantioselective Catalysis

Matthieu Raynal[*,‡] and Anton Vidal-Ferran[†,§]

*Sorbonne Université, CNRS, Institut Parisien de Chimie Moléculaire, Equipe Chimie des Polymères, Paris 75005, France
†ICREA, Passeig Lluís Companys, 23, Barcelona E-08010, Spain & Department of Inorganic and Organic Chemistry, University of Barcelona, C. Martí i Franquès 1-11, Barcelona E-08028, Spain
‡matthieu.raynal@sorbonne-universite.fr
§anton.vidal@icrea.cat

Table of Contents

List of Abbreviations	60
I. Introduction	61
II. Supramolecular Regulation of Bidentate Ligands in Metal Complexes	63
A. Supramolecular Regulation with Aza-Crown-Ether Motifs as the Regulation Sites	64
B. Supramolecular Regulation with Linear Polyether Motifs as Regulation Sites	65

III. Induction of Chirality by Chiral Regulating Agents 70
   A. Bidentate Ligands with a Remote Chiral Regulating Agent 71
   B. Ion-paired Chiral Ligands 76
   C. Supramolecular Helical Catalysts 80
IV. Supramolecular Regulation in Organocatalysis 85
V. Conclusions and Outlook 90
VI. Acknowledgments 91
VII. References 92

List of Abbreviations

acac	acetylacetonato
BArF	tetrakis[3,5-bis(trifluoromethyl)phenyl]borate
BTA	benzene-1,3,5-tricarboxamide
b/l	branched-to-linear ratio
BINOL	[1,1'-binaphthalene]-2,2'-diol
DIPEA	N,N-diisopropylethylamine
EAS	enantioselective allylic substitution
EHF	enantioselective hydroformylation
C	catalyst
ee	enantiomeric excess
er	enantiomeric ratio
L	ligand
nbd	norbornadiene
P	product
S	substrate, starting material or reactant
S/C	substrate-to-catalyst ratio
RA	regulation agent
rt	room temperature
Salen	2,2'-(($1E, 1'E$)-(ethane-1,2-diylbis(azanylylidene))bis(methanylylidene)diphenol
TADDOL	2,2-dialkyl-$\alpha, \alpha, \alpha', \alpha'$-tetraaryldioxolane-4,5-dimethanol
TOF	turnover frequency
VANOL	3,3'-diphenyl-[2,2'-binaphthalene]-1,1'-diol

I. Introduction

The use of supramolecular interactions in catalysis has undergone major growth in the last 15 years and has contributed to major advances in the field of enantioselective catalysis. Supramolecular interactions have been applied in enantioselective catalysis in a number of ways.

In some cases, supramolecular interactions have been exploited to construct the skeleton of the desired chiral catalyst itself. This is done by attaching two (or more) building blocks that contain the functional groups desired for catalysis, on one hand, and the motifs needed for the supramolecular assembly, on the other hand. These methodologies have enabled the synthesis of libraries of structurally diverse supramolecular catalysts with greater ease than by standard covalent chemistry.[1–5] In other examples, supramolecular interactions have been exploited to facilitate a good three-dimensional stereochemical match between substrate(s), reagent(s), and catalytic system, thus enhancing the stereoselectivity of the corresponding transformations. Molecular recognition of the substrate by the catalytic system in a specific site facilitates substrate preorganization with an overall enhancement of the (stereo)selectivity of the corresponding transformations.[6–8] In other cases, enantioselective supramolecular catalysts mimic enzyme catalysis by focusing on a *host-guest approach*: the catalytic system brings the substrate(s) and reagent(s) together with an overall reduction of the entropic costs associated with the process and enhancement of the reaction rates. Moreover, the reduction of the degrees of freedom of the substrate(s) and/or reagent(s) upon binding to the catalytic system often translates into increased stereoselectivities.[9,10] These topics will not be discussed in detail in this text, as they have been reviewed recently in a comprehensive manner.[11–14] The specific purpose of this chapter will be to highlight the effectiveness of well-defined additives in the supramolecular regulation of the catalytic activity.

Supramolecular regulation of the catalytic site has proven to be indeed a useful approach in enantioselective catalysis. Regulation refers to "the addition of a chemical species (hereinafter referred to as regulation agent [RA]) that is not directly involved in the catalytic site but interacts with the assembled ligand (or catalytic system) through reversible interactions, providing a change in the three-dimensional structure of the catalyst with respect to that obtained with other regulation agents." This strategy generally seeks to override one of the intrinsic limitations of enantioselective catalysts: their lack of generality. It is well known indeed that structural changes in the substrate(s) and/or reagent(s) often translate into a loss of enantioselectivity and, consequently, structural variations of the catalytic system are required for improving the stereochemical outcome of the reaction. Supramolecular regulation has proven to be an efficient strategy to produce libraries of enantioselective catalysts, where the whole set of catalysts preserve the main structural characteristics, but incorporate subtle structural differences at the catalytic site, which are related to the RA used. Overall, one of the members of a particular library of catalysts is capable of adapting to the requirements of a given substrate, providing the highest performance in terms of enantioselectivity for this particular substrate.[15]

The generation of libraries of enantioselective catalysts by the regulation strategy has been made following two approaches:

- *Supramolecular regulation onto a preformed chiral catalyst.* Catalysts belonging to this category possess a suitable site for enantioselective catalysis and a distal regulation site capable of interacting with the RA. The library resulting from the binding of an array of diverse RAs preserves most of the structural characteristics and, at the same time, incorporates structural peculiarities that depend on the size and shape of the RA employed. This approach is generally inspired by the process of allosteric regulation in nature, in which the activity of an enzyme is modified by binding an effector molecule at a site other than the enzyme's active site.[16,17]
- *Supramolecular regulation onto a prochiral ligand or catalyst.* In terms of generating catalyst diversity, the principle of this approach is similar to the one summarized above with one crucial difference: the initial catalyst is achiral (or chiral but poorly selective) in the absence of the RA and the addition of this external chemical species induces a suitable chiral environment to the assembled catalytic site. This strategy infers efficient chirality transfer through noncovalent interactions, that is, chirogenesis,[18] which eventually leads to the formation of a major enantiomer in an enantioselective catalytic reaction. In this chapter,

only small-sized and discrete RAs will be covered. For the use of biomacromolecules as chiral scaffolds for enantioselective reactions, the reader is referred to recent reviews[19–21] and to chapter "DNA-based enantioselective catalysis; past, present and future" of this book.

The discussion that follows is divided into three different sections according to the nature of the catalytic system and the aforementioned approaches. Accordingly, supramolecular regulation of enantioselective catalysts will be discussed in Section II, whereas chirality induction to metal catalysts will be mentioned in Section III. The fourth section will be dedicated to supramolecular regulation of enantioselective organic catalysts. This text does not intend to be a comprehensive summary of all articles published in the literature on supramolecular regulation in enantioselective transformations due to space limitations. This text will disclose the most representative examples on this area of research. It has to be mentioned also that examples in which the RA or additive is directly involved in the coordination sphere of the metal have not been summarized. The most developed strategy in this context has been to employ chiral counterions and, for more information on this topic, the reader is referred to reviews[22,23] as well as to the chapter "Chiral counterions in enantioselective organometallic catalysis" in this book.

II. Supramolecular Regulation of Bidentate Ligands in Metal Complexes

This section highlights relevant examples of supramolecular regulation onto an already formed bidentate metal complex, functioning as an enantioselective catalyst in an organic transformation of interest. Supramolecularly regulated enantioselective catalysts belonging to this category possess a catalytic site with a well-defined three-dimensional structure that already incorporates the required stereogenic elements for controlling the stereochemical outcome of the reaction. Moreover, these enantiopure catalysts also possess a distal regulation site containing a supramolecular motif capable of interacting with an external RA *via* supramolecular interactions. Reported regulation sites are mainly based on polyether functional groups of various sizes and topology (cyclic or linear), with variable number of oxygen atoms and the presence or absence of other heteroatoms. These regulation sites offer a wide range of binding capacities to RAs of different size and types, as these binding motifs are excellent hosts for alkali metal and ammonium-based cations.[24–26] In this way, the interaction between the enantiopure catalyst and an array of structurally diverse RAs translates into the generation of a library of catalysts that preserve most of the structural characteristics but also incorporate specific structural features (*i.e.* spatial environment around the active site) that finely depend on the size and shape of the external agent employed. Supramolecular regulations involving major alterations of the structural or functional features of the catalysts are outside the scope of this section.

A. Supramolecular Regulation with Aza-Crown-Ether Motifs as the Regulation Sites

Ito and coworkers pioneered the field of supramolecular regulation of the catalytic activity by considering palladium-based catalysts, derived from enantiopure bisphosphine ligands, in enantioselective allylic substitutions (EASs) (Figure 1).[27] The distal regulation site incorporated an aza-crown-ether motif

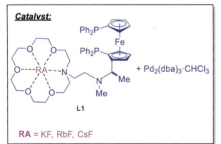

Figure 1. Supramolecularly regulated catalysts for EASs.[27]

and the principle underlying the regulation mechanism relied on secondary interactions between the reacting nucleophile (metal enolates derived from **S1** or **S2**) and the aza-crown-ether motif. It was envisaged that the inclusion complex between the enolate and the aza-crown-ether motif would translate into a closer proximity between the nucleophile and the allyl group bound to the palladium center, with an overall improvement of the reaction rate and stereocontrol. Palladium catalysts were formed *in situ* from equimolar amounts of $Pd_2(dba)_3 \cdot CHCl_3$ and ligand **L1**, and combined then with an array of alkali metal fluorides. The authors envisaged that metal fluorides, with cationic components of varying size, could operate as RAs and influence the relative arrangement of the nucleophile and the electrophilic allyl group. Ligand **L1** was found to be optimal in terms of enantioselectivity for the alkylation of allyl acetate with α-nitro ketones *via* palladium catalysis (the reader is referred to the original publication[27] for further details on the choice of the chiral ligand, including the size and nature of the crown ether site). The authors reported that the combined use of **L1**, the palladium precursor, and RbF as RA provided higher conversion rates and enantioselectivities than those obtained in the absence of RA or with either KF or CsF. The regulation effects are summarized in the Table to Figure 1 (for KF, compare entries 1 *vs* 2 for 2-nitrocyclohexanone **S1**, and 4 *vs* 5 for 2-nitropropanoate **S2**; for CsF, see entries 3 *vs* 2 and 6 *vs* 5). Thus, the better complementarity between RbF and the distal monoaza-18-crown-6 regulation site translated into positive regulation effects in terms of enantioselectivity.

B. Supramolecular Regulation with Linear Polyether Motifs as Regulation Sites

After the seminal and elegant example of regulation of the activity and stereoselectivity in enantioselective catalysts containing an (aza)-crown-ether motif in their regulation site summarized in the previous section, numerous designs of related supramolecularly regulated catalysts for diverse chemistries have been published and comprehensively reviewed.[11,12,15,28] The discussion that follows highlights only examples in which the final enantioselectivities surpass 80% enantiomeric excess (ee) (enantiomeric ratio [er] >92.5:7.5) and regulation effects are higher than 25% ee (*i.e.* the increase in the ee is of at least 25% for a given RA, with respect to the nonregulated catalyst). It should also be noted that this section highlights only studies where the regulation ability of a catalyst is demonstrated for a set of RAs on an array of substrates.

Following the work of Fan, He, and coworkers on a linear polyether-based supramolecularly regulated catalysts for enantioselective hydrogenations,[29] Vidal-Ferran and coworkers reported the use of supramolecularly regulated α,ω-bisphosphites in rhodium-promoted enantioselective hydroformylations (EHFs) (Figure 2).[30] With the aim of identifying the optimal ligand geometry, the effects of structural variations in the backbone of the ligand have been investigated. This led to the discovery of a highly performing supramolecular bisphosphite ligand (**L2**)[31] for the hydroformylation of linear alkenes, and a variant of this ligand (**L3**)[32] for the hydroformylation of heterocyclic olefins. Both ligands incorporate polyether-based regulation sites, which consist of a four-ethyleneoxy-containing linear chain for **L2**, and two two-ethyleneoxy-containing linear chains separated by a configurationally stable (R_a)-[1,1'-binaphthalene]-2,2'-diol (or (R_a)-BINOL) motif for **L3** (see Figure 2(a)). Alkali metal and ammonium salts were considered as potential RAs, as these salts are known to be good polyether binders in organic solvents. Regarding the anionic component of the RAs, tetrakis[3,5-bis(trifluoromethyl)phenyl]borate (hereafter referred to as BArF) turned out to be the best option, as the corresponding catalysts provided higher branched-to-linear (b/l) ratios than the tetrafluoroborate or perchlorate salts, for instance.[31] Moreover, the adequate solubility profile of BArF salts in solvents suitable for hydroformylation reactions (*e.g.* mixtures of toluene and tetrahydrofuran) was also considered advantageous for further studies.

Catalytic studies focused first on evaluating the design principle, namely, whether the addition of a RA would be able to tune the catalytic activity of the rhodium complexes derived from **L2**. The Table to Figure 2 summarizes the effects of various alkali metal and ammonium BArF salts on the outcome of the rhodium-catalyzed EHF of vinyl acetate (**S3**, Figure 2(b)). Of the whole series of RAs tested, RbBArF gave the best results: 99% conversion, complete regioselectivity (>99:1 in favor of the branched product) and perfect enantioselectivity (99% ee; entry 4, Table to Figure 2). Addition of RbBArF provided the most positive regulation

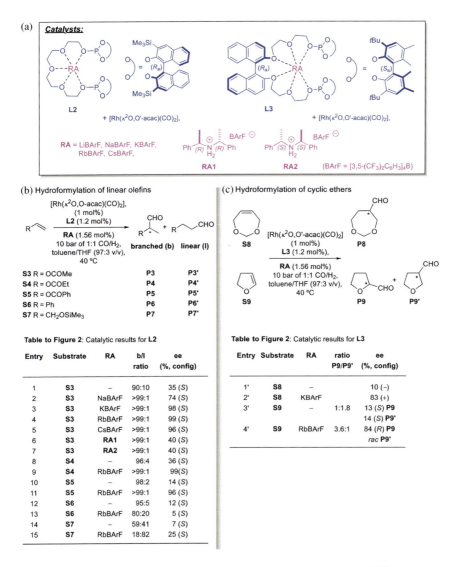

Figure 2. Supramolecularly regulated catalysts for EHFs.[31,32]

effect, with an impressive 64% enhancement of the ee, with KBArF and CsBArF also providing very high enantioselectivities (98% and 96% ee, respectively; entries 3 and 5, Table to Figure 2). In contrast, when enantiomerically pure ammonium salts **RA1** and **RA2** were used, the selectivity of the rhodium-based catalytic system derived from **L2** worsened (entries 6 and 7, Table to Figure 2). The high catalytic activity of the catalytic system developed by Vidal-Ferran and coworkers was demonstrated by increasing the substrate-to-catalyst (S/C) ratio in the hydroformylation of vinyl acetate from 100 (standard screening conditions) up to 1000 (*i.e.* 0.1 mol% of catalyst), with almost no decrease in the catalytic activity (99% conversion, b/l ratio >99:1 and 97% ee in favor of (*S*)-**P3**).[31] The addition of RbBArF not only increased the ee, but also mediated a *ca.* four-fold increase in the reaction rate (with a S/C ratio = 1000, at 50% conversion, turnover frequency [TOF] = 40 h^{-1} in the absence of RbBArF and TOF = 177 h^{-1} in the presence of RbBArF).[31]

Rhodium complexes derived from **L2**, combined with RAs, also provided high enantioselectivities in the hydroformylation of other linear alkenes. For instance, the combination of [Rh(κ^2O, O-acac)(CO)$_2$] as rhodium precursor, **L2**, and RbBArF provided the highest performing catalyst for the hydroformylation of **S4**, **S5**, and **S7** (entries 9, 11, and 15 in the Table to Figure 2) with enhancements in the ees of 63%, 82%, and 18%, respectively. Unfortunately, the regulation approach did not work for styrene **S6** (entries 12 and 13), as the use of RAs led to lower regio- and enantioselectivities.

Mechanistic studies on the hydroformylation reaction were performed with continuous monitoring of the evolution of reaction components employing a nuclear magnetic resonance (NMR) flow probe. The reaction mixture was recirculated through the reaction vessel and the NMR spectrometer with simultaneous monitoring of the starting material, product, and the metal hydride intermediates. Analysis of the reaction profile revealed a first-order dependence of the reaction rate with respect to the catalyst and the alkene, which ultimately indicates that olefin-hydride insertion is the rate- and stereo-determining step in the catalytic cycle.[33] Insight into the structural changes induced by the RA was gained through computational studies, which revealed that the ionic radius of the RA cation correlated with the width of the P–Rh–P angle (from 113° in the absence of RA to *ca.* 122° for Cs$^+$).[31] Moreover, these studies revealed that the increase in the enantioselectivity resulted from adaptation of the P–Rh–P angle to its optimal value.[31]

The same regulation approach was applied to the highly EHF of heterocyclic olefins employing **L3** as the optimal ligand, and alkali metal BArF salts as RAs (Figure 2c). Ligand **L3** incorporates a (R_a)-BINOL motif at the regulation site and two 3,3'-di-*t*butyl-[1,1'-biphenyl]-2,2'-diol-derived phosphites at the catalytic site. High enantioselectivities were observed in the hydroformylation of the heterocyclic olefins **S8** and **S9** in Figure 2.[32] Starting from 4,7-dihydro-1,3-dioxepine **S8**, the presence of KBArF improved both the conversion rates (from 23% to >99%) and the enantioselectivity, with a remarkable 73% increase of the ee, reaching 83% at 40°C (compare entries 1' and 2'). The high catalytic activity translated into the possibility of performing the hydroformylation reaction even at room temperature (rt), without detrimental effects on the conversion rate (conversion >99%) and with excellent ee (93% ee). To the best of our knowledge, this is the highest ee reported for this substrate.

KBArF was also the RA of choice for achieving high enantioselectivities in the hydroformylation of 2,3-dihydrofuran **S9**. As shown in Figure 2, entries 3' vs 4', addition of KBArF changed the **P9:P9'** product ratio (α-aldehyde vs β-aldehyde), leading mainly to the α-aldehyde **P9** with much higher enantioselectivity with respect to the parent catalyst (84% vs 13% ee).[32]

As a continuation of their research on supramolecular regulation, Fan, He, and coworkers have reported the first example of a regulated catalyst for the enantioselective Henry reactions (Figure 3).[34] The designed catalyst features the enantiopure ligand **L4** and incorporates two chromium(III)-salen moieties (for the salen motif, see gray fragment in Figure 3) separated by a polyether chain. Fan, He, and coworkers envisaged that supramolecular interactions between the polyether chain of the α,ω-bis(metallosalen) derivative and a RA would bring the two chromium(III)-salen units together. This would generate a bimetallic catalyst that could operate via a cooperative mechanism, such as previously reported for enantioselective Henry reactions. The authors assessed the new catalyst in the Henry reaction between 2-methoxybenzaldehyde (**S10a**) and nitromethane as model substrates in the presence of an array of alkali metal BArF salts as RAs. As summarized in the Table to Figure 3, the size of the cation in the RA affected the catalytic activity with the best results in terms of activity and enantioselectivity being obtained with KBArF (entries 2–5). When the amount of KBArF was increased up to two equivalents with respect to the amount of α,ω-bis(metallosalen) derivative, almost complete conversion and an enhancement of up to 6% in the ee were observed for **S10a** (up to 93% ee, compare entries 4, 6, and 7).

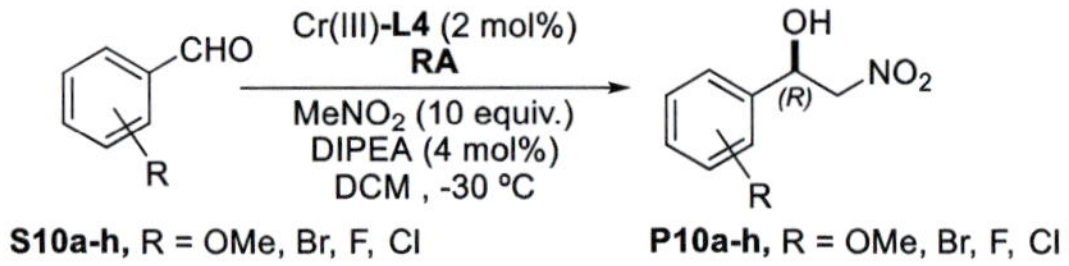

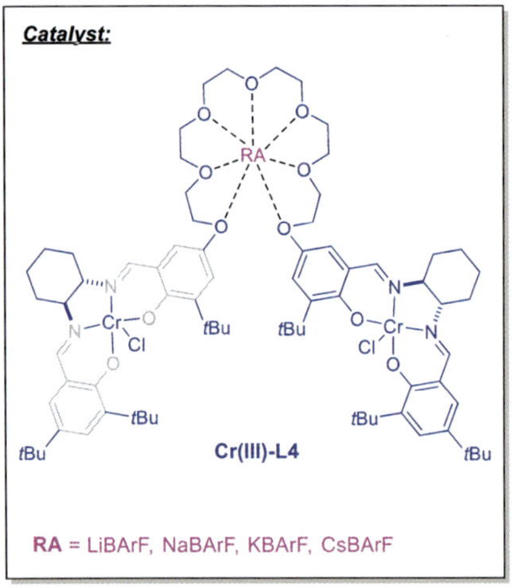

Figure 3. Catalytic results

Entry	Substrate, R	RA (mol%)	ee (%, config)
1	S10a, 2-OMe	–	87 (R)
2	S10a, 2-OMe	LiBArF (2)	89 (R)
3	S10a, 2-OMe	NaBArF (2)	89 (R)
4	S10a, 2-OMe	KBArF (2)	91 (R)
5	S10a, 2-OMe	CsBArF (2)	89 (R)
6	S10a, 2-OMe	KBArF (1)	90 (R)
7	S10a, 2-OMe	KBArF (4)	93 (R)
8	S10b, 2-Br	–	77 (R)
9	S10b, 2-Br	KBArF (4)	86 (R)
10	S10c, 2-F	–	91 (R)
11	S10c, 2-F	KBArF (4)	90 (R)
12	S10d, 2-Cl	–	80 (R)
13	S10d, 2-Cl	KBArF (4)	96 (R)
14	S10e, 3-OMe	–	53 (R)
15	S10e, 3-OMe	KBArF (4)	84 (R)
16	S10f, 3-Cl	–	66 (R)
17	S10f, 3-Cl	KBArF (4)	84 (R)
18	S10g, 4-F	–	77 (R)
19	S10g, 4-F	KBArF (4)	85 (R)
20	S10h, 4-Cl	–	74 (R)
21	S10h, 4-Cl	KBArF (4)	83 (R)

Figure 3. Supramolecularly regulated catalysts for enantioselective Henry reactions.[34]

With the optimized reaction conditions in hand, the authors expanded the substrate scope to a set of structurally diverse aromatic aldehydes, **S10b–h**. This study revealed that the RA affects the outcome of the reaction at a major extent, both in terms of conversion and enantioselectivity, in the case of the *meta*-substituted benzaldehydes **S10e** and **S10f** (see entries 14–17). A positive regulation effect of up to 31% ee, that is, from 53% to 84% ee, has been observed for **S10e**.

III. Induction of Chirality by Chiral Regulating Agents

A main question that arose from the emergence and development of efficient enantioselective metal catalysts during the second half of the twentieth century was related to the optimal position and nature of the stereogenic elements in the ligand system, with respect to the metal center. The paradigm quickly moved away from "the closer, the better," leading to the development of efficient chiral ligands displaying nonstereogenic coordinating centers and chiral backbones. Although these enantiopure ligands are quite easily available, their limited modularity remains a common drawback. As elegantly illustrated in the previous section, this drawback can be somewhat circumvented by the addition of an achiral regulating agent. In a different approach, the noncovalent binding of a chiral RA in the backbone of a prochiral ligand can be used to induce a significant level of enantioselectivity in catalytic reactions, as shown in the following examples.

A. Bidentate Ligands with a Remote Chiral Regulating Agent

Bidentate phosphines are key ligands in enantioselective metal catalytic systems, and privileged ones are usually constructed from axially chiral 1,1′-binaphthyls, chiral diols (*e.g.* BINOL, 2,2-dialkyl-α, α, α', α'-tetraaryldioxolane-4,5-dimethanol [TADDOL]) or amino acids. Adapting their covalent backbone to enlarge the substrate scope for a given catalytic reaction, usually faces significant synthetic hurdles. To circumvent this issue, combinatorial approaches have been considered in which bidentate ligands are built *in situ* from mixtures of chiral monodentate species. Thus, engaging mixtures of monodentate phosphine ligands containing recognition sites will generate libraries of (supramolecular) bidentate ligands at a far lower preparation cost.[35–38]

In an alternative approach, van Leeuwen and coworkers engaged a chiral inducer as an integrative building block of a supramolecular catalyst made of monodentate achiral phosphines.[39] The designed phosphines, **L**, are ditopic ligands featuring Schiff base units as the additional coordination sites. The catalytic system is assembled as follows (Figure 4a): a chiral diol (acting as the chiral RA) and two molecules of the ditopic ligand are bonded to the same titanium center, through the exclusive coordination of the Schiff base part of the ligand. The free phosphorus atoms are engaged then in the coordination to the catalytic metal center. The different building blocks are assembled *in situ*, with [Ti(O*i*Pr)$_4$] and [Rh(nbd)$_2$BF$_4$] acting as the template and catalytic metal precursors,

(a)

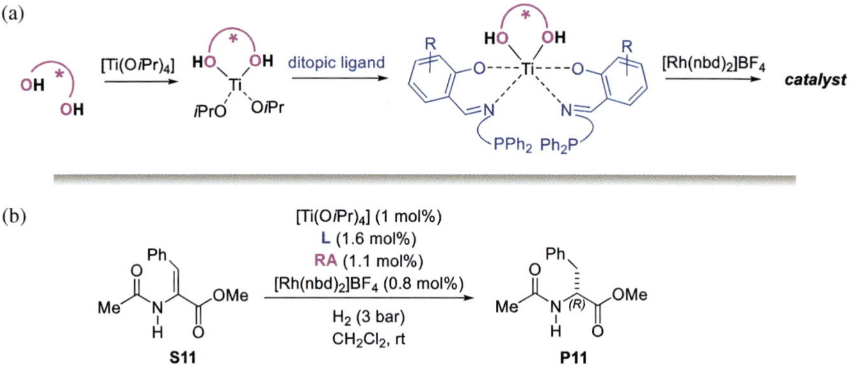

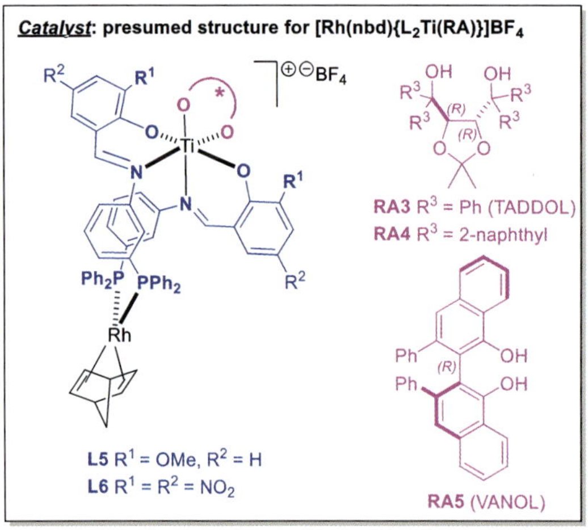

Table to Figure 4: Catalytic results

Entry	L	RA	ee (%)
1	L5	RA3	86.9
2	L5	RA4	80.5
3	L5	RA5	80.5
4	L6	RA3	9.1
5	L6	RA4	-49.5
6	L6	RA5	18.0
7	L5	RA3	92.0[a]

[a] **S11**/[Rh] = 1000

Figure 4. Supramolecular bidentate ligands and catalysts featuring a remote chiral diol as the RA: (a) Preparation; (b) rhodium-catalyzed hydrogenations of **S11** and postulated catalyst structure.[39]

respectively. It was anticipated that the free isopropanol generated upon addition of the chiral diol and the ditopic ligand to [Ti(O*i*Pr)$_4$] imparts certain reversibility to the system, thus providing the same final catalytic species whatever the order of addition of the different building blocks. These catalysts were tested in the rhodium-catalyzed hydrogenation of methyl (Z)-α-acetamidocinnamate (**S11**, Figure 4). Combinations of ditopic ligands, differing by the nature of the groups located at the *ortho* and *para* position of the phenol unit, and a variety of diols were screened. For the majority of the ditopic ligands tested, the highest enantioselectivities were obtained with TADDOL (**RA3**), the 2-naphthyl analogue of TADDOL (**RA4**) and 3,3′-diphenyl-[2,2′-binaphthalene]-1,1′-diol (VANOL) (**RA5**) as the chiral diols.

However, the selectivities vary significantly as a function of the ligand/diol pairs. Thus, opposite enantiomers of the final product were obtained when combining **L5** and **L6** with the same chiral regulating agent, **RA4**, highlighting the large structural diversity spanned by the assembled catalysts (entries 2 and 5, Table to Figure 4b). Under optimized conditions, the combination of **L5** and **RA3** furnished **P11** with 92% ee and the same catalyst was found to be the most efficient of the whole library for two other substrates. The precatalyst embedding **RA4** as the chiral diol and **L5** was isolated under the form of [Rh(nbd){(**L5**)$_2$Ti(**RA4**)}] BF$_4$. NMR analysis indicates a well-defined C_1-symmetric rhodium complex with the templated phosphine acting as a bidentate ligand (see a representation of the complex in Figure 4b). The clean formation of a single species demonstrates an efficient transfer of chirality from the diol to the Ti center (adopting a Δ or ∧ configuration) and then to the chelate ring formed by phosphine coordination to the rhodium center (relative orientation of the *meta*-substituted aryl groups). Importantly, the fact that the bridgehead protons of the norbornadiene are diastereotopic further substantiates the efficient transfer of chiral information from the diol located more than 14 Å away. Such a long-range induction of chirality remains rare in enantioselective catalysis with the notable exception of catalysts adopting a helical structure (see one example below).[40–42]

Reek and coworkers developed a class of diphosphine ligands, whose selectivity stems from a chiral anion located in the ligand backbone.[43] Here, the covalent scaffold of the ligand is equipped with a diamido-diindolylmethane anion receptor, which has a strong ability to bind carboxylate anions ($K_a > 10^5$ M^{-1} in CD$_2$Cl$_2$). In the initially designed ligand (**L7**, Figure 5), the metal binding group (PPh$_2$ moieties) and the anion bonding site were connected through a phenyl group, on which they were located in relative *para* positions. The X-ray structure of the rhodium precatalyst [Rh(nbd)**L7**]BF$_4$ showed that the BF$_4$ anion is bound to the receptor site and that the PPh$_2$ moieties are coordinated to the Rh center in a relative *cis* orientation. The distance between the anion and the rhodium center (B⋯Rh = 5.86 Å) is significantly shorter than the distance

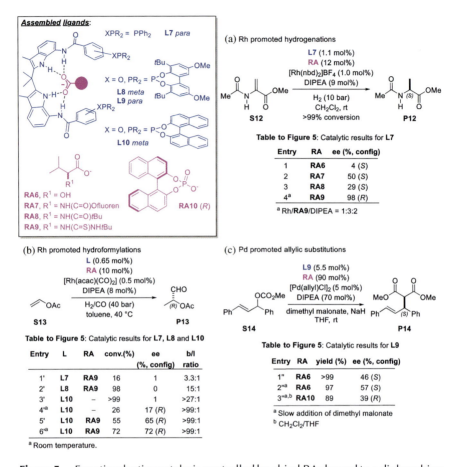

Figure 5. Enantioselective catalysis controlled by chiral RAs bound to a diphosphine backbone.[43–45]

between the chiral diol and the Rh center in the complex displayed in Figure 4. Simple enantiopure RAs, such as α-hydroxy acids, α-amino acids, and their derivatives, readily replace the BF_4 anion and the dissymmetric nature of the resulting complexes was evidenced by the fact that the N–H protons and the P atoms became diastereotopic. It was anticipated that the structural unit of the regulating agent that comes in close proximity to the metal complex will play a major role in the catalytic process, reminiscent of the role of cofactors in enzyme catalysis.

Catalytic mixtures composed of **L7**, [Rh(nbd)$_2$]BF$_4$, a variety of chiral acids and *N*, *N*-diisopropylethylamine (DIPEA) were evaluated in the hydrogenation of ethyl 2-acetamidoacrylate (**S12**, Figure 5). *N*-substituted amino acids provided moderate enantioselectivity (*e.g.* **RA7**, **RA8**), whereas the amino acid featuring a *tert*-butyl thiourea unit (**RA9**) gave a hit (98% ee, Table to Figure 5). Control experiments and density functional theory (DFT) calculations provided a precise molecular model that rationalized the preferential formation of the (*R*) enantiomer of **P12**: the carboxylate anion of **RA9** is bound to the recognition site while its thiourea carbonyl function is engaged into a hydrogen bonding interaction with the amide N–H function of **S12**. The drastically lower enantioselectivity observed for the **L7·RA9** combination in the hydrogenation of alkenes lacking N–H protons substantiates the importance of this hydrogen bonding interaction in determining the selectivity of the reaction.

It is noteworthy that the assembled catalysts are prepared *in situ* allowing, in principle, large libraries of catalysts to be prepared and tested rapidly through combinatorial approaches and deconvolution strategies. Remarkably, a series of experiments in which mixtures of RAs compete for the rhodium complex of **L7** identified **RA9** as the best cofactor (in terms of enantioselectivity), similarly to what was obtained by assessing each assembled catalyst individually. This was attributed to the stronger, preferential interaction between the bonding pocket of **L7** and **RA9**, compared to the other RAs, involving that the most selective catalyst is also the most abundant in solution.

The potential of these self-assembled catalysts was probed for other transition metal-catalyzed reactions. Ligand **L7**, as well as two newly designed ligands with phosphite functions, was tested that display either *tropos* (configurationally labile biaryl motif, **L8**) or *atropos* (configurationally stable biaryl group, **L10**) backbones. These ligands were evaluated in the rhodium-catalyzed hydroformylation of various alkenes including vinyl acetate.[44] With **RA9**, **L7** and **L8** yielded racemic aldehydes (entries 1'-2', Table to Figure 5). However, **RA9** drastically enhanced the enantioselectivity provided by the chiral diphosphine ligand **L10** in the EHF of functionalized alkenes (Δee = +55% in the hydroformylation of vinyl acetate at rt, compare entries 4' and 6' in Table to Figure 5).

Interestingly, **RA9** proved again to be the most efficient chiral inducer among a large library of chiral acids.

Ligand **L9**, with a *tropos* phosphite unit in *para* position of the aryl linker, was tested in combination with various RAs in palladium-catalyzed allylic alkylation reactions.[45] Here, the (*S*)-2-hydroxy-3-methylbutanoate, **RA6,** and the enantiopure phosphate anion **RA10** provided the best enantioselectivity (entries 1″–3″, Table to Figure 5). The ability of the chiral anions to induce a chiral conformation to the otherwise flexible biaryl phosphite unit was demonstrated through NMR analyses of the respective Pd complexes coordinated to fumaronitrile (NC–CH = CH–CN, serving as a model for the *E*-olefin produced by the alkylation reaction). Importantly, in these Pd-catalyzed reactions, no specific interaction occurs between the RA and the substrate and, as such, the essential role of the RA is to control the chirality of the ligand, similarly to the remote chiral diol effects in the aforementioned Ti/Rh-heterobinuclear complexes (Figure 4).

B. Ion-paired Chiral Ligands

Monodentate phosphine ligands have gained much interest in enantioselective catalysis, in part because of their simpler structure relatively to diphosphines. Likewise, a mixture of monodentate ligands could generate metal complexes with two different (hetero complexes) or identical ligands (homo complexes) and such modulation of the catalytic structure proved to be beneficial in a range of reactions.[46] However, in conventional approaches, phosphines with stereogenic P centers or stereogenic centers located next to the phosphorus atom are employed, thus limiting the modularity of the resulting catalysts.

With the objective of addressing this issue, Ooi and coworkers precisely designed a new class of supramolecular chiral ligands in which a quaternary ammonium ion is embedded in the scaffold of an achiral phosphine ligand. The onium ion will interact with a chiral binaphtholate or a chiral phosphate anion thus generating an ion-paired chiral ligand (Figure 6a and 6b).[47]

Structural determination of a member of this family (**L11•RA11**) revealed that the phosphine lone pair is oriented toward the ion-pairing

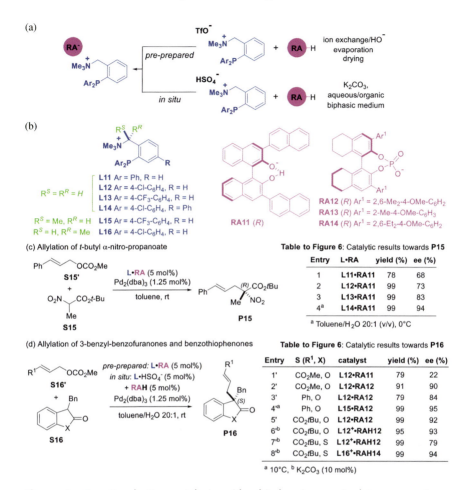

Figure 6. Enantioselective catalysis with chiral anions paired to ammonium-functionalized phosphine ligands.[47–49] (a) Preparation of the ion-paired ligands; (b) structures of the assembled ligands; (c and d) catalytic tests.

site and consequently variations in the structure of both the P-ligand and the chiral anion are expected to affect significantly the performance of the resulting metal complexes. This was indeed demonstrated in the Pd-catalyzed enantioselective allylation of α-nitrocarboxylate **S15** with cinnamyl carbonate **S15'** (Figure 6c). (*R*)-3,3′-disubstituted BINOL **RA11** was selected as the chiral inducer and subtle modifications of the chemical structure of the achiral ligand provided a hit (**L14•RA11**, entry 4, Table to

Figure 6, 94% ee). The same combination displayed excellent levels of enantiocontrol for reactions employing other cinnamyl carbonates (91–97% ee).

For the allylation of 3-benzylbenzofuranone **S16** (X = O, Figure 6d), both components of the catalyst had to be tuned and catalyst optimization was required when the nature of the electrophilic partner **S16'** was changed.[48] **L12•RA12** and **L15•RA12** are the most efficient combinations for methoxycarbonyl- and phenyl-substituted allyl carbonates, respectively (R^1 = CO_2Me or Ph, entries 1'–4', Table to Figure 6). When allyl carbonates with disubstituted double bonds were used as the electrophilic partners, the stereochemistry of the double bond of the allylated product is controlled by finely tuning the structure of the catalyst.[50] The desired *E*-configured product was obtained almost exclusively in high enantioselectivity by employing the combination of **L13** and **RA13**.

In the catalytic tests given earlier, the ion-paired ligands were actually prepared *prior* to the catalytic experiments. In further studies, a practical method for the *in situ* generation of these ion-paired ligands was established through ion metathesis between the hydrogen sulfate salt of the ammonium-phosphine ligand and the chiral phosphoric acid in the presence of K_2CO_3 under biphasic conditions (Figure 6a).[49] The *pre-prepared* (**L12•RA12**) and *in situ*-generated catalysts (**L12$^+$/RAH12**) provided similar selectivity in the palladium-catalyzed allylation of 3-benzylbenzofuranone **S16** (X = O) (92% vs 93% ee, entries 5'–6', Table to Figure 6). Thus, combinatorial screening of ion-paired ligand libraries became easily accessible and was expected to be crucial given the high sensibility of the Pd-catalyzed EAS reactions to the structure of the substrates. For instance, the combination **L12$^+$/RAH12**, best performing in the allylation of 3-benzylbenzofuranone (**S16**, X = O), proved to be less efficient in the case of 3-benzylbenzothiophenone (**S16**, X = S, 79% ee, entry 7', Table to Figure 6). An iterative deconvolution strategy was adopted to screen 144 combinations of ion-paired ligands, from which 16 experiments only were necessary to identify the optimal combination (**L16$^+$/RAH14**, 94% ee, entry 8', Table to Figure 6). The individual evaluations of all the possible 144 combinations also highlighted **L16$^+$/RAH14** as the best one suggesting that heteroleptic bisphosphine–palladium complexes, potentially generated during the combinatorial screening, did not prevent the identification of the best ion-paired ligand.

Phipps and coworkers have disclosed a related concept in which the location of the charges has been reversed (Figure 7).[51] Indeed, a bipyridine ligand was functionalized with a sulfonate anion, enabling cations to be used as the chiral RAs. Unlike their anionic counterparts, chiral cations have been scarcely employed in enantioselective transition metal catalysis and their general use will unlock an important domain of the (chiral) chemical space. The catalytic performance of the bipyridine-sulfonate **L17**, combined with the dihydroquinine-derived chiral cations **RA15** or **RA16**, was evaluated in the iridium-catalyzed enantioselective desymmetrization of benzhydrylamide **S17** via C–H borylation. **RA16** provided the monoborylated product with excellent enantioselectivity (96% ee) after optimization of the reaction conditions (Table to Figure 7) and the scope was expanded to other benzhydrylamides as well as to diaryl phosphinamides. The high regioselectivity for borylation at the *meta*

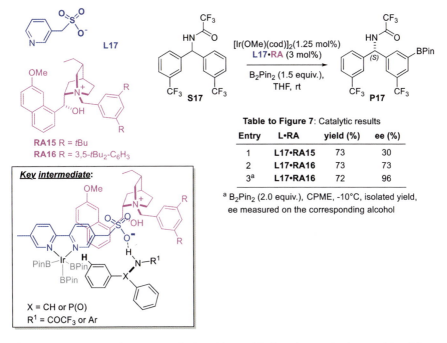

Figure 7. Enantioselective catalysis with a chiral cation paired to a bipyridine ligand.[51] The ligand–substrate interaction in the key intermediate offers a rationale for the selective borylation at the *meta* position.

position of the arene was attributed to the existence of a hydrogen bond between the substrate and the sulfonate group of the ligand, as represented in Figure 7. Hence, this catalytic system combines high regioselectivity, by means of ligand–substrate interactions, and chiral discrimination of enantiotopic sites located far from the prostereogenic center, thanks to the chiral RA. A control experiment was performed in which the ion-paired chiral ligand was tentatively generated *in situ* by mixing the tetrabutylammonium salt of **L17** with the bromide salt of **RA16**. The enantioselectivity dropped to 58% ee, as a result of an incomplete counterion exchange.

The few studies summarized in this section demonstrate that the design of building blocks for ion-paired ligands is relatively simple and its generalization to other families of ligands is expected in the upcoming years. *In situ* generation of catalysts will enable a faster screening of ion-paired ligand libraries and identification of the optimal pairs, similarly to the experiments on onium ligand/chiral anion combinations mentioned earlier (Figure 6).[50]

C. Supramolecular Helical Catalysts

Most of the catalysts reported for enantioselective reactions display ligand-embedded stereogenic centers, stereogenic metals, as well as axial chirality or planar chirality in their scaffolds. Inherently chiral ligands, that is, ligands devoid of these chirality elements, have recently emerged as an appealing class of enantioselective catalysts since they exhibit structural properties that drastically differ from those of classical chiral ligands. A recent example is provided by mechanically planar chiral rotaxane-type ligands.[52] In this context, catalysts built on helical scaffolds have attracted most of the attention in part because significant progresses have been made toward their (enantioselective) synthesis.[53] Consequently, P-containing groups have been attached to a variety of helical scaffolds including helicenes and helical polymers.[54] These enantioselective catalysts usually combine, in addition to their inherent helical chirality, point chirality for controlling the optical purity of the helices and may contain a stereogenic P center; all these elements may contribute to a different extent to the enantiodiscrimination process.[55] However, a limited number of catalysts based on *covalent polymers* has also been developed, for which enantioselectivity is controlled exclusively by the inherent chirality of the helical scaffold.[40,42] This particular class of enantioselective catalysts presents several key features: (i) the enantioselectivity is related to the optical purity of the helices, (ii) the configuration of the main enantiomer of the product depends on the handedness of the helices, (iii) the proximity between the catalytic sites and the helical structure and the related steric effects can enhance the catalytic performance, and (iv) the configuration of the polymer main chain can be controlled by chemical triggers, the amount of which can be substantially lower than the amount of monomers present in the polymer (chirality amplification,[53] *vide infra*). Extending this concept to *supramolecular polymers*, that is, macromolecules spontaneously assembled from relatively simple monomers through noncovalent interactions,[56] would facilitate the construction of highly tunable enantioselective catalysts.

With this objective in mind, Raynal and coworkers decided to anchor PPh$_2$ groups at the periphery of the helical assemblies formed by benzene-1,3,5-tricarboxamide (BTA) monomers. BTA monomers assemble cooperatively to yield long one-dimensional helical structures by a combination of hydrogen bonding and aromatic π-interactions.[57] Supramolecular polymeric ligands, and the corresponding metal complexes, are expected to result from the assembly process as represented in Figure 8a.

In the initial experiments, the stereogenic centers, required for twisting the helices in a preferential direction, and the PPh$_2$ group (as a metal binding site) were located on different arms of the same BTA molecule (**L18**, Figure 8).[58] The respective precatalysts were prepared by mixing the phosphorus-functionalized BTA monomers in the presence of the metal precursor, namely rhodium(I) complexes, in a relatively polar solvent. The final helical catalytic assemblies were formed upon evaporation of this solvent and dissolution of the mixtures in an apolar medium.

The combination of **L18** and [Rh(cod)$_2$BArF] was tested as the catalyst in the enantioselective hydrogenation of dimethyl itaconate **S18** (Figure 8b). It provided the hydrogenation product **P18** with 82% ee (entry 1, Table to Figure 8). The selectivity of the reaction is remarkable considering that the rhodium center and the chiral centers in **L18** are separated by 12 covalent bonds. Controlled catalytic experiments and analytical data actually indicated that the selectivity stems from the

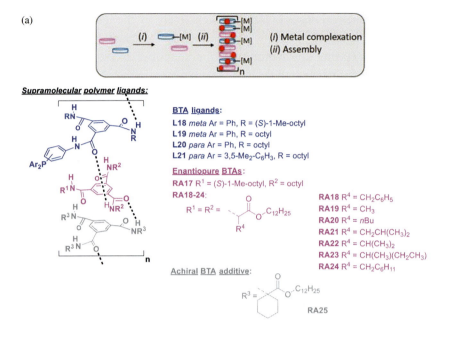

Figure 8. Supramolecular helical catalysts.[58–63] For clarity, a single array of hydrogen bonds is shown in the schematic representation of the supramolecular helical ligand.

inherently chiral nature of the assemblies formed by **L18**, not from isolated **L18** molecules. Thus, the rhodium centers are postulated to bridge two consecutive BTA units of the helically chiral assemblies, through their PPh$_2$ functions.

These inaugural results hold out the possibility of optimizing the catalyst structure by simply mixing BTA-based ligands and phosphine-free, chiral BTA monomers, the latter playing the role of the RAs. The selectivity is increased upon the addition of **RA17** to **L18** (88% ee, entry 2, Table to Figure 8) as the result of their association in the same helical structure. **RA17** is likely to improve the net helicity of the coassemblies, that is, the formation of one-handed helices.

Finally, mixing achiral BTA ligands and enantiopure BTAs (as the RAs) appeared as the simplest way to induce chirality in these systems. Achieving high level of enantioinduction with these catalytic mixtures is not trivial since the achiral ligand may operate either in a chiral or in an achiral environment and any ligands located in an achiral one will significantly lower the selectivity of the reaction. Indeed, the combination of the achiral BTA ligand **L19** and **RA17** provided **P18** with a modest 31% ee (entry 3, Table to Figure 8), indicating that further optimization of the helical structure of the catalyst is required to improve the selectivity.

The chemical structures of the BTA components were varied as follows: (i) the relative position of the PPh$_2$ group and the amide function was set to either *meta* or *para* in the achiral BTA ligands and (ii) the enantiopure BTAs **RA18-24** were prepared from α-amino esters (NHR^2 = NH-CHR^4-CO$_2$C$_{12}$H$_{25}$),[64,65] thus providing a facile access to a library of RAs, whose members differ by the nature of the group R^4 attached to the stereogenic center. The ease of preparation of the catalytic mixtures enabled the screening of several BTA combinations in the rhodium-catalyzed hydrogenation of itaconate **S18** (Figure 8b, entries 4–10, Table to Figure 8)[59] and in the copper-catalyzed hydrosilylation of **S19** (Figure 8c).[60] In these experiments, the RA component is in equimolar amount or in slight excess to the BTA ligand. In the hydrogenation reaction, the combination of the achiral ligand **L19** and (*S*)-**RA23** equals the chiral BTA ligand **L18** in terms of selectivity (85 ± 7% ee, entry 10, Table to Figure 8), thus highlighting the feasibility of reaching high levels of stereocontrol with RA as the unique source of chirality. Also, enantiomeric RAs provided opposite enantiomers of the product, as a result of their ability to induce a preferential handedness to the helical catalyst (entries 6–7, Table to Figure 8).[59] In the Cu-promoted hydrosilylation of *p*-nitroacetophenone **S19**, **L20**, and (*S*)-**RA24** proved to be the optimal combination providing **P19** with 58% ee (entry 7', Table to Figure 8).[60–62]

Subsequently, the influence of a substoichiometric amount of the RA on the outcome of the catalytic reaction was probed. The amount of (*S*)-**RA23** could be decreased down to one-fourth of that of **L19** (*i.e.* one RA for two rhodium atoms) without deteriorating the enantioselectivity of the hydrogenation reaction (78% ee, entry 11, Table to Figure 8).[59] The RA acts as "sergeant" and imposes its preferential handedness to three achiral ligands on average, the latter behaving as the "soldiers." Chirality amplification through the "sergeants-and-soldiers" effect[53,66] is well established in supramolecular polymers adopting a helical configuration, and this study demonstrated its implementation in the context of an enantioselective catalytic reaction.

The effect was modest in the hydrosilylation reaction in Figure 8c, since an equimolar fraction of **L20** and (*S*)-**RA24** ($f_{RA} = 50\%$) was necessary to reach the optimal selectivity (58% ee, entry 7'). However, in that case, the addition of an achiral BTA monomer (**RA25**) was found to enhance by two orders of magnitude the sergeants-and-soldiers effect and consequently a 50% ee in **P19** was obtained with as little as 0.25% of the (*S*)-**RA24** monomer (entry 10', Table to Figure 8).[63] Importantly, no selectivity was observed under the same conditions in the absence of the achiral additive **RA25** (entry 9', Table to Figure 8). Further tuning of the structure of the BTA ligand (**L21**) and optimization of the reaction conditions furnished the hydrosilylation product **P19** with 90% ee, even though only 0.25% of (*S*)-**RA24** was present in the supramolecular helices (entry 12', Table to Figure 8). The increased level of chirality amplification in presence of **RA25** was assigned to its ability to stabilize and decrease the amount of defects in the coassemblies, that is, the number of helix reversals. In such a three-component system, not only an enantiopure RA but also an achiral RA have to be combined to promote the formation of single-handed helices and to finely tune the structure of the supramolecular helical catalyst.

The screening of different L/RA combinations (see Tables to Figure 8) showed the strong influence of the chemical structure of the RAs on the enantioselectivity. Thus, with the aim of harnessing the precise role of the RAs in these supramolecular systems, the optical purity, composition, and structure of the helical coassemblies embedding different types of RA were precisely determined.[62] It was demonstrated that **RA21** and **RA24** almost fully intercalate into the helical stacks of **L20** thus generating long helical assemblies (l > 100 nm), which, in turn, provide the optimal selectivity in the hydrosilylation reaction. On the contrary, **RA18** and **RA19** behave as chain cappers, yielding short and poorly helically biased catalysts. **RA20** plays both the role of intercalator and chain capper, thus explaining the fact that this RA provides significant but not optimal selectivity. This study indicates that the RA acts as a remote chiral inducer, located in the second coordination sphere of the metal complex, which both biases the twist of the helices and affects their length. It remains to be demonstrated whether close contacts might occur between the RA and the catalytic site within the helical coassemblies, thus directly affecting the first coordination sphere of the metal catalyst.

IV. Supramolecular Regulation in Organocatalysis

As previously discussed, tuning the spatial environment around the active sites of synthetic catalysts is a task that has been addressed in a number of ways for transition metal-based catalysts. The discussion that follows highlights three relevant examples on how regulation of the catalytic site has been accomplished for organocatalysts by employing ionic- and hydrogen bonding supramolecular interactions.

In 2009, Bella and coworkers reported an elegant strategy for regulating the activity of an organocatalyst for the formal cycloaddition of aldehydes to cyclic enones *via* supramolecular interactions.[67] The authors developed a strategy in which the starting linear aldehyde is activated toward electrophilic attack by formation of an enamine with the secondary amino group of proline (**C1**) used as the catalyst. Although proline alone turned out to be inefficient in this transformation, the addition of a tertiary amine (Et_3N) translated into the formation of the product in good yield, though with low stereocontrol (low diastereo- and enantioselectivity; entries 1 and 2, Table to Figure 9). The use of quinine **RA26** provided a modest increase in the enantioselectivity, with the diastereoselectivity remaining unaltered with respect to that obtained with triethylamine

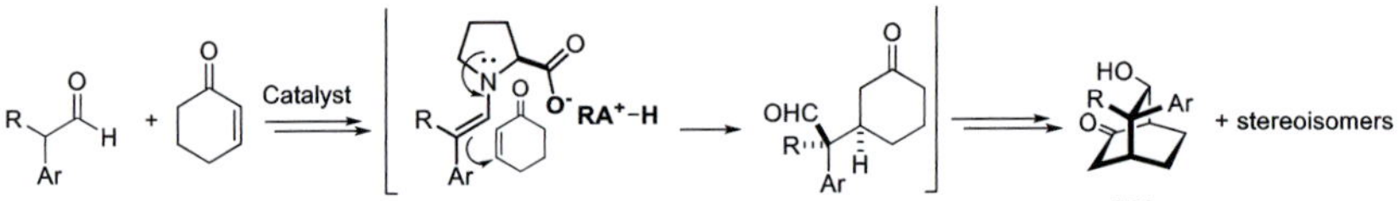

S20 **S20'** **P20**

R = H, Me; Ar = Ph, 4-Br-C_6H_4, 4-Cl-C_6H_4, 4-MeO-C_6H_4, 2-naphthyl (6 examples)

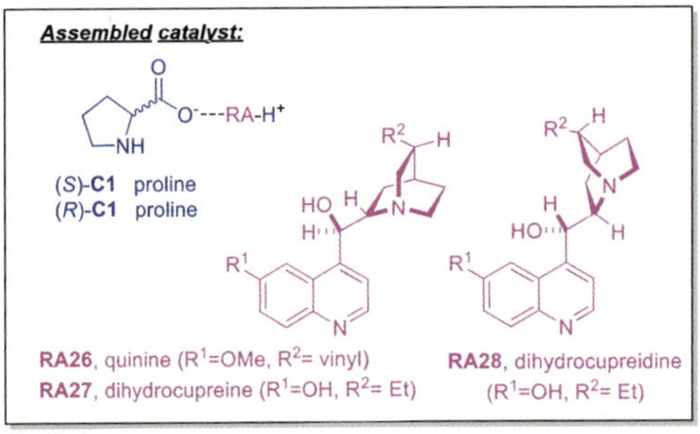

Table to Figure 9: Catalytic results (R = H, Ar = Ph)

Entry	Catalyst	Yield (%)	dr[a]	ee (%)
1	(*S*)-**C1**, –	–	–	–
2	(*S*)-**C1**, Et_3N	76	1.5:1	20
3	(*S*)-**C1**, **RA26**	75	1.5:1	33
4	(*S*)-**C1**, **RA27**	85	1.6:1	64
5	(*S*)-**C1**, **RA28**	83	1.8:1	66
6	(*R*)-**C1**, **RA27**	80	2.3:1	–68
7	(*R*)-**C1**, **RA28**	75	2.2:1	–51

[a] Diastereomeric ratio **P20**:**P20'**

P20' =

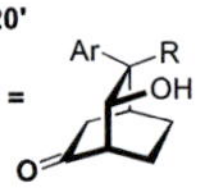

Figure 9. Supramolecularly regulated catalysts for the formal [4+2] cycloaddition of aldehydes to cyclic enones.[67]

(entry 3, Table to Figure 9). After screening other *cinchona* alkaloids such as **RA27** and **RA28**, the yield and enantioselectivity could be increased. Dihydrocupreidine (**RA28**) led to the highest level of stereocontrol (diastereomeric ratio [dr] = 1.8:1 with a 66% ee for the most abundant diastereoisomer; entry 5, Table to Figure 9).

The authors were also interested in assessing the effect of the absolute configuration of the RAs in the outcome of the reaction. For the sake of simplicity, the authors combined unnatural proline (*i.e.* the (*R*) enantiomer) with the *cinchona*-based derivatives already studied (**RA26**–**RA28**). Dihydrocupreine (**RA27**) provided the highest level of diastereo- and enantioselectivity in the reaction (dr = 2.3:1, and a –68% ee for the major diastereoisomer; entry 6, Table to Figure 9). Interestingly, an inversion of the absolute configuration of the product was observed when using unnatural proline as the catalyst (entries 6 and 7, Table to Figure 9), which clearly indicates that, although the absolute configuration of both the catalyst and the RAs influence the stereoselectivity of the process, the stereochemical outcome of the reaction in terms of absolute configuration of the final product is determined by the catalyst and not by the chiral RA.

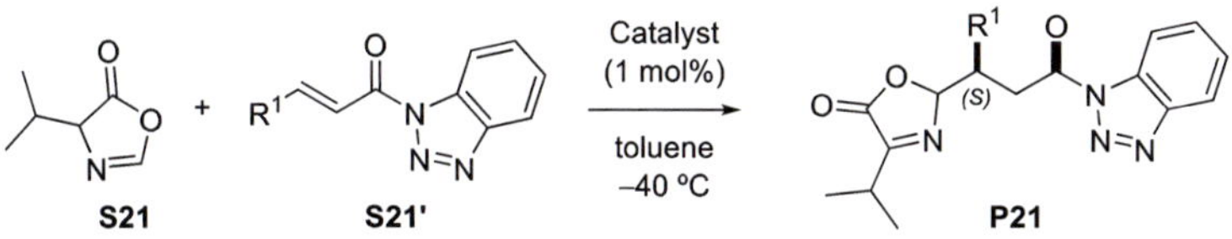

R¹ = Ph, 4-MeO-C₆H₄, 4-Br-C₆H₄, 2-Me-C₆H₄, 3-Br-C₆H₄, 1-naphthyl, 2-furyl, Me, Me(CH₂)₄, Ph(CH₂)₂, cyclohexyl (11 examples)

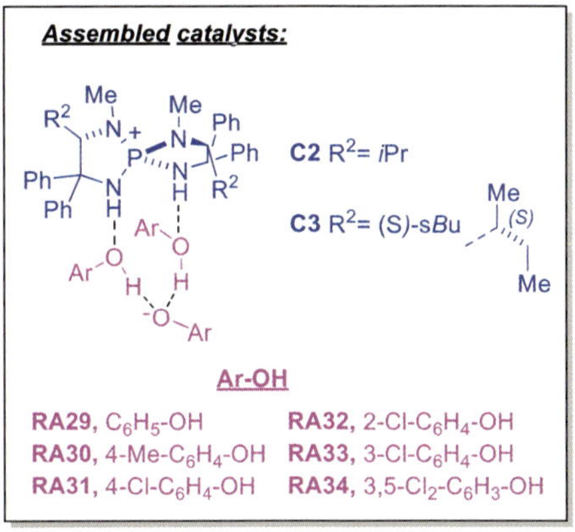

Table to Figure 10 Catalytic results (R^1 = Ph)

Entry	Catalyst	Yield (%)	ee (%)
1	C2, RA29	99	60
2	C2, RA30	96	58
3	C2, RA31	97	75
4	C2, RA32	94	63
5	C2, RA33	93	70
6	C2, RA34	92	80
7[a]	C2, RA34	99	87
8[a]	C3, RA34	95	95

[a] 10-fold increase in concentration

Assembled catalysts:

C2 R^2 = *i*Pr
C3 R^2 = (S)-*s*Bu

Ar-OH

RA29, C₆H₅-OH
RA30, 4-Me-C₆H₄-OH
RA31, 4-Cl-C₆H₄-OH
RA32, 2-Cl-C₆H₄-OH
RA33, 3-Cl-C₆H₄-OH
RA34, 3,5-Cl₂-C₆H₃-OH

Figure 10. Supramolecularly regulated catalysts for the addition of azlactones to α,β-unsaturated amides.[68]

Ooi and coworkers have reported the combined use of enantiopure tetraaminophosphonium salts with phenols and phenoxides to form a hydrogen bond–assembled chiral organic ion pair that promotes the enantioselective conjugate addition of azlactone **S21**, an acyl anion equivalent, to the α,β-unsaturated ester surrogates **S21'** (Figure 10).[68] The distinctive feature of Ooi's supramolecular assembly is that its ability to bias the stereochemical outcome of the reaction is not only controlled by the enantiopure tetraaminophosphonium component (see fragment in blue in Figure 10) but also by the achiral phenols and phenoxides that produce the organic ion pair (see fragment in magenta in Figure 10). In this regard, the achiral component can be considered as the RA in these supramolecular enantioselective catalysts.

The chlorides of the enantiopure phosphoniums **C2** and **C3** were readily synthesized from the corresponding α-amino acids, *L*-valine, and *L*-isoleucine, respectively. Anion exchange with hydroxide followed by neutralization with a phenol rendered the target supramolecular catalysts, whose structure was unambiguously determined by means of X-ray diffraction analysis. The final assembly incorporated one tetraaminophosphonium cation, two phenols, and one phenoxide anion, which formed a 10-membered cyclic hydrogen bonding network (see Figure 10).

Interestingly, the chiral information of the cationic unit was effectively transmitted by the two phenol molecules to the remote position held by the phenoxide anion. This anion is readily replaced then by the enolate of azlactone **S21**, which initiates the catalytic cycle. Finally, the supramolecular complexes prepared by Ooi and coworkers proved to be efficient catalysts for the enantioselective conjugate addition of the enolate of **S21** to the α,β-unsaturated acyl-benzotriazole **S21'** (entries 1–6, Table to Figure 10). The stereoselectivity of the reaction could be tuned through structural modification of the achiral phenolic component, with the highest enantioselectivity being obtained with 3,5-dichlorophenol (**RA34**, 80% ee, entry 6, Table to Figure 10). Interestingly, an increased concentration of the reactants led to improved enantioselectivity (87% ee, compare entries 6 and 7, Table to Figure 10) due to an increased propensity for molecular association in solution.

The structural modification of the chiral cationic moiety, by changing the R^2 substituent from *i*Pr in **C2** to (*S*)-*s*Bu in **C3**, was crucial for increasing the enantioselectivity of the reaction (95% ee, entry 8, Table to Figure 10). Finally, the scope of α,β-unsaturated acyl-benzotriazoles **S21'** was also expanded, with excellent yields and enantioselectivities ranging from 93% to 98% ee (11 examples).[68]

Clarke and coworkers elegantly demonstrated that regulation strategies can be used in enantioselective organocatalytic Michael additions of ketone enolates to nitroalkenes (Figure 11).[69] These authors envisaged that 2-amino-1,8-naphthyridines (as featured in enantiopure **C4** and **C5**) and 1-(1*H*-imidazol-2-yl)ureas (see compounds **RA35** and **RA36**) would be highly complementary recognition motifs through hydrogen bonding interactions. Thus, these structural units were chosen for achieving precise self-assembly by hydrogen bonding toward the target catalysts. Catalyst (*S*)-**C4** combines the 2-amino-1,3-naphthyridine unit with a chiral pyrrolidin-2-yl ring responsible for enantioselective catalysis. The (imidazol-2-yl)urea additives **RA35** (R = H) and **RA36** (R = *t*Bu) functioned as RAs

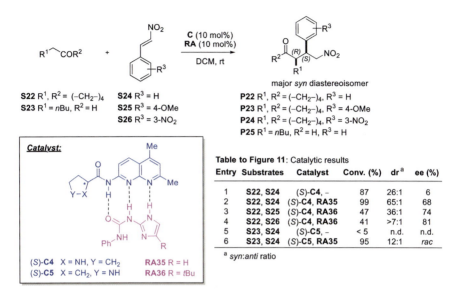

Figure 11. Supramolecularly regulated catalysts for enantioselective Michael addition reactions.[69]

by tuning the steric environment at the catalytic site. Clarke and coworkers observed that the supramolecular assemblies arising from the combination of the two components were more active in mediating Michael additions than the proline-derived catalyst **C4** alone. Thus, for instance, in the enantioselective Michael addition of cyclohexanone **S22** to β-nitrostyrene **S24**, the use of **RA35**, combined to catalyst **C4**, led to an enhancement in both the diastereoselectivity (from a 26:1 to a 65:1 *syn* to *anti* ratio of diastereomers) and the enantioselectivity (from 6% to 68% ee), as summarized in entries 1 and 2 of the Table to Figure 11. Interestingly, the use of the RA **RA36** instead of **RA35** led to the highest diastereo- and enantio-selectivities in the Michael additions to 4-methoxy- and 3-nitro substituted β-nitrostyrenes (substrates **S25** and **S26**; entries 3 and 4 in Table to Figure 11).

The Michael addition of the linear aldehyde **S23** to β-nitrostyrene required the combination of the RA **RA35** and the new pyrrolidin-3-yl-containing catalyst (*S*)-**C5**, instead of the original catalyst (*S*)-**C4**. Although a high diastereoselectivity was attained with this organocatalyst,

(*S*)-**C5•RA35** (dr = 12:1 in favor of the *syn* diastereoisomer), the final product was obtained as a racemic mixture (entries 5 and 6 in Table to Figure 11).

V. Conclusions and Outlook

This chapter highlights different strategies for generating libraries of supramolecularly regulated enantioselective catalysts. In each library of catalysts, the components retain most of the backbone structural features, yet they incorporate, at the same time, changes at their active sites that are determined by the structural characteristics of an external molecule (or RA) that is employed.

Two different categories of supramolecularly regulated enantioselective organometallic catalytic systems have been developed. The first type possesses a catalytic site with a well-defined, asymmetric three-dimensional structure suitable for enantioselective catalysis and a distal regulation site containing a supramolecular motif capable of interacting with the RA. The interactions between cyclic (aza)-crown-ether or polyether motifs, as supramolecular regulation sites, and alkali metal salts or ammonium derivatives as RAs have been exploited for the easy generation of libraries of enantioselective catalysts. These libraries have made it possible to adapt the geometric features of the catalyst to the requirements of a given substrate, so as to reach high enantioselectivity in palladium-promoted allylic substitutions, rhodium-promoted hydroformylations, and chromium-promoted Henry reactions, among other transformations. This strategy for the regulation of the active site has not been restricted to transition metal–based transformations, since supramolecularly regulated enantioselective organocatalysts have been reported for the formal [4+2] cycloadditions of aldehydes with enones, as well as for the addition of carbon nucleophiles to α,β-unsaturated derivatives.

The second category of supramolecularly finely regulated catalytic systems is made of catalysts that are generated from achiral (or prochiral) catalytic units. The three-dimensional structure of the catalytic site is not well defined in the absence of a RA and, in this elegant approach, addition of an enantiopure RA triggers chirogenesis processes involving chirality transfer from the RA to the backbone containing the catalytic site. Overall, bidentate ligands, ion-paired ligands, and helical catalysts have been prepared following this approach and have been efficiently used for an array of structurally diverse substrates in enantioselective hydrogenation, hydroformylation, allylic substitution, borylation, and hydrosilylation reactions. It is noteworthy to mention that the role played by the enantiopure RA can vary significantly from one system to another, and even for the same system as a function of the substrate. Therefore, it is not surprising that RAs have been named as chiral remote inducers,[39] chiral cofactors,[43] and "sergeants"[59] in the original publications quoted in this chapter. In some cases, the role of the RA is restricted to its ability to induce a preferential chiral conformation to the metal catalytic centers, whereas in other cases, its ability to interact with the substrate and/or to trigger a significant structural change to the whole catalytic system have been clearly demonstrated.

The reader is reminded that the selected examples presented here are not meant to be exhaustive; rather they are intended to be representative of key concepts in this area of research. We believe that the concepts that have been summarized and discussed in this review will pave the way to the design and preparation of novel supramolecularly regulated catalysts with ever greater efficiency (*i.e.* turnovers, chemo- and stereoselectivities). In addition, it will be important to determine whether the role of the RA can be extended to other catalytic purposes, for example, to switching the enantioselectivity of a given catalyst in real time in order to efficiently achieve stereodivergent processes.[62]

VI. Acknowledgments

M.R. gratefully acknowledges cooperation and support of all collaborators and coworkers involved in the research projects dealing with "supramolecular helical catalysts" and whose names are mentioned in the related publications. M.R. also thanks the French Agence Nationale de la Recherche, project ANR-13-BS07-0021 (SupraCatal) and project ANR-17-CE07-0002 (AbsoluCat), for financial support. A.V. thanks MINECO (PID2020-115658GB-I00) for the financial support.

VII. References

1. Breit, B. *Angew. Chem. Int. Ed.* **2005**, *44*, 6816–6825.
2. Goudriaan, P. E.; van Leeuwen, P. W. N. M.; Birkholz, M.-N.; Reek, J. N. H. *Eur. J. Inorg.Chem.* **2008**, *2008*, 2939–2958.
3. Meeuwissen, J.; Reek, J. N. H. *Nat. Chem.* **2010**, *2*, 615–621.
4. Carboni, S.; Gennari, C.; Pignataro, L.; Piarulli, U. *Dalton Trans.* **2011**, *40*, 4355–4373.
5. Ohmatsu, K.; Ooi, T. *Tetrahedron Lett.* **2015**, *56*, 2043–2048.
6. Dydio, P.; Reek, J. N. H. *Chem. Sci.* **2014**, *5*, 2135–2145.
7. Fanourakis, A.; Docherty, P. J.; Chuentragool, P.; Phipps, R. J. *ACS Catal.* **2020**, *10*, 10672–10714.
8. Mote, N. R.; Chikkali, S. H. *Chem. Asian J.* **2018**, *13*, 3623–3646.
9. Tan, C. X.; Chu, D. D.; Tang, X. H.; Liu, Y.; Xuan, W. M.; Cui, Y. *Chem. — Eur. J.* **2019**, *25*, 662–672.
10. Li, X. Z.; Wu, J. G.; He, C.; Meng, Q. T.; Duan, C. Y. *Small* **2019**, *15*, 1804770.
11. Raynal, M.; Ballester, P.; Vidal-Ferran, A.; van Leeuwen, P. W. N. M. *Chem. Soc. Rev.* **2014**, *43*, 1660–1733.
12. Raynal, M.; Ballester, P.; Vidal-Ferran, A.; van Leeuwen, P. W. N. M. *Chem. Soc. Rev.* **2014**, *43*, 1734–1787.
13. van Leeuwen, P. W. N. M. Ed., *Supramolecular Catalysis*. Wiley-VCH: Weinheim, **2008**.
14. van Leeuwen, P. W. N. M.; Raynal, M. Supramolecular Strategies for Efficient Catalysis in Water. In *Supramolecular Chemistry in Water*; Kubik, S., Ed. Wiley VCH: Weinheim, **2018**.
15. Vaquero, M.; Rovira, L.; Vidal-Ferran, A. *Chem. Commun.* **2016**, *52*, 11038–11051.
16. Perutz, M. F.; Fermi, G.; Luisi, B.; Shaanan, B.; Liddington, R. C. *Acc. Chem. Res.* **1987**, *20*, 309–321.
17. Grandori, R.; Lavoie, T. A.; Pflumm, M.; Tian, G.; Niersbach, H.; Maas, W. K.; Fairman, R.; Carey, J. *J. Mol. Biol.* **1995**, *254*, 150–162.
18. Escárcega-Bobadilla, M. V.; Kleij, A. W. *Chem. Sci.* **2012**, *3*, 2421–2428.
19. Boersma, A. J.; Megens, R. P.; Feringa, B. L.; Roelfes, G. *Chem. Soc. Rev.* **2010**, *39*, 2083–2092.
20. Park, S.; Sugiyama, H. *Angew. Chem. Int. Ed.* **2010**, *49*, 3870–3878.
21. Hyster, T. K.; Ward, T. R. *Angew. Chem. Int. Ed.* **2016**, *55*, 7344–7357.
22. Phipps, R. J.; Hamilton, G. L.; Toste, F. D. *Nat. Chem.* **2012**, *4*, 603–614.
23. Mahlau, M.; List, B. *Angew. Chem. Int. Ed.* **2013**, *52*, 518–533.
24. Pedersen, C. J. *Angew. Chem. Int. Ed. Engl.* **1988**, *100*, 1021–1027.
25. Cram, D. J. *Angew. Chem. Int. Ed. Engl.* **1988**, *100*, 1009–1020.
26. Steed, J. W. *Coord. Chem. Rev.* **2001**, *215*, 171–221.
27. Sawamura, M.; Nakayama, Y.; Tang, W.-M.; Ito, Y. *J. Org. Chem.* **1996**, *61*, 9090–9096.

28. Yoo, C.; Dodge, H. M.; Miller, A. J. M. *Chem. Commun.* **2019**, *55*, 5047–5059.
29. Li, Y.; Ma, B.; He, Y.; Zhang, F.; Fan, Q.-H. *Chem. — Asian J.* **2010**, *5*, 2454–2458.
30. Mon, I.; Jose, D. A.; Vidal-Ferran, A. *Chem. — Eur. J.* **2013**, *19*, 2720–2725.
31. Vidal-Ferran, A.; Mon, I.; Bauzà, A.; Frontera, A.; Rovira, L. *Chem. — Eur. J.* **2015**, *21*, 11417–11426.
32. Rovira, L.; Vaquero, M.; Vidal-Ferran, A. *J. Org. Chem.* **2015**, *80*, 10397–10403.
33. Martínez-Carrión, A.; Howlett, M. G.; Alamillo-Ferrer, C.; Clayton, A. D.; Bourne, R. A.; Codina, A.; Vidal-Ferran, A.; Adams, R. W.; Burés, J. *Angew. Chem. Int. Ed.* **2019**, *58*, 10189–10193.
34. Ouyang, G.-H.; He, Y.-M.; Fan, Q.-H. *Chem. — Eur. J.* **2014**, *20*, 16454–16457.
35. Ding, K.; Du, H.; Yuan, Y.; Long, J. *Chem. — Eur. J.* **2004**, *10*, 2872–2884.
36. Moteki, S. A.; Takacs, J. M. *Angew. Chem. Int. Ed.* **2008**, *47*, 894–897.
37. Meeuwissen, J.; Kuil, M.; van der Burg, A. M.; Sandee, A. J.; Reek, J. N. H. *Chem. — Eur. J.* **2009**, *125*, 10272–10279.
38. Wieland, J.; Breit, B. *Nat. Chem.* **2010**, *2*, 832–837.
39. van Leeuwen, P. W. N. M.; Rivillo, D.; Raynal, M.; Freixa, Z. *J. Am. Chem. Soc.* **2011**, *133*, 18562–18565.
40. Suginome, M.; Yamamoto, T.; Nagata, Y.; Yamada, T.; Akai, Y. *Pure Appl. Chem.* **2012**, *84*, 1759–1769.
41. Le Bailly, B. A. F.; Byrne, L.; Clayden, J. *Angew. Chem. Int. Ed.* **2016**, *55*, 2132–2136.
42. Li, Y.; Bouteiller, L.; Raynal, M. *ChemCatChem* **2019**, *11*, 5212–5226.
43. Dydio, P.; Rubay, C.; Gadzikwa, T.; Lutz, M.; Reek, J. N. H. *J. Am. Chem. Soc.* **2011**, *133*, 17176–17179.
44. Bai, S. T.; Kluwer, A. M.; Reek, J. N. H. *Chem. Commun.* **2019**, *55*, 14151–14154.
45. Theveau, L.; Bellini, R.; Dydio, P.; Szabo, Z.; van der Werf, A.; Sander, R. A.; Reek, J. N. H.; Moberg, C. *Organometallics* **2016**, *35*, 1956–1963.
46. Reetz, M. T. *Angew. Chem. Int. Ed.* **2008**, *47*, 2556–2588.
47. Ohmatsu, K.; Ito, M.; Kunieda, T.; Ooi, T. *Nat. Chem.* **2012**, *4*, 473–477.
48. Ohmatsu, K.; Ito, M.; Kunieda, T.; Ooi, T. *J. Am. Chem. Soc.* **2013**, *135*, 590–593.
49. Ohmatsu, K.; Hara, Y.; Ooi, T. *Chem. Sci.* **2014**, *5*, 3645–3650.
50. Ohmatsu, K.; Ito, M.; Ooi, T. *Chem. Commun.* **2014**, *50*, 4554–4557.
51. Genov, G. R.; Douthwaite, J. L.; Lahdenpera, A. S. K.; Gibson, D. C.; Phipps, R. J. *Science* **2020**, *367*, 1246–1251.
52. Heard, A. W.; Goldup, S. M. *Chem* **2020**, *6*, 994–1006.
53. Yashima, E.; Ousaka, N.; Taura, D.; Shimomura, K.; Ikai, T.; Maeda, K. *Chem. Rev.* **2016**, *116*, 13752–13990.
54. Megens, R. P.; Roelfes, G. *Chem. — Eur. J.* **2011**, *17*, 8514–8523.
55. Yavari, K.; Aillard, P.; Zhang, Y.; Nuter, F.; Retailleau, P.; Voituriez, A.; Marinetti, A. *Angew. Chem. Int. Ed.* **2014**, *53*, 861–865.
56. De Greef, T. F. A.; Smulders, M. M. J.; Wolffs, M.; Schenning, A. P. H. J.; Sijbesma, R. P.; Meijer, E. W. *Chem. Rev.* **2009**, *109*, 5687–5754.

57. Cantekin, S.; de Greef, T. F. A.; Palmans, A. R. A. *Chem. Soc. Rev.* **2012**, *41*, 6125–6137.
58. Raynal, M.; Portier, F.; van Leeuwen, P. W. N. M.; Bouteiller, L. *J. Am. Chem. Soc.* **2013**, *135*, 17687–17690.
59. Desmarchelier, A.; Caumes, X.; Raynal, M.; Vidal-Ferran, A.; van Leeuwen, P. W. N. M.; Bouteiller, L. *J. Am. Chem. Soc.* **2016**, *138*, 4908–4916.
60. Zimbron, J. M.; Caumes, X.; Li, Y.; Thomas, C. M.; Raynal, M.; Bouteiller, L. *Angew. Chem. Int. Ed.* **2017**, *56*, 14016–14019.
61. Li, Y.; Caumes, X.; Raynal, M.; Bouteiller, L. *Chem. Commun.* **2019**, *55*, 2162–2165.
62. Martínez-Aguirre, M. A.; Li, Y.; Vanthuyne, N.; Bouteiller, L.; Raynal, M. *Angew. Chem. Int. Ed.* **2021**, *60*, 4183–4191.
63. Li, Y.; Hammoud, A.; Bouteiller, L.; Raynal, M. *J. Am. Chem. Soc.* **2020**, *142*, 5676–5688.
64. Desmarchelier, A.; Raynal, M.; Brocorens, P.; Vanthuyne, N.; Bouteiller, L. *Chem. Commun.* **2015**, *51*, 7397–7400.
65. Desmarchelier, A.; Alvarenga, B. G.; Caumes, X.; Dubreucq, L.; Trouffard, C.; Tessier, M.; Vanthuyne, N.; Idé, J.; Maistriaux, T.; Beljonne, D.; Brocorens, P.; Lazzaroni, R.; Raynal, M.; Bouteiller, L. *Soft Matter* **2016**, *12*, 7824–7838.
66. Palmans, A. R. A.; Meijer, E. W. *Angew. Chem. Int. Ed.* **2007**, *46*, 8948–8968.
67. Bella, M.; Schietroma, D. M. S.; Cusella, P. P.; Gasperi, T.; Visca, V. *Chem. Commun.* **2009**, 597–599.
68. Uraguchi, D.; Ueki, Y.; Ooi, T. *Science* **2009**, *326*, 120–123.
69. Fuentes, J. A.; Lebl, T.; Slawin, A. M. Z.; Clarke, M. L. *Chem. Sci.* **2011**, *2*, 1997–2005.

© 2022 World Scientific Publishing Company
https://doi.org/10.1142/9789811248436_0004

4 Chiral Counterions in Enantioselective Organometallic Catalysis

Louis Fensterbank,* Cyril Ollivier,* Antoine Roblin* and Marion Barbazanges*,†

*Sorbonne Université, CNRS, Institut Parisien de Chimie Moléculaire, UMR 8232, 4 Place Jussieu, 75005 Paris, France
†marion.barbazanges@sorbonne-universite.fr

Table of Contents

List of Abbreviations	96
I. Introduction	97
II. Reactions Involving Formation of C–H Bonds	102
III. Reactions Involving Formation of C–Heteroatom Bonds	103
A. Heterocyclizations *via* Hydroaminations or Hydroalkoxylations	103
B. Diamination of Alkenes	109
C. Overman Rearrangement	110
IV. Reactions Involving Formation of C–C Bonds	111
A. Cycloisomerization Reactions	111
B. Tandem Cycloisomerization/Addition Reactions	114

C. [2+2+2] Cycloaddition Reactions		116
D. Hydrovinylation of Olefins		119
V. Conclusion		120
VI. Acknowledgments		120
VII. References		121

List of Abbreviations

ACDC	asymmetric counterion–directed catalysis
Ar	aryl
atm	atmosphere
BINAP	2,2′-bis(diphenyl phosphino)-1,1′-binaphthyl
BIPHEP	1,1′-bis(phosphanyl)biphenyl
Bu	butyl
cod	cyclooctadiene
Cy	cyclohexyl
Cyp	cyclopentyl
DCE	dichloroethane
DFT	density functional theory
DIPAMP	1,2-bis[(2-methoxyphenyl)phenylphosphino]ethane
DM-BIPHEP	(1,1′-Biphenyl-2,2′-diyl)bis[bis(3,5-dimethylphenyl)phosphine]
DM-BINAP	(1,1′-Binaphthalene-2,2′-diyl)bis[bis(3,5-dimethylphenyl)phosphine]
DOSY	diffusion-ordered spectroscopy
dppb	1,4-bis(diphenylphosphino)butane
dppm	bis(diphenylphosphino)methane
EWG	electrowithdrawing group
EXAFS	extended X-ray absorption fine structure
HRMS	high-resolution mass spectroscopy
MS	molecular sieves
MTBE	methyl *tert*-butyl ether
NHC	*N*-heterocyclic carbene
NMR	nuclear magnetic resonance
rt	room temperature
Tf	CF_3SO_2

Tol-BINAP	(2,2′-bis(di-*p*-tolylphosphino)-1,1′-binaphthyl
TPGS	DL-alpha-tocopherol methoxypolyethylene glycol succinate
TRIP	3,3′-Bis(2,4,6-triisopropylphenyl)-1,1′-binaphthyl-2,2′-diyl phosphate
TRISPHAT	[Tetrabutylammonium] [Δ-tris(tetrachloro-1,2-benzenediolato)phosphate(V)]
Ts	tosyl
X^{*-}	chiral counterion

I. Introduction

The control of chirality is essential in many fields, whether for therapeutic molecules or for molecular materials and even in macromolecules, since it ensures the delivery of materials with more defined properties, thus favoring the designed uses. Although Mother Nature is intrinsically a chiral world, it handles the control of chirality with daunting but also inspiring ease. The synthetic chemist can take advantage of this knowledge and design, *de novo*, tools for asymmetric induction. One of the first thinkers on asymmetric induction is certainly Pasteur when he claimed in 1860, as translated by Henri Kagan, that "[…] one needs to use dissymmetric forces, to have recourse to solenoids, to dissymmetric movements of light, to the action of substances themselves dissymmetric […]" to prepare chiral substances (dissymmetric being understood today as chiral).[1] What is amazing in Pasteur's vision is that he put at the same level asymmetric physical forces (chiral polarized light, magnetic field) over chemical asymmetric induction, whereas he had already evidenced the chirality of tartrates. While Le Bel and Van't Hoff nurtured the thoughts of that time, it is probably Marckwald who settled the ground of asymmetric synthesis in 1904 by coining the term itself: "Asymmetric syntheses are those reactions which produce optically active substances from symmetrically constituted compounds, with the intermediate use of optically active materials, but with the exclusion of all analytical processes."[2] The term was coined on the occasion of the first asymmetric reaction that corresponds to the enantioselective decarboxylation of a malonic acid derivative. It is nearly a century later (in 2001) that a Nobel Prize was awarded to Sharpless, Knowles, and

Noyori for the discovery of catalytic enantioselective reactions. A common feature of the rewarded works of these scientists is the use, as catalysts, of transition metal complexes bearing chiral ligands. This asymmetric homogenous catalysis approach is by far the most chosen one and it corresponds now to a mainstream practice with current applications in industrial manufacturing. Nevertheless, it remains intrinsically highly challenging to find a proper ligand for running asymmetric catalysis since it needs to satisfy a dual role: first, providing the adequate electronic configuration and steric environment to promote the elementary steps of the catalytic cycle and second, fixing the highest possible asymmetric induction. Finding an optimal match between these two goals often reveals complex and this set of constraints generally translates into a tedious screening of various types of ligands. An alternative to the chiral ligands approach is conceivable when dealing with cationic organometallic catalysts. It consists in assorting the cationic metallic moiety with a chiral counterion. The main advantage of this strategy, coined as asymmetric counteranion–directed catalysis (ACDC) by Benjamin List, resides in the fact that no sophisticated ligands are required in that case, it just needs to set the right electronics around the metal and, then, the screening for asymmetric induction will rely on a single component: the chiral anion (Figure 1).[3]

Certainly, the ground works of Lacour in the 2000s on the chiral [tetrabutylammonium] [Δ-tris(tetrachloro-1,2-benzenediolato)phosphate(V)] (TRISPHAT) anion have paved the way for important developments in asymmetric catalysis with chiral anions (Figure 2).[4] TRISPHAT is an organic salt.

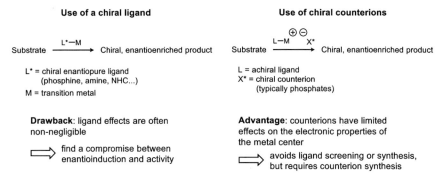

Figure 1. Chiral ligand strategy *vs* chiral counterion strategy.

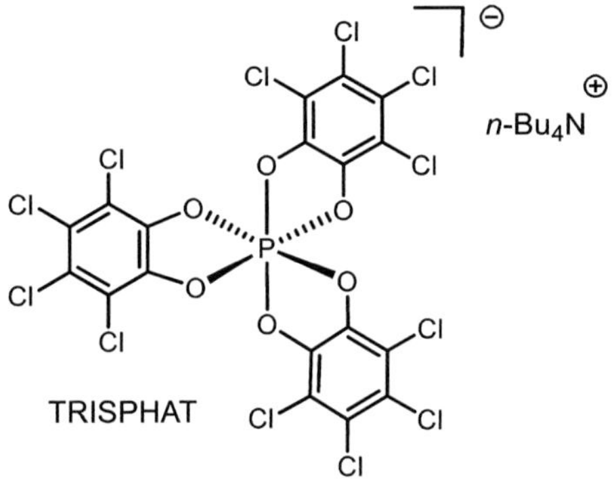

Figure 2. A TRISPHAT chiral salt.

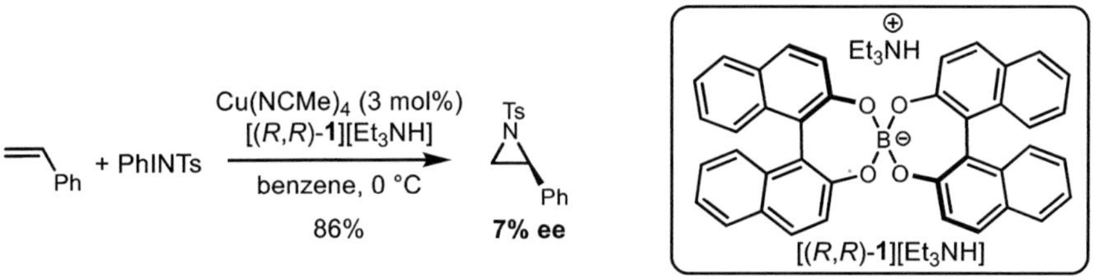

Scheme 1. Seminal study toward asymmetric catalysis using a borate chiral anion.[6]

The anion features a phosphorus(V) bonded to three tetrachlorocatecholate ligands. This anion can be resolved into two enantiomers that are optically stable. The main use of TRISPHAT has consisted of applications as a chiral shift reagent for cations. Notably, it enables resolution of nuclear magnetic resonance (NMR) spectra by forming diastereomeric ion pairs with chiral substrates.

In 2000, Arndtsen and coworkers reported a rare example where the bis(binaphthol)borate anion 1, initially designed by Yamamoto,[5] could be used as a chiral anion with some achiral Cu(I) complexes to promote the enantioselective asymmetric aziridination of styrene (Scheme 1). Although the target product was obtained in only 7% enantiomeric excess (ee), this example validated that chiral anions could be valuable agents in asymmetric catalysis.[6]

Chiral binaphthyl phosphoric acids, first developed by Terada[7] and Akiyama,[8] have been intensively used as chiral catalysts and cocatalysts.[9]

In 2006, List introduced the highly selective 3,3-bis(2,4,6-triisopropylphenyl)-1,1'-binaphthyl-2,2'-diyl phosphoric acid (abbreviated as "H-TRIP").[10,11] The same year, at the occasion of the development of an asymmetric organocatalytic transfer hydrogenation of α,β-unsaturated aldehydes, List also proposed the term "asymmetric counteranion–directed synthesis."[12] In 2009, Lacour[4] pointed that the use of the term "counteranion" should be avoided and "counterion" is the proper term. So, in this chapter, we will refer to "ACDC" and "counterion." It is in 2007 that List first demonstrated the relevance of the use of the TRIP chiral phosphate in an organometallic catalyzed reaction.[13] The asymmetric α-allylation of aldehydes was developed upon mixing an aldehyde and an allylic amine, in the presence of 3.0 mol% of Pd(PPh$_3$)$_4$ and 1.5 mol% H(R)-TRIP. Allylated aldehydes were obtained in high yields with excellent ees. This was rationalized by the initial formation of an enammonium–TRIP ionic pair. Upon reaction with Pd(0), a tightly associated assembly of a cationic π-allyl palladium complex, the enamine and the TRIP anion is formed. It reacts then through a nucleophilic addition, leading to the asymmetric formation of a C–C bond and delivering an α-allylated iminium cation that gets hydrolyzed to liberate the desired aldehyde (Scheme 2).

At this time, Krische also reported the enantioselective hydrogenative coupling of 1,3-enynes to heterocyclic aromatic aldehydes and ketones *via*

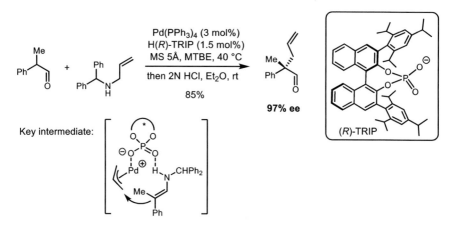

Scheme 2. Enantioselective Pd/Brønsted acid-catalyzed direct α-allylation of aldehydes.[13]

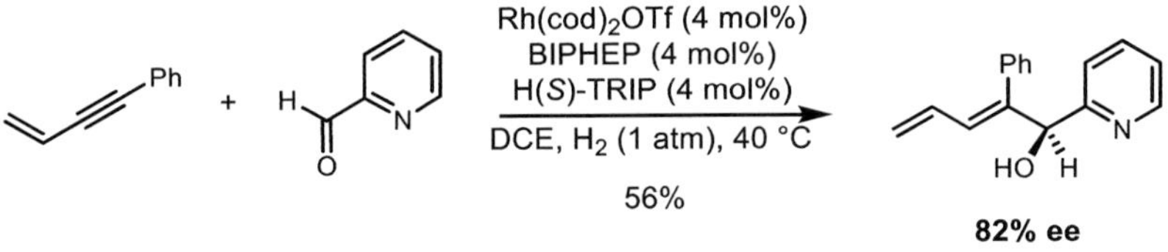

Scheme 3. Hydrogenative coupling of 1,3-enynes to carbonyls using rhodium/phosphate ion pair.[14]

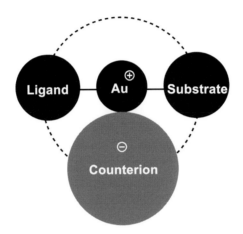

Figure 3. Schematic view of cationic gold(I) intermediates in catalysis.

rhodium catalysis in the presence of a TRIP anion, which is formed after *in situ* deprotonation of H-TRIP by a pyridine (Scheme 3).[14]

Probably the most iconic example about the use of chiral anionic phosphates was reported in 2007 by Toste in the context of gold(I)-catalyzed asymmetric hydroxycyclization reactions.[15] The choice of gold(I) catalysis was fully meaningful, since the challenge is particularly high in this case. Because of the linear coordination imposed by Au(I), which sets the ligand and the substrate at 180°, and the degree of rotational freedom around the ligand–Au axis, finding suitable ligands for asymmetric induction always necessitates focused design efforts (Figure 3).[16] In contrast, a bulky anion with restricted mobility should be able to fix discriminating transition states.

Thanks to the use of gold chloride complexes, combined with Ag[(S)-TRIP], excellent ees could be reached in several hydroxycyclizations and

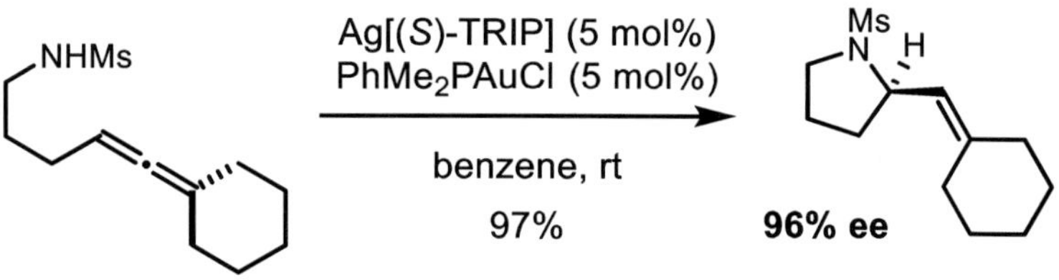

Scheme 4. Intramolecular allene hydroamination using a gold(I)/phosphate ion pair.[15]

intramolecular hydroaminations of allenes. Here, the silver phosphate generates a cationic species from the (PhMe$_2$P)AuCl complex, which features a "simple," achiral phosphine ligand (Scheme 4).

Several excellent reviews have been published on the topic of catalysis with chiral anions.[3,9,17] Herein, this monograph will strictly focus on transformations entailing full atom economy, transiting by π-activation of C=C or C≡C bonds and in which chiral species are involved in catalytic amounts.

II. Reactions Involving Formation of C–H Bonds

In 2010, Leitner, Klankermayer, and coworkers reported the only atom economic C–H bond creation involving π-activation of an olefinic C=C bond, in the chiral counterion strategy, by using the chiral borate **1** counterion as the source of stereoinduction (Scheme 5).[18] They disclosed the enantioselective hydrogenation of dimethyl itaconate, by using enantiopure borate ammonium salt [*S*,*S*-**1**][Et$_3$NH] in the presence of a racemic cationic rhodium complex, [(BINAP)Rh(cod)][BF4] (BINAP = *rac*-2,2'-bis(diphenylphosphino)-1,1'-binaphthyl). By using this system under hydrogen pressure, after probable counterion metathesis, the desired reduced product was isolated in ees increasing with the amount of borate introduced, ranging from 11% ee (for **1**/[Rh] = 2,6/1) to 57% ee (for **1**/[Rh] = 60/1]).

Interestingly, solvents screening supported a true ion-pair mechanism, since the presence of ethanol as the polar cosolvent totally inhibited the stereoinduction and led to a racemic product.

Further experiments have been carried out to enlighten the stereoinduction process. Thus, when [(*rac*-BINAP)Rh(cod)][BF$_4$] was treated with

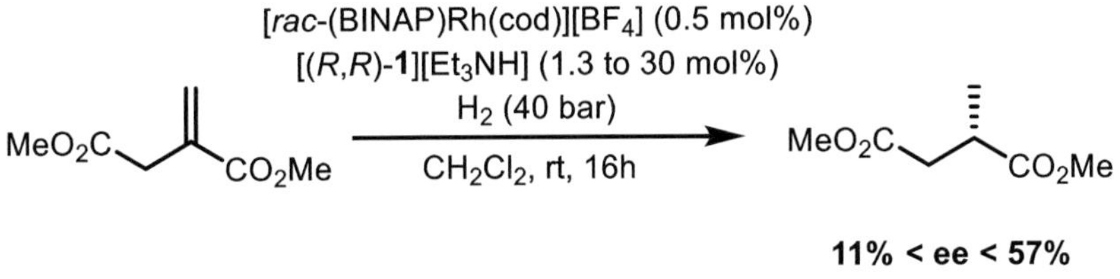

Scheme 5. Alkene hydrogenation using a rhodium/borate ion pair.[18]

Reaction conditions shown: [rac-(BINAP)Rh(cod)][BF$_4$] (0.5 mol%), [(R,R)-1][Et$_3$NH] (1.3 to 30 mol%), H$_2$ (40 bar), CH$_2$Cl$_2$, rt, 16h. Substrate: MeO$_2$C–C(=CH$_2$)–CO$_2$Me → product MeO$_2$C–CH(CH$_3$)–CO$_2$Me. 11% < ee < 57%.

the borate/ammonium salt [(R,R)-1][Et$_3$NH], ^{31}P NMR analysis disclosed the formation of two diastereoisomeric complexes, presumably [(R-BINAP)Rh(cod)][(R,R)-1] and [(S-BINAP)Rh(cod)][(R,R)-1]. Kinetic studies enlightened a *match* and a *mismatch* pairing between BINAP and the counterion, where the *mismatch* pair led to a deactivation process that controlled the stereochemical outcome (chiral poisoning). Thus, the (S)-BINAP/(R,R)-1 *match* pair, which gives an (S)-configured product, catalyzed the reaction faster than the (R)-BINAP/(R,R)-borate *mismatch* pair, explaining why the *rac*-BINAP/(R,R)-1 combination led to the (S)-adduct predominantly.

III. Reactions Involving Formation of C–Heteroatom Bonds

A. Heterocyclizations *via* Hydroaminations or Hydroalkoxylations

The significance and increasing popularity of the electrophilic metal-catalyzed cycloisomerizations of allene derivatives, which allow the formation of carbon–heteroatom bonds in a single step, are evident from the number of publications that appeared in the literature. High degrees of chemo-, regio- and stereoselectivity have been attained in these reactions and some successful examples have been evidenced where the control of the enantioselectivity is achieved with chiral anions instead of chiral ligands.

As detailed in the Introduction, in 2007, Toste and coworkers published a study dealing with gold-mediated asymmetric intramolecular hydroalkoxylations, hydroaminations, and hydrocarboxylations of allenes,

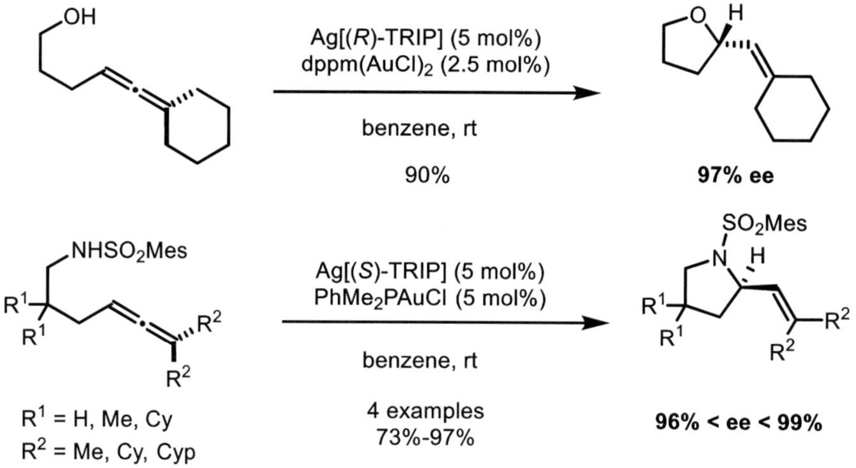

Scheme 6. Intramolecular allenol hydroalkoxylation and allene–sulfonamide hydroamination using gold(I)/phosphate ion pairs.[15]

based on TRIP anion as the chiral source.[15] The authors showed that the chiral counterion–directed strategy can achieve very high levels of enantioselectivity in these reactions where the chiral ligand strategy revealed in most cases ineffective: the use of TRIP in either (di)cationic mononuclear or dinuclear gold catalysts generated products in 90–99% ee (Scheme 6). Thus, almost unprecedented in metal catalysis at that time, the use of chiral anions resulted in exquisite enantioselectivities, whether they were combined with an achiral or a chiral ligand as an amplification source (*match* effect).

This opportunity of chiral amplification was clearly demonstrated with the involvement of gold complexes featuring the (R)-TRIP chiral anion and (S)-BINAP [or (S,S)- 1,2-bis[(2-methoxyphenyl)phenylphosphino]ethane (DIPAMP)] as the chiral ligand for the cycloisomerization of allene carboxylates [or allenols] as typified in Scheme 7, Eq. 1.[15] A more environmentally friendly and efficient process has been developed then by Lipshutz and coworkers by running these reactions in aqueous micellar medium. Upon treatment with the substituted (R)-BIPHEP (BIPHEP = MeO-1,1′- bis(phosphanyl)biphenyl)(AuCl)$_2$/[Ag(R)-TRIP]$_2$ ion pair in Scheme 7, allene carboxylates underwent gold-catalyzed lactamization with excellent results. The catalytic activity and the enantioselectivity

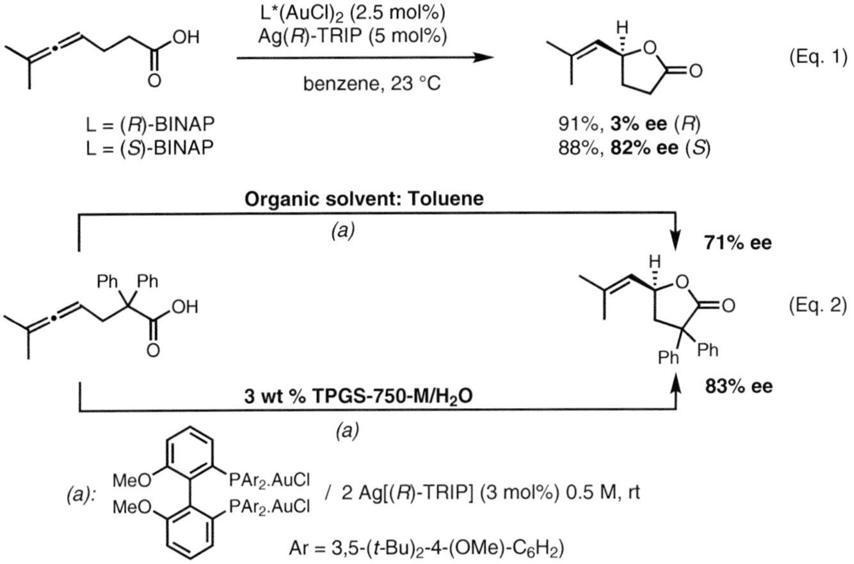

Scheme 7. Synergistic effects between chiral ligands and chiral anions: gold-promoted intramolecular hydrocarboxylation of allene carboxylic acids in organic solvent (Eq. 1)[15] and in aqueous medium (Eq. 2).[19]

in aqueous micelles showed to be better than in organic solvents and both the catalyst and the surfactant could be recycled at least five times. The increase of enantioselectivity may be explained by the hydrophobicity of the nanomicelle core, which improves the interactions between the cationic gold complex and the phosphate anion (Scheme 7, Eq. 2).[19]

In 2012, Hii and Togni developed chiral monodental ferrocenylphosphine ligands possessing a planar chirality. Used in allenol cyclization in collaboration with a chiral phosphate counterion, the latter led to enantioenriched tetrahydrofuranes in ees up to 47% when *match* pair was used.[45]

In 2010, Toste and coworkers extended the scope of these reactions to the cyclizations of allene hydroxylamines *via* hydroalkoxylation or hydroamination of the allene unit. Here, the catalytic system combining the achiral complex (dppm)AuCl2 (dppm = bis(diphenylphosphino)methane) (6 mol%) associated to the chiral silver phosphate Ag[(S)-TRIP] (6 mol%) was found to be highly effective in terms of yield and asymmetric induction for the synthesis of isoxazolidines by intramolecular hydroalkoxylation. It is important to note also that the synergic action of the chiral complex (S, S)-DIPAMP(AuCl)$_2$ (3 mol%) and the chiral silver phosphate Ag[(S)-TRIP] (6 mol%) has demonstrated once again its value by promoting the asymmetric synthesis of tetrahydrooxazines through hydroamination, with very high enantioselectivity.[20]

At the meantime, Mikami and coworkers developed a strategy to prepare new enantiopure axially chiral (Ar-BIPHEP)–gold(I) complexes (BIPHEP = 1,1'-bis(phosphanyl)biphenyl) by resolution of the dinuclear cationic complexes with a binaphthol-derived phosphate or N-triflyl phosphoramide as the chiral anions (X*$^-$, Scheme 8). For instance, the thermodynamically favored diastereoisomer ((S)-DM-BIPHEP)Au$_2$((S)-X*)$_2$

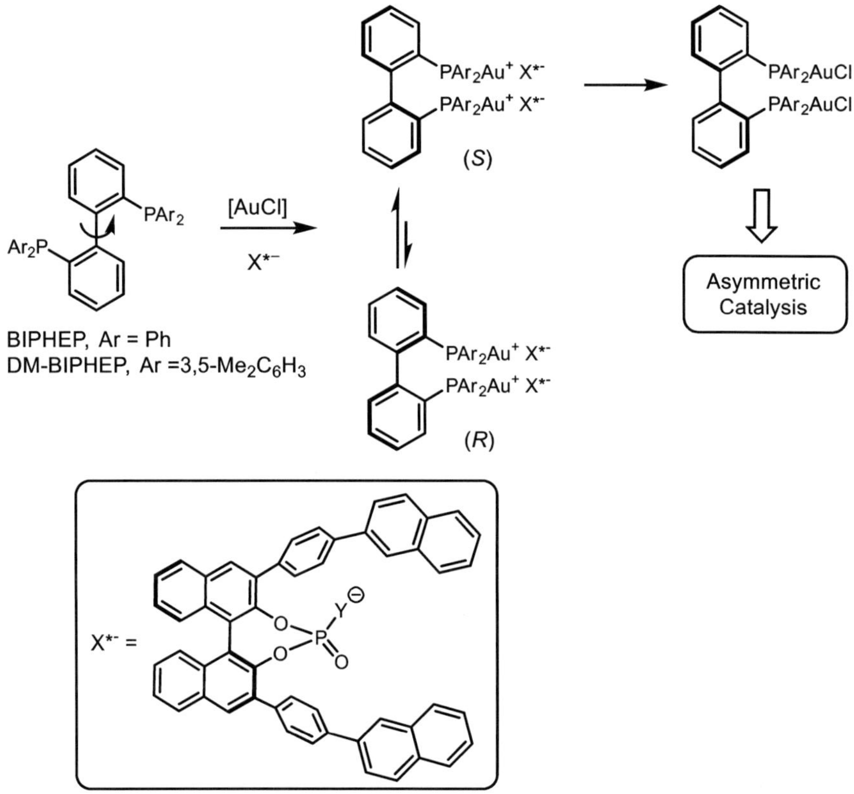

Scheme 8. Resolution of axially chiral (BIPHEP)–gold complexes with binaphthol-derived phosphates (Y = O) or N-triflyl phosphoramides (Y = NTf). Adapted from Ref. 23, with permission from Wiley.

was preferentially formed and was converted then into the enantiopure (S)-DM-BIPHEP(AuX)$_2$ complex that could be successfully used in enantioselective catalysis, when combined either with achiral or chiral counterions. The X-ray structure of the bimetallic complex has revealed a Au–Au distance of 3.10 Å, suggesting a Au–Au interaction.[21–23]

The concept of chiral amplification was investigated by using the enantiopure dinuclear BIPHEP–gold complexes in the presence of the chiral silver binaphthol–derived phosphates, AgX*, as chiral anions sources. Depending on the number of silver salt equivalents, the authors obtained either the monocationic (1 equivalent) or the dicationic (2 equivalents) digold complex. Tested in asymmetric allenol hydroalkoxylation reactions, the catalytic activity and enantioselectivity of the monocationic complex proved to be better than that of the dicationic complex (Scheme 9, Eq. 1). The best results were reached with the combination (R)-DM-BIPHEP(AuCl)$_2$/Ag[(S)-X*], for Ar′ = 4-(4-tBuC$_6$H$_4$)C$_6$H$_4$ (Scheme 9, Eq. 2). Finally, application to the intramolecular enantioselective hydroamination of an N-alkenyl urea with the catalytic system (S)-DM-BIPHEP(AuCl)$_2$/Ag[(S)-X*] (Ar′ = 4-(2-naphthyl)C$_6$H$_4$) provided the cyclized product in excellent yield but only moderate ee (Scheme 9, Eq. 3).[21]

In 2015, Toste and Zi extended the ACDC strategy to the efficient desymmetrisation of 1,3-diol. In implementing intramolecular hydroalkoxylation of allenes in the presence of bisgold/phosphate ion pair, they got access to multisubstituted oxygen heterocycles incorporating a quaternary stereocenter in both high enantio- and diastereoselectivity. Once again, the gold/phosphate ratio has a critical impact on the enantioselectivity. However, it is worthy to mention that this time, lower ees are obtained when a 2:1 Au/phosphate ratio is used.[46]

More recently, Czekelius and coworkers have reported an elegant application of the chiral anion strategy to the asymmetric desymmetrization of a 1,4-diynamide for the preparation of a key intermediate in the total synthesis of (+)-mesembrine. They showed that cycloisomerization of the 1,4-diynamide catalyzed by a cationic gold(I) complex associated with the chiral phosphate TRIP, (t-Bu$_3$P)Au[(R)-TRIP], led to the methylene pyrrolidine with a quaternary stereocenter in 3-position in 76% yield and 70% ee (Scheme 10).[24]

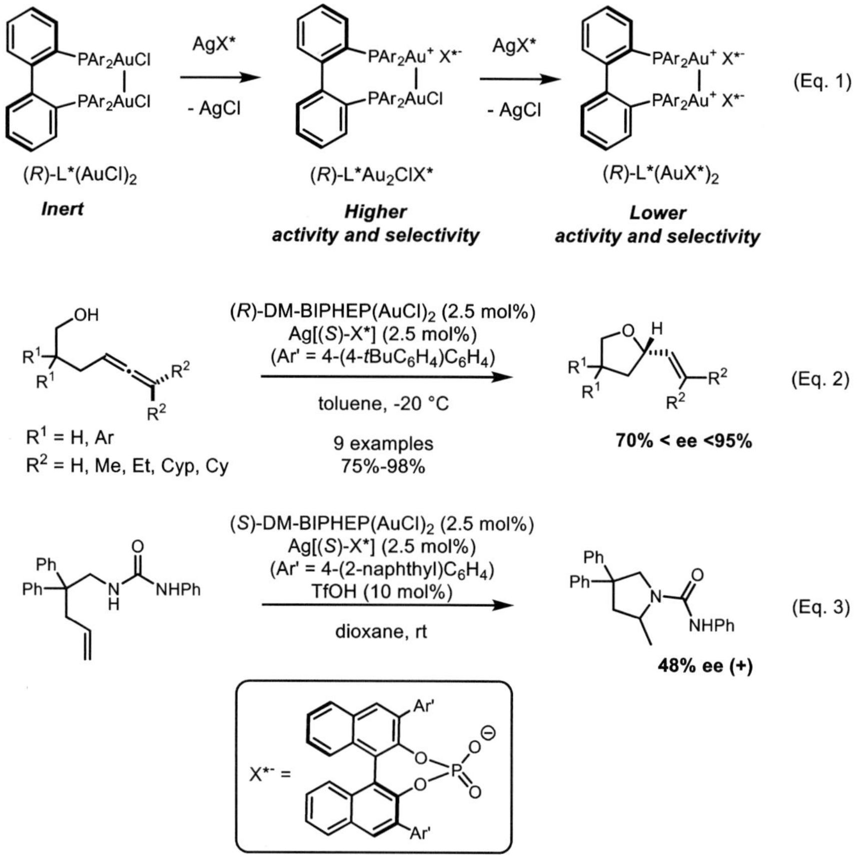

Scheme 9. Monocationic and dicationic dinuclear gold complexes for chiral amplification (Eq. 1). Adapted from Ref. 24, with permission from Wiley. Applications to the intramolecular hydroalkoxylation of allenols (Eq. 2)[23] and hydroamination of an N-alkenyl urea (Eq. 3).[21]

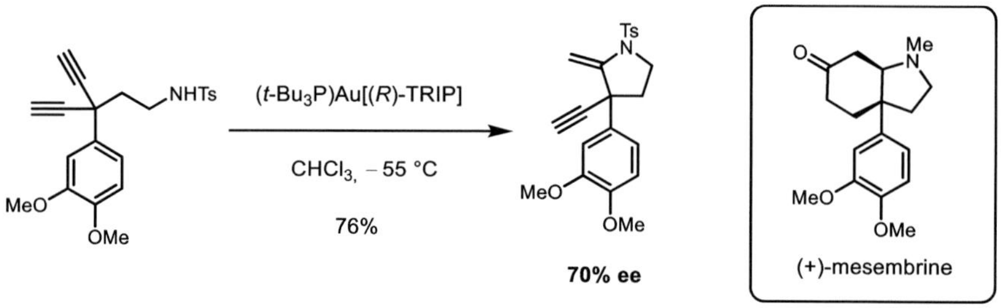

Scheme 10. Enantioselective desymmetrization of a 1,4-diynamide using a gold(I)/phosphate.[24]

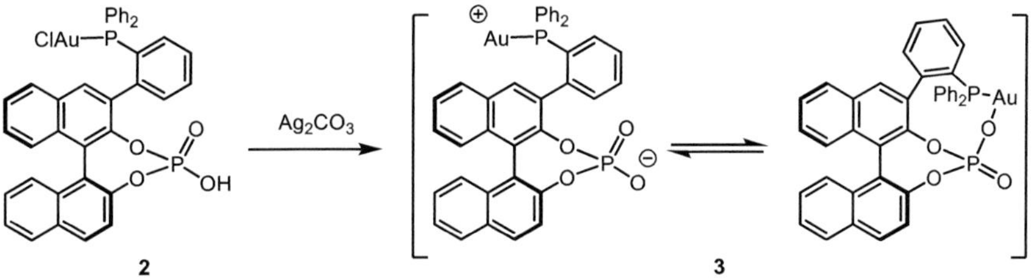

Scheme 11. Formation of gold phosphates through deprotonation of a phosphine-phosphoric acid ligand.[27]

In all these gold(I)-catalyzed heterocyclizations, the question of the true nature of the phosphate, that is, a counterion or an X-type ligand, remains. Several pieces of evidence support a covalent bonding between gold and phosphate at the resting state, whether in the solid state (Au–O distance in X-ray crystal structures[21,22,25]) or in solution (molar extinction coefficient and long-range NMR coupling[25]). In addition, in 2012 Nguyen addressed this issue through extended X-ray absorption fine structure (EXAFS) analysis. Focusing on the (Ph_3P)Au[1,1'-binaphthalene-2,2'-diyl phosphate] complex, the authors observed a significant bonding degree between the cationic gold and the phosphate, and ^{31}P NMR analysis confirmed the lack of mobility of the phosphate that remains tightly fixed to the gold cation.[26] However, the counterionic character of the phosphate was supported by the strong influence of the solvent polarity on the ee of some catalytic reactions,[15] as well as by the excellent ee obtained with bisgold complexes. In addition, in 2020, Guinchard, Marinetti, and coworkers designed a new gold complex **2** with a tethered phosphoric acid unit (Scheme 11).[27] ^{31}P NMR analysis of the putative corresponding gold/phosphate complex **3** showed no long-range coupling thus pleading in this case for a gold/phosphate bond more ionic than covalent.

B. Diamination of Alkenes

Since Arndsten's pioneering work on the copper-catalyzed asymmetric aziridination of styrene with PhNITs as the nitrene precursor and bis(binaphthol)borate **1** as the sole chiral source (see Introduction,

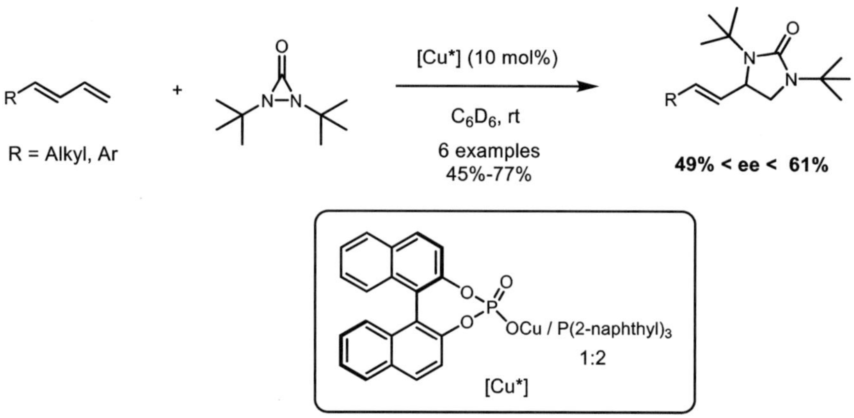

Scheme 12. Catalytic asymmetric diamination of conjugated dienes using a copper(I)/phosphate.[28]

Scheme 1),[6] new environmentally friendly enantioselective processes to form carbon–nitrogen bond have been developed, including the diamination of alkenes. In 2009, Shi and coworkers have devised new chiral copper(I) complexes capable of mediating diamination of conjugated dienes and trienes.[28] 1,3-Di-*tert*-butyldiaziridinone was used as the nitrogen source (Scheme 12). A chiral copper(I) phosphate [Cu*] obtained from mesitylcopper(I), tris(2-naphthyl)phosphine (1:2), and a BINOL-derived phosphoric acid provided the best results as a catalyst (10 mol%), in terms of both yields and selectivity. This transformation, whose mechanism is supposed to be radical, represents a big synthetic challenge. The functionalization occurs selectively at the terminal position of the conjugated dienes but the ees are only moderate (49–61%). Many questions remain unanswered, such as the molecular structure of the catalyst, and whether the phosphate acts here as a counterion or a ligand.

C. Overman Rearrangement

Another study using the ACDC concept for carbon–nitrogen bonds formation was conducted by List and coworkers in 2011 and was applied to the so-called Overman rearrangement.[29] A catalytic amount of a chiral palladium complex has been formed *in situ* from the commercially available dimeric chiral palladacycle [(*S*)-**[Pd]**Cl]$_2$ in Scheme 13 (1 mol%) and the

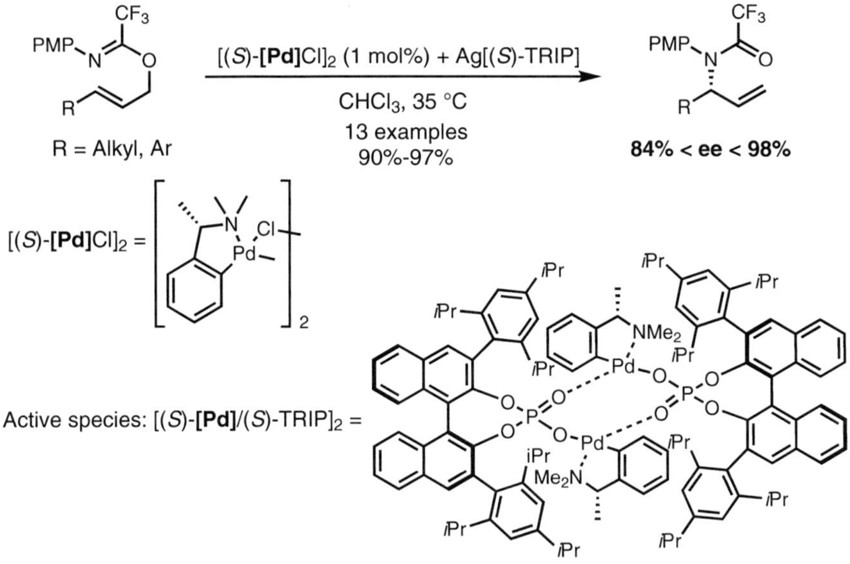

Scheme 13. Asymmetric Overman rearrangement using a palladium(II)/phosphate ion pair.[29]

chiral phosphate Ag[(S)-TRIP] (2 mol%, *match* pair). The mixture has been shown to efficiently catalyze the conversion of allylic imidates to allylic amine derivatives. Good to excellent yields and enantiomeric ratios were achieved. It has been demonstrated also that the chiral palladacycle reacts with the chiral phosphate to yield the [(S)-**Pd**]/(S)-TRIP]$_2$ dimer in Scheme 13, which provided comparable results in catalytic tests. The authors thus suggested that this complex should be the actual catalyst, whether preformed or generated *in situ*. To confirm the structure of the active species, imidazole was added to the dimer [(S)-**Pd**]/(S)-TRIP]$_2$, so as to access a monomeric palladium complex whose structure could be established by X-ray diffraction studies.

IV. Reactions Involving Formation of C–C Bonds

A. Cycloisomerization Reactions

The next challenge within the chiral counterion strategy was the enantioselective formation of carbon–carbon bonds. Of particular interest were

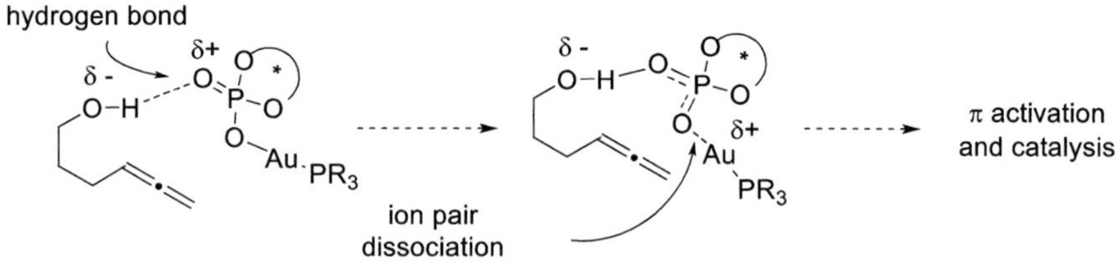

Scheme 14. Postulated mechanism for Au–O bond dissociation in Toste's allene heterocyclizations.

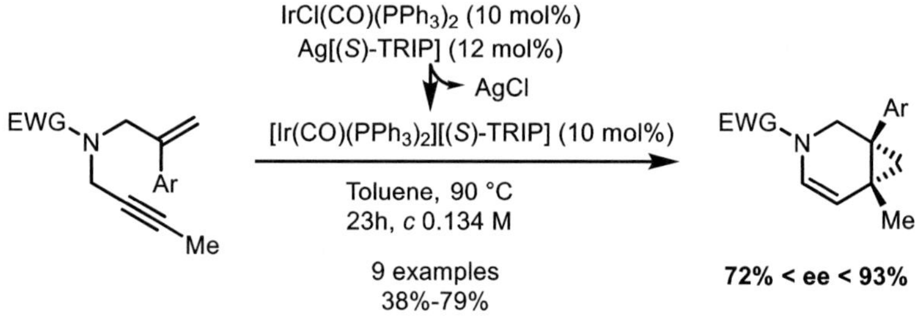

Scheme 15. Enyne cycloisomerization using an iridium/phosphate ion pair.[30]

the cycloisomerization reactions that are known to convert polyunsaturated substrates into various complex cyclic adducts under metal catalysis, including gold catalysis. However, the previously developed gold–phosphate complexes that led to exceptional results in the enantioselective Toste's allene heterocyclizations (see Scheme 6)[15] did not give satisfying results in initial studies on the cycloisomerization of 1,6-enyne-type substrates.[25,30] On one hand, no catalytic activity was observed in apolar solvents. This is probably due to the covalent nature of the gold–phosphate bond, which needs to be dissociated to generate the catalytically active species. This dissociation step can be favored by substrates, such as alcohols, that feature H-bonding functions (Scheme 14), while it is precluded with other nonfunctionalized substrates.

On the other hand, introduction of a polar cosolvent led to restoration of the reactivity, the polar cosolvent probably assisting the ion-pair dissociation process, but racemic products were isolated in all cases.[25,30]

In 2011, Aubert, Fensterbank, Amouri, and coworkers disclosed the first enantioselective cycloisomerization process based on the counterion strategy (Scheme 15).[30] To enforce dissociation of the strong phosphate metal covalent bonding, they moved from linear gold(I) complexes toward square planar Vaska's-type iridium complex. According to density functional theory (DFT) calculation, the steric hindrance between the bulky phosphate and the ligands in relative *cis* positions promotes bond dissociation, leading to the cationic iridium/anionic phosphate ion pair. Thus, the cycloisomerization of 1,6-enynes took place in toluene at 90°C and led to optically enriched bicyclo-[4.1.0]hept-2-enes in ees ranging between 73% and 93% when the catalyst was generated *in situ* by treatment of the Vaska's complex with Ag[(*S*)-TRIP]. In this case, the chiral counterion strategy outperformed the chiral ligand–based approach, as the use of Tol-BINAP as the chiral ligand in a similar iridium catalyst was reported to provide ees in the range 44–74%.[31] Once again, the influence of the solvent revealed crucial as the addition of methanol as a polar cosolvent led to racemic adducts.

Infrared (IR) and ^{31}P NMR experiments confirmed that neither loss of ligands nor change in the *trans* geometry of the Vaska's complex took place during the reaction. DFT calculations suggested a 6-*endo* cyclization mechanism, involving an ion-pair-dissociated transition state, in which the phosphate anion binds the substrate through hydrogen bonding, while the cationic iridium coordinates to the alkyne moiety (Figure 4).

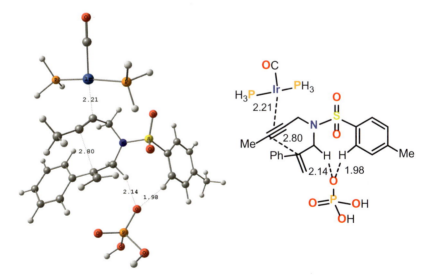

Figure 4. Postulated transition state from DFT calculations on the model catalyst [Ir(CO)(PH$_3$)$_2$][OP(O)(OH)$_2$].[31] Reproduced from Ref. 30 with permission from Wiley.

B. Tandem Cycloisomerization/Addition Reactions

A cycloisomerization/nucleophilic addition sequence was reported in 2011 by Toste and coworkers, using copper as the cationic metal and a chiral phosphate counterpart (Scheme 16).[32] These reactions involve the intramolecular cycloisomerization of propargylic ketones displaying a cyclohexenone unit into the corresponding furans, by treatment with Cu[(S)-TRIP]$_2$. The rearrangement generates an electron-deficient intermediate that undergoes enantioselective addition of indole and affords the desired products in ees ranging from 73% to 93%. Based on kinetic studies and high-resolution mass spectroscopy (HRMS) experiments, the authors suggested that a single phosphate is involved in the stereodetermining step, acting either as a cuprate X-type ligand or as an anion associated to the carbocationic intermediate.

Interestingly, the silver salt itself catalyzes the reaction, albeit with a decreased ee (74% ee when Ag[(S)-TRIP] was used as the catalytic species). In 2018, Marinetti and Betzer studied the same tandem reaction further and developed a new chiral paracyclophane-based phosphate silver

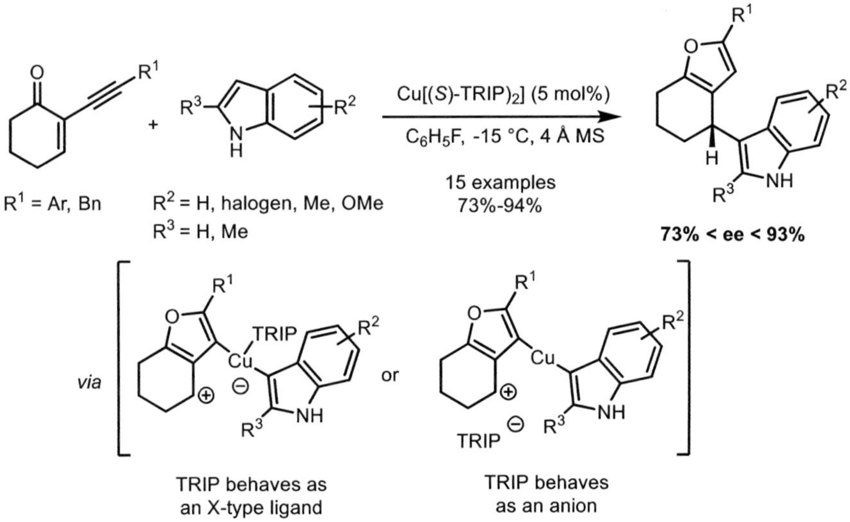

Scheme 16. Tandem cycloisomerization–indole addition reactions catalyzed by a copper/phosphate ion pair.[32]

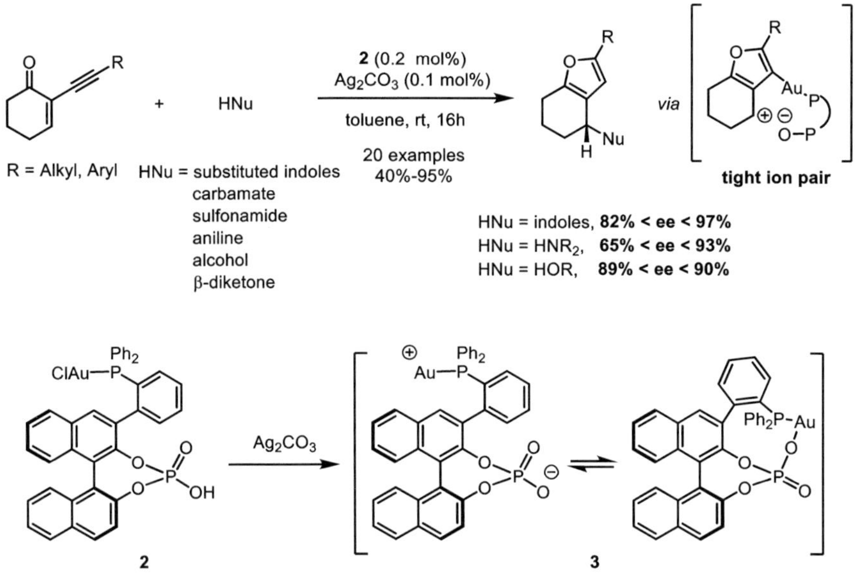

Scheme 17. Cycloisomerization–nucleophilic addition reactions catalyzed by a tethered gold/phosphate ion pair.[27]

complex, which, in conjunction with the use of dichloroethylene as the solvent, allowed to get the products in higher ees.[33] As the ee dropped when using an *N*-methylindole as the nucleophile, it was postulated that H-bonding between the phosphate and the indole can be involved in the enantioselection process in this case.

Development of a gold-catalyzed version of this reaction was successfully achieved in 2020 by Guinchard and Marinetti, thanks to the design of a bifunctional ligand that prefigures and facilitates the formation of a chiral ion pair.[27] The precatalyst is the phosphoric acid-tethered phosphine gold chloride **2**. The active metallic species **3** was generated by concomitant chloride abstraction and phosphoric acid deprotonation in the presence of silver carbonate. The gold catalyst with the tethered phosphate counterion allowed to reach high ees, whereas the use of the analogous, nontethered complex (PPh$_3$)Au[(*S*)-TRIP] led to only 24% ee. Noteworthy, when weakly basic nucleophiles were used, no silver salt was required to generate the active species. Indeed, in this case, **2** may react with the basic nucleophile, resulting in the spontaneous and reversible formation of the active gold phosphate catalyst **3** by HCl abstraction.

The gold complex **3** revealed to be a highly versatile catalyst for this tandem transformation, yielding substituted furans in high yields and ees by using a variety of nucleophiles, ranging from substituted indoles to carbamates, sulfonamides, anilines, alcohols, and enols (Scheme 17).

Both experiments and mechanistic considerations supported by DFT calculations allowed to rule out a stereodetermining H-bonding between the phosphate and the indole *N*-group. The authors propose that "the geometrical constraints enforced by tethering gold(I) to phosphate create a more organized spatial arrangement and enable the excellent stereocontrol" *via* the tight ion pair between the phosphate and the prochiral carbon center.

C. [2+2+2] Cycloaddition Reactions

Most commonly, enantioselective [2+2+2] cycloaddition reactions are performed by using a cationic Rh(I) complex associated with a bidental chiral ligand. Thanks to the use of axially chiral bidentate phosphines, this strategy revealed extremely efficient, but it remains still quite substrate dependent.

In 2013, Aubert, Fensterbank, and Ollivier have applied the chiral counterion strategy to this class of reactions. They disclosed indeed a rhodium-based [2+2+2] cycloaddition process leading to the enantioselective synthesis of chiral pyridinones, by which the control of axial chirality was carried out for the first time via the ACDC approach (Scheme 18).[34] This strategy revealed complementary to the chiral ligand–based approach, affording high ees in reactions where chiral ligands failed, and vice versa.[35]

Axially chiral pyridin-2(*1H*)-ones were obtained *via* [2+2+2]-cycloadditions of diynes with *ortho*-substituted arylisocyanates. Starting from the [Rh(cod)Cl]$_2$ (cod=cyclooctadiene) complex and 1,4-bis(diphenylphosphino)butane (dppb) as the ligand of choice, once again the (*S*)-TRIP phosphate turned out to be a suitable counterion to control the enantioselectivity. By performing the reaction in dichloromethane at 80°C (sealed tube), the desired pyridinones were obtained in ees ranging from 58% to 82%, from *o*-alkoxy-substituted arylisocyanates, thus overcoming, in this case, the chiral ligand strategy.[35] On the other hand, with arylisocyanates displaying an alkyl or a halogen substituent in *ortho* position, the

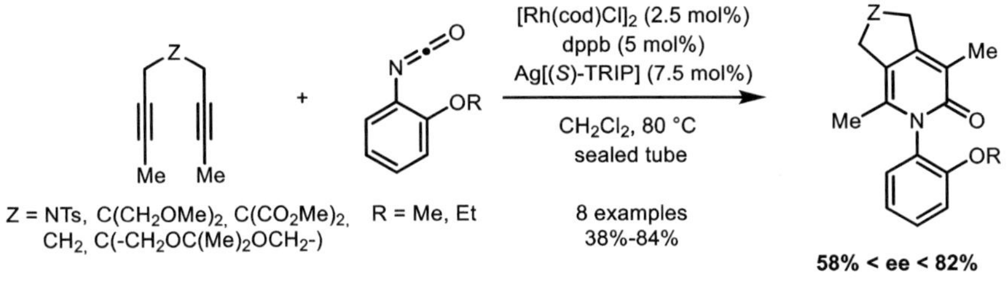

Scheme 18. [2+2+2] Cycloadditions of diynes to isocyanates promoted by rhodium/phosphate ion pairs.[34]

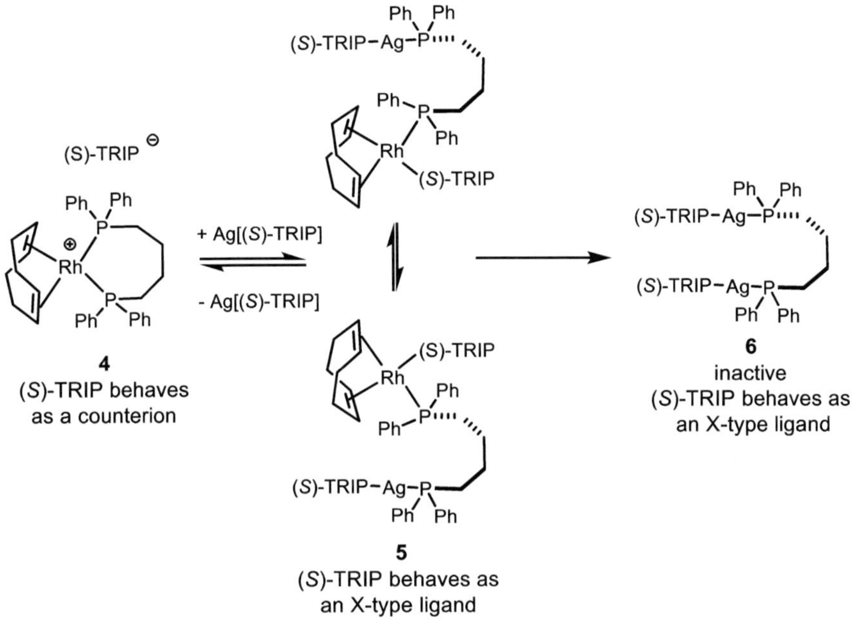

Scheme 19. Rhodium species identified in the reaction mixture.[36]

selectivity dropped, and the chiral ligand strategy revealed more relevant for these substrates.

In a following publication, the authors have highlighted two catalytic species involved in this reaction.[36] The major one, **4**, is a monometallic [Rh(cod)dppb][(S)-TRIP] ion pair (Scheme 19). Diffusion-ordered spectroscopy (DOSY) 2D-NMR experiments confirmed the ionic character of the phosphate anion in this species, since the cationic rhodium and its

(S)-TRIP counterion proved to move independently. A second species, **5**, appears when more than 1 equivalent of Ag[(S)-TRIP] is introduced, and is in equilibrium with complex **4**. Complex **5** revealed to be a rhodium–silver bimetallic species arising from opening of the chelate rhodium–dppb complex by the excess of Ag[(S)-TRIP]. Its [(S)-TRIP]Rh(cod)[PPh$_2$(CH$_2$)$_4$PPh$_2$]Ag[(S)-TRIP] structure was confirmed by NMR and mass spectrometry. Here the phosphate acts as an X-type ligand, as proven by the long-distance ^{31}P-^{109}Rh coupling observed in ^{31}P-^{109}Rh 2D HMBC and the ^{103}Rh chemical shift. Complex **5** evolves with time into the catalytically inactive complex **6**, dppb(Ag[(S)-TRIP])$_2$, showing that Ag[(S)-TRIP] totally displaces the dppb ligand from its rhodium complex (Scheme 19). It is worthy to mention that, in presence the presence of **5**, the ee increased significantly, as shown in Scheme 19.

Thus, in this work, the authors showed that the phosphate anion can act both as a counterion and as a ligand in rhodium complexes, and that these species may be in equilibrium.

Within the same studies, the double stereodifferentiation process was also evaluated, by using a chiral ligand associated with a chiral phosphate counterion.[37] However, by switching from dppb to chiral BINAP-type ligands, the conditions for the formation of active catalysts had to be tuned again: a prior hydrogenation step revealed crucial, as well as prolonged prestirring at room temperature. In the case of a *match* pairing, the double

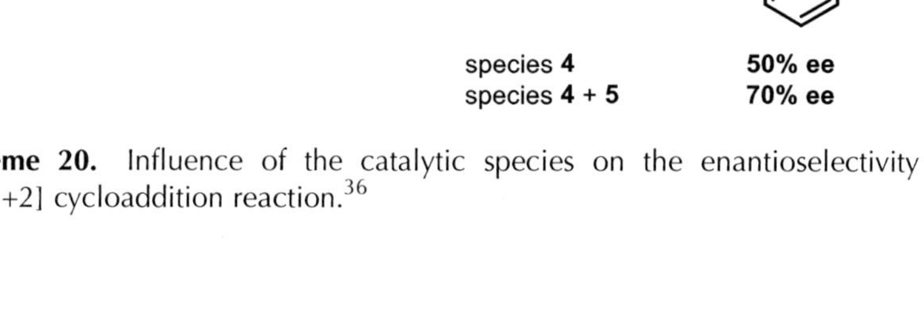

Scheme 20. Influence of the catalytic species on the enantioselectivity of a [2+2+2] cycloaddition reaction.[36]

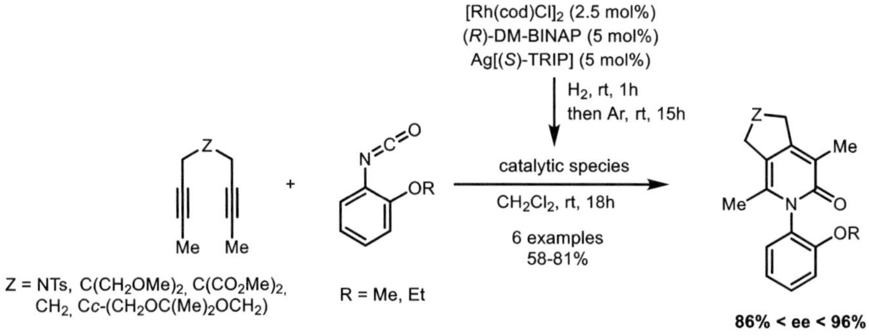

Scheme 21. Rhodium promoted [2+2+2] cycloadditions under double stereodifferentiating conditions.[37]

stereochemical information provided by the (R)-DM-BINAP ligand and the (S)-TRIP counterion rose the ees. It gave better results than the chiral counterion strategy or the chiral ligand strategy separately, provided that an ether substituent is introduced on the arylisocyanate (Scheme 21).

D. Hydrovinylation of Olefins

In 2011, List and coworkers have investigated the metal-promoted hydrovinylation of olefins by an ACDC approach.[38] Whereas cationic nickel derivatives failed to afford any enantioselectivity in spite of the presence of a chiral phosphate or disulfonimide counterion, the use of cationic ruthenium complexes revealed successful. Interestingly, ruthenium is rarely used to perform hydrovinylation reactions and the development of asymmetric versions is still challenging with this metal.[39,40]

Once again, (S)-TRIP revealed to be the counterion of choice to perform the reaction. The [HRu(CO)(PR′$_3$)$_2$][(S)-TRIP] ion pair was generated *in situ*, by treatment of HRuCl(PR′$_3$)$_2$(CO) with Ag[(S)-TRIP]. It allowed the enantioselective hydrovinylation of styrene derivatives in benzene, at 8°C, giving high yields and regioselectivity toward the branched products. Some enantioselectivity was observed, as the products were obtained in ees ranging from 34% to 44% (Scheme 22).

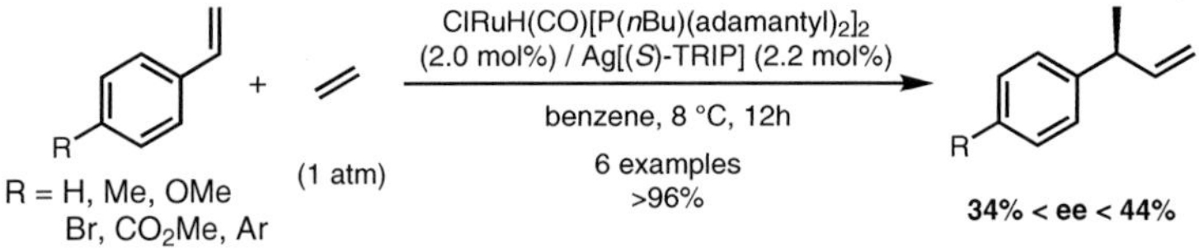

Scheme 22. Olefin hydrovinylation reactions catalyzed by a ruthenium/phosphate ion pair.[38]

V. Conclusion

Although the chiral counterion strategy first appeared as a fascinating, "magical" tool to avoid ligand and metal-related geometry constraints, finally the ACDC approach has revealed to be complementary to other strategies — such as the use of chiral ligands or of chiral ionic liquids — for the efficient stereocontrol in organometallic catalysis. Mostly, the atropochiral TRIP phosphate (Scheme 2) has been used successfully so far, although the alternative design of bifunctional phosphine-phosphate ligands has demonstrated good potential for selected applications. The true nature of the chiral phosphate, namely does it behave as a counterion or an X-type ligand, remains a debatable question that has to be investigated for each transformation independently.

Beyond the frame of this short review focused on catalytic processes involving π-activation of C=C and C≡C bonds, many other reports deal with the chiral counterion concept in organometallic catalysis, as it has been applied to a variety of transformations, atom-economic or not, from epoxidations to C–H activations or again hydrogenations.[3,9,17] As an additional example, it can be mentioned that chiral metal phosphates also display significant potential in polymerization reactions for applications in materials chemistry. Thus, Leibfarth and Teator elegantly used a titanium/phosphate ion pair to control the stereoselectivity of the cationic polymerization of vinyl ethers into isotactic polymers.[41] Moreover, research in the domain of ACDC also motivates studies on other aspects of coordination chemistry, where the counterion will be associated, for instance, to charged ligands.[42–44]

VI. Acknowledgments

The authors thank Sorbonne Université and the CNRS.

VII. References

1. Kagan, H. B.; Gopalaiah, K. *New J. Chem.* **2011**, *35*, 1933–1937.
2. (a) Marckwald, W. *Ber. Dtsch. Chem. Ges.* **1904**, *37*, 349–354. (b) Marckwald, W. *Ber. Dtsch. Chem. Ges.* **1904**, *37*, 1368–1370.
3. Mahlau, M.; List, B. *Angew. Chem., Int. Ed.* **2013**, *52*, 518–533.
4. Lacour, L.; Moraleda, D. *Chem. Commun.* **2009**, *2009*, 7073–7089.
5. Ishihara, K.; Miyata, M.; Hattori, K.; Tada, T.; Yamamoto, H. *J. Am. Chem. Soc.* **1994**, *116*, 10520–10524.
6. Llewellyn, D. B.; Adamson, D.; Arndtsen, B. A. *Org. Lett.* **2000**, *2*, 4165–4168.
7. Uraguchi, D.; Terada, M. *J. Am. Chem. Soc.* **2004**, *126*, 5356–5357.
8. Akiyama, T.; Itoh, J.; Yokota, K.; Fuchibe, K. *Angew. Chem. Int. Ed.* **2004**, *43*, 1566–1568.
9. (a) Parra, A.; Reboredo, Martín Castro, A. M.; Alemán, J. *Org. Biomol. Chem.* **2012**, *10*, 5001–5020; (b) Parmar, D.; Sugiono, E.; Raja, S.; Rueping, M. *Chem. Rev.* **2014**, *114*, 9047–9153; (c) Fang, G.-C.; Cheng, Y.-F.; Zhang-Long Yu, Z.-L.; Li, Z.-L.; Liu, X.-Y. *Top. Curr. Chem.* **2019,** *377,* 23.
10. Seayad, J.; Seayad, A. M.; List, B. *J. Am. Chem. Soc.* **2006**, *128*, 1086–1087.
11. Klussmann, M.; Ratjen, L.; Hoffmann, S.; Wakchaure, V.; Goddard, R.; List, B. *Synlett* **2010**, *2010*, 2189–2192.
12. Mayer, S.; List, B. *Angew. Chem. Int. Ed.* **2006**, *45*, 4193–4195.
13. Mukherjee, S.; List, B. *J. Am. Chem. Soc.* **2007**, *129*, 11336–11337.
14. Komanduri, V.; Krische, M. J. *J. Am. Chem. Soc.* **2006**, *128*, 16448–16449.
15. Hamilton, G. L.; Kang, E. J.; Mba, M.; Toste, F. D. *Science* **2007**, *317*, 496–499.
16. Zi, W.; Toste, F. D. *Chem. Soc. Rev.* **2016**, *45*, 4567–4589.
17. (a) Ávila, E. P.; Amarante, G. W. *ChemCatChem* **2012**, *4*, 1713–1721; (b) Mahlau, M.; List, B. *Isr. J. Chem.* **2012**, *52*, 630–638; (c) Phipps, R.; Hamilton, G.; Toste, F. *Nat. Chem.* **2012**, *4*, 603–614; (d) Mahlau, M.; List, B. *Angew. Chem. Int. Ed.* **2013**, *52*, 518–533.
18. Chen, D.; Sundararaju, B.; Krause, R.; Klankermayer, J.; Dixneuf, P. H.; Leitner, W. *ChemCatChem* **2010**, *2*, 55–57.
19. Handa, S.; Lippincott, D. J.; Aue, D. H.; Lipshutz, B. H. *Angew. Chem. Int. Ed.* **2014**, *53*, 10658–10662.
20. LaLonde, R. L.; Wang, Z. J.; Mba, M.; Lackner, A. D.; Toste, F. D. *Angew. Chem. Int. Ed.* **2010**, *49*, 598–601.
21. Kojima, M.; Mikami, K. *Synlett* **2012**, *23*, 57–61.
22. Aikawa, K.; Kojima, M.; Mikami, K. *Angew. Chem., Int. Ed.* **2009**, *48*, 6073–6077.
23. Aikawa, K.; Kojima, M.; Mikami, K. *Adv. Synth. Catal.* **2010**, *352*, 3131–3135.
24. (a) Spittler, M.; Lutsenko, K.; Czekelius, C. *J. Org. Chem.* **2016**, *81*, 6100–6105. (b) Mourad, A. K.; Leutzow, J.; Czekelius, C. *Angew. Chem. Int. Ed.* **2012**, *51*, 11149–11152.
25. Raducan, M.; Moreno, M.; Bour, C.; Echavarren, A. M. *Chem. Commun.* **2012**, *48*, 52–54.

26. Nguyen, B. N.; Adrio, L. A.; Barreiro, E. M.; Brazier, J. B.; Haycock, P.; Hii, K. K. (M.); Nachtegaal, M.; Newton, M. A.; Szlachetko, J. *Organometallics* **2012**, *31*, 2395–2402.
27. Zhang, Z.; Smal, V.; Retailleau, P.; Voituriez, A.; Frison, G.; Marinetti, A.; Guinchard, X. *J. Am. Chem. Soc.* **2020**, *142*, 3797–3805.
28. Zhao, B.; Du, H.; Shi, Y. *J. Org. Chem.* **2009**, *74*, 8392–8395.
29. Jiang, G.; Halder, R.; Fang, Y.; List, B. *Angew. Chem. Int. Ed.* **2011**, *50*, 9752–9755.
30. Barbazanges, M.; Augé, M.; Moussa, J.; Amouri, H.; Aubert, C.; Desmarets, C.; Fensterbank, L.; Gandon, V.; Malacria, M.; Ollivier, C. *Chem. Eur. J.* **2011**, *17*, 13789–13794.
31. Shibata, T.; Kobayashi, Y.; Maekawa, S.; Toshida, N.; Takagi, K. *Tetrahedron* **2005**, *61*, 9018–9024.
32. Rauniyar, V.; Wang, Z. J.; Burks, H. E.; Toste, F. D. *J. Am. Chem. Soc.* **2011**, *133*, 8486–8489.
33. Force, G.; Lock Toy Ki, Y.; Isaac, K.; Retailleau, P.; Marinetti, A.; Betzer, J.-F. *Adv. Synth. Catal.* **2018**, *360,* 3356–3366.
34. Augé, M.; Barbazanges, M.; Tran, A. T.; Simonneau, A.; Elley, P.; Amouri, H.; Aubert, C.; Fensterbank, L.; Gandon, V.; Malacria, M.; Moussa, J.; Ollivier, C. *Chem. Commun.* **2013**, *49*, 7833–7835.
35. Tanaka, K.; Takahashi, Y.; Suda, T.; Hirano, M. *Synlett* **2008**, *2008*, 1724–1728.
36. Barbazanges, M.; Caytan, E.; Lesage, D.; Aubert, C.; Fensterbank, L.; Gandon, V.; Ollivier, C. *Chem. Eur. J.* **2016**, *22*, 8553–8558.
37. Augé, M.; Feraldi-Xypolia, A.; Barbazanges, M.; Aubert, C.; Fensterbank, L.; Gandon, V.; Kolodziej, E.; Ollivier, C. *Org. Lett.* **2015**, *17*, 3754–3757.
38. Jiang, G.; List, B. *Chem. Commun.* **2011**, *47*, 10022–10024.
39. Wang, Q.-S.; Xie, J.-H.; Li, W.; Zhu, S.-F.; Wang, L.-X.; Zhou, Q.-L. *Org. Lett.* **2011**, *13*, 3388–3391.
40. Hirano, M. *ACS Catal.* **2019**, *9*, 1408–1430.
41. Teator A. J.; Leibfarth, F. A. *Science* **2019**, *363*, 1439–1443.
42. (a) Ohmatsu, K.; Ito, M.; Kunieda, T.; Ooi, T. *Nat. Chem.* **2012**, *4*, 473–477; (b) Ohmatsu, K.; Ito, M.; Kunieda, T.; Ooi, T. *J. Am. Chem. Soc.* **2013**, *135*, 590–593; (c) Ohmatsu, K.; Hara, Y.; Kusano, Y.; Ooi, T. *Synlett* **2016**, *27*, 1047–1050.
43. Rueping, M.; Koenigs, R. M. *Chem. Commun.* **2011**, *47*, 304–306.
44. (a) Li, C.; Wang, C.; Villa-Marcos, B.; Xiao, J. *J. Am. Chem. Soc.* **2008**, *130*, 14450–14451. (b) Li, C.; Villa-Marcos, B.; Xiao, J. *J. Am. Chem. Soc.* **2009**, *131*, 6967–6969.
45. Barreiro, E. M.; Broggini, D. F. D.; Adrio, L. A.; White, A. J. P.; Schwenk, R.; Togni, A.; Hii, K. K. (M.) *Organometallics* 2012, 31, 3745–3754
46. Zi, W.; Toste, F. D. Angew. Chem. Int. Ed. 2015, 54, 14447–14451

© 2022 World Scientific Publishing Company
https://doi.org/10.1142/9789811248436_0005

5 Enantioselective Catalysis by Nonmetal Frustrated Lewis Pairs

Armen Panossian,*,† Julien Bortoluzzi* and Frédéric R. Leroux*

*Université de Strasbourg, Université de Haute-Alsace, CNRS, LIMA UMR 7042, 25 Rue Becquerel F-67000 Strasbourg, France
†armen.panossian@unistra.fr

Table of Contents

List of Abbreviations	124
I. Introduction	124
II. Mechanistic Aspects of Enantioselective Catalysis by FLPs	126
III. FLP Catalysts Featuring Chiral Lewis Acid Moieties	128
A. Boron-based Chiral Lewis Acids	128
B. Phosphorus-based Chiral Lewis Acids	141
IV. FLP Catalysts Featuring Chiral Lewis Base Moieties	142
V. Chiral Intramolecular FLP Catalysts	148
A. Phosphorus/Boron-based Chiral Intramolecular FLPs	149
B. Nitrogen/Boron-based Chiral Intramolecular FLPs	150
C. Oxygen/Boron-based Chiral Intramolecular FLPs	151

VI.	Rationalization of Enantioselectivity	151
VII.	Conclusion	155
VIII.	Acknowledgments	156
IX.	References	156

List of Abbreviations

Alk	alkyl group
Ar	aryl group
Ar^F	fluorinated aryl group
9-BBN	9-borabicyclo[3.3.1]nonane
BCF	tris(pentafluorophenyl)borane
Bpin	pinacolatoboranyl
Bn	benzyl
Boc	*tert*-butyloxycarbonyl
ee	enantiomeric excess
EWG	electron-withdrawing group
FLP	frustrated Lewis pair
Hal	halogen
HetAr	heteroaryl group
LA	Lewis acid
LB	Lewis base
Mes	mesityl
Napht	naphthyl
PG	protecting group
PMP	1,2,2,6,6-pentamethylpiperidine
TBAF	tetrabutylammonium fluoride
Tf	trifluoromethanesulfonyl

I. Introduction

The term frustrated Lewis pair (FLP) describes the pair made of a Lewis acid and a Lewis base that are unable to associate via a dative bond, due to strong steric hindrance, and retains therefore the Lewis acidic and basic properties of its components. The term was coined in 2007[1] although examples of such impeached adducts were already known in literature.

FLP chemistry truly emerged following a seminal paper by Stephan and coworkers demonstrating the reversible activation of molecular hydrogen by a nonmetal-based FLP, where phosphorus plays the role of the Lewis basic center and boron the acidic one.[2] In this reaction, the base and the acid act cooperatively to cleave dihydrogen heterolytically. In this respect, the method represents a suitable alternative to the traditional activation of H_2 by transition metal catalysts (Scheme 1). The work of Stephan and coworkers opened the way to many catalytic applications of FLPs, a diverse and blossoming area of chemistry, which was reviewed on many occasions.[3–6]

Surprisingly, despite the constant growth of the amount of work on FLPs, the catalytic applications of chiral, nonracemic FLPs are still rare and almost exclusively restricted to reduction reactions, with particular focus on hydrogenations, as it will be shown in this chapter. This is most probably due to the inherent strong interest and utmost importance of hydrogenation reactions in both academia and industry. Hydrogenation reactions were conducted up to then almost exclusively by means of

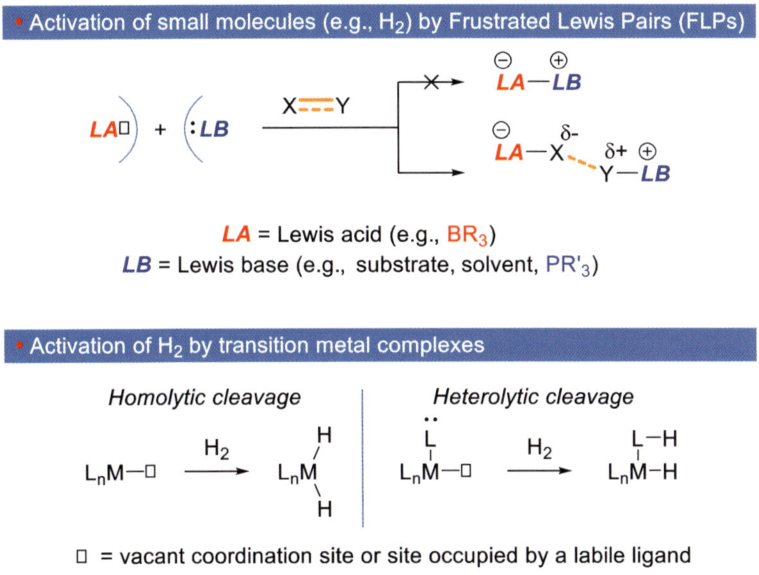

Scheme 1. The concept of frustrated Lewis pairs (FLPs) and the heterolytic activation of small molecules.

transition metal catalysts that have known decades of considerable success, due to their versatility, high efficiency at very low loadings, user-friendliness, and modularity of ligand design. The revolutionary discovery that nonmetal-based FLPs could catalyze hydrogenations and their enantioselective variants attracted intense synthetic and theoretical focus. However, it may also have somewhat eclipsed other enantioselective processes with high synthetic potential. Also, only a narrow range of chiral FLP structures is available so far, which mostly rely on chiral boranes, as it will be shown in the following sections.

A few reviews on enantioselective catalysis by FLPs have been published.[7–10] This chapter will complete the picture and provide perhaps the reader with another viewpoint.

II. Mechanistic Aspects of Enantioselective Catalysis by FLPs

As mentioned previously, most reports on enantioselective catalysis by FLPs deal with enantioselective hydrogenation and the related hydrosilylation and hydroboration reactions. It is therefore important to recall some features of such reduction reactions, as enlightened by mechanistic, experimental, and theoretical studies.

The details of H_2 and hydrides activation and release by FLPs are still the subject of theoretical debate; however, the following general picture can be given for the reduction process (Scheme 2).[11–20] The FLPs exist in solution as dissociated pairs, which can "associate" however into the so-called "encounter complexes." Obviously, these complexes do not result from donation of the lone pair of the Lewis base to the Lewis acid but from noncovalent weak interactions. These encounter complexes have been evidenced experimentally.[21,22] They are able to activate heterolytically the key hydride source (molecular hydrogen, hydrosilane, or hydroboronic species), with the Lewis acid capturing the hydride and the Lewis base evolving into an onium species. The resulting ion pair in turn activates the unsaturated substrate while delivering the hydride and finally releasing the reduced product and the FLP. It should be noted that in hydrogenation reactions, the Lewis basic component of the FLP actually plays the role of a Brønsted base, as well as that the rate-determining step

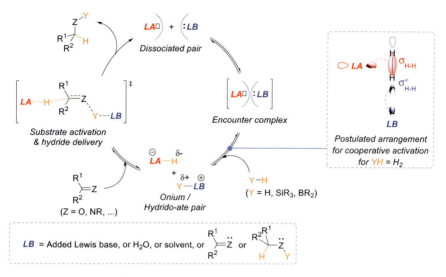

Scheme 2. Simplified mechanistic view of FLP-catalyzed reductions (LA = Lewis acid, LB = Lewis base).

appears to be the assembly of reactants (acid, base, and H_2), not the cleavage nor the formation of the bonds.[23]

However, this oversimplified mechanistic view is often complexified by the noninnocent role played by either the substrate, the product, the solvent, or trace moisture. Indeed, it was demonstrated notably for the hydrosilylation of imines that both the substrate and the reduction product can act as or produce Lewis bases upon silylium or proton exchange, which are competent for the activation of the hydride donor and subsequent steps.[24,25] Similarly, ketone hydrosilylations catalyzed by tris(pentafluorophenyl)borane (BCF), in the absence of added Lewis base, are known to imply activation of the silane by both the borane and the substrate acting itself as the Lewis base.[26–28] This process is referred to as the Piers hydrosilylation, which predates the use of FLPs but can be seen as a particular case of FLP catalysis. Other FLP-type mechanisms have also been proposed for reduction reactions carried out in wet solvents, where the solvent and water molecules may have a critical role. Thus, in the presence of BCF, water may act as the Lewis base and activate H_2 cooperatively with BCF. Alternatively, the water–BCF adduct may behave as a strong Brønsted acid and activate ketone-type substrates, while H_2

activation will take place then under the combined effects of the solvent, as the Lewis base, and the activated carbonyl itself, as the Lewis acid.[29–31]

In addition, one should also be aware that perfluoroarylboranes, which are the most used Lewis acids in FLP chemistry, may undergo nucleophilic attack by Lewis bases at their fluorinated *para*-carbons. These reactions convert the FLPs into intramolecular Lewis pairs, which are also competent in H_2 activation.[32] Last but not least, activation of H_2 can also be effected by non-frustrated, i.e. at least partially associated Lewis pairs.[33] All these issues taken together, it is often difficult to ascertain the nature of the catalyst and its effective behavior as a canonical FLP, let alone vocabulary issues on the proper use of the term "frustrated Lewis pair."

This issue takes on additional importance in enantioselective catalysis by chiral FLPs, where the aim of reaching high stereoinduction might be compromised by the undefined nature of the true catalyst. Accordingly, and due to the young age of the field, efforts toward enantioselective catalysis by FLPs almost rely on trial-and-error approaches and systematic screening of chiral auxiliaries. Also, rationalization of stereoinduction in the case of effective enantioselective catalysts is still rare.

Additionally, enantioselective methods in FLP chemistry almost exclusively concern reduction reactions whose enantio-discriminating step is presumably the transfer of hydride from the Lewis acid. Consequently, most of the chiral FLP-type catalysts developed so far rely on chiral Lewis acid components and, to a lesser extent, on intramolecular FLPs featuring a chiral tether.

Yet, several interesting strategies have already been explored and highly efficient systems have already been developed in the young field of enantioselective catalysis by chiral FLPs. These efforts will be described hereafter.

III. FLP Catalysts Featuring Chiral Lewis Acid Moieties

A. Boron-based Chiral Lewis Acids

The first example of enantioselective catalysis by chiral FLPs was described as a proof of concept by Klankermeyer and coworkers.[34] In this single example, *N*-(1-phenylethylidene)aniline was hydrogenated with

13% ee by means of the chiral perfluoroarylborane **1** derived from (+)-α-pinene, initially described by Piers and coworkers (Table 1, entry 1).[35] The easy strategy to access this chiral borane — namely hydroboration of (+)-α-pinene by the Piers' borane $HB(C_6F_5)_2$ — inspired the synthesis of other chiral boranes. Klankermayer and coworkers similarly prepared (+)-camphor-derived boranes and used them as catalysts, either without added Lewis base or in combination with a phosphine (tri-*tert*-butylphosphine or trimesitylphosphine). Both the hydrogenations and the hydrosilylations of imines were considered (Table 1, entries 2–3, and Scheme 3).[36,37] For reactions carried out without added base, it is assumed that either the starting imine or the final product plays the role of co-catalyst (autocatalysis). In these pioneering studies, it was shown that good levels of enantioselectivity can be attained, with up to 83% ee in hydrogenation reactions and up to 87% ee in hydrosilylations. But maybe more interestingly, these studies highlighted the possible impact of an added Lewis

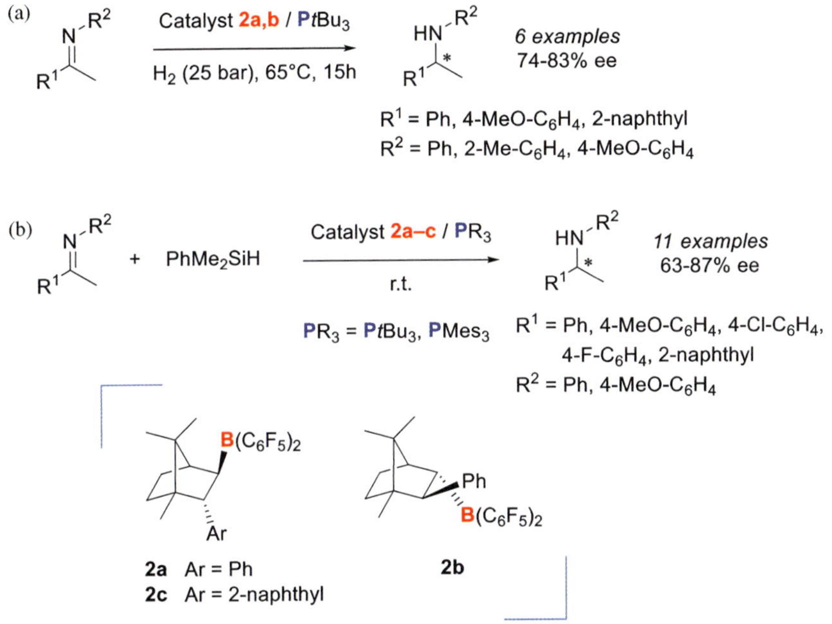

Scheme 3. Hydrogenation and hydrosilylation of imines promoted by FLPs involving camphor-derived chiral boranes.[36,37]

base, namely a phosphine, not only on the catalytic activity but also on the enantioselectivity induced by the chiral Lewis acid. One could presume indeed that the large distance between the Lewis base and the reacting center of the substrate, together with the high flexibility of the transition state (See Scheme 2), would preclude a dramatic influence by the Lewis base. The study showed that, in the hydrosilylation of N-(1-phenylethylidene)aniline ($R^1 = R^2 = Ph$), borane **2c** alone quickly produced the expected amine (quantitative yield after 2 h), but with almost no enantioselectivity, whereas the FLPs **2c**/PMes$_3$ and **2c**/PtBu$_3$ proved to be slower (55%–67% yields after 4 days) but much more enantioselective (63% and 79% ee, respectively).

Much interestingly, it was recently shown by Repo, Pápai, and coworkers that substituting the phenyl ring of borane **2a** with two *tert*-butyl groups positively affects the properties of the catalyst; indeed, added phosphine was no longer required, and the imine hydrogenation products were produced under milder conditions and with higher enantioselectivities by means of borane **2d** alone (up to 96% ee, Table 1, entry 4).[38]

The only drawback of the initial approach by Klankermayer and coworkers was the limited modularity of the structure of the chiral borane. This issue was ingeniously tackled by Du and coworkers who reported, in a series of papers, the use of the chiral bis-boranes **3** produced in situ by hydroboration of a library of tens of 3,3′-disubstituted binaphthyl-2,2′-dienes or -diynes (Table 1, entries 5–15, and Scheme 4), as well as a pyrrolidine-derived diene (bis-borane **4**, entry 16 in Table 1), with the Piers' borane.[39–50] Although the carbon backbone of such catalysts is made heavy by the bulky substituents needed to extend the asymmetric environment of the binaphthyl in the vicinity of the active boron centers, this smart strategy allowed to achieve very high levels of enantioselectivity in the direct or transfer hydrogenation and hydrosilylation of unsaturated substrates, including aza-heterocycles. Interestingly, for the reduction of oxygenated substrates, namely silyl enol ethers,[40,41] 1,2-dicarbonyl compounds,[45] and ketones,[50] a phosphine was required as a Lewis base to ensure high activity and/or enantioselectivity.

Du's binaphthyl-2,2′-diene-derived chiral bis(boranes) **3** were also elegantly employed by Wasa and coworkers for an enantioselective process differing from the usual reduction reactions. In this work, the authors

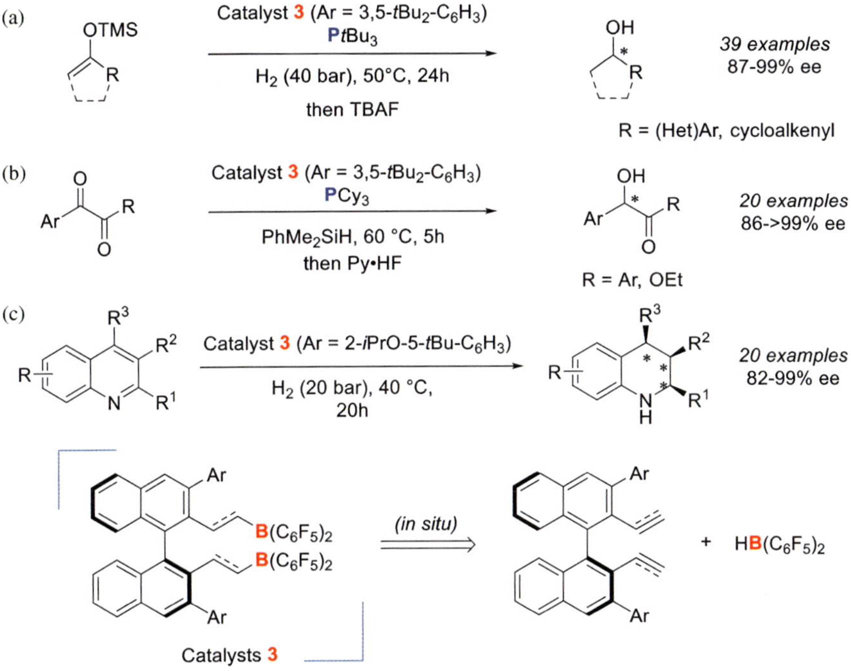

Scheme 4. Hydrogenation and hydrosilylation reactions promoted by axially chiral binaphthyl-derived diboranes **3**.[39–50]

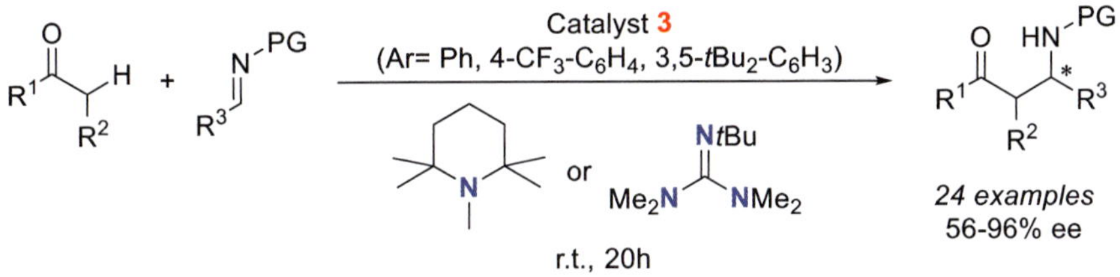

Scheme 5. FLP-promoted Mannich reactions with chiral borane catalysts **3**.[51]

used the chiral boranes in combination with 1,2,2,6,6-pentamethylpiperidine (PMP) or 2-*tert*-butyl-1,1,3,3-tetramethylguanidine (Barton's base) to catalyze Mannich reactions between α-H carbonyl compounds and aldimine derivatives with high levels of enantio- and diastereoselectivity (Table 1, entry 17 and Scheme 5).[51] Notably, in this process, as in hydrogenations, the Lewis base actually behaves as a Brønsted base.

Following the same strategy of using Piers' borane to generate chiral boranes from chiral olefins, and in an effort to simplify the procedure, Du and coworkers developed a simple approach to chiral alkenes starting from a library of chiral BINOLs and *gem*-bis(chloromethyl)ethylene. The resulting olefins were converted into the corresponding boranes **5** (Table 1, entry 18).[52] The borane catalysts **5** were effective in the hydrogenation of *N*-aryl imines and provided good enantioselectivities but slightly lower ee than those obtained with catalysts **3**[39] (4 examples with 78%–89% ee, 12 examples with 60%–72% ee). Interestingly, the two series of catalysts were tentatively compared in preliminary computational studies, showing that, in the case of identical substituents at the backbone, compound **5** bears more positive charge at the Lewis acidic boron atom. It also displays a more acute dihedral angle at the naphthyl moiety, compared to **3** that bears shorter but more flexible spacers between the binaphthyl unit and the boron centers.

Other chiral dienes were also used as precursors for bis-boranes. Peng, Wang, and coworkers reported on the hydroboration of chiral bicyclo[3.3.0]octa-2,6-dienes, which interestingly furnished the diastereoisomeric bis-boranes **6a** and/or **6b** depending on the reaction temperature. The latter were then used successfully in the hydrogenation of *N*-aryl 1-alkyl,1-aryl ketimines (Table 1, entry 19).[53] Very recently, the same series of diboranes was employed in combination with *N*-methylpiperidine to effect the highly enantioselective FLP-catalyzed vinylogous Mannich reactions of enones (45 examples, 82%–96% ee; Table 1, entry 20 and Scheme 6), as a nice complement to the work of Wasa and coworkers (*vide supra*).[54]

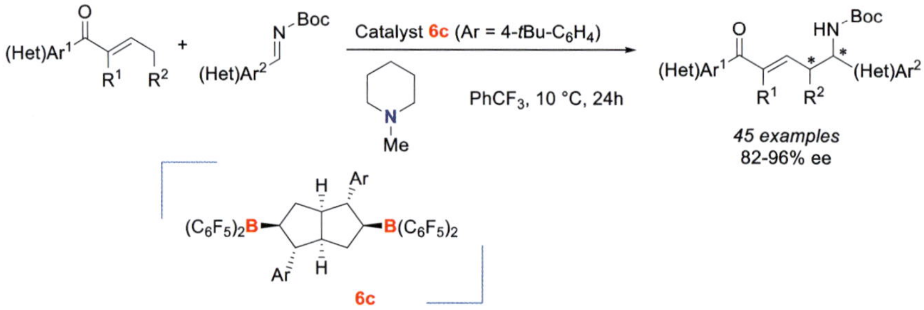

Scheme 6. FLP-promoted Mannich reaction of enones involving diborane **6c**.[54]

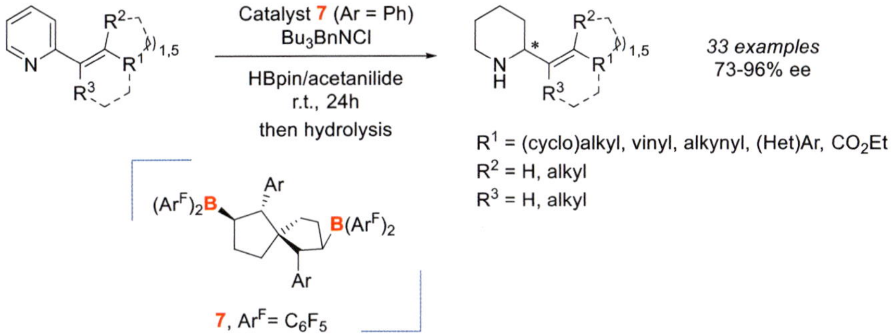

Scheme 7. FLP-promoted enantioselective reduction of pyridines with the chiral spiro-diborane catalyst **7**.[55,56]

Alternately, Wang and coworkers developed chiral bis-boranes from spiro [4.4] dienes, by relying on the increased steric bulk around the boron centers to hopefully increase the stereocontrol in catalytic reactions (entries 21-22).[55,56] This time, only one isomer of the desired bis-borane **7** was formed upon hydroboration. These catalysts were applied to the hydrogenation of 2-subtituted quinolines and the reduction of 2-vinylpyridines (Scheme 7) by means of a pinacolborane/acetanilide or pinacolborane/diphenylamine system as formal hydrogen sources. In both cases, the targeted double bonds were efficiently reduced with high enantioselectivity, without affecting other unsaturated units and functional groups. Notably, for the reduction of 2-(het)aryl-substituted quinolines, addition of tris[3,5-bis(trifluoromethyl)phenyl]phosphine as Lewis base was beneficial and led to increased yields.

All of the chiral boranes discussed previously were produced by the hydroboration of chiral alkenes by the Piers' borane. However, the retrohydroboration and degradation of the catalyst might be a side-reaction.[57] For this reason among others, other chiral boranes were developed for FLP applications where the retro-hydroboration reaction is unlikely. Thus, Oestreich and coworkers developed the axially chiral binaphthoborepines **8** for the hydrosilylations of ketones and imines (Table 1, entries 23-24). Initially, when using an unsubstituted binaphthyl backbone (R=H), only modest enantioinduction was achieved (up to 62% ee).[24,58,59] However, upon introduction of phenyl substituents in positions 3 and 3′ of the binaphthyl, very high enantioselectivities were obtained in the

Table 1. Boron-based chiral Lewis acids and FLPs used as enantioselective catalysts.

	Lewis acids	Added Lewis bases	Reactions	ee %
1	**1**	None	Hydrogenation of imine PhN = C(Me)Ph[34]	*1 substrate* 13% ee
2	**2a** or **2b**	P*t*Bu$_3$	Hydrogenation of imines ArN = C(Me)Ar'[36]	*6 substrates* 74%–83% ee
3	**2a** or **2c**	None or PMes$_3$ or P*t*Bu$_3$	Hydrosilylation of imines ArN = C(Me)Ar'[37]	*11 substrates* 63%–87% ee
4	**2d**	None	Hydrogenation of imines RN = C(Alk)Ar'[38]	*8 substrates* 33%–96% ee

Enantioselective Catalysis by Nonmetal FLPs 135

	Reaction	Base	Substrates / ee
5	Hydrogenation of imines PhN = C(Me)Ar[39]	None	19 substrates, 74%–89% ee
6	Hydrosilylation of silyl enol ethers[40,41]	PtBu₃	39 substrates, 87%–99% ee
7	Hydrosilylation of 1,2-dicarbonyl compounds[45]	PCy₃	20 substrates, 86%–99% ee
8	Hydrosilylation of chromones and flavones[48]	None	8 substrates, 11%–32% ee
9	Hydrogenation of a cis-1,2-diimine[44]	None	1 substrate, 10% ee
10	Hydrogenation of quinolines[43]	None	20 substrates, 82%–99% ee
11	Hydrogenation of quinoxalines[42]	None	14 substrates, 67%–96% ee
12	Hydrogenation of 1,8-naphthyridines[46]	None	14 substrates, 14%–74% ee
13	Hydrogenation of 2H-1,4-benzoxazines[47]	None	6 substrates, 30%–42% ee
14	Transfer hydrogenation of imines[49]	None	8 substrates, 16%–38% ee
15	Hydrosilylation of ketones[50]	PtBu₃	27 substrates, 81%–99% ee
16	Hydrogenation of imine PhN = C(Me)Ph[39]	None	1 substrate, 41% ee
17	Mannich reactions[51]	PMP or tBuNC(NMe₂)₂	24 substrates, 56%–96% ee

3 (42 boranes), R = H, Me

4 (92% ee)

3: R = H, Ar = Ph, p-CF₃-C₆H₅, 3,5-(tBu)₂C₆H₃

(Continued)

Table 1. (Continued)

	Lewis acids	Added Lewis bases	Reactions	ee %
18	**5** (12 catalysts)	None	Hydrogenation of imines[52]	20 substrates 45%–89%
19	**6a** (4 catalysts) or **6b** (2 catalysts)	None	Hydrogenation of imines[53]	33 substrates 3%–95% ee
20	**6c** Ar = 4-tBu(C_6H_4)	N-methyl-piperidine	Vinylogous Mannich reactions of enones[54]	45 substrates 82%–96% ee

#	Catalyst	Additive	Reaction	Results
21	**7** (5 catalysts)	None or (3,5-(F$_3$C)$_2$C$_6$H$_3$)$_3$P	Hydrogenation of quinolines[55]	45 substrates 87%–99% ee
22			Reduction of 2-vinylpyridines with pinacolborane[56]	33 substrates 73%–96% ee
23	**8** (R = H or Ph)	None	Hydrosilylation of ketones[58,60]	16 substrates 14%–99% ee
24		None	Hydrosilylation of imines[24,59]	5 substrates 33%–62%
25	**9**	None	Hydrogenation of imine PhN = C(Me)Ph[61]	1 substrate 0% ee
26	**10**	None	Hydrosilylation of acetophenone[62]	1 substrate 20% ee

(Continued)

Table 1. (Continued)

Lewis acids	Added Lewis bases	Reactions	ee %
(structures 11, 12, 13, 14, 15, 27)	None	Hydrogenation of imine PhN = C(Me)Ph[63]	1%–20% ee

28	![compound 16]	None	Hydrosilylation of imines RN = C(CH$_2$R')Ar[64]	8 substrates 28%–86% ee
29	⊖NTf$_2$ or {Al[OC(CF$_3$)$_3$]$_4$}	None	Hydrogenation of imines RN = C(CH$_2$R')Ph[64]	4 substrates 2%–80% ee
30	![compound 17] ⊖B(C$_6$F$_5$)$_4$	None	Hydrosilylation of imine PhN = C(Me)Ph[65]	20% ee

hydrosilylation of ketones with PhSiH$_3$ under mild conditions (neat, room temperature: 9 examples with 80%–99% ee).[60] As an alternative to centrally and axially chiral boranes, Paradies and coworkers introduced the planar chiral [2.2]paracyclophane-derived borane **9** (entry 25). Despite a promising chiral architecture, this borane produced a racemic product in the sole reported catalytic test, namely the hydrogenation of *N*-(1-phenylethylidene)aniline.[61]

As an auspicious strategy to overcome the limitations of chiral, strongly Lewis acidic, neutral perfluoroaryl-boranes, several groups have investigated positively charged borane-Lewis base adducts (R$_2$B/Lewis base$^{(+)}$). These species can be viewed either as neutral, trivalent boranes displaying a positively charged substituent, or as divalent borenium salts stabilized by neutral donor ligands. A first example of chiral catalysts of this class for FLP-type applications was described by Jäkle and coworkers. The authors used a planar chiral ferrocene-derived borenium salt (**10**) to catalyze the Piers hydrosilylation of a few ketones, obtaining the desired products with a modest 20% ee in the best case (Table 1, entry 26).[62] Later, in a collaborative work by Eisenberger, Melen, Crudden, Stephan, and others, several chiral carbene-stabilized boreniums were studied as catalysts in the hydrogenation of imines.[63] The main interest of the strategy is the structural modularity and easy availability of the catalysts that are generated by the complexation of chiral carbenes with commercially available hydroboranes (BH$_3$, 9-BBN or diisopinocampheylborane), followed by hydride abstraction with tritylium tetrakis(pentafluorophenyl) borate. Unfortunately, despite the diversity of structures tested, **11–15**, only low enantioselectivity was achieved in the hydrogenation of *N*-(1-phenylethylidene)aniline as the only studied substrate (entry 27, ee < 20%).

However, still convinced of the high potential of this strategy, Fuchter and coworkers later reexamined the method involving catalyst **15**, by varying several parameters, especially the reductant (a silane instead of dihydrogen), the counterion of the borenium catalyst **16**, the substituents of the imine substrate and the solvent. In their hands, although *N*-(1-phenylethylidene) aniline again furnished a modest 30% ee, other imines, namely imines with *N*-alkyl substituents (benzyl, butyl, cyclohexyl, cyclopentyl, isopropyl) were reduced with up to 86% ee (entries 28-29).[64]

In parallel to this work, Chuzel and coworkers took advantage of a chiral dihydroborane–NHC complex of their design and transformed it, in a few steps, into the NHC-stabilized chiral borenium **17**, which afforded a moderate 20% ee in the hydrosilylation of the ubiquitous *N*-(1-phenylethylidene) aniline substrate (entry 30).[65]

B. Phosphorus-based Chiral Lewis Acids

In parallel to their intense efforts to develop borane-based FLP chemistry, Stephan and coworkers also embarked with others in the quest for other strong Lewis acids that would be able to take part in FLP-type chemistry. Among the most successful ones, fluorophosphonium cations showed broad reaction scope and structural modularity. Their interest also resides in their elegant and simple access from the corresponding trivalent phosphines by an *Umpolung* strategy via an oxidative difluorination/fluoride abstraction sequence.[66,67] However, only one study on a chiral fluorophosphonium has been reported so far, by the combined efforts of Oestreich, Stephan, and coworkers.[68] These authors assessed the binaphthophosphepine-derived fluorophosphonium cations **18** as catalysts for the hydrosilylation of ketones (Table 2, entry 1). Unexpectedly, the products were always obtained as racemates, which led the authors to question the true nature of the catalyst. Finally, they proposed that the reaction is actually catalyzed by an achiral silylium cation, whose formation is triggered by the fluorophosphonium salt.

A more successful series of phosphorus-based Lewis acid catalysts has been described recently by Speed and co-workers[69] within studies on the use of chiral diazaphospholenes as catalysts for reductive processes.[70–73] Speed and coworkers synthesized a series of diazaphosphenium cations **19** in two to three steps from chiral amines and applied them to the hydrosilylation and hydroboration of cyclic and acyclic imines, obtaining high levels of enantioselectivity (Table 2, entry 2). The modularity of the synthetic access to these catalysts looks very promising. The mechanism proposed by the authors is analogous to that of the borane-catalyzed hydrosilylations of imines: the productive reaction pathway implies (1) bonding of the imine to the hydride source — hydrosilane or hydroborane — rather than to the B- or P-based Lewis acids, followed by (2) hydride capture by the Lewis acid,

Table 2. Phosphorus-based chiral Lewis acids used as enantioselective catalysts.

	Lewis acids	Added Lewis bases	Reactions	Comments
1	**18** (R = Ph or C_6F_5), with $B(C_6F_5)_4^-$ counterion	None	Hydrosilylation of ketones[68]	4 substrates 0% ee
2	**19a**, **19b**, **19c**, **19d** (all with TfO$^-$ counterion)	None	Hydrosilylation and hydroboration of imines[69]	24 substrates 60%–94% ee

which (3) finally delivers the hydride to the imine activated by the silylium (or proton,[24] see part II of this chapter) or pinacolborenium ions linked to its nitrogen atom (Scheme 8).

IV. FLP Catalysts Featuring Chiral Lewis Base Moieties

As already stated earlier, reductions are arguably the most studied reactions under FLP catalysis. The transfer of the hydride from the Lewis acid to the substrate is likely to be the stereo-discriminating step in such reductions. Consequently, very little effort has been devoted to the development of chiral catalysts where chirality is borne by the Lewis base component only. Yet, a few studies have been reported.

As chiral phosphines are so efficient stereoinducers in asymmetric organometallic catalysis, and particularly in hydrogenations, it seems

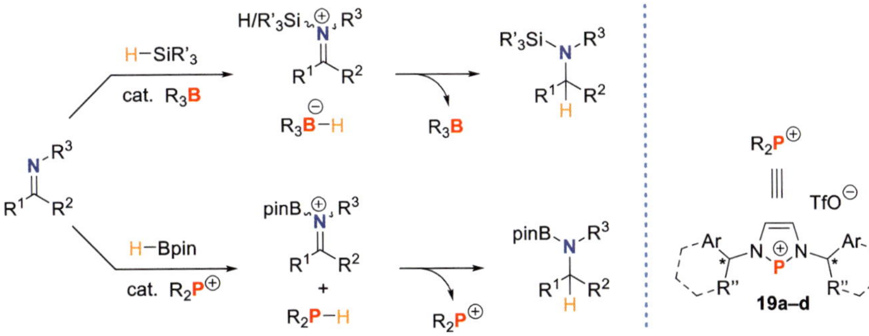

Scheme 8. Mechanisms of the borane-catalyzed hydrosilylation of imines and diazaphosphenium-catalyzed hydroboration of imines.[69]

natural that Stephan and coworkers decided to assess some of them (namely (R)-BINAP **20**, (S,S)-DIOP **21** and (S,S)-Chiraphos **22**) as chiral Lewis bases in FLP-catalyzed hydrogenations of imines.[74] Although the enantiomeric excesses were null to modest (Table 3, entry 1), this was the first example of an enantioselective process involving FLP catalysts with chirality located on the Lewis base only. The rationale for stereoinduction remains unknown.

Following this work, Panossian and coworkers assessed the axially chiral biphenyl-derived monophosphines **23** as chiral Lewis bases, in combination with BCF, in the hydrosilylation of acetophenone and the corresponding N-phenyl imine. Despite reasonable catalytic activities, only racemic products were obtained, presumably due to the large distance between the chiral Lewis base and the functional group to be reduced, as well as to the flexibility of the transition state (Table 3, entry 2).[75] The strategy of using chiral base-derived FLP catalysts proved more successful in studies where nitrogen-based Lewis/Brønsted bases were used, as shown hereafter.

Choosing, once again, to look for other reactions than reductions, Wasa and coworkers reported on the α-amination of α-H acetophenone derivatives (tetralone, chromanones, etc.) with azodicarboxylates, catalyzed by a combination of BCF as Lewis acid and various chiral amines (Scheme 9). The initial mechanistic hypothesis involved the activation of the ketone by BCF, followed by its enolization promoted by the chiral base. The resulting chiral ammonium salt would activate then the

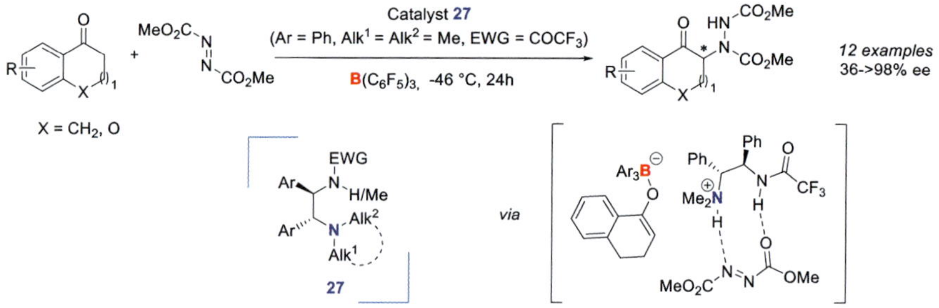

Scheme 9. FLP-promoted α-amination of α-tetralones, chromanones and other cyclic ketones.[76]

azodicarboxylate electrophile by hydrogen bonding and would thus effect stereoinduction at the enolate addition step. After an initial screening, it appeared that monoamine **24**, aminoether **25** and (−)-sparteine (**26**) only led to racemates, whereas amines **27** and **28** bearing an additional hydrogen bond donor group provided high enantioselectivities. The authors rationalized this successful outcome by both a more rigid transition state and a better activation of the electrophile that results from the double hydrogen-bonding interaction between the chiral ammonium-amine and the azodicarboxylate. All in all, very high enantioselectivites were achieved in many cases (Table 3, entries 4–5).[76]

Interestingly, the challenging hydrogenation of ketones by FLPs could be achieved by Meng, Du and coworkers with up to 87% ee when using a catalyst composed of chiral bis- or mono-oxazolines (**29** and **30** respectively) and a tris(fluoroaryl)borane (Scheme 10). This achievement is remarkable, not only due to the notorious difficulty of hydrogenating ketones by means of FLP catalysts, in comparison with unsaturated nitrogen-based substrates, but also because it validates successfully the strategy of using chiral Lewis base/achiral Lewis acid combinations, which seemed to be difficult, or perhaps even doomed to failure, in previously attempted H_2-, hydrosilane-, or hydroborane-mediated reductions. The authors also extended their work to the conjugate hydrogenation of tetralone-derived enones (Scheme 10(b)) and chromones with the same type of catalytic system, obtaining higher enantioselectivities (up to 95% ee) (Table 3, entries 6–10).[77]

Table 3. Chiral Lewis bases used in enantioselective catalysis by FLPs.

	Lewis bases	Lewis acid	Reactions	Comments
1	**20**, **21**, **22**	$B(C_6F_5)_3$	Hydrogenation of PhN=C(Me)Ph[74]	1 substrate 0%–25% ee
2	**23**, **24**, **25**	$B(C_6F_5)_3$	Hydrosilylation PhCOMe and PhN=C(Me)Ph[75]	2 substrates 0% ee
3	**26**	$B(C_6F_5)_3$	α-Amination of α-tetralone[76]	1 substrate ≤4% ee

(Continued)

Table 3. (Continued)

	Lewis bases	Lewis acid	Reactions	Comments
4	**27** (10 amines) EWG = CO_2tBu, $COCF_3$, $COCCl_3$, SO_2CF_3 Ar = Ph, 4-(MeO)C_6H_4, 2-Cl-C_6H_4, 1-Napht	$B(C_6F_5)_3$	α-Amination of α-tetralone[76]	1 substrate 60%–94% ee
5	**27** (Ar = Ph, Alk^1 = Alk^2 = Me, EWG = $COCF_3$)	$B(C_6F_5)_3$	α-Amination of α-tetralone derivatives[76]	12 substrates 36–>98% ee
6	**29** (2 bases)	$B(C_6F_5)_3$	Hydrogenation of 4-CF_3-C_6H_4-C(O)Me[77]	1 substrate 41%–60% ee

Enantioselective Catalysis by Nonmetal FLPs 147

	Catalyst structure	Borane	Reaction	Scope
7	**30** (11 bases) with R¹, R², R³ substituents	B(C$_6$F$_5$)$_3$	Hydrogenation of 4-CF$_3$-C$_6$H$_4$-C(O)Me[77]	1 substrate 7%–72% ee
8	**30a** (tBu, Ph)	B(4-HC$_6$F$_4$)$_3$	Hydrogenation of ketones[77]	29 substrates 50%–87% ee
9	**30b** (CEt$_3$, Ph)	B(4-HC$_6$F$_4$)$_3$	Hydrogenation of α-arylidene-tetralones[77]	34 substrates 63%–92% ee
10		B(4-HC$_6$F$_4$)$_2$Ar (Ar = 3,4,5-H$_3$C$_6$F$_2$)	Hydrogenation of chromones[77]	16 substrates 33%–95% ee

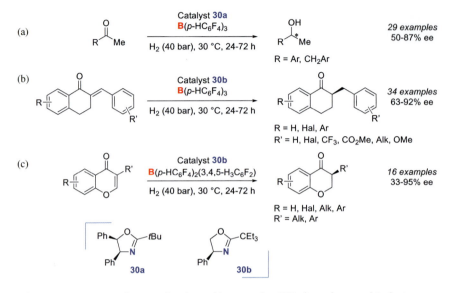

Scheme 10. 1,2- and 1,4-reduction of ketones by FLPs based on a chiral nitrogen base.[77]

V. Chiral Intramolecular FLP Catalysts

The strategies described previously based on chiral acid/achiral base or achiral acid/chiral base combinations offer the advantage of being, in theory, highly modular and allowing the efficient screening of a variety of FLPs for the fine-tuning of the catalyst. On the other hand, the use of intramolecular chiral FLPs can be envisioned but would require a much heavier synthetic work, which would preclude easy modulation. However, this alternative strategy has demonstrated other benefits. Indeed, with the acid and the base being both connected to the chiral backbone, there might be a higher chance for stereoinduction from the catalyst. Moreover, since in some reactions, such as hydrogenations, the rate-determining step involves the concomitant encounter of the acid, the base, and one of the reactants (see Scheme 2),[23] using an intramolecular FLP would accelerate this step and lead to more efficient catalytic processes hence to milder reaction conditions. Hereafter are described examples of chiral intramolecular FLPs used in catalysis.

A. Phosphorus/Boron-based Chiral Intramolecular FLPs

Following their work on FLPs based on camphor-derived boranes (*vide supra*), Klankermayer and coworkers applied the same synthetic strategy to a camphor-derived phosphine-borane. After a multistep synthesis, the intramolecular FLP **31** was isolated as its H_2 adduct, that is, the corresponding hydrophosphonium/hydridoborate zwitterion, which showed excellent stability as it could even be purified by column chromatography, recovered, and recycled for further catalytic experiments. The catalyst was able to hydrogenate a small series of imines with ee's of 70%–76% (Table 4, entry 1).[78]

Erker and coworkers on the other hand applied their extensive know-how in FLP chemistry to the synthesis of planar chiral ferrocene-derived P/B FLPs **32** and **33**. In a first paper, they demonstrated the feasibility of asymmetric induction by their structures; however, ee's were modest, presumably due to the flexibility and the length of the spacer between the key boron center and the chiral ferrocene unit (Table 4, entry 2).[79] Then, in a second paper, they introduced a phenyl substituent α to boron, thus adding a stereocenter in close vicinity to boron. Accordingly, they managed to improve ee's up to 69% (entry 3).[80]

Table 4. Phosphorus- and boron-based chiral intramolecular FLPs used in catalysis.

	Chiral intramolecular FLP	Reactions	Comments
1	**31** (camphor-derived, $B(C_6F_5)_2$ / $PtBu_2$)	Hydrogenation of imines ArNH = C(Me)Ar'[78]	8 substrates 70%–76% ee
2	**32** (ferrocene, $B(C_6F_5)_2$ / $PMes_2$)	Hydrogenation of imines RNH = C(Me)Ar'[79]	3 substrates 2%–26% ee
3	**33** (ferrocene, Ph, $B(C_6F_5)_2$ / $PMes_2$)	Hydrogenation of imines PhN = C(Me)Ar[80]	6 substrates 42%–69% ee

B. Nitrogen/Boron-based Chiral Intramolecular FLPs

Although the use of phosphorus-derived bases in intramolecular FLPs is advantageous in terms of synthesis, as the phosphorous moiety can be introduced straightforwardly by substitution reactions on easily available phosphorous electrophiles, the sensitivity of phosphines to oxygen is a common drawback. Accordingly, using nitrogen-derived bases instead could be beneficial.

Having in hand a simple and quick synthetic access, Repo and coworkers devised a series of intramolecular B/N FLPs, two of which (**34** and **35**) were highly enantioenriched. The Lewis basic sites were cyclic chiral amines, whose conformation was determined by a carbon stereocenter that led to a defined stereogenicity at nitrogen. The authors assessed these catalysts in the hydrogenation of a couple of imines and a quinoline, obtaining low enantioselectivities (Table 5, entry 1).[81]

The same group later reported on another system (**36**), based on an axially chiral 1,1′-binaphthyl core, with the Lewis acidic and basic sites on the 2 and 2′ positions. In this case, both the boron and nitrogen centers would benefit equally well from the close chirality of the backbone, which is, moreover, rather rigid and would lead therefore to better-defined transition states. Furthermore, in hydrogenation reactions, the spatial proximity of the boron and nitrogen centers would allow easier dihydrogen

Table 5. Nitrogen- and boron-based chiral intramolecular FLPs used in catalysis.

	Chiral intramolecular FLP	Reactions	Comments
1	**34**, **35**	Hydrogenation of imines and of a quinoline[81]	4 substrates 17%–37% ee
2	**36**	Hydrogenation of imines $R^1(R^2CH_2)C = N(R^3)$[82]	6 substrates 32%–83% ee
3		Hydrogenation of enamines[82]	4 substrates 47%–99% ee

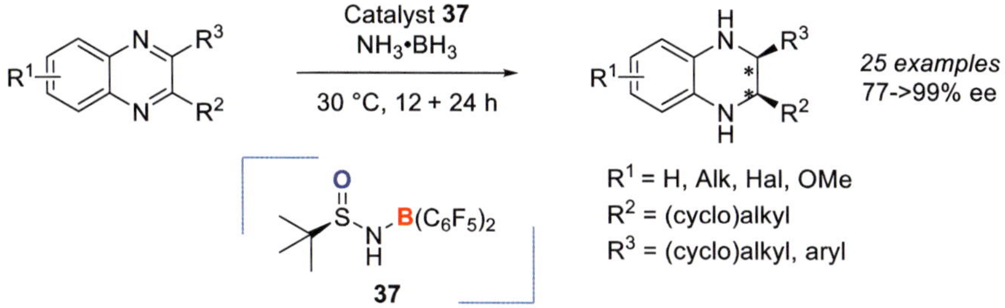

Scheme 11. Transfer hydrogenation of quinoxalines with ammonia-borane catalyzed by Du's N-boranylsulfinamide FLP.[83–86]

activation, thus an overall faster reduction process. When employed in the hydrogenation of a small series of imines and enamines, catalyst **36** afforded ee's, which were highly dependent on the substrate structure, yet reaching sometimes optimal levels. Very interestingly, the above-mentioned structural features of the catalyst indeed led to high activity under unusually mild conditions for FLP-catalyzed hydrogenations — 2 bars of H_2, catalyst loading of 2.5 to 7.5 mol%, room temperature, reaction times as short as 0.5 h (Table 5, entries 2–3).[82]

C. Oxygen/Boron-based Chiral Intramolecular FLPs

Last but not least, one should mention another elegant strategy explored by Du and coworkers consisting in the in situ generation of a chiral intramolecular FLP upon complexation of Piers' borane with (S)-*tert*-butanesulfinamide, followed by release of dihydrogen to the substrate. The resulting intramolecular B/O FLP **37** proved efficient and enantioselective in the transfer hydrogenation of imines, quinoxalines (Scheme 11), indoles and β-aminocinnamates, using ammonia-borane as formal dihydrogen source (Table 6). A possible drawback is the long time required in certain cases for the in situ generation of the catalyst from the sulfinamide-$HB(C_6F_5)_2$ complex.[83–86]

VI. Rationalization of Enantioselectivity

We already emphasized on the youth of the field of enantioselective catalysis by chiral FLPs and on the somewhat limited number of chiral

Table 6. Chiral sulfinamide-borane FLPs used as enantioselective catalysts.

Intramolecular FLP	Reactions	Comments
1	Transfer hydrogenation of imines ArN = C(Me)Ar' (stoichiometric version)[83]	13 substrates 73%–96% ee
2	Transfer hydrogenation of imines (4-NC-C_6H_4) N = C(Me)Ar with $NH_3 \cdot BH_3$[83]	17 substrates 84%–95% ee
3	Transfer hydrogenation of 2-alkyl-3-arylquinoxalines[84]	7 substrates 77%–86% ee (*cis* isomer)
4	Transfer hydrogenation of 2,3-dialkylquinoxalines[84]	18 substrates 89–>99% ee (*trans* isomer)
5	Transfer hydrogenation of enamino esters[85] ArNH-C(Ar') = CH(CO_2R)	35 substrates 41%–91% ee
6	Transfer hydrogenation of 2-substituted and 2,3-disubstituted indoles[86]	24 substrates 0%–90% ee

(Structure **37**: tBu–S(=O)–N(H)–B($C_6F_5)_2$)

FLP structures, relying usually on chiral Lewis acids. These reasons may account also for the very few attempts to rationalize enantioselectivity. To the best of our knowledge, only five papers tentatively address this issue.[38,69,77,82,87]

Repo, Pápai, and coworkers used computational studies to gain insight in the elementary steps of the reductions catalyzed by the binaphthyl-derived FLP **36,** as well as to explain the sense of enantioinduction. Calculations appeared to confirm that hydride transfer from the hydridoborate to the protonated enamine is the rate- and stereo-determining step (Scheme 12); however, the authors could not clearly identify the most favored transition state among several of very close energies, due to a competition between steric repulsion and noncovalent attractive interactions.[82]

The same authors recently carried out extensive computational work to rationalize the enantioselectivity of imine hydrogenations by means of Klankermayer's boranes derived from pinene and camphor, **1** and **2a,b**, as well as for related boranes of their own design including **2d** (Table 1, entries 1–4). Several valuable conclusions could be drawn from this work. First, the activation of dihydrogen by the combination of borane **2a** and the imine substrate was shown to be much easier than activation by the borane/tri-*tert*-butylphosphine

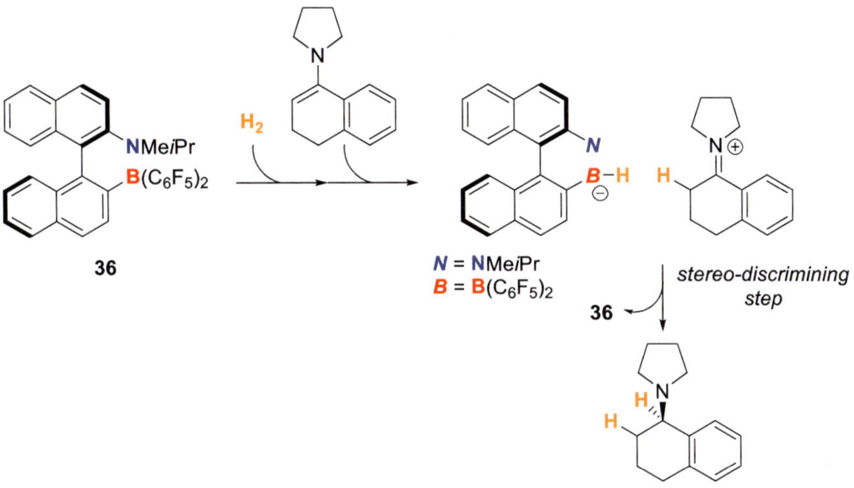

Scheme 12. Hydrogenation of enamines by means of FLP catalyst **36**.[38]

pair, due to high steric hinderance by the bulky phosphine. Second, the stereo-discriminating step was confirmed to be the delivery of the hydride from the hydridoborate to the protonated imine and was shown to be governed by the Curtin–Hammett principle. Indeed, several interconverting isomers of close energy were identified for the hydridoborate/iminium intermediate, the most stable of which was not the one with the lowest activation barrier in the irreversible hydride delivery. Third, energy differences between isomeric intermediates or transition states were caused by the combinations of noncovalent interactions such as π–π stacking (between C_6F_5 groups of the borane and Ph rings of the imine), CH•••π interactions (between the CH_3 group of the imine and the Ph substituent of the borane backbone), and Ph-Ph interactions (between the Ph substituent of nitrogen and the Ph substituent of the borane) (Scheme 13). Gratifyingly, the authors could closely reproduce computationally the experimental enantiomeric excess and absolute configuration of the product, and accordingly they could optimize the design of the borane to reach higher selectivity.[38]

Concerning another type of imine reduction, in their work on chiral diazaphosphenium Lewis acid catalysts, Speed and coworkers proposed a simple model based on minimization of steric hindrance to rationalize the sense of stereoinduction (Scheme 14).[69]

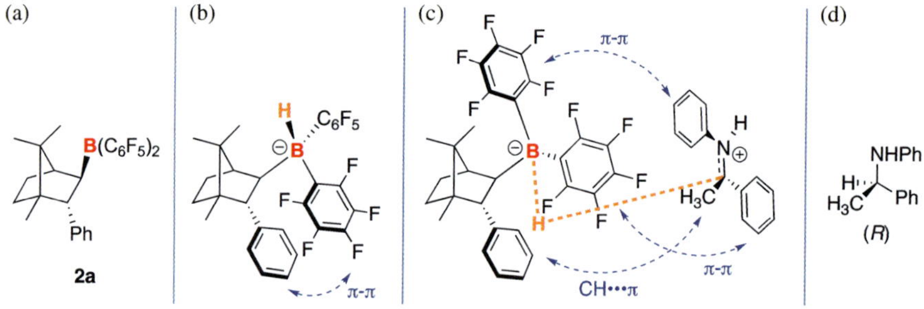

Scheme 13. Simplified rationale for the enantioselectivity of imine hydrogenations by means of borane **2a**. (a) Borane **2a**; (b) most stable conformer of the borohydride; (c) most stable transition state (distances between the borohydride and the iminium are represented as exceedingly large for easier reading); (d) absolute configuration of the amine product.[38]

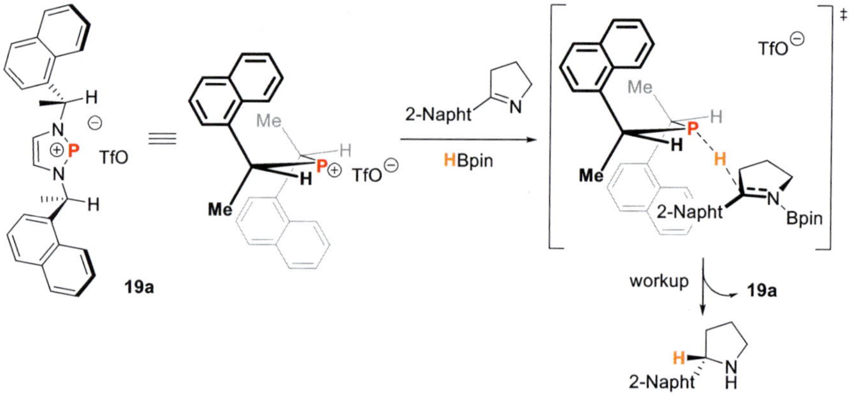

Scheme 14. Stereochemical model for catalyst **19a** by Speed and coworkers.[69]

Du, Meng, and coworkers also used computations to model the hydrogenation of ketones by chiral oxazoline-based FLPs (Scheme 10). According to this study, the FLP activates H_2 following the same pathway depicted in Scheme 2. The resulting ammonium/hydridoborate intermediate activates ketones by means of H-bonding between the ammonium hydrogen and the carbonyl oxygen of the substrate as well as by CH···F interactions and π-stackings. However, the stereo-discriminating step might be different depending on the substrate (Scheme 15). For 4′-(trifluoromethyl)acetophenone

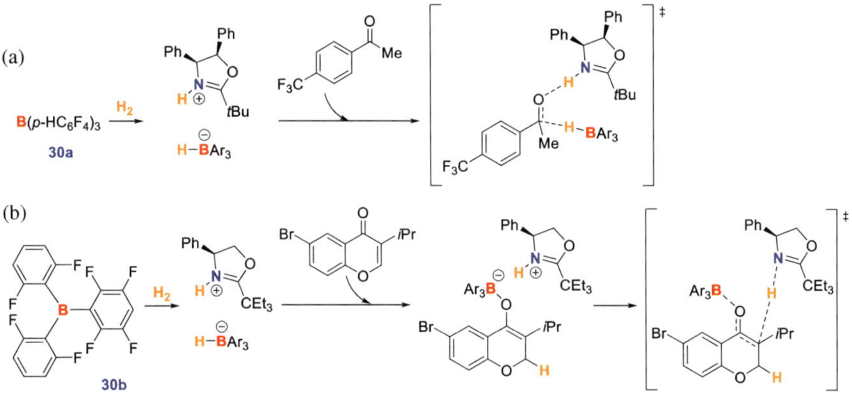

Scheme 15. Hydrogenation of ketones and chromones by means of FLP catalysts based on chiral oxazolines **30**.[39]

(Scheme 15(a)), the stereo-discriminating step would be the concerted proton and hydride delivery and the preferred transition state is the one that displays the larger number of CH···F interactions, the other transition state being higher in energy by 5.5 kcal/mol. On the other hand, in the hydrogenation of 6-bromo-3-*iso*propylchromone (Scheme 15(b)), the stereo-discriminating step would be the protonation of the boron enolate (generated upon hydride delivery to the enone) by the *N*-H oxazolinium. A 2.2 kcal/mol difference was found between the transition states leading to the opposite enantiomers.[77]

Finally, a recent computational study by Platts and coworkers approached the issue of enantioselective catalysis from the opposite angle. The authors indeed conceived potential FLPs based on a chiral backbone and tried to predict the stereochemical outcome of the reactions. However, these calculations have not been validated experimentally yet.[87]

VII. Conclusion

The young field of enantioselective catalysis by FLPs has already achieved some remarkable results despite the relatively small number of studies. Very high enantioselectivities and high activities have been attained notably in the hydrogenations of simple imines, ketones, and cyclic enones, in hydrosilylation reactions, and in aminations of cyclic ketones. FLPs based on amine–borane pairs constitute so far the most successful series, while efficient phosphorus-based catalysts remain to be developed. For historical reasons, the most studied catalytic systems are based on chiral, highly Lewis acidic boranes as the key components of the FLP. Confronted with the difficult synthesis or handling of such boranes, chemists have devised various strategies, whose peculiar assets are either the modularity of the catalyst, the simple preparation, or the shift of chirality to the Lewis base, to cite a few. Other reactions than the "traditional" reductions of unsaturated substrates have now just started to be developed under enantioselective catalysis by FLPs. Again, the field is still young, but it holds promises of innovative advances.

VIII. Acknowledgments

The authors thank the French Agence Nationale pour la Recherche (ANR) (grant number ANR-18-CE07-0007-01), the CNRS, and the Université de Strasbourg for financial support.

IX. References

1. McCahill, J. S.; Welch, G. C.; Stephan, D. W. *Angew. Chem. Int. Ed.* **2007**, *46*, 4968–4971.
2. Welch, G. C.; San Juan, R. R.; Masuda, J. D.; Stephan, D. W. *Science* **2006**, *314*, 1124–1126.
3. Stephan, D. W.; Erker, G. *Angew. Chem. Int. Ed.* **2010**, *49*, 46–76.
4. Erker, G.; Stephan, D. W. *Frustrated Lewis Pairs I. Top Curr Chem.* Springer Berlin Heidelberg, **2013**; Vol. 332.
5. Erker, G.; Stephan, D. W. *Frustrated Lewis Pairs II. Top Curr Chem.* Springer Berlin Heidelberg, **2013**; Vol. 334.
6. Stephan, D. W.; Erker, G. *Angew. Chem. Int. Ed.* **2015**, *54*, 6400–6441.
7. Feng, X.; Du, H. *Tetrahedron Lett.* **2014**, *55*, 6959–6964.
8. Shi, L.; Zhou, Y.-G. *ChemCatChem* **2015**, *7*, 54–56.
9. Meng, W.; Feng, X.; Du, H. *Acc. Chem. Res.* **2018**, *51*, 191–201.
10. Meng, W.; Feng, X.; Du, H. *Chin. J. Chem.* **2020**, *38*, 625–634.
11. Piers, W. E.; Marwitz, A. J. V.; Mercier, L. G. *Inorg. Chem.* **2011**, *50*, 12252–12262.
12. Rokob, T. A.; Hamza, A.; Papai, I. *J. Am. Chem. Soc.* **2009**, *131*, 10701–10710.
13. Schulz, F.; Sumerin, V.; Heikkinen, S.; Pedersen, B.; Wang, C.; Atsumi, M.; Leskelä, M.; Repo, T.; Pyykkö, P.; Petry, W.; Rieger, B. *J. Am. Chem. Soc.* **2011**, *133*, 20245–20257.

14. Pu, M.; Privalov, T. *J. Chem. Phys.* **2013**, *138*, 154305.
15. Rokob, T. A.; Bakó, I.; Stirling, A.; Hamza, A.; Pápai, I. *J. Am. Chem. Soc.* **2013**, *135*, 4425–4437.
16. Zeonjuk, L. L.; Vankova, N.; Mavrandonakis, A.; Heine, T.; Röschenthaler, G. V.; Eicher, J. *Chem. Eur. J.* **2013**, *19*, 17413–17424.
17. Pu, M.; Privalov, T. *ChemPhysChem.* **2014**, *15*, 3714–3719.
18. Skara, G.; Pinter, B.; Top, J.; Geerlings, P.; De Proft, F.; De Vleeschouwer, F. *Chem. Eur. J.* **2015**, *21*, 5510–5519.
19. Daru, J.; Bakó, I.; Stirling, A.; Pápai, I. *ACS Catal.* **2019**, *9*, 6049–6057.
20. Heshmat, M.; Privalov, T. *J. Phys. Chem. A.* **2018**, *122*, 7202–7211.
21. Rocchigiani, L.; Ciancaleoni, G.; Zuccaccia, C.; Macchioni, A. *J. Am. Chem. Soc.* **2014**, *136*, 112–115.
22. Brown, L. C.; Hogg, J. M.; Gilmore, M.; Moura, L.; Imberti, S.; Gartner, S.; Gunaratne, H. Q. N.; O'Donnell, R. J.; Artioli, N.; Holbrey, J. D.; Swadzba-Kwasny, M. *Chem. Commun.* **2018**, *54*, 8689–8692.
23. Houghton, A. Y.; Autrey, T. *J. Phys. Chem. A.* **2017**, *121*, 8785–8790.
24. Hermeke, J.; Mewald, M.; Oestreich, M. *J. Am. Chem. Soc.* **2013**, *135*, 17537–17546.
25. Blackwell, J. M.; Sonmor, E. R.; Scoccitti, T.; Piers, W. E. *Org. Lett.* **2000**, *2*, 3921–3923.
26. Sakata, K.; Fujimoto, H. *J. Org. Chem.* **2013**, *78*, 12505–12512.
27. Parks, D. J.; Blackwell, J. M.; Piers, W. E. *J. Org. Chem.* **2000**, *65*, 3090–3098.
28. Houghton, A. Y.; Hurmalainen, J.; Mansikkamaki, A.; Piers, W. E.; Tuononen, H. M. *Nature Chem.* **2014**, *6*, 983–988.
29. Heshmat, M.; Privalov, T. *Chem. Eur. J.* **2017**, *23*, 11489–11493.
30. Heshmat, M.; Privalov, T. *J. Phys. Chem. B.* **2018**, *122*, 8952–8962.
31. Heshmat, M.; Privalov, T. *Chem. Eur. J.* **2017**, *23*, 9098–9113.
32. Marwitz, A. J.; Dutton, J. L.; Mercier, L. G.; Piers, W. E. *J. Am. Chem. Soc.* **2011**, *133*, 10026–10029.
33. Mane, M. V.; Vanka, K. *ChemCatChem.* **2017**, *9*, 3013–3022.
34. Chen, D.; Klankermayer, J. *Chem. Commun.* **2008**, 2130–2131.
35. Parks, D. J.; Piers, W. E.; Yap, G. P. A. *Organometallics.* **1998**, *17*, 5492–5503.
36. Chen, D.; Wang, Y.; Klankermayer, J. *Angew. Chem. Int. Ed.* **2010**, *49*, 9475–9478.
37. Chen, D.; Leich, V.; Pan, F.; Klankermayer, J. *Chem. Eur. J.* **2012**, *18*, 5184–5187.
38. Hamza, A.; Sorochkina, K.; Kótai, B.; Chernichenko, K.; Berta, D.; Bolte, M.; Nieger, M.; Repo, T.; Pápai, I. *ACS Catal.* **2020**, *10*, 14290–14301.
39. Liu, Y.; Du, H. *J. Am. Chem. Soc.* **2013**, *135*, 6810–6813.
40. Wei, S.; Du, H. *J. Am. Chem. Soc.* **2014**, *136*, 12261–12264.
41. Ren, X.; Li, G.; Wei, S.; Du, H. *Org. Lett.* **2015**, *17*, 990–993.
42. Zhang, Z.; Du, H. *Angew. Chem. Int. Ed.* **2015**, *54*, 623–626.
43. Zhang, Z.; Du, H. *Org. Lett.* **2015**, *17*, 2816–2819.
44. Zhu, X.; Du, H. *Org. Lett.* **2015**, *17*, 3106–3109.
45. Ren, X.; Du, H. *J. Am. Chem. Soc.* **2016**, *138*, 810–813.
46. Wang, W.; Feng, X.; Du, H. *Org. Biomol. Chem.* **2016**, *14*, 6683–6686.

47. Wei, S.; Feng, X.; Du, H. *Org. Biomol. Chem.* **2016**, *14*, 8026–8029.
48. Ren, X.; Han, C.; Feng, X.; Du, H. *Synlett.* **2017**, *28*, 2421–2424.
49. Wang, Q.; Chen, J.; Feng, X.; Du, H. *Org. Biomol. Chem.* **2018**, *16*, 1448–1451.
50. Liu, X.; Wang, Q.; Han, C.; Feng, X.; Du, H. *Chin. J. Chem.* **2019**, *37*, 663–666.
51. Shang, M.; Cao, M.; Wang, Q.; Wasa, M. *Angew. Chem. Int. Ed.* **2017**, *56*, 13338–13341.
52. Liu, X.; Liu, T.; Meng, W.; Du, H. *Org. Biomol. Chem.* **2018**, *16*, 8686–8689.
53. Tu, X.-S.; Zeng, N.-N.; Li, R.-Y.; Zhao, Y.-Q.; Xie, D.-Z.; Peng, Q.; Wang, X.-C. *Angew. Chem. Int. Ed.* **2018**, *57*, 15096–15100.
54. Tian, J.-J.; Liu, N.; Liu, Q.-F.; Sun, W.; Wang, X.-C. *J. Am. Chem. Soc.* **2021**, *143*, 3054–3059.
55. Li, X.; Tian, J.-J.; Liu, N.; Tu, X.-S.; Zeng, N.-N.; Wang, X.-C. *Angew. Chem. Int. Ed.* **2019**, *58*, 4664–4668.
56. Tian, J.-J.; Yang, Z.-Y.; Liang, X.-S.; Liu, N.; Hu, C.-Y.; Tu, X.-S.; Li, X.; Wang, X.-C. *Angew. Chem. Int. Ed.* **2020**, *59*, 18452–18456.
57. Lindqvist, M.; Axenov, K.; Nieger, M.; Räisänen, M.; Leskelä, M.; Repo, T. *Chem. Eur. J.* **2013**, *19*, 10412–10418.
58. Mewald, M.; Fröhlich, R.; Oestreich, M. *Chem. Eur. J.* **2011**, *17*, 9406–9414.
59. Mewald, M.; Oestreich, M. *Chem. Eur. J.* **2012**, *18*, 14079–14084.
60. Süsse, L.; Hermeke, J.; Oestreich, M. *J. Am. Chem. Soc.* **2016**, *138*, 6940–6943.
61. Greb, L.; Paradies, J. Paracyclophane Derivatives in Frustrated Lewis Pair Chemistry. In *Top Curr Chem*, Springer Berlin Heidelberg, **2013**; Vol. 334, pp. 81–100.
62. Chen, J.; Lalancette, R. A.; Jäkle, F. *Chem. Commun.* **2013**, *49*, 4893–4895.
63. Lam, J.; Gunther, B. A.; Farrell, J. M.; Eisenberger, P.; Bestvater, B. P.; Newman, P. D.; Melen, R. L.; Crudden, C. M.; Stephan, D. W. *Dalton Trans.* **2016**, *45*, 15303–15316.
64. Mercea, D. M.; Howlett, M. G.; Piascik, A. D.; Scott, D. J.; Steven, A.; Ashley, A. E.; Fuchter, M. J. *Chem. Commun.* **2019**, *55*, 7077–7080.
65. Aupic, C.; Abdou Mohamed, A.; Figliola, C.; Nava, P.; Tuccio, B.; Chouraqui, G.; Parrain, J.-L.; Chuzel, O. *Chem. Sci.* **2019**, *10*, 6524–6530.
66. Bayne, J. M.; Stephan, D. W. *Chem. Soc. Rev.* **2016**, *45*, 765–774.
67. Stephan, D. W. *Angew. Chem. Int. Ed.* **2017**, *56*, 5984–5992.
68. Süsse, L.; LaFortune, J. H. W.; Stephan, D. W.; Oestreich, M. *Organometallics* **2019**, *38*, 712–721.
69. Lundrigan, T.; Welsh, E. N.; Hynes, T.; Tien, C.-H.; Adams, M. R.; Roy, K. R.; Robertson, K. N.; Speed, A. W. H. *J. Am. Chem. Soc.* **2019**, *141*, 14083–14088.
70. Adams, M. R.; Tien, C.-H.; McDonald, R.; Speed, A. W. H. *Angew. Chem. Int. Ed.* **2017**, *56*, 16660–16663.
71. Miaskiewicz, S.; Reed, J. H.; Donets, P. A.; Oliveira, C. C.; Cramer, N. *Angew. Chem. Int. Ed.* **2018**, *57*, 4039–4042.
72. Reed, J. H.; Donets, P. A.; Miaskiewicz, S.; Cramer, N. *Angew. Chem. Int. Ed.* **2019**, *58*, 8893–8897.
73. Reed, J. H.; Cramer, N. *ChemCatChem.* **2020**, *12*, 4262–4266.

74. Stephan, D. W.; Greenberg, S.; Graham, T. W.; Chase, P.; Hastie, J. J.; Geier, S. J.; Farrell, J. M.; Brown, C. C.; Heiden, Z. M.; Welch, G. C.; Ullrich, M. *Inorg. Chem.* **2011**, *50*, 12338–12348.
75. Fer, M. J.; Cinqualbre, J.; Bortoluzzi, J.; Chessé, M.; Leroux, F. R.; Panossian, A. *Eur. J. Org. Chem.* **2016**, 4545–4553.
76. Shang, M.; Wang, X.; Koo, S. M.; Youn, J.; Chan, J. Z.; Yao, W.; Hastings, B. T.; Wasa, M. *J. Am. Chem. Soc.* **2017**, *139*, 95–98.
77. Gao, B.; Feng, X.; Meng, W.; Du, H. *Angew. Chem. Int. Ed.* **2020**, *59*, 4498–4504.
78. Ghattas, G.; Chen, D.; Pan, F.; Klankermayer, J. *Dalton Trans.* **2012**, *41*, 9026–9028.
79. Wang, X.; Kehr, G.; Daniliuc, C. G.; Erker, G. *J. Am. Chem. Soc.* **2014**, *136*, 3293–3303.
80. Ye, K. Y.; Wang, X. W.; Daniliuc, C. G.; Kehr, G.; Erker, G. *Eur. J. Inorg. Chem.* **2017**, 368–371.
81. Sumerin, V.; Chernichenko, K.; Nieger, M.; Leskelä, M.; Rieger, B.; Repo, T. *Adv. Synth. Catal.* **2011**, *353*, 2093–2110.
82. Lindqvist, M.; Borre, K.; Axenov, K.; Kótai, B.; Nieger, M.; Leskelä, M.; Pápai, I.; Repo, T. *J. Am. Chem. Soc.* **2015**, *137*, 4038–4041.
83. Li, S.; Li, G.; Meng, W.; Du, H. *J. Am. Chem. Soc.* **2016**, *138*, 12956–12962.
84. Li, S.; Meng, W.; Du, H. *Org. Lett.* **2017**, *19*, 2604–2606.
85. Zhao, W.; Feng, X.; Yang, J.; Du, H. *Tetrahedron Lett.* **2019**, *60*, 1193–1196.
86. Zhao, W.; Zhang, Z.; Feng, X.; Yang, J.; Du, H. *Org. Lett.* **2020**, *22*, 5850–5854.
87. Sharma, G.; Newman, P. D.; Melen, R.; Platts, J. A. *J. Theor. Comput. Chem.* **2020**, *19*, 2050009.

© 2022 World Scientific Publishing Company
https://doi.org/10.1142/9789811248436_0006

6 Organocatalytic and Asymmetric Processes via P^{III}/P^V Redox Cycling

Charlotte Lorton[*] and Arnaud Voituriez[*,†]

[*]Université Paris-Saclay, CNRS, Institut de Chimie des Substances Naturelles, UPR 2301, Gif-sur-Yvette 91198, France
[†]arnaud.voituriez@cnrs.fr

Table of Contents

List of Abbreviations	162
I. Introduction	162
II. Toward the Development of Enantioselective CWRs	163
A. Catalytic Wittig Reactions	163
1. Initial studies and prerequisites for the development of phosphine-catalyzed P^{III}/P^V redox reactions	163
2. Further optimization and development of CWR	168
B. Asymmetric Wittig Reactions	172
1. Asymmetric Wittig–type olefination reactions using stoichiometric amounts of phosphorus reagents	172
2. Enantioselective CWRs	174

	3. Application in total synthesis	175
III.	Tandem Reactions Including Wittig and aza-Wittig Reactions	176
	A. Tandem Michael Addition/Wittig Olefination	177
	1. A catalytic tandem reaction	177
	2. Catalytic and asymmetric Michael addition/Wittig reactions	180
	B. Tandem Staudinger/aza-Wittig Reaction	183
	1. Catalytic tandem processes	183
	2. Catalytic and asymmetric Staudinger/aza-Wittig reaction	184
IV.	The Catalytic Mitsunobu Reaction	186
	A. Introduction to the Mitsunobu Reaction	186
	B. Catalytic Mitsunobu Reactions *via* P^{III}/P^{V} Processes	188
	C. Catalytic Mitsunobu Reactions *via* P^{V}/P^{V} Processes	189
V.	Conclusion and Outlook	191
VI.	Acknowledgments	192
VII.	References	192

List of Abbreviations

CWR	catalytic Wittig reaction
DAAD	dialkyl acetylenedicarboxylate
DFT	density functional theory
DIAD	diisopropyl azodicarboxylate
DMSO	dimethyl sulfoxide
PMHS	polymethylhydrosiloxane
RDS	rate-determining step
TMDS	tetramethyldisiloxane

I. Introduction

Phosphines play a key role as reagents in a variety of impactful transformations such as the Wittig, Staudinger, Mitsunobu reactions, and many others.[1] Despite their usefulness, these reactions possess a major drawback, namely the concomitant formation of a stoichiometric quantity of phosphine oxide, which often complicates the purification in large-scale processes and globally decreases the atom economy in these reactions. The formation of the thermodynamically favored P=O bond is at the same time the driving force of these reactions and their Achille's heel.

In view of an optimal synthetic approach, it was envisaged to simplify the purification step of these processes by using several methodologies, that is, among others: (a) the use of water-soluble phosphines,[2] (b) the use of polymer-supported phosphines,[3] and (c) the use of fluorinated phosphines and rapid solid phase extraction over fluorous silica.[4] In addition to all these, with the aim of developing more environmentally friendly methods, a substoichiometric amount of phosphine should be ideally employed. To this end, in the last decade, catalytic strategies based on P^V/P^V redox neutral process involving activation of phosphine oxides with isocyanates or oxalyl chloride[5] and a strategy involving *in situ* reduction of the phosphine oxide (P^{III}/P^V redox catalysis) have been used worldwide by several research groups. Consequently, with the latter strategy, it is now possible to render catalytic in phosphine many phosphine-promoted transformations, with the use of silanes as *in situ* reducing agents.[6] In addition to simplify the purification process, compared to transformations involving stoichiometric amounts of phosphines, these methodologies now allow the development of catalytic and asymmetric reactions.

In this chapter, the catalytic Wittig reaction (CWR) and the general requirements to develop P^{III}/P^V catalytic processes will be presented first. In the last two sections, (asymmetric) tandem transformations and the catalytic Mitsunobu reaction will be summarized.

II. Toward the Development of Enantioselective CWRs

A. Catalytic Wittig Reactions

1. *Initial studies and prerequisites for the development of phosphine-catalyzed P^{III}/P^V redox reactions*

Among the phosphine-promoted transformations, the Wittig reaction, discovered in 1953 by Georg Wittig,[7] is one of the most venerable methods for the synthesis of carbon–carbon double bonds from aldehydes or ketones. For this olefination reaction, Wittig received the Nobel Prize in 1979 for the "development of the use of phosphorus compounds as an important reagent in organic synthesis." Among the outstanding applications of this reaction in industry, the large-scale synthesis of vitamin A is especially remarkable.[8] However, this reaction suffers from some drawbacks, the principal one being the formation of stoichiometric amounts of phosphine oxide waste. It raises therefore the general issue of the regeneration of phosphines by *in situ* reduction of phosphine oxides, and the related chemoselectivity concerns.

The mechanism of the CWR can be divided into the following elementary steps: (i) formation of a phosphonium salt by addition of the trivalent phosphine to the halide **A**; (ii) deprotonation of intermediate **B** with a base to generate the phosphonium ylide **C**, followed by (iii) reaction with an aldehyde to furnish the olefin and the phosphine oxide as by-product. Finally, (iv) the phosphine should be regenerated in the presence of a reducing agent, by *in situ* reduction of the phosphine oxide (Scheme 1).

The first Wittig reaction involving a substoichiometric amount of phosphine was developed in 2009 by C. J. O'Brien and coworkers by using 10 mol% of 3-methyl-1-phenylphospholane 1-oxide **P1** (see Scheme 2) as precatalyst, diphenylsilane as reducing agent, and sodium carbonate as inorganic base for the *in situ* formation of the phosphorus

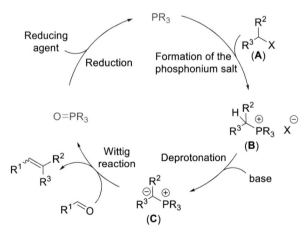

Scheme 1. Proposed mechanism for CWR.

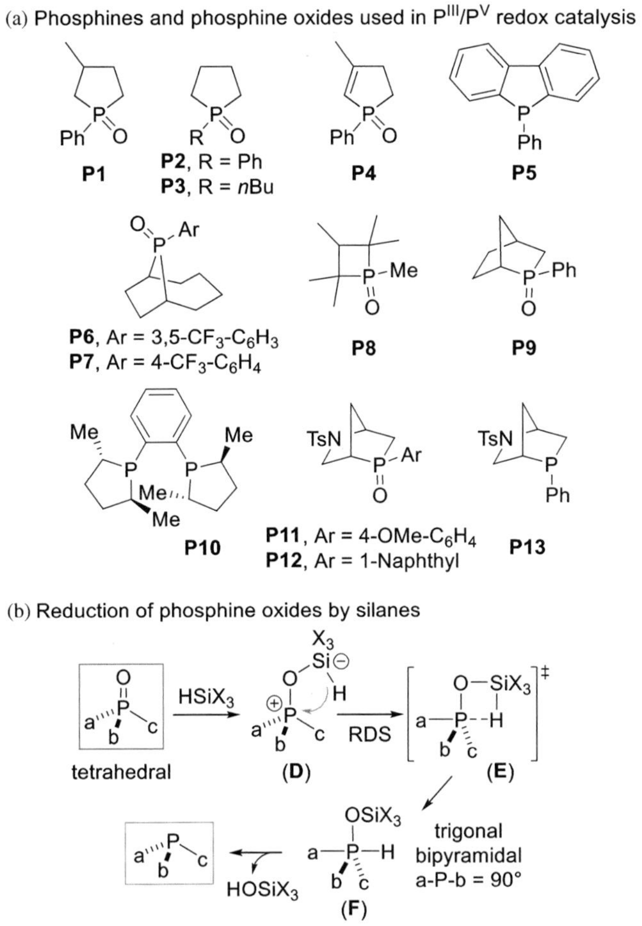

Scheme 2. (a) Phosphines and phosphine oxides for P^{III}/P^V redox catalysis and (b) postulated reduction mechanism.

ylide.[9] After 24 h in toluene at 100°C, the desired olefinic derivatives were isolated in overall good yields and moderate to excellent *E/Z* isomeric *ratios* (Scheme 3). The reaction applies to both aromatic and aliphatic aldehydes and to bromoalkyl derivatives with at least one electron-deficient substituent (ester, arylketone, or cyano derivatives). This first catalytic transformation is a cornerstone toward the development of more sustainable phosphine-mediated processes and gives an opportunity to the

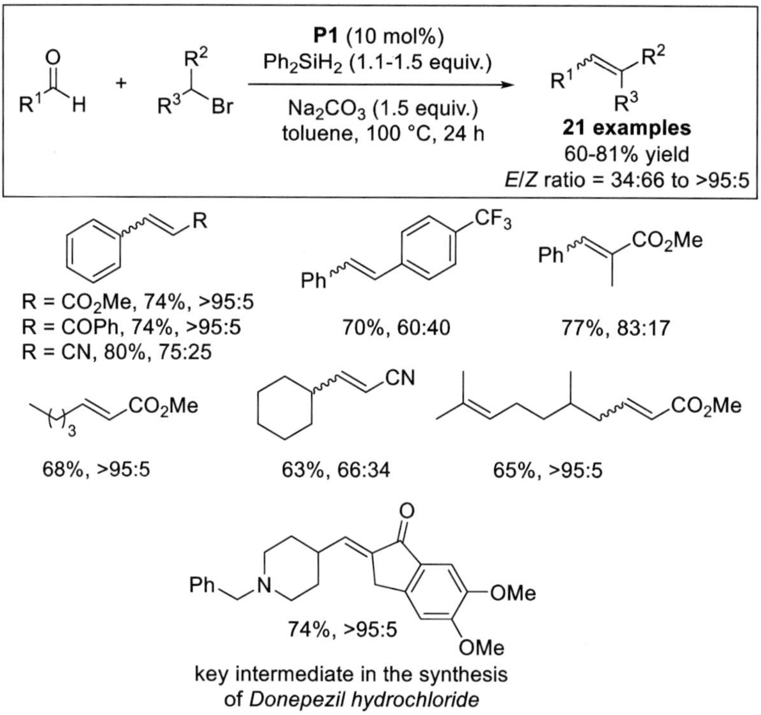

Scheme 3. Development of the first CWR.[9]

organic chemist community to imagine synthetic strategies using Wittig reactions, without concerns about the separation of the phosphine oxide from the desired products. As a representative example, a key intermediate in the synthesis of *donepezil hydrochloride*, a drug used in the palliative treatment of mild Alzheimer's diseases, has been synthesized in 74% yield on multidecagram scale.

These initial results highlight some guidelines to develop new catalytic processes. Three major items, that is, the phosphine, the reducing agent, and the additives have to be finely tuned and chosen to successfully reach this goal:

(1) Concerning the phosphine, the right balance has to be found between the nucleophilicity of the trivalent phosphine (to ensure its alkylation) and the facility to reduce the corresponding phosphine oxide (for a

catalytic process). Representative examples of phosphine oxides used in P^{III}/P^{V} redox catalysis are shown in Scheme 2(a). On one hand, acyclic electron-rich phosphine, such as tributylphosphine, might possess good nucleophilicity, but the corresponding phosphine oxides are very difficult to reduce. On the other hand, cyclic phosphines, such as phospholanes **P1-3**, phospholene **P4,** or dibenzophosphole **P5**, possess quite good nucleophilicity and the corresponding phosphine oxides are more easily reducible by silanes than acyclic phosphine oxides.

To understand the ease by which cyclic phosphine oxides can be reduced, the accepted mechanism for the reduction step must be considered.[10] It has been calculated by density functional theory (DFT) that the rate-determining step (RDS) of the reduction is the hydride transfer from the silicon atom to the phosphorus center.[10e] In this step, the pseudo-tetrahedral phosphorus atom of intermediate **D** becomes trigonal bipyramidal (Intermediate **F**, Scheme 2(b)). The energetic barrier related to this geometrical transformation is lowered by the adequate bond angles at phosphorus imposed by the cyclic backbone in the five-membered rings (a-P-b angles close to the ideal 90°). This is why research groups involved in this field have generally chosen these cyclic phosphines (including chiral phosphines) because they represent the ideal balance between nucleophilicity and facility to realize the P^{III}/P^{V} redox cycling. As a representative example, it can be mentioned that O. Kwon focused her studies on a series of chiral phosphabicyclo[2.2.1]heptanes derived from L-hydroxyproline. Phosphines **P11-13** display a-P-b angles of about 93° and are easily reducible in mild conditions (20–40°C).[11]

(2) Regarding to the reducing agent, the best compromise between reactivity and chemoselectivity was found with the use of silanes. Indeed, if we consider the Wittig reaction, the presence in the reaction mixture of both aldehydes, bromoalkyls, olefins, and phosphine oxide raises the key question of the chemoselectivity of the reduction. This prevents the use of strong reducing agents such as aluminum hydrides.[12] Therefore, phenylsilane ($PhSiH_3$) and diphenylsilane (Ph_2SiH_2) are the most used reducing agents because they possess this key feature of being selective for the reduction of the P=O bond. It is also possible to use tetramethyldisiloxane (TMDS) or polymethylhydrosiloxane

(PMHS) as reductants,[13] but their polymeric structure decreases their reactivity, resulting in the necessity to heat the reaction medium or to use additives.

(3) To facilitate the reduction of phosphine oxides and thus the whole catalytic process, some additives have been considered. Either organometallic complexes, such as Ti(OiPr)$_4$,[14] or organic compounds can be used to activate the reducing agent. Due to sustainable chemistry and efficiency concerns, most of the time substoichiometric amounts of organic additives such as carboxylic acids (PhCO$_2$H and pNO$_2$-C$_6$H$_4$CO$_2$H)[15] or bis(4-nitrophenyl)phosphate [(pNO$_2$-C$_6$H$_4$O)$_2$PO$_2$H][16] were used. This last additive induces *in situ* formation of a bifunctional silyl–phosphate catalyst, which simultaneously activates the silane and the phosphine oxide. This explains the beneficial effect of such additive.

After the development of the first CWR that provided key knowledge on the prerequisites for catalytic processes, improvements were made and the reaction scope was broadened significantly.

2. *Further optimization and development of CWR*

Following the pioneering work above, several improvements of the CWR have been disclosed by the O'Brien group. From a practical and sustainable point of view, heating the reaction mixture at 100°C for 24 h is quite inappropriate, and this is the reason why a room temperature CWR was developed later, thanks to the use of additives. Indeed, O'Brien and coworkers realized that the use of "aged" benzaldehyde, contaminated with trace amounts of benzoic acid, promotes the phosphine oxide reduction step, which was in most of the case the RDS of the whole process.[15] A simple comparison of reactions with or without benzoic acid showed a drastic increase of the conversion rates, from <30% to >70%. As an additional effect, this additive also potentially increases the solubility of the phosphonium salt intermediate, **B** in Scheme 1. In the presence of catalytic amounts of both 1-butylpholane 1-oxide **P3** and 4-nitrobenzoic acid, many different olefination reactions occurred at room temperature (16 examples, 61–91% yield, up to >95/5 isomeric *ratio*, Scheme 4(a)).

In further studies, to widely open the scope of the CWRs, substrates that generate stabilized ylides (pK_a = 8–12 in DMSO, see Scheme 3) have

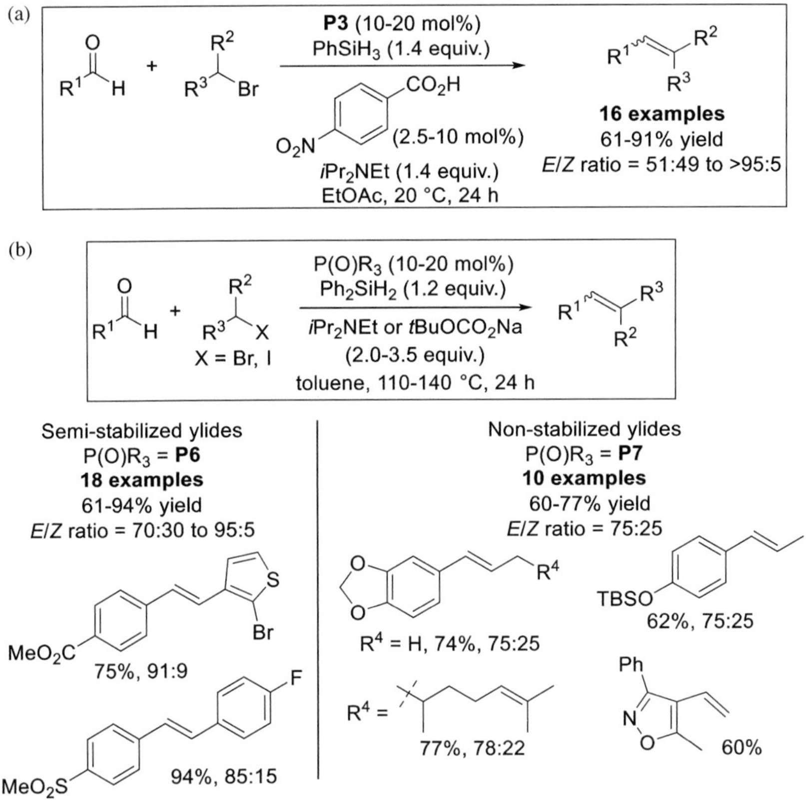

Scheme 4. (a)[15] Room temperature CWRs and (b) reactions involving semi- and nonstabilized ylides.[17]

been replaced by substrates leading to semistabilized (pK_a = 16–18 in DMSO) and nonstabilized (pK_a = 22–25 in DMSO) ylides.[17] In order to make possible such trickier Wittig reactions, two improvements have been made, namely the use of electron-poor bicyclic phosphines such as **P6** and **P7**, and the application of a masked base, sodium *tert*-butyl carbonate. This carbonate slowly delivers *in situ* sodium *tert*-butoxide (*t*-BuONa, pK_a = 20 in DMSO), which enables the generation of nonstabilized phosphorus ylides and the subsequent Wittig reactions, whereas the stronger bases usually employed in the literature, such as sodium *bis*(trimethylsilyl)amide (NaHMDS) or *n*-butyllithium, would not be compatible with the presence of organosilanes or alkylhalides in the reaction mixture. On the other side, the use of bicyclic phosphines where the phosphorus atom bears electron-deficient aromatic groups, such as 3,5-*bis*(trifluoromethyl)phenyl and 4-(trifluoromethyl)phenyl substituents, allowed to decrease the pK_a value of the ylide precursors **B** (Scheme 1). Finally, the fine tuning of the phosphine oxide and the reaction conditions allowed the synthesis of variously substituted olefins, starting either from benzyl bromides (18 examples, 61–94% yield, up to 95/5: *E*/*Z* ratio, Scheme 4(b), left) or from alkyl iodides (10 examples, 60–77% yield, 75/25: *E*/*Z* ratio, Scheme 4(b), right).

Another possible drawback of this CWR is the pretty high catalytic loading (i.e. 10 mol %) that is required. To answer this concern, Werner and coworkers used a phosphetane oxide precatalyst, **P8** (Scheme 2(a)). This four-membered phosphacycle, whose use had been popularized by Radosevich's group in other P(III)/P(V) redox cycling catalytic transformations,[18] proved to give outstanding results in the CWR at low catalyst loading, working even at 25°C (25 examples, 54–97% yield, up to >99/1: *E*/*Z* ratio, Scheme 5).[19] Only 2 mol % of phosphine was utilized, without the need of Brønsted acid additives such as carboxylic acids. Moreover, the reaction could be performed starting from alkyl chlorides, bromides, and iodides. In this process, the RDS proved to be the reduction of the phosphetane oxide.

In 2015, Werner and coworkers developed an interesting olefination method that does not require the use of a base to *in situ* form phosphorus ylides. This reaction, which can be seen as a "base free Wittig reaction,"

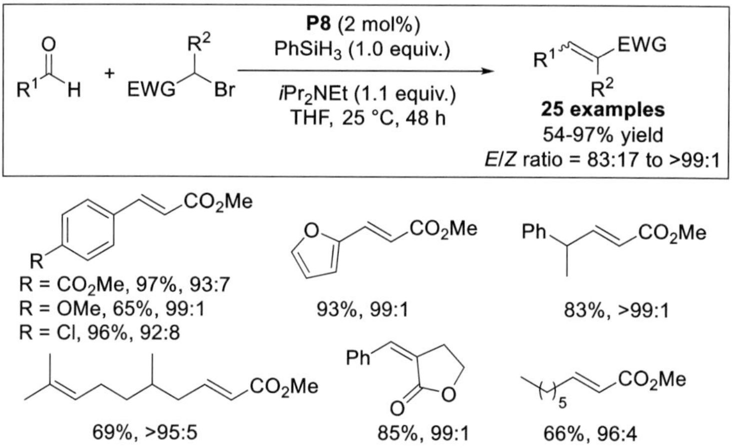

Scheme 5. CWRs promoted by phosphetane oxide **P8** at low catalytic loading.[19]

starts with the addition of the trivalent phosphine to a Michael acceptor, such as diethyl maleate **G**, to furnish the zwitterionic species **H**. After a [1,2]-*H*-shift, an ylide **I** is formed and react with diverse aldehydes *via* a Wittig process, to form the desired olefin and phosphine oxide as a by-product (Scheme 6(a)). The use of a silane as reducing agent allows to recover the trivalent phosphine. Overall, this catalytic process proved to

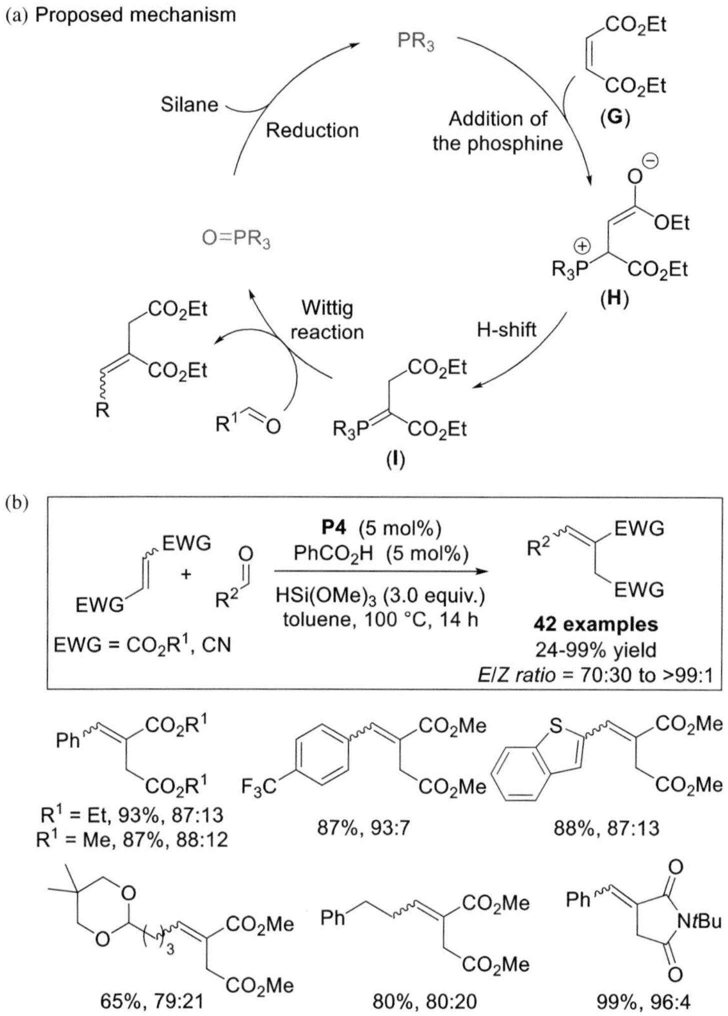

Scheme 6. Base free CWR.[20]

be very efficient for the synthesis of substituted succinate derivatives (42 examples, up to 99% yield, *E/Z ratio*: 70/30 to >99/1, Scheme 6(b)).[20] The optimal reaction conditions employed the phosphole catalyst **P4** (5 mol%), in the presence of trimethoxysilane and catalytic amounts of benzoic acid to facilitate the reduction of the phosphine oxide.

B. Asymmetric Wittig Reactions

Asymmetric Wittig-type transformations *via* desymmetrization of carbonyl derivatives is a long-standing quest in organic synthesis. To develop such processes, the first approaches involved the use of either chiral substrates or stoichiometric quantities of chiral phosphorus compounds. Some work in this field will be detailed in the next paragraph. Later on, as we have just mentioned throughout the preceding paragraph, many solutions have been developed in the literature in order to establish new standards in the CWR. In particular, although it is hardly imaginable to develop efficient asymmetric reactions with large quantities of very expensive chiral phosphines, the possibility to use substoichiometric amounts of the phosphorus reagent opens now the way to the development of catalytic and enantioselective processes.

1. *Asymmetric Wittig–type olefination reactions using stoichiometric amounts of phosphorus reagents*

Different strategies have been proposed since the 1960s for the development of asymmetric Wittig–type reactions,[21] such as the reaction of chiral phosphorus ylides or chiral ketones, the kinetic resolution of either racemic ylides or racemic carbonyl compounds, and the desymmetrization of prochiral dicarbonyl substrates by means of chiral phosphorus reagents. In the context of this chapter, we will specifically develop the most significant results of the latter strategy.

In 1980, Trost and Curran developed the first asymmetric Wittig reaction *via* the desymmetrization of the 2-methylcyclopentane-1,3-dione in Scheme 7(a), with the use of stoichiometric amounts of chiral trivalent phosphines (Scheme 7(a)).[22] In this intramolecular reaction, the best result in terms of enantiomeric excess (ee) was obtained with the

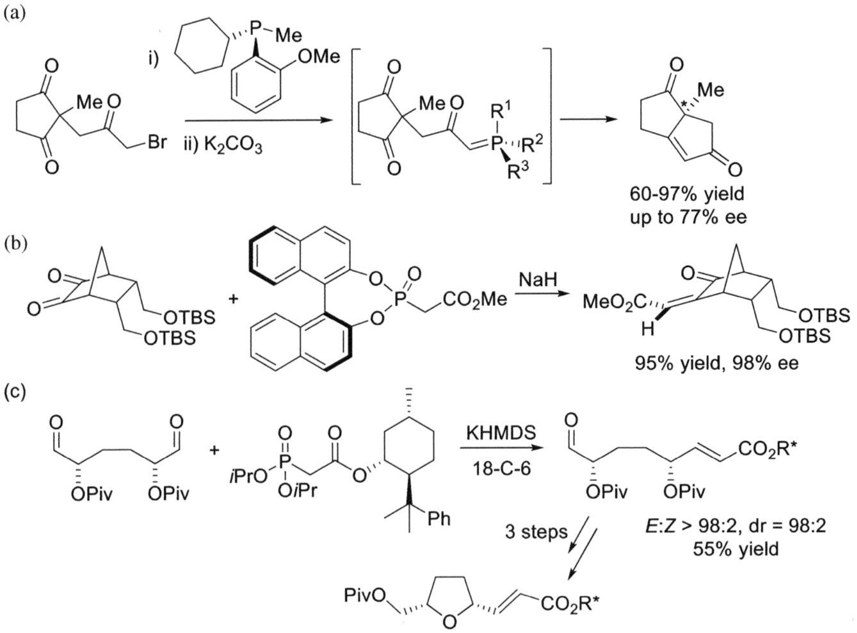

Scheme 7. Asymmetric olefinations with stoichiometric amounts of chiral phosphorus reagents: (a)[22] Wittig reactions and (b)[23] and (c) Horner–Wadsworth–Emmons–type reactions.[24]

cyclohexyl-*O*-anisyl-methyl phosphine (*S*)-CAMP, with up to 77% ee at −10°C. The enantioenriched bicyclic product is a key building block in the total synthesis of natural products.

After these preliminary results in intramolecular Wittig olefination reactions, the desymmetrization of prochiral ketones was investigated also *via* Horner–Wadsworth–Emmons–type olefinations. Thus, for instance, stoichiometric amounts of a chiral binaphthol–derived phosphonate were reacted with the bicycle [2.2.1] heptane-2,3-dione shown in Scheme 7(b). After *in situ* formation of the carbanionic intermediate, the corresponding Z-olefin was isolated in 95% yield and 98% ee.[23]

The asymmetric Horner–Wadsworth–Emmons reaction has been then extended by Rein and coworkers to the olefination of the *meso*-dialdehyde in Scheme 7(c), for the preparation of tetrahydrofuran derivatives. The mono-olefination of the *bis*-pivalic ester of 2,5-dihydroxyhexanedial

(Piv = pivaloyl = C(O)*t*Bu) gave the desired product in 55% yield, with almost complete *E* selectivity and a 98:2 diastereoisomeric *ratio*. Consecutively, the enantioenriched 2,5-*cis*-disubstituted tetrahydrofuran derivative was isolated after three additional steps.[24]

2. *Enantioselective CWRs*

In 2014, Werner and coworkers developed the first enantioselective catalytic Wittig–type transformation (Scheme 8).[25] The reaction starts with the *in situ* formation of the phosphonium salt of chiral phosphines and the corresponding ylide after deprotonation with either sodium carbonate or butylene oxide. The ylide will then react preferentially by intramolecular Wittig reaction with one of the two prochiral ketone functions, thus revealing a quaternary stereogenic center in the product. The phosphine oxides produced by the Wittig reaction are reduced then *in situ* with silanes. Unfortunately, this catalytic desymmetrization of prochiral diketones did not allow to obtain, at the same time, decent yields and ees. Indeed, using conventional heating (conditions **A**), the best result in terms of ee was obtained with (*S*, *S*)-Me-DuPhos **P10**. Even if the low conversion rate (<10% NMR yield) reminds more of a stoichiometric than a catalytic reaction, the tetrahydropentalene-1,5-dione derivative was obtained in about 90% ee. More interestingly, using microwave conditions in dioxane at 150°C for 2 h, with phenylsilane as reducing agent and butylene oxide as masked base, 39% isolated yield and 62% ee were obtained. This result paved the way to other work in

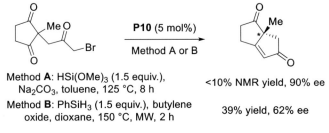

Scheme 8. The first asymmetric CWR.[25]

the field of P(III)/P(V) redox catalysis that will be more efficient in terms of both yields and enantioselectivities, particularly in the case of tandem processes.

To allow access to chiral compounds with excellent enantioselectivities and decent yields, the chiral phosphine must have several important properties. First, the chiral phosphine must possess a good nucleophilicity and to be reactive enough with the substrates engaged in the different transformations and the corresponding phosphine oxide P(V) should be easily reduced in order to regenerate the catalyst in its P(III) active form. Furthermore, the chiral catalyst must be easily modulated, with different substituents having different electronic and/or steric properties in order to optimize the chiral induction. The consideration of all these criteria, including the fine optimization of the reaction conditions and the design of new chiral phosphines, should allow in the future bringing this new catalytic methodology into the toolbox of the organic chemist.

3. *Application in total synthesis*

One of the most convincing ways to validate the robustness of a catalytic methodology is the application in total synthesis. Thus, after careful optimization of the reaction conditions, Christmann, Tantillo, and coworkers used the asymmetric and catalytic methodology developed by Werner in the synthesis of two natural products, the dichrocephones A and B (Scheme 9).[26,27]

The key step in the synthesis is a CWR on a prochiral diketone, by which the first quaternary stereogenic center was installed. The optimal conditions used catalytic amounts of (R, R)-Me-Duphos **P10** as the chiral phosphine, phenylsilane as reducing agent, and butylene oxide as masked base. After 20 h at 150°C in a sealed tube, the product was isolated in 60% yield and 96% ee. After vinyl cuprate addition and ring-closing metathesis, the propellane core was formed in very good yield.[26] Then, several synthetic steps allowed to isolate the targeted compound. However, the NMR data of the synthetic sample did not match with those of the natural product. Thus, suspecting an incorrect assignment of the natural product, the authors reversed the stereochemistry of one of the stereogenic centers

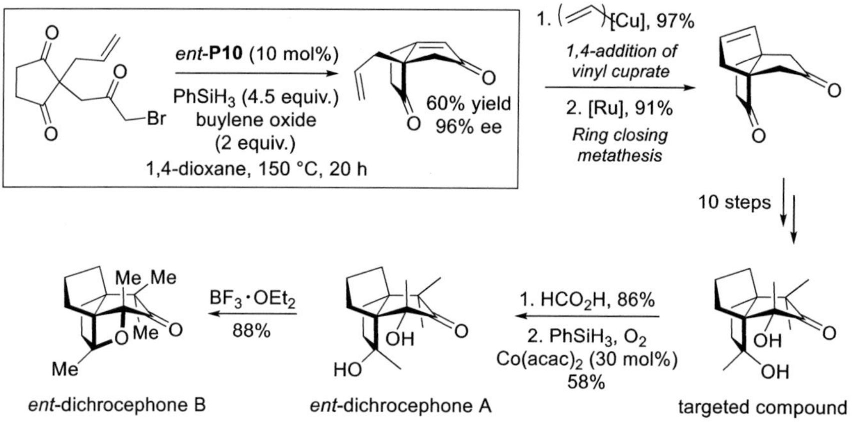

Scheme 9. Total synthesis of dichrocephones A and B *via* a catalytic enantioselective Wittig reaction.[26]

and isolated the other diastereoisomer, which turned out to be indeed the natural product, dichrocephone A. In only one additional step, it was possible to isolate another natural product, dichrocephone B. Thus the catalytic and asymmetric Wittig reaction has been used elegantly in total synthesis and the work shown here demonstrates a certain maturity of this methodology, which will certainly be used again in the future in the synthesis of natural products.

III. Tandem Reactions Including Wittig and aza-Wittig Reactions

After the discovery and the exemplification of catalytic Wittig olefinations involving phosphonium salts as the ylide precursors, this chapter will explore the development of new tandem reactions including Wittig olefinations as the final chemical step, in which, however, the phosphorus ylides are formed *via* either Michael additions or Staudinger reactions. A tandem reaction is defined as a process that includes at least two successive transformations, forming at least two new bonds, and so that further transformations cannot take place without the previous step. To do this, no additional reagents and/or catalyst must be added during the reaction.

Hereafter, we will present first Michael addition/CWR sequences, including the catalytic and asymmetric version of this reaction, before switching to tandem Staudinger/aza-Wittig reactions.

A. Tandem Michael Addition/Wittig Olefination

1. *A catalytic tandem reaction*

The tandem Michael addition/Wittig olefinations using stoichiometric amounts of phosphines have been described in the litterature in the 1960s. These reactions have been designed initially as suitable alternatives to classical Wittig olefinations for the synthesis of 3*H*-pyrrolizines. Initially, Schweizer and Light disclosed the synthesis of 3*H*-pyrrolizines from 1*H*-pyrrole-2-carbaldehyde and vinylphosphonium salts, in the presence of sodium hydride.[28] Later on, in order to prevent the use of stoichiometric amounts of base and alkyl halides for the preparation of phosphorus ylides, Yavari and coworkers have cleverly developed a tandem reaction for the formation of fused pyrrolizines (pyrrolo[1,2-*a*]indoles), which involves indole-2-carboxaldehydes and dialkyl acetylenedicarboxylate (DAAD) as starting materials, in the presence of triphenylphosphine (Scheme 10(a)).[29] According to the proposed mechanism (Scheme 10(b)), the tandem reaction starts with the addition of the phosphine to the DAAD substrate to form the zwitterionic species **J**. This intermediate subsequently deprotonates indole-2-carboxaldehyde, to give the vinylphosphonium salt **K**. Addition of the conjugate base of the indole to **K** generates *in situ* a phosphorus ylide **L**, which accomplishes an intramolecular Wittig olefination to form, after migration of the double bond, the final product **M** and triphenylphosphine oxide.

To avoid the use of huge amounts of triphenylphosphine and to simplify the purification process, Voituriez and coworkers carried out this transformation using only 5 mol% of phosphine **P4**, thanks to the *in situ* reduction of phosphine oxide with phenylsilane (Scheme 10(c)).[30] This catalytic protocol was applied to the synthesis of a wide range of pyrrolo[1,2-*a*]indoles (14 examples, 70–98% yield) and pyrrolizines (4 examples, 83–96% yield).

Interestingly, the scope of this catalytic methodology could be extended to the synthesis of many nitrogen-containing heterocycles

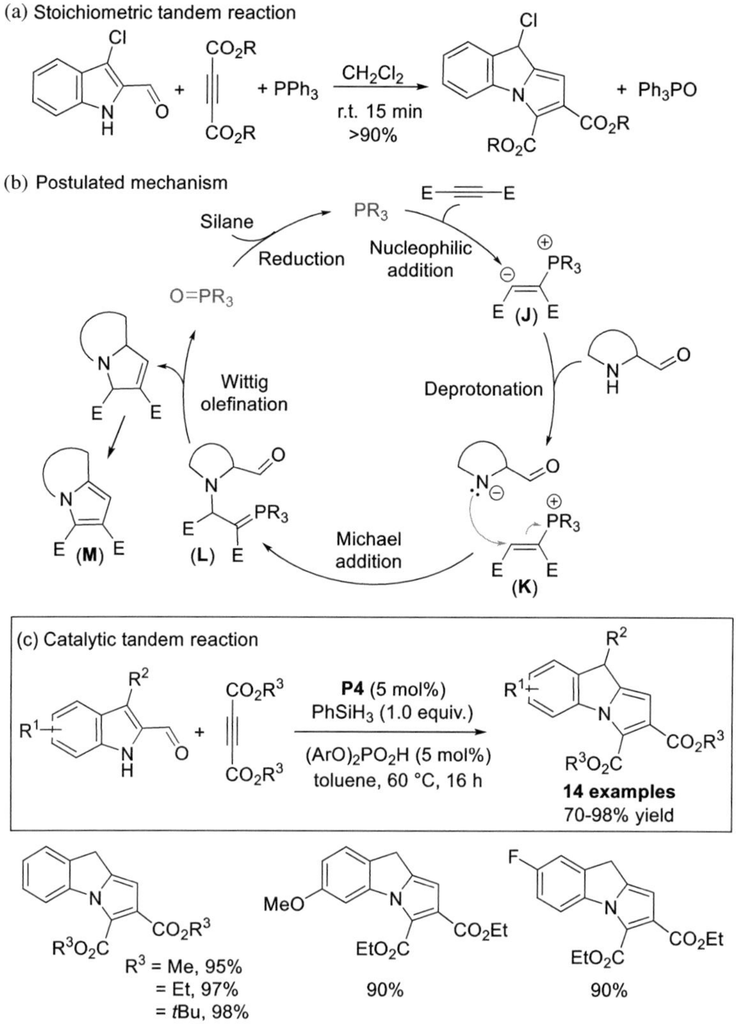

Scheme 10. Tandem aza-Michael addition/Wittig reaction: (a) stoichiometric reaction,[29] (b) postulated mechanism, and (c) catalytic variant.[30]

(Scheme 11).[31] Indeed, the generalization of this tandem process requires only the use of substrates having both a function that can be deprotonated (here an N–H function) and a carbonyl function, such as an aldehyde or an ester, for the intramolecular Wittig reaction. With 5 mol% of precatalyst **P4**, 5 mol% of *bis*(4-nitrophenyl)phosphate to facilitate the reduction

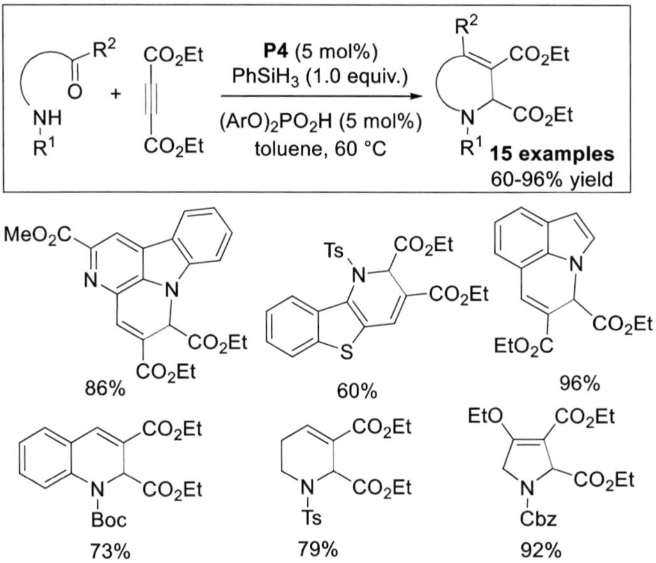

Scheme 11. Synthesis of nitrogen-containing heterocycles *via* catalytic aza-Michael–Wittig tandem reactions.[31]

of the phosphine oxide with phenylsilane, and the DAAD reagent, many amino-carbaldehydes and amino-esters were cyclized into the corresponding 4*H*-pyrrolo[3,2,1-*ij*]quinolone, 4*H*-indolo[3,2,1-*de*][1,5]naphthyridine, tetrahydropyridine, and 1,2-dihydroquinoline derivatives (15 examples, 60–96% yield). In this study, the authors also attempted to develop the catalytic and asymmetric synthesis of 2,5-dihydro-1*H*-pyrrole derivatives, with the use of chiral phosphines. Good isolated yields were obtained; however, the enantioselectivity was unfortunately very low (<30% ee). Finally, the extension of this catalytic tandem reaction to the synthesis of carbocycles such as cyclopentenone derivatives has been also exemplified, starting from butan-2,3-dione or 1-phenylpropane-1,2-dione and DAAD.

This tandem reaction has been extended by Kwon and coworkers to other substrates than DAAD, namely allenoates (Scheme 12).[32] The mechanism remains similar to that mentioned in Scheme 10, except that the phosphine adds to the central carbon of the allenoate, generating a zwitterionic species that can be protonated by the sulfonamide. Subsequently,

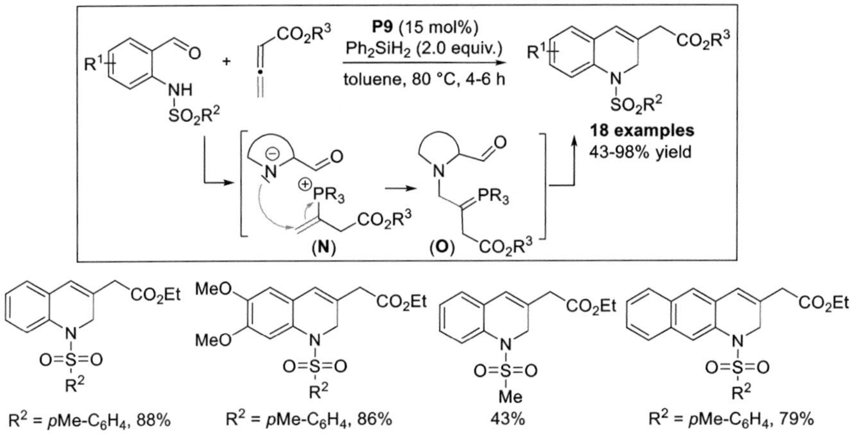

Scheme 12. Use of allenoates in a catalytic tandem aza-Michael–Wittig reaction.[32]

addition of the sulfonamide on the vinyl phosphonium salt **N** allows the *in situ* formation of a phosphorus ylide **O**, which can react in an intramolecular Wittig reaction to form the desired alkyl 2-(1,2-dihydroquinolin-3-yl) acetates (18 examples, 43–98% yield). In this work, Kwon and coworkers demonstrated that the newly developed bridged bicyclic phosphine oxide **P9** was easily reduced in the presence of diphenylsilane, without degradation of the allenoate. This reduction step is the RDS of this catalytic transformation, as often in phosphine redox catalysis.

Voituriez and coworkers have explored the analogous synthesis of pyrrolo[1,2-*a*]indoles from *1H*-indole-2-carbaldehydes with the use of allenoates as substrates. These reactions gave excellent yields using stoichiometric amounts of phosphine but only moderate results in P^{III}/P^V redox catalysis.[33]

2. Catalytic and asymmetric Michael addition/Wittig reactions

In 2019, Voituriez and coworkers developed the first catalytic and enantioselective tandem process including a Michael addition, followed by an intramolecular Wittig reaction.[34] This study focused on the reaction that converts DAAD and activated 1,3-dicarbonyl compounds, such as 4,4,4-trifluoro-1-arylbutane-1,3-diones,[29c,d] into cyclobutenes (Scheme 13). The

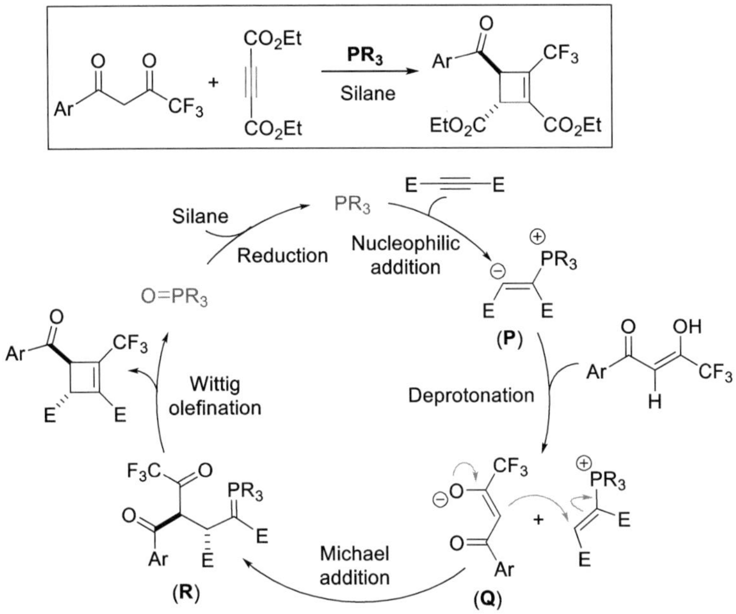

Scheme 13. Synthesis of cyclobutenes *via* tandem Michael–Wittig reactions.[29d,34]

reaction starts with the formation of the zwitterion **P** and its protonation by the 4,4,4-trifluoro-1-arylbutane-1,3-dione substrates. The resulting enolate **Q** adds on the vinylphosphonium salt to generate the ylide **R**. After a Wittig reaction, the chiral cyclobutenes are delivered and the phosphine oxide is reduced *in situ* by the silane to regenerate the phosphine catalyst.

The real challenge in this study was to find the ideal phosphine that will meet the criteria imposed by the development of both a redox catalytic cycle and an asymmetric transformation (Scheme 14). As already mentioned in Chapter II.B.2., for the enantioselective Wittig reaction, to find the right balance between reactivity, enantioselectivity, and facility to reduce the phosphine oxide was not obvious. The use of commercially available HypPhos chiral phosphines **P11-12** (Scheme 2),[11] developed by O. Kwon in 2014, contributed greatly to the success of this catalytic process. In the end, numerous (trifluoromethyl)cyclobutene derivatives have been synthesized starting from simple 4,4,4-trifluoro-1-arylbutane-1,3-diones and DAAD as substrates, with 5 mol% of chiral phosphine **P11**, arylphosphate, and phenylsilane as reducing agent (26 examples, 52–99% yield, up

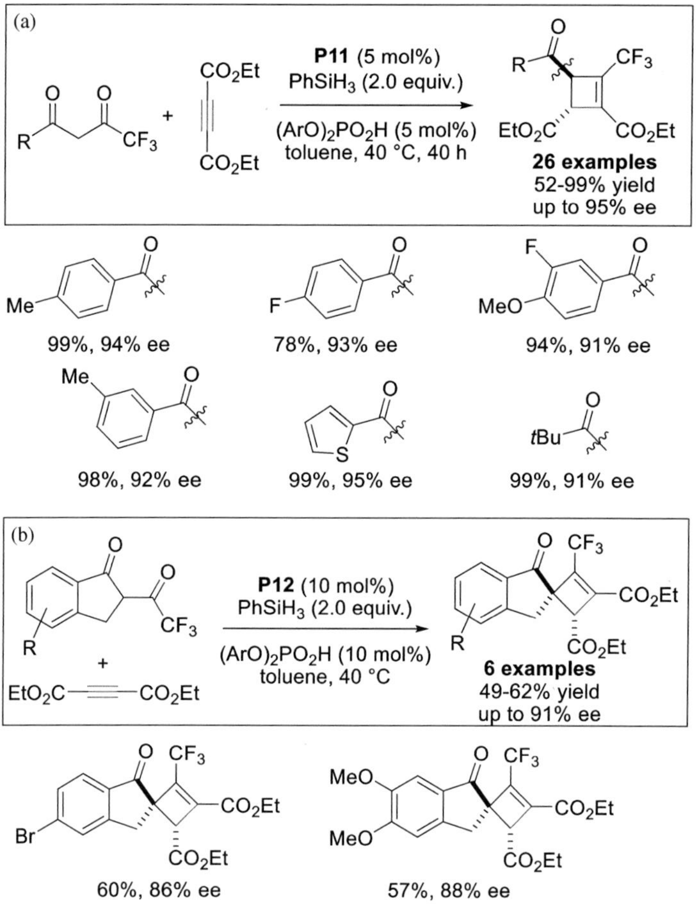

Scheme 14. Synthesis of fluorinated cyclobutenes *via* asymmetric redox P^{III}/P^V catalysis.[34]

to 95% ee, Scheme 14(a). It is noteworthy that both excellent yields and ees have been obtained on a large number of variously substituted substrates. It is now conceivable to isolate gram-scale quantities of enantioenriched cyclobutenes (3 mmol scale, 1.11 g, 99% yield, 90% ee), starting with only 2 mol% of **P11** (23 mg). Furthermore, the method has been applied to the synthesis of fluorinated spiro[3.4]octanones (6 examples, 49–62% yield, up to 91% ee, Scheme 14(b)). The synthesis of such chiral backbones is worthy of interest, especially for applications in medicinal chemistry and total synthesis.[35]

B. Tandem Staudinger/aza-Wittig Reaction

1. *Catalytic tandem processes*

In 2013, van Delft and coworkers developed a Staudinger/aza-Wittig tandem reaction using substoichiometric amounts of phosphines.[36] Starting from arylazides bearing a suitable ester function, these intramolecular reactions led to benzoxazoles (Scheme 15(a)). The postulated mechanism starts with the reaction of the trivalent phosphine with the azido function of the substrate, **S**, to form an iminophosphorane intermediate **T** (Scheme 15(b)). The iminophosphorane reacts then with the carbonyl function of the substrate *via* an aza-Wittig reaction to give the benzoxazoles. From the reactions in Scheme 15(a), the desired benzoxazoles were obtained in

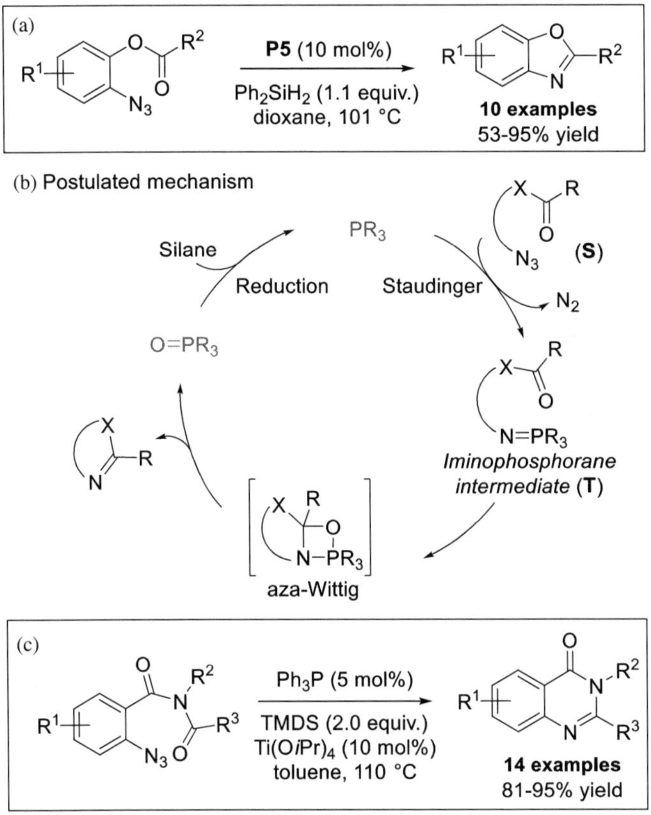

Scheme 15. Catalytic Staudinger/Aza-Wittig tandem transformations.[36,37]

high yields (10 examples, 53–95% yield). The appropriate phosphine for this transformation is the 5-phenyl-5H-benzo[b]phosphindole **P5**, in the presence of diphenylsilane as reducing agent, for reactions carried out at 101°C in dioxane. With this reaction conditions, the competitive Staudinger reduction of the azides RN_3 into primary amines (RNH_2) *via* the reduction of the iminophosphorane intermediate is decreased. By using the same tandem reaction, some benzodiazepines (4 examples, 60–88% yield) were also isolated.

In 2014, Ding and coworkers applied this transformation to the synthesis of quinazolinone derivatives (14 examples, 81–95% yield, Scheme 15(c)).[37] Interestingly, TMDS, a waste of the silicon industry, could be used as reducing agent, in combination with titanium isopropoxide [Ti(O*i*Pr)$_4$] as Lewis acid to activate *in situ* the phosphine oxide and overall facilitate its reduction to trivalent phosphine. For this reaction, only 5 mol% of triphenylphosphine has been used in toluene, at 110°C. The proof of concept of the feasibility of such catalytic tandem reactions opened the way to the development of asymmetric processes, which will be presented hereafter.

2. *Catalytic and asymmetric Staudinger/aza-Wittig reaction*

Before mentioning catalytic and asymmetric processes, it is important to recontextualize this original work in the general context of asymmetric aza-Wittig reactions. In 2006, Marsden and coworkers[38a] developed the first asymmetric aza-Wittig process, *via* the formation of a chiral iminophosphorane intermediate from enantiopure phosphorus derivatives (Scheme 16). Starting from prochiral 1,3-dicarbonyls with pendant azide functions, in the presence of 1.2 equivalents of a chiral oxazaphospholidine or a diazaphospholidine, the corresponding hexahydroquinolinones were isolated after *N*-acetyl protection. This desymmetrization process occurred in good yields and up to 60% ee. Later, the enantioselectivity has been increased to 84% ee with the use of a *P*-stereogenic phosphine.[38b]

Very interestingly, Kwon and coworkers developed in 2019 the catalytic and asymmetric desymmetrization of prochiral azido-1,3-diketones *via* a P^{III}/P^{V} process (Scheme 17).[39] In the presence of 20 mol% of the chiral phosphine *endo*-phenyl-HypPhos **P13**, phenylsilane as reducing

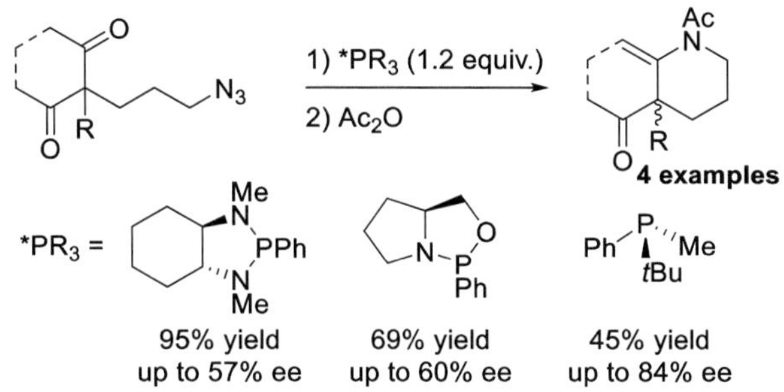

Scheme 16. Asymmetric Staudinger-aza-Wittig tandem reactions with stoichiometric amounts of phosphorus reagents.[38]

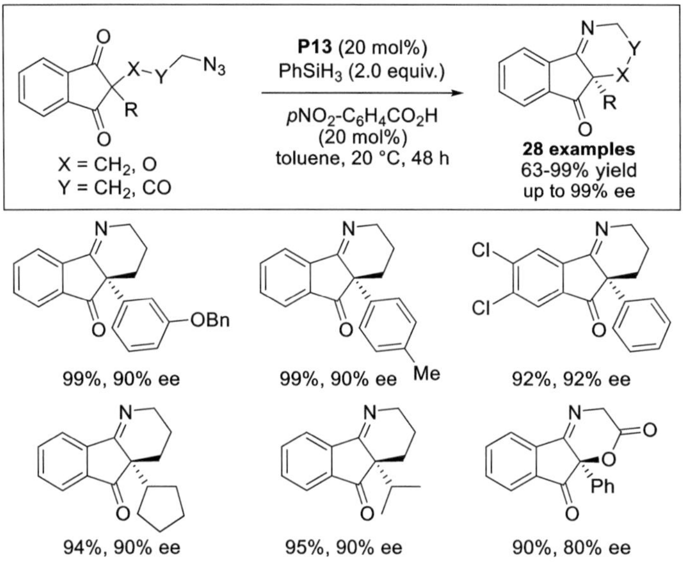

Scheme 17. Catalytic and enantioselective Staudinger-aza-Wittig tandem reaction, via a P^{III}/P^V redox process.[39]

agent and 2-nitro-benzoic acid as additive in toluene at room temperature, various nitrogen-containing heterocycles were isolated (28 examples, 63–99% yield, up to 99% ee). These structures of interest display indanopiperidine, dihydrooxazinone, and dihydrooxazine backbones that are frequently encountered in bioactive molecules. Mechanistic investigations by DFT calculation allowed to rationalize the origin of the enantioselectivity in this tandem process and to highlight the major role of the carboxylic acid additive both in accelerating the reduction of phosphine oxide, and in increasing the enantioselectivity induced by the chiral phosphine **P13**.

Overall, the recent concomitant development of these and other highly enantioselective tandem processes is expected to give a boost to the field and gives a glimpse of the possible future developments in asymmetric P^{III}/P^{V} catalysis.

IV. The Catalytic Mitsunobu Reaction

A. Introduction to the Mitsunobu Reaction

The Mitsunobu reaction has been discovered in 1967.[40] This substitution reaction allows to convert primary or secondary alcohols into other functionalized compounds (amines, esters, ethers etc.) by substitution with a nucleophile (Scheme 18). Among all suitable pronucleophiles, (thio)carboxylic acids, (thio)phenols, functionalized amines, and sulfonamides can be mentioned, allowing notably the formation of C–O, C–N, and C–S bonds. This reaction would be performed ideally by reacting an alcohol

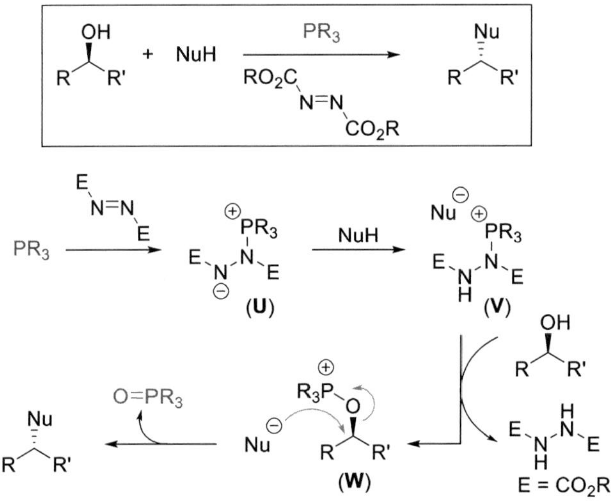

Scheme 18. Mechanism of the Mitsunobu reaction.

and a nucleophile, with the formation of water as the only byproduct. In reality, for both kinetic and thermodynamic reasons, this transformation needs prior activation of the alcohol function. Indeed, the nucleophilic substitutions on alcohols with different pronucleophiles (NuH), promoted by trivalent phosphines, require the use of dialkyl azodicarboxylate reagents. More specifically, this transformation begins with a nucleophilic addition of the trivalent phosphine on the dialkyl azodicarboxylate to generate a zwitterionic species **U**, called the "Huisgen zwitterion" (Scheme 18). The latter can act as a base to deprotonate the pronucleophile NuH, which most often possesses a $pk_a < 15$, to form the intermediate **V**. After that, the addition of the alcohol on **V** gives the phosphonium salt **W**. Finally, this intermediate reacts with the nucleophile *via* a bimolecular nucleophilic substitution (S_N2) to form the C–Nu bond with complete inversion of stereochemistry of the stereogenic center.

Thus, the Mitsunobu reaction is an important tool in organic chemistry to create C–Nu bonds with a good stereoselectivity under mild, neutral conditions, and typically at room temperature. However, this efficient transformation suffers from a major drawback, namely the concomitant formation of phosphine oxide ($R_3P=O$) and hydrazine (E–NH–NH–E) as wastes, which complicates the purification and makes this process an atom-uneconomical reaction. Considering the additional safety hazards associated with dialkyl azodicarboxylates, this is why the Mitsunobu reaction is seldom used in process chemistry and manufacturing.[41] To solve these problems, several innovative reagents have been developed that can be easily removed by liquid–liquid or solid–liquid extractions to facilitate the purification step, but the ideal solution would be to develop a Mitsunobu reaction catalytic in both phosphine and azodicarboxylate reagents (E–N=N–E).[42] To date, the development of such fully catalytic process is, however, difficult to achieve because it involves the addition in the same flask of a reducing agent (reduction of the phosphine oxide to trivalent phosphine) and an oxidant (oxidation of E–NH–NH–E to E–N=N–E). Nevertheless, a few solutions have emerged in the literature to make this reaction either catalytic in phosphine or catalytic in azocarboxylate reagent. In the context of this chapter, only Mitsunobu reactions catalytic in phosphine will be developed.[43] For the purpose of using a phosphine catalyst in this transformation, it is necessary to distinguish the catalytic P^{III}/P^{V} redox processes from the recently reported catalytic P^{V}/P^{V} processes.

B. Catalytic Mitsunobu Reactions *via* P^{III}/P^{V} Processes

During the Mitsunobu transformation, stoichiometric amounts of phosphine oxide are formed. The first strategy to render this process catalytic in phosphine has been to develop a P^{III}/P^{V} redox process, with the use of an external reducing agent to *in situ* reduce the phosphine oxide to the active trivalent phosphine. The first catalytic Mitsunobu P^{III}/P^{V} redox process has been disclosed in 2010 in an O'Brien patent (Scheme 19(a)).[44] Starting from benzyl alcohol and 4-nitrobenzoic acid in the presence of diisopropyl azodicarboxylate (DIAD), 20 mol% of 3-methyl-1-phenylphospholane

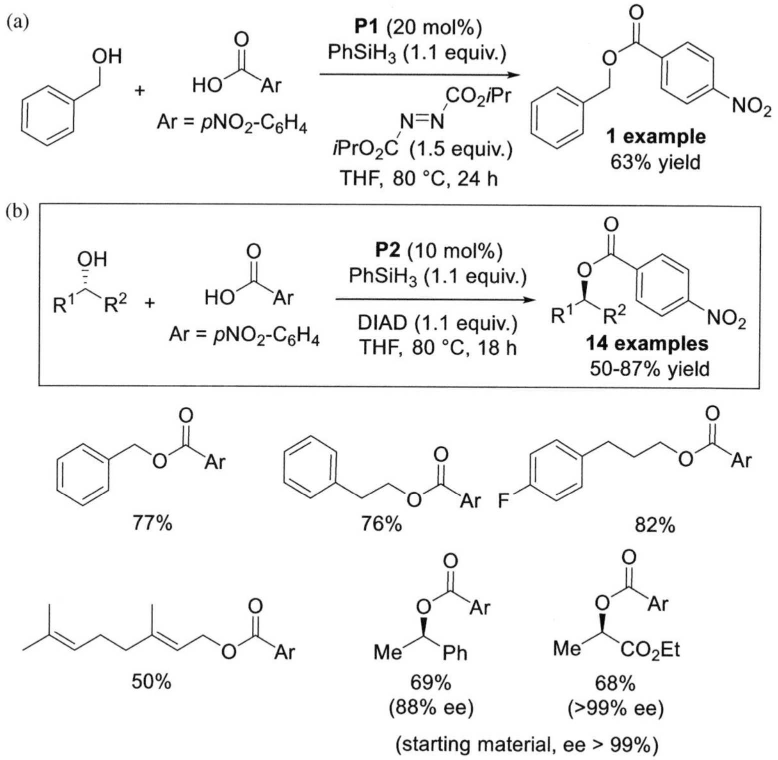

Scheme 19. Phosphine-catalyzed Mitsunobu reactions.[44,45]

1-oxide **P1** as precatalyst and phenylsilane as reducing agent, the target compound has been isolated in 63% yield.

In 2015, this pioneering work has been extended by Aldrich and Buonomo to other substrates including secondary chiral alcohols (14 examples, 50–87% yield, Scheme 19(b)). This led notably to the formation of enantioenriched 1-phenylethyl benzoate (88% ee) and 1-ethoxy-1-oxopropan-2-yl benzoate (99% ee) from the corresponding enantioenriched alcohols.[45] The inversion of the stereochemistry of the chiral secondary alcohols in the corresponding ester products was fully predictable and confirms the classical S_N2 mechanism of the Mitsunobu reaction, even using catalytic conditions. Unfortunately, and despite the initial claims of this study, the fully catalytic Mitsunobu transformation, catalytic both in phosphine and azodicarboxylate, still remains to be developed.[46] That said, thanks to this work it is now possible to easily access chiral esters *via* a Mitsunobu reaction, by using only 10 mol% of phosphine. However, such P^{III}/P^V redox processes require a stoichiometric quantity of reducing agent to regenerate *in situ* the catalyst. To overcome this, an alternative P^V/P^V-based approach has recently emerged in the literature, in which the oxidation state of phosphorus is invariant all through the catalytic cycle.

C. Catalytic Mitsunobu Reactions *via* P^V/P^V Processes

In 2019, an innovative method for the nucleophilic substitution reactions on alcohols was developed by Denton and coworkers.[47] Several primary and secondary alcohols have been converted into the corresponding esters and amides in the presence of 10 mol% of a phosphine oxide catalyst, with 35–97% yield and up to 99% ee (Scheme 20(a)). The originality of this strategy is based on the design of the phosphorus catalyst, which is a phosphine oxide bearing a phenol function. This phosphine oxide is activated *in situ* by the acidic pronucleophile (mainly NuH = 2,4-dinitrobenzoic acid), to undergo a cyclization/dehydration step (Scheme 20(b)). The corresponding oxyphosphonium salt **Y** is generated and undergoes then addition of the alcohol to form an alkoxyphosphonium/nucleophile ion pair **Z** (Scheme 20(b)).

Here, **Z** is the classical intermediate encountered in the Mitsunobu reaction and the subsequent nucleophilic substitution (S_N2) leads to the

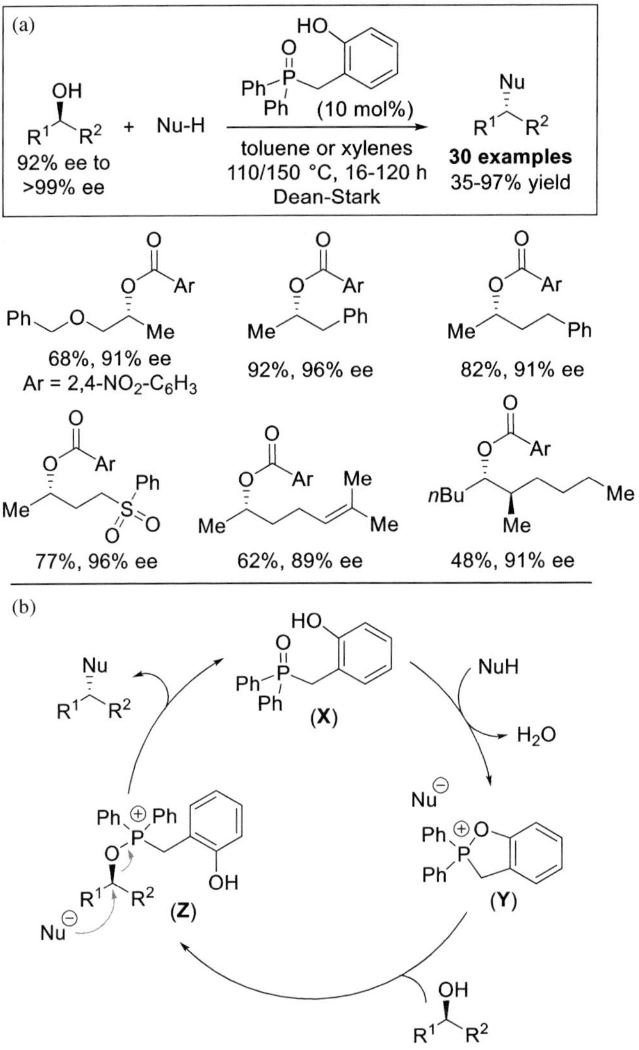

Scheme 20. Catalytic nucleophilic substitutions on alcohols *via* P^V/P^V catalysis.[47]

desired products and the phosphine oxide **X**, which can be reengaged in a catalytic cycle. Thus, during this redox neutral process, the phosphorus oxidation state remains unchanged, avoiding the use of an external reducing agent. It should be noted that substrates hardly compatible with nucleophilic trivalent phosphines, such as alkyl halides or azides, are well tolerated under these operating conditions, contrary to the "classical" Mitsunobu conditions. Furthermore, enantiopure alcohols gave the corresponding substitution products with good stereoselectivity, *via* inversion of configuration of the stereogenic center. This new strategy is limited to the use of very acidic pronucleophiles ($pK_a < 4$), which facilitate the formation of the activated catalyst species **Y**. Finally, the use of a Dean–Stark and heating to 110–150°C in toluene or xylenes is mandatory to trap water and to form intermediate **Y**. Further computational studies[48] by Houk and coworkers highlighted that the RDS of this catalytic process is the addition of the nucleophile on the alkoxyphosphonium salt **Z** and that this step could presumably be accelerated by using (difluoro(2-hydroxyphenyl) methyl)phosphine oxide as catalyst.

Overall, whichever the method used, that is, the redox P^{III}/P^V catalysis or the redox neutral P^V/P^V transformation, the "reaction mass efficiency" has been significantly improved in these reactions. This parameter simply expresses the efficienty with which the mass of the substrates ends up in the final product.[42b,49] In the same transformation, it has been evaluated to 21% in typical stoichiometric Mitsunobu reactions, 27% in the P^{III}/P^V process, and 65% in the P^V/P^V catalysis.

V. Conclusion and Outlook

Many different applications of phosphine catalysis *via* P^{III}/P^V (asymmetric) redox cycling have been outlined in this chapter. These reactions typically involve the *in situ* reduction of phosphine oxides to regenerate the P(III) catalysts. The usefulness of the method has been clearly demonstrated. However, it is still legitimate to ask whether the use of a stoichiometric amount of reducing agent in place of a large amount of phosphine is a more sustainable alternative. This pivotal question can be answered in three main points: (1) first of all, in processes that are catalytic in phosphine, the purification step is greatly simplified because the silanol/siloxane waste (coming from the silane reducing agent) can be easily separated from the crude mixture with a simple liquid–liquid extraction. (2) Furthermore, in 2013 Huijbregts and coworkers compared the "life cycle assessment" of the stoichiometric Wittig reaction to its catalytic version, and showed that the use of a catalytic amount of phosphine offers significant environmental improvements in terms of cumulative energy demand and greenhouse gas emissions.[50] (3) Finally, catalytic processes could also unlock a major bolt in phosphine-mediated reactions, namely the development of asymmetric transformations. Thus, it is now possible to consider the use of small amounts of chiral phosphines in enantioselective catalytic reactions, whereas it was previously inconceivable to use huge quantities of these expensive phosphines, as required in stoichiometric transformations.

To further improve the reaction scope and the versatility of the P^{III}/P^{V} redox processes, it will be possible, in the future, to work in different directions. Some work has to be done for the design of new cyclic (chiral) phosphines to facilitate the reduction of the corresponding phosphine oxide precatalyst and maximize the efficiency in terms of isolated yields, catalyst turnover, and enantioselectivity. The development of more reactive — but still chemoselective — silanes is also expected. Morover, the extended use of more sustainable reducing agent, such as TMDS or PMHS, that is, silicon industry waste products, is also a great concern. The implementation of these (asymmetric) innovative strategies on a larger scale and in industry would be a major step forward for process chemists.

VI. Acknowledgments

A.V. is personally grateful to students and researchers who worked in his group on P(III)/P(V) redox catalysis projects: Dr. Kévin Fourmy, Dr. Nidal Saleh, Charlotte Lorton, Xu Han, Dr. Thomas Castanheiro, Antoine Roblin, and Romain Losa. The authors acknowledge support from the "Centre National de la Recherche Scientifique" (CNRS), Paris-Saclay University, the "Agence Nationale de la Recherche" (ANR-13-JS07-0008), and CHARMMMAT Labex (ANR-11-LABX-0039).

VII. References

1. (a) Valentine, D. H.; Hillhouse, J. H. *Synthesis* **2003**, *3*, 317–334. (b) Xu, S.; He, Z. *RSC Adv.* **2013**, *3*, 16885–16904. (c) Karanam, P.; Reddy, G. M.; Koppolu, S. R.; Lin, W. *Tetrahedron Lett.* **2018**, *59*, 59–76.
2. Bottaro, J. C. *Synth. Commun.* **1985**, *15*, 195–199.

3. Guino, M.; Hii, K. K. *Chem. Soc. Rev.* **2007**, *36*, 608–617.
4. Dandapani, S.; Curran, D. P. *Tetrahedron* **2002**, *58*, 3855–3864.
5. (a) Marsden, S. P.; McGonagle, A. E.; McKeever-Abbas, B. *Org. Lett.* **2008**, *10*, 2589–2591. (b) An, J.; Denton, R. M.; Lambert, T. H.; Nacsa, E. D. *Org. Biomol. Chem.* **2014**, *12*, 2993–3003.
6. (a) Voituriez, A.; Saleh, N. *Tetrahedron Lett.* **2016**, *57*, 4443–4451. (b) Guo, H. C.; Fan, Y. C.; Sun, Z. H.; Wu, Y.; Kwon, O. *Chem. Rev.* **2018**, *118*, 10049–10293. (c) Longwitz, L.; Werner, T. *Pure Appl. Chem.* **2019**, *91*, 95–102.
7. Wittig, G.; Geissler, G. *Liebigs Ann.* **1953**, *580*, 44–57.
8. Rocha, D. H. A.; Pinto, D. C. G. A.; Silva, A. M. S. *Eur. J. Org. Chem.* **2018**, 2443–2457.
9. (a) O'Brien, C. J.; Tellez, J. L.; Nixon, Z. S.; Kang, L. J.; Carter, A. L.; Kunkel, S. R.; Przeworski, K. C.; Chass, G. A. *Angew. Chem. Int. Ed.* **2009**, *48*, 6836–6839. (b) O'Brien, C. J.; Nixon, Z. S.; Holohan, A. J.; Kunkel, S. R.; Tellez, J. L.; Doonan, B. J.; Coyle, E. E.; Lavigne, F.; Kang, L. J.; Przeworski, K. C. *Chem. Eur. J.* **2013**, *19*, 15281–15289.
10. (a) Horner, L.; Balzer, W. D. *Tetrahedron Lett.* **1965**, 1157. (b) Naumann, K.; Zon, G.; Mislow, K. *J. Am. Chem. Soc.* **1969**, *91*, 2788. (c) Naumann, K.; Zon, G.; Mislow, K. *J. Am. Chem. Soc.* **1969**, *91*, 7012. (d) Marsi, K. L. *J. Org. Chem.* **1974**, *39*, 265–267. (e) Krenske, E. H. *J. Org. Chem.* **2012**, *77*, 3969–3977. (f) Kirk, A. M.; O'Brien, C. J.; Krenske, E. H. *Chem. Commun.* **2020**, *56*, 1227–1230.
11. Henry, C. E.; Xu, Q. H.; Fan, Y. C.; Martin, T. J.; Belding, L.; Dudding, T.; Kwon, O. *J. Am. Chem. Soc.* **2014**, *136*, 11890–11893.
12. (a) Herault, D.; Duc Hanh, N.; Nuel, D.; Buono, G. *Chem. Soc. Rev.* **2015**, *44*, 2508–2528. (b) Podyacheva, E.; Kuchuk, E.; Chusov, D. *Tetrahedron Lett.* **2019**, *60*, 575–582.
13. Lenstra, D. C.; Lenting, P. E.; Mecinović, J. *Green Chem.* **2018**, *20*, 4418–4422.
14. Harris, J. R.; Haynes, II, M. T.; Thomas, A. M.; Woerpel, K. A. *J. Org. Chem.* **2010**, *75*, 5083–5091.
15. O'Brien, C. J.; Lavigne, F.; Coyle, E. E.; Holohan, A. J.; Doonan, B. J. *Chem. Eur. J.* **2013**, *19*, 5854–5858.
16. Li, Y.; Lu, L.-Q.; Das, S.; Pisiewicz, S.; Junge, K.; Beller, M. *J. Am. Chem. Soc.* **2012**, *134*, 18325–18329.
17. Coyle, E. E.; Doonan, B. J.; Holohan, A. J.; Walsh, K. A.; Lavigne, F.; Krenske, E. H.; O'Brien, C. J. *Angew. Chem. Int. Ed.* **2014**, *53*, 12907–12911.
18. (a) Nykaza, T. V.; Harrison, T. S.; Ghosh, A.; Putnik, R. A.; Radosevich, A. T. *J. Am. Chem. Soc.* **2017**, *139*, 6839–6842. (b) Nykaza, T. V.; Ramirez, A.; Harrison, T. S.; Luzung, M. R.; Radosevich, A. T. *J. Am. Chem. Soc.* **2018**, *140*, 3103–3113.
19. Longwitz, L.; Spannenberg, A.; Werner, T. *ACS Catal.* **2019**, *9*, 9237–9244.
20. (a) Schirmer, M.-L.; Adomeit, S.; Werner, T. *Org. Lett.* **2015**, *17*, 3078–3081. (b) Schirmer, M.-L.; Adomeit, S.; Spannenberg, A.; Werner, T. *Chem. Eur. J.* **2016**, *22*, 2458–2465.
21. Rein, T.; Pedersen, T. M. *Synthesis* **2002**, 579–594.

22. (a) Trost, B. M.; Curran, D. P. *J. Am. Chem. Soc.* **1980**, *102*, 5699–5700; (b) Trost, B. M.; Curran, D. P. *Tetrahedron Lett.* **1981**, *22*, 4929–4932.
23. (a) Tanaka, K.; Ohta, Y.; Fuji, K.; Taga, T. *Tetrahedron Lett.* **1993**, *34*, 4071–4074. (b) Tanaka, K.; Watanabe, T.; Ohta, Y.; Fuji, K. *Tetrahedron Lett.* **1997**, *38*, 8943–8946.
24. (a) Kann, N.; Rein, T. *J. Org. Chem.* **1993**, *58*, 3802–3804. (b) Tullis, J. S.; Vares, L.; Kann, N.; Norrby, P.-O.; Rein, T. *J. Org. Chem.* **1998**, *63*, 8284–8294. (c) Vares, L.; Rein, T. *Org. Lett.* **2000**, *2*, 2611–2614.
25. Werner, T.; Hoffmann, M.; Deshmukh, S. *Eur. J. Org. Chem.* **2014**, *2014*, 6630–6633.
26. Schmiedel, V. M.; Hong, Y. J.; Lentz, D.; Tantillo, D. J.; Christmann, M. *Angew. Chem. Int. Ed.* **2018**, *57*, 2419–2422.
27. For a more general approach of the asymmetric synthesis of carbocyclic propellanes, see: Schneider, L. M.; Schmiedel, V. M.; Pecchioli, T.; Lentz, D.; Merten, C.; Christmann, M. *Org. Lett.* **2017**, *19*, 2310–2313.
28. (a) Schweizer, E. E.; Light, K. K. *J. Am. Chem. Soc.* **1964**, *86*, 2963. (b) Schweizer, E. E.; Light, K. K. *J. Org. Chem.* **1966**, *31*, 870–872.
29. (a) Yavari, I.; Adib, M.; Sayahi, M. H. *J. Chem. Soc. Perkin Trans.* **2002**, *1*, 1517–1519. (b) Esmaeili, A. A.; Kheybari, H. *J. Chem. Res.* **2002**, *9*, 465–466. (c) For other PPh$_3$-mediated transformations, see: Yavari, I.; Asghari, S. *Tetrahedron* **1999**, *55*, 11853–11858. (d) Mosslemin, M. H.; Yavari, I.; Anary-Abbasinejad, M.; Nateghi, M. R. *Synthesis* **2004**, *7*, 1029–1032.
30. Saleh, N.; Voituriez, A. *J. Org. Chem.* **2016**, *81*, 4371–4377.
31. (a) Saleh, N.; Blanchard, F.; Voituriez, A. *Adv. Synth. Catal.* **2017**, *359*, 2304–2315. (b) Han, X.; Saleh, N.; Retailleau, P.; Voituriez, A. *Org. Lett.* **2018**, *20*, 4584–4588.
32. Zhang, K.; Cai, L. C.; Yang, Z. Y.; Houk, K. N.; Kwon, O. *Chem. Sci.* **2018**, *9*, 1867–1872.
33. Lorton, C.; Voituriez, A. *J. Org. Chem.* **2018**, *83*, 5801–5806.
34. Lorton, C.; Castanheiro, T.; Voituriez, A. *J. Am. Chem. Soc.* **2019**, *141*, 10142–10147.
35. Ding, A. S.; Meazza, M.; Guo, H.; Yang, J. W.; Rios, R. *Chem. Soc. Rev.* **2018**, *47*, 5946–5996.
36. (a) van Kalkeren, H. A.; te Grotenhuis, C.; Haasjes, F. S.; Hommersom, C. A.; Rutjes, F. P. J. T.; van Delft, F. L. *Eur. J. Org. Chem.* **2013**, 7059–7066. (b) In 2008, Marsden developed a catalytic P^V/P^V redox neutral aza-Wittig process, starting from isocyanates, see 5a.
37. (a) Wang, L.; Wang, Y.; Chen, M.; Ding, M.-W. *Adv. Synth. Catal.* **2014**, *356*, 1098–1104. For other examples, see: (b) Wang, L.; Xie, Y.-B.; Huang, N.-Y.; Yan, J.-Y.; Hu, W.-M.; Liu, M.-G.; Ding, M.-W. *ACS Catal.* **2016**, *6*, 4010–4016. (c) Ren, Z.-L.; Liu, J.-C.; Ding, M.-W. *Synthesis* **2017**, *49*, 745–754.
38. (a) Lertpibulpanya, D.; Marsden, S. P.; Rodriguez-Garcia, I.; Kilner, C. A. *Angew. Chem. Int. Ed.* **2006**, *45*, 5000–5002. (b) Headley, C. E.; Marsden, S. P. *J. Org. Chem.* **2007**, *72*, 7185–7189.

39. Cai, L.; Zhang, K.; Chen, S.; Lepage, R. J.; Houk, K. N.; Krenske, E. H.; Kwon, O. *J. Am. Chem. Soc.* **2019**, *141*, 9537–9542.
40. (a) Mitsunobu, O.; Yamada, M.; Mukaiyama, T. *Bull. Chem. Soc. Jpn* **1967**, *40*, 935–939. (b) Mitsunobu, O.; Yamada, M. *Bull. Chem. Soc. Jpn* **1967**, *40*, 2380–2382. For reviews, see: (c) Mitsunobu, O. *Synthesis*, **1981**, *1*, 1–28. (d) Swamy, K. C. K.; Kumar, N. N. B.; Balaraman, E.; Kumar, K. V. P. P. *Chem. Rev.* **2009**, *109*, 2551–2651.
41. Carey, J. S.; Laffan, D.; Thomson, C.; Williams, M. T. *Org. Biomol. Chem.* **2006**, *4*, 2337–2347.
42. (a) Fletcher, S. *Org. Chem. Front.* **2015**, *2*, 739–752. (b) Beddoe, R. H.; Sneddon, H. F.; Denton, R. M. *Org. Biomol. Chem.* **2018**, *16*, 7774–7781.
43. For reactions catalytic in azocarboxylate, see: (a) But, T. Y. S.; Toy, P. H. *J. Am. Chem. Soc.* **2006**, *128*, 9636–9637. (b) Hirose, D.; Taniguchi, T.; Ishibashi, H. *Angew. Chem. Int. Ed.* **2013**, *52*, 4613–4617.
44. O'Brien, C. J. "Catalytic Wittig and Mitsunobu reactions" PCT Int. Appl, *WO2010*/118042A2.2010, **2010**.
45. Buonomo, J. A.; Aldrich, C. C. *Angew. Chem. Int. Ed.* **2015**, *54*, 13041–13044.
46. Hirose, D.; Gazvoda, M.; Košmrlj, J.; Taniguchi, T. *Org. Lett.* **2016**, *18*, 4036–4039.
47. Beddoe, R. H.; Andrews, K. G.; Magné, V.; Cuthbertson, J. D.; Saska, J.; Shannon-Little, A. L.; Shanahan, S. E.; Sneddon, H. F.; Denton, R. M. *Science* **2019**, *365*, 910–914.
48. Zou, Y.; Wong, J. J.; Houk, K. N. *J. Am. Chem. Soc.* **2020**, *142*, 16403–16408.
49. Curzons, A. D.; Constable, D. J. C.; Mortimer, D. N.; Cunningham, V. L. *Green Chem.* **2001**, *3*, 1–6.
50. van Kalkeren, H. A.; Blom, A. L.; Rutjes, F. P. J. T.; Huijbregts, M. A. *J. Green Chem.* **2013**, *15*, 1255–1263.

© 2022 World Scientific Publishing Company
https://doi.org/10.1142/9789811248436_0007

7. Recent Advances in the Use of Three-dimensional-printed Devices in Organic Synthesis Including Enantioselective Catalysis

Sergio Rossi,* Alessandra Puglisi,*
Laura Maria Raimondi* and Maurizio Benaglia*,†

*Dipartimento di Chimica, Università degli Studi di Milano,
Via Golgi 19, Milano 20133, Italy
†maurizio.benaglia@unimi.it

Table of Contents

List of Abbreviations	198
I. Introduction	199
II. 3D-Printed Devices in Nonstereoselective Reactions	207
A. Catalytic 3D-Printed Devices in Batch Reactions	209
B. 3D-Printed Devices for In-flow Reactions	223
1. Reactions under heterogeneous conditions	224
2. Reactions under homogeneous conditions	226
C. 3DP for Drugs and Pharmaceuticals Applications	236

III. 3D-Printed Devices in Stereoselective Reactions	240
IV. Conclusion	244
V. Acknowledgments	245
VI. References	245

List of Abbreviations

μRD	microrobotic deposition
2PP	two-photon polymerization
3D	three-dimensional
3DP	three-dimensional printing
AAPTS	[3-(2-aminoethylamino)propyl]trimethoxysilane
ABS	acrylonitrile-butadiene-styrene copolymer
APTS	(3-aminopropyl)trimethoxysilane
BIS	beam interference solidification
BPM	ballistic particle manufacturing
CAD	computer-aided design
CLIP	continuous liquid interface printing
CSTR	continuous stirred tank reactor
CuAAC	copper-catalyzed azide–alkyne cycloaddition
DBU	1,8-diazabiciclo[5.4.0]undec-7-ene
DCM	dichloromethane
DIPEA	N, N-Diisopropyl-N-ethylamine
DIW	direct ink qriting
DLP	digital light processing
DMAP	N, N-Dimethylaminopyridine
DMD	direct metal deposition
DMF	dimethylformamide
DMLM	direct metal laser melting
DWA	direct write assembly
EBM	electron beam melting
FDA	Food and Drug Administration
FDM	fused deposition modeling
HIPS	high-impact polystyrene
HIS	holographic interference solidification
HME	hot melt extrusion

ID	internal diameter
IJP	inkjet printing
LENS	laser-engineered net shaping
LOM	laminated object manufacturing
LPD	laser powder deposition
LPT	liquid thermal polymerization
MJM	multijet modeling
MOF	metal-organic framework
OBR	oscillatory baffled reactor
PC	polycarbonate
PE	polyethylene
PEG	polyethylene glycol
PEGDA	polyethylene glycol diacrylate
PET	polyethylene terephthalate
PLA	polylactic acid
PLP	pyridoxal phosphate
POM	polyoxymethylene
PP	polypropylene
p-TSA	*para*-Toluenesulfonic acid
RC	robocasting
SBS	styrene-butadiene-styrene copolymer
SFP	solid foil polymerization
SGC	solid ground curing
SLA	stereolithography apparatus
SLM	selective laser melting
SLS	selective laser sintering
SNAr	nucleophilic aromatic substitution
STL	standard tessellation language (stereolithography)

I. Introduction

Additive manufacturing, which commonly refers to three-dimensional (3D) printing technology, has found fruitful applications in a wide range of fields and in the last few years it has kindly made its way through our everyday life. From the manufacture of objects for clothes manufacturing and jewelry products,[1,2] three-dimensional printing (3DP) has spread to almost all fields of application, and nowadays, the realization of medical implants[3] or 3D-printed dental appliances has become a common procedure to realize highly customized and personalized products.[4,5] The ease of the 3DP process combined with its low manufacturing cost and with the possibility to realize objects with intricate internal structures (far away from sites where they were originally designed) represent its key to success. 3DP is the perfect representative technology of the third industrial revolution[6] and represents the evolution of the standard mass production. Mass production involves an investment in manufacturing facilities necessary to efficiently produce copies of the same object. This business is sustainable until demand continues to exist for a particular product; in this situation, production costs continuous decreasing with a consequent increment of profit margins. However, with this approach, products cannot be sold until they are produced, and an excess of production compared to the request could generate waste. 3DP process cannot compete with the numbers of pieces produced by the mass production, but it offers the possibility to model and customize objects before any manufacturing has taken place since the production step is on client demand. Once the electronic file of the desired object has been purchased, its production is made locally, and no stock, no shipping, and no waste are produced. There are wide number of product categories far away from the mass production (categories that include products made on a limited or personalized scale) where the benefits of 3DP are making the difference, as the aerospace and aircraft industries,[7,8] oral drug productions,[9] biomedical engineering,[10] and food fabrication.[11,12]

The revolution induced by 3DP has been made possible, thanks to the amazing advances in chemical science, focused mainly in the creation of many different materials that can undergo the 3DP process. Different materials means different printing techniques; the most important 3DP techniques are stereolithography (STL), fused deposition modeling (FDM), selective laser sintering (SLS), selective laser melting (SLM), digital light processing (DLP), continuous liquid interface printing (CLIP), robocasting (RC) (also known as direct-ink writing [DIW] and, less frequently, as direct-write assembly [DWA] or micro robotic deposition [μRD]), two-photon polymerization (2PP), laminated object manufacturing (LOM), solid ground curing (SGC), ballistic particle manufacturing (BPM), and inkjet printing (IJP). Not all of these have directly applied in organic synthesis, and the precise principles, advantages, and limitations are beyond the scope of this chapter. However, a nonexhaustive summary of their properties is reported in Table 1.

Given the many different 3DP techniques, it is not easy to define what a "3D printing process" is. Any of various layer-by-layer processes used to make a 3D solid object starting from a digital file could be considered a 3D-printed process. Regardless the type of 3DP technology used, the steps involved in the realization of the final object are always the same[30] and consist in the following:

1. Description of the geometry of the desired object using a computer-aided design (CAD) software
2. Conversion of the geometry in a STereo Lithography interface format (STL format). This is a specific codification in which the geometry is described as a raw, unstructured triangulated surface (using unit normal and vertices of triangles) in a 3D Cartesian coordinate system. Due to these characteristics, STL files contain no scale information.
3. Performing the "slicing" process, which consists in the conversion of a 3D object model into specific instructions for the printer. This step is very important and it is strictly correlated to the physical characteristic of the 3D printer employed. In this step, all the required printer parameters need to be defined (layer thickness, resolution, temperature, etc.) as well as the desired dimensions of the object. The slicer software divides the object as a stack of flat 2D layers, and writes encoded instructions that are essential for the correct movements of the 3D printer extruder, fixation laser printer plates, and so on. A new file is generated, which needs to be transferred to the printer.
4. Setup of the 3D printer (loading of the printable material, technical calibration) and get the printing process started. If everything has been properly set up, the printer will fabricate the whole object layer-by-layer.
5. Final post printing process, which consists in the removal of printing supports generated during the printing and brushing off any residual material.

Table 1. Comparison of different 3DP techniques.

Process type	Techniques	Build materials	Description	Advantages	Disadvantages	Printing resolution	Ref
Laser based	STL DLP SGC Liquid thermal polymerization (LTP) Beam interference solidification (BIS) Holographic interference solidification (HIS)	Curable resins	A light source (laser, led or high-powered ultraviolet [UV] lamp) is employed to photocatalyze or thermally initiate the polymerization of the curable resin.	High speed and high resolution, good accuracy.	High cost of resins. Small dimension of objects. SGC presents high waste production	>50 μm	10–15
	2PP	Photoresist blends	Two light sources are employed to promote a two-photon absorption and subsequent polymerization process of the building material	Very high resolution. The laser radiation is not harmful for cells.	Highly expensive	100–800 nm	15–17

					14,15, 18–20
SLM, also known as direct metal laser melting (DMLM) or laser powder bed fusion (LPBF)	Metal powders (aluminum alloys, steels, titanium alloys, nickel super alloys) and refractory materials	A high power-density laser is used to melt, to fuse or sin-tering together building materials	High speeds and a wide variety of materials having different characteristics in terms of strength, durability, and functionality	Grainy surface finish due to powder particle size. Final objects present porous surfaces. Long process time required	Depending on the material. Typically >1 mm SLS and SLM 20–150 μm
Laser powder deposition (LPD)					
Laser-engineered net shaping (LENS)					
Direct metal deposition (DMD)					
Selective laser cladding (SLC)					
SLS	Polymers such as poly-amides, polystyrenes, thermoplastic elastomers and polyary-letherketones				

(Continued)

Table 1. (Continued)

Process type	Techniques	Build materials	Description	Advantages	Disadvantages	Printing resolution	Ref
Extrusion processes	FDM	Acrylonitrile-butadiene-styrene (ABS), polylactic acid (PLA), polyethylene terephthalate (PET), styrene-butadiene-styrene (SBS), high-impact polystyrene (HIPS), Primalloy, nylon, polyoxymethylene (POM), polyethylene (PE), polypropylene (PP), polycarbonate (PC)	A filament of material is melted in a heated head and deposited on a built plate. Objects are realized by layer-to-layer approach	High scalability and flexibility in terms of building materials	Low quality of the final object	Resolution depends from nozzle. 50–300 μm	21–24

RC, also known as DIW or DWA or μRD	Colloidal slurries, dense ceramics, and composites.	The paste-like material is extruded from a small nozzle while the nozzle is moved across a platform	Short production runs Compatible with biological applications	Post printed process often required	30–500 μm	20,25
Material jetting (powder based) IJP 3DP Multijet modeling (MJM) BPM Thermojet	Thermo-polymer Materials (wax, acrylates), metals, plastics, and ceramics in particle forms	Ejecting liquid-phase materials in droplet form through print head nozzles onto the build object in order to bind the powder into a solid object. MMJ involves the use of UV light to polymerize droplets	High positioning accuracy. Small droplet size. Useful for tissue engineering. Requires minimal post-processing. Minimal power consumption. Ability to perform in microgravity and vacuum environments.	Porous features In some cases, the liquid content must be evaporated prior to use the final object High price of resin and printers. Low viscosity of resin required.	IJP 1–50 μm	20,26,27

(Continued)

Table 1. (Continued)

Process type	Techniques	Build materials	Description	Advantages	Disadvantages	Printing resolution	Ref
Material adhesion	LOM Solid foil polymerization (SFP)	Plastics, composites, metal foil, ceramics, paper, polymer film	Pre-cut foils with a heat-sensitive adhesive on one side which are compacted together and heated to activate a curing process	Ability to produce larger-scaled models Use of inexpensive materials Requires no support structures	Not ideal for complex geometries Lower accuracy compared other techniques	Limited to the foils thickness	28
Electron beam	Electron beam melting (EBM)	Metals	A high-voltage electron beam is used as energy source to melt the metal powder.	Allow to realize fully dense, void free, and extremely strong parts. High manufacturing speed.	Granular surfaces Less accuracy compared to laser beam	>60 mm	29

3DP techniques can be formally classified into two distinct groups: the first includes all techniques in which an object is generated starting from patternable (polymerizable) materials, whereas the second group includes all other methodologies in which a material is extruded through a nozzle onto a platform. Even if in both cases objects will be fabricated in a layer-by-layer approach, they will result in different physical and chemical properties that are essentially defined by the technology employed for their production[15] and by the material used.[31,32]

Although 3DP was initially used to evaluate prototypes, its diffusion, supported by a thriving open-source community, has reached also chemical science. In this field, 3DP technology was often combined with other enabling technologies for the realization of tissues,[33-38] analytical detectors,[39-41] electrodes,[42,43] and micro- and macro fluidic devices.[22,44-54] Other applications are in separation sciences,[55] adsorbers,[56] airlift crystallizers,[57] and to the realization of supplies for laboratories[58-60] and basilar instrument for chemical education purpose.[61-64] Moreover, in the last two years, chemists have started to combine 3DP technology with organic compounds for the development of "3D-printed catalysts," which have been employed in organic synthesis.[65,66] In this chapter, only the very recent advances in the application of 3DP technology in organic synthesis will be discussed, with a particular focus on the implementation of the technology in batch chemistry as well as under continuous-flow conditions, in the synthesis of drugs and pharmaceuticals and in the stereoselective synthesis of chiral molecules in the last three years (2018–2020).

II. 3D-Printed Devices in Nonstereoselective Reactions

3DP is one of the most promising technology that allows to fabricate multifunctional and effective mixers and geometrically optimized reactors that can be employed in chemical transformations.[20,22] In a further improvement, by adding an active ingredient in the composition of the printable material, with due caution, it is possible to generate "3D-printed catalysts" with high control of the catalyst distribution.[67]

The initial, simpler approach developed in this sense was the creation of 3D-printed reactionware. One of the first examples of 3D-printed reactors was reported by Cronin and coworkers, who described the fabrication of a reusable 3D-printed reactionware by FDM, in which reagents were included during preprogrammed pauses of the printing process. The particular shape of the reactor, easily obtained by the 3DP process, was fundamental to perform in real-time electrochemical and spectroscopic analysis of phenanthridine-based heterocycles.[41,68] With this example, Cronin showed that 3DP technology was ready to create ad hoc devices for chemical transformations. One year later, with the same approach, Cronin and coworkers reported the realization of four sealed cube reactionwares sequentially connected with circular channels; the system was used to perform a multistep reaction sequence without any pumps with a minimal handling by the operator. Acrolein and a cyclopentadiene were involved in a sequential (i) Diels–Alder cyclization for the synthesis of the corresponding adduct that, in the presence of aniline (preloaded in the reactor) was subjected to (ii) imine formation followed by (iii) C=N reduction promoted by triethylsilane and Pd/C.[69] Starting from this point, many other examples of 3D-printed reactors were reported, such as the realization of sealed 3D-printed hydrothermal reactionware employed in the synthesis of metal-organic frameworks (MOFs),[70,71] 3D-printed PP tube for coupling reactions[72] or 3D-printed nuclear magnetic resonance (NMR) tubes and cuvettes for the study of cross-coupling reactions by magnetic resonance and infrared spectroscopy.[73] However, since these works have been widely described in details in precedent reviews,[65,74] they will not be herein discussed.

More recently, Cronin and coworkers reported a methodology for the translation of synthetic procedures from bench scale to a step-by-step workflow in custom 3D-printed reactionwares.[75] The full synthesis, workups, and corresponding purification processes of the racemic version of three active pharmaceutic ingredients (baclofen, lamotrigine, and zolimidine) were performed using a system of integrated cartridges 3D printed in PP. The synthetic sequence of each product was split into a series of bespoke reaction modules designed to accomplish a specific chemical process. The geometry of each single module is a composition of three different parts (top part, central part, bottom) that are selected from a CAD library and combined by the software to produce a wide range of 3D-printable module geometries. These geometries are specifically designed for a particular synthetic process and reaction parameters (e.g., solvent volumes and number of inputs and outputs) and each different

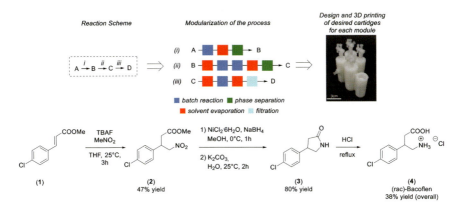

Scheme 1. Schematic representation of the multistep synthesis of (rac)-baclofen performed in 3D-printed reactionware. Adapted from Ref. 75 with permission from American Association for the Advancement of Science (AAAS), Washington, DC.

module can be connected to another one by a "plug-and-play" approach. The synthesis of (rac)-baclofen starts from the commercially available methyl 4-chloro-cinnamate **1**, which undergoes a Michael addition of nitromethane, generating compound **2**. This compound was then subjected to the nickel-catalyzed reductive lactamization for the synthesis of **3** that in presence of hydrochloric acid gave the hydrochloride salt **4** in 38% overall yield. The synthesis, formally composed by three steps, was modularized in 12 individual processing operations (Scheme 1) that were incorporated into five reactionware devices.

According to the authors, the advantages of this approach are related to (i) the simplification of the operations required to perform the reaction sequence, since the participation of the operator is minimal compared to the traditional bench synthesis and to the (ii) lower level of technical skills required to perform the synthesis. However, at the moment, there is a lack in the specific regulatory framework necessary to validate the synthesis of pharmaceutical products according to this modular approach.

A. Catalytic 3D-Printed Devices in Batch Reactions

The realization of 3D-printed devices containing an embedded or a bounded (organo)catalyst offers the possibility to control architectures and pore size dimensions of the final object, which will be employed to promote chemical transformations. Intrinsic advantages of this approach are (i) the easy removal of the 3D-printed catalytic species at the end of the reaction and (ii) the simplification of the purification process. In addition, from a theoretical point of view, the recovered 3D-printed device could be also reused for further transformations although, in some cases, a reactivation process of the catalyst could be required in order to maintain high performances. The immobilization of catalytic species on solid reusable supports is one of the challenging methods to develop cleaner and sustainable processes, and in recent years, 3D-printed devices based on doped polymers, carbon materials,[76] metal and metal oxides,[77] or zeolites[78,79] have received considerable attention in catalytic transformations.[67,80] Advantages of this type of devices can be summarized in easy recovery and recycling procedures and simplification of the workup process. One example was reported in 2016 by Sotelo, Gil, and coworkers who reported the fabrication of a robust, efficient, and reusable copper heterogeneous inorganic catalytic system to promote Cu-mediated coupling reactions.[81] A Cu/Al_2O_3 catalytic system with a woodpile porous structure was synthetized starting from a colloidal gel ink composed of Al_2O_3 ceramic powder, polymer binders, and $Cu(NO_3)_2$ 2.5 H_2O. This mixture was extruded to build a 3D woodpile structure using a RC 3D printer (diameter nozzle = 410 μm) and sintered at 1400°C for 24 h in a conventional furnace. With this thermal treatment, solvents and polymers employed during the ink preparation were removed and the structure resulted in an increased mechanical strength. The final cylindrical woodpile structure obtained presents 10 mm diameter and 12 mm height, an open porosity of 57% with a copper loading of 2.3 wt%. This functionalized device was successfully employed as catalyst in the synthesis of N-aryl substituted imidazoles, benzimidazoles, and N-aryl amides starting from iodobenzenes (Scheme 2).

One year later, the same group showed that the simple 3D-printed woodpile device made of Al_2O_3 is able to act as a Lewis acid catalyst in the synthesis of biologically active 3,4-dihydropyrimidin-2(1H)-ones and 1,4-dihydropyridines under solvent-free conditions, upon microwave irradiation, starting from a mixture of aromatic aldehydes and dicarbonyl compounds (Scheme 3).[82] In this case, the 3D-printed Al_2O_3 woodpile structure was sintered at 1500°C in order to reduce the number of

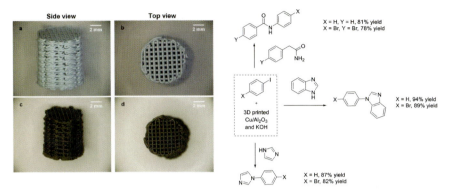

Scheme 2. (a and b) Cu/Al$_2$O$_3$ woodpile porous structure air dried and (c and d) after sintering process and its uses in Cu-catalyzed Ullmann reactions. Adapted from Ref. 81 with permission from Elsevier.

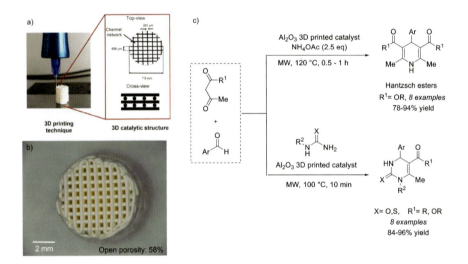

Scheme 3. (a) Structure of the Al$_2$O$_3$ woodpile porous device. (b) Al$_2$O$_3$ woodpile devices after sintering process. (c) Biginelli and Hantzsch reactions catalyzed by 3D-printed Al$_2$O$_3$. Adapted from Ref. 82 with permission from Elsevier.

hydroxyl groups present in the alumina structure, with consequent increase of the Lewis acidic character.

In 2018, Sotelo, Gil, and coworkers reported a combination of RC 3D-printed heterogeneous copper and palladium woodpiles devices to promote multicatalytic and multicomponent reactions.[83] A silica-based Pd and Cu heterogeneous monolithic catalysts were synthetized by 3DP process to perform heterogeneous multicatalytic multicomponent reactions based on a copper alkyne-azide cycloaddition, followed by a palladium catalyzed cross-coupling reaction for the one-pot synthesis of variously substituted 1,2,3-triazoles (Scheme 4). In this case, the synthetic approach was different from the previously reported: the unfunctionalized 3D woodpile structure was obtained from a colloidal gel ink made by SiO_2 ceramic powder and polymer binders (poly(vinyl butyral-co-vinyl alcohol-co-vinyl acetate) and polyethylene glycol [PEG]) by 3DP approach.

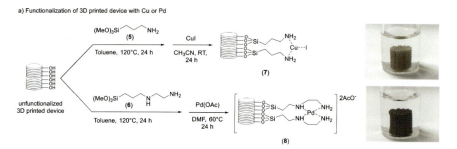

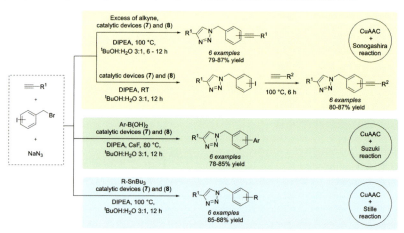

Scheme 4. 3D-Printed heterogeneous Cu and Pd catalytic devices **7** and **8** in multicatalytic multicomponent reactions. Adapted from Ref. 83 with permission © 2017 American Chemical Society.

The resulting device consists of a 10-mm diameter cylinder composed by 40 layers; each layer is formed by parallel rods (rod diameter = 410 μm) and each layer was rotated by 90° from the previous one. After the sintering process (1500°C), the device was reacted with (3-aminopropyl) trimethoxysilane (**5**, APTS) or [3-(2-aminoethylamino)propyl]trimethoxysilane (**6**, AAPTS) in order to introduce free NH_2 groups on the surface of the device. A treatment with CuI or $Pd(OAc)_2$ gave the desired copper or palladium 3D monolithic devices **7** and **8** (Scheme 4). The presence of catalytic species only on the surface of the monolith represents a key point in the minimization of the cost production; in this way, all the metal employed should be accessible by the reactants thus avoiding the waste of metal inside the mass of the device.

After a preliminary investigation of the reaction conditions, the Cu-functionalized device **7** was used in combination with the Pd-functionalized device **8** to promote the copper-catalyzed azide–alkyne cycloaddition (CuAAC), followed by a palladium-catalyzed cross-coupling reaction. Using an orbital shaker, sodium azide, iodo benzyl bromides, and alkynes were combined in the presence of **7** to generate the corresponding triazole, which in the presence of catalyst **8** and another coupling partner (an organotin reagent, a boronic acid or a second molecule of alkyne) led to the formation of high functionalized triazoles.

Using both catalytic devices simultaneously, the "click reactions" and Sonogashira coupling occurred satisfactorily within 6–12 h, leading to the formation of substituted triazoles in high yields. In the same manner, cycloadditions combined with Stille and Suzuki coupling reactions allowed to obtain the desired products in comparable yields. After completing the reactions, each monolithic catalyst was easily removed from the reaction mixture and, after a wash/dry cycle, was reused at least 10 times without a noticeable drop in the catalytic activity, although some changes in the oxidation state were observed. Investigations on the metal leaching reveal that only 0.07% for device **7** and 0.03% for device **8** of the initially added catalyst was released.

In 2019, Coelho and coworkers reported that functionalized triazoles can be synthetized using a combination of a catalytic 3D-printed devices, ferritic copper(I) magnetic nanoparticles and 3D-printed PP capsules.[84] In this approach, a catalytic triad based on Cu^{2+}, Cu^{1+}, and $Pd^{(0)}$ catalytic species was employed to promote the synthesis of substituted benzyl-1,2,3-triazoles, based on a tandem sequence of Chan–Lam azidation/copper alkyne–azide cycloaddition/Suzuki reaction. Initially, a PP capsule containing a copper(II) species supported on 1,5,7-triazabicyclo [4.4.0]dec-5-ene-polystyrene [PS-TBD-Cu$^{(II)}$] was realized using a FDM 3D printer. PS-TBD-Cu$^{(II)}$ was loaded into an empty capsule (which served as a protective envelope) during a pause of the 3DP process. This capsule was sealed with a permeable PP membrane, which ensured the interaction between the PS-TBD-Cu$^{(II)}$ resin and the reagents, but at the same time it prevented the leaching of the catalyst. Pd$^{(0)}$-functionalized 3D-printed SiO$_2$ monolithic device **9** was synthetized by RC approach starting from a SiO$_2$ ink gel. This process was performed in two stages: first, the 3D-printed woodpile structure was realized by RC, then the surface of the device was functionalized using a mixture of polyimide resin and Pd(OAc)$_2$. The printed monolith consisted of a 10-mm diameter cylinder structure composed by 40 layers formed by parallel rods (rod diameter = 410 μm), with 1.6 mg of palladium distributed on the monolith surface. Magnetic Cu$_2$O/chitosan–Fe$_3$O$_4$ (Cu$^{(I)}$-CS-Fe$_3$O$_4$) nanoparticles were also synthetized following a standard protocol. These three catalysts were loaded sequentially in a reaction vessel in the presence of sodium azide, iodophenylboronic acids, arylboronic acids, and alkynes for the synthesis of 1-([1,1'-biphenyl]-4-yl)-*1H*-1,2,3-triazoles (Scheme 5). The azidation reaction between iodophenylboronic acid and NaN$_3$, promoted by the presence of the PS-TBD-Cu$^{(II)}$ capsule, generated intermediate **A**. After 3 h of reaction time, the capsule was removed, followed by a concomitant addition of Cu$^{(I)}$-CS-Fe$_3$O$_4$ nanoparticles and the desired alkyne. Under these conditions, intermediate **A** reacted with the alkyne to generate the intermediate 1,2,3-iodotriazoles **B** in high yields, as single products, in 4–5 h. After this time, the nanoparticles were scavenged from the reaction medium by applying an external magnet. A new boronic acid and the Pd-functionalized 3D-printed device **9** were subsequently added, and the Suzuki transformation took place efficiently, leading to the formation of the desired compounds after 2–6 h.

The reusability of all the catalytic devices was effective for at least 10 runs without a noticeable drop in the product yield or catalytic activity. Investigations on the leaching of active metal species during the reactions demonstrated that only 0.0008% from the initial Cu content in the

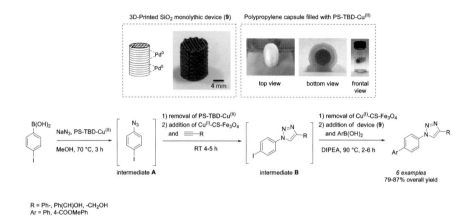

Scheme 5. $Cu^{(II)}/Cu^{(I)}/Pd^{(0)}$ multicatalytic sequential one-pot reactions for the synthesis of substituted 1,2,3-triazoles. Adapted from Ref. 84 with permission © 2019 American Chemical Society.

nanoparticles and 0.004% of the initial Pd on the monolithic catalyst was released in solution. Leaching of Cu by the PS-TBD-$Cu^{(II)}$ catalyst confined in the capsule was almost negligible.

One year later, Coelho and coworkers adopted the same RC technology to realize a Pd-based silica monolithic catalyst **10**, which was integrated in a PP reactor realized by FDM 3D-printing technology.[85] The monolithic device was obtained by RC printing of a SiO_2 ink gel; after sintering, the unfunctionalized device was immersed in a hot solution of K_2PdCl_4, to give a stable deposition of palladium nanoparticles on silica surface via strong electrostatic adsorption. The monolith was then treated with $NaBH_4$ in order to reduce $Pd^{(II)}$ to $Pd^{(0)}$ and after a drying process, it was transferred into the PP vial reactor. This PP-capped reactor containing device **10** was employed in Sonogashira and Suzuki reactions (Scheme 6). Suzuki reaction between haloarenes and boronic acids was conducted in the presence of potassium carbonate and a 2:1 mixture of $^tBuOH:H_2O$ as solvent, under orbital stirring, at 90°C, for up to 8 h. Reaction products were obtained in high yields, and no purification was necessary, since they frequently crystallized in the reactor walls, after cooling at room temperature. The catalytic device (**10**) instead was easily removed using tweezers. The Sonogashira coupling between haloarenes and simple terminal

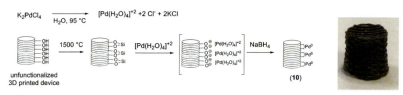

Scheme 6. Impregnated palladium on silica monolith in Sonogashira and Suzuki reactions performed in 3D-printed PP vessels. Adapted from Ref. 85 with permission from John Wiley & Sons, Inc.

alkynes was performed in the presence of *N, N*-diisopropyl-*N*-ethylamine (DIPEA) as base and dimethylformamide (DMF) as solvent and also in this case, the desired adduct was obtained in a short time and with slightly lower yields compared to the Suzuki reaction. Recycling studies demonstrated that 3D-printed device **10** can be reused at least six times without significant loss of catalytic performance, with 0.06% metal released after the first run, even if traces of some soluble palladium species were identified in the solution.

Functionalized 3D-printed monolithic structures have been investigated in different applications than coupling reactions. Xiong, Zheng, Liu, and coworkers reported the synthesis of 3D-printed devices functionalized with different organic bases that were employed in the synthesis of 2-aryliden-malononitriles, direct precursors of the corresponding α,β-unsaturated carbonyl compounds.[86] An unfunctionalized woodpile structure **11** was realized by RC approach, starting from a colloidal ink powder composed by SiO_2 powder, sodium alginate, and ethanol (radius 14 mm × height 14 mm, filling pitch 1.6 mm). After the sintering process at 1500°C, the surface of the device was activated using H_2O_2 under heating, and then reacted with 3-chloropropyltriethoxysilane. The new device **12** was loaded with different amines to generate the 3D-printed functionalized monolithic devices **13–17** (Scheme 7).

A further improvement in the realization of catalytic 3D-printed devices was reported by Hilton *et al.*, who filed a patent application disclosing a 3D-printed, catalyst impregnated plastic stirrer holder, as a sort of "catalytic stirrer" to be employed in different chemical reactions.[87,88] In this approach, as a proof of concept, a photo-curable resin was mixed with a catalyst and used to realize 3D-printed devices according to a SLA approach. Stirrer bead holders were printed starting from a resin

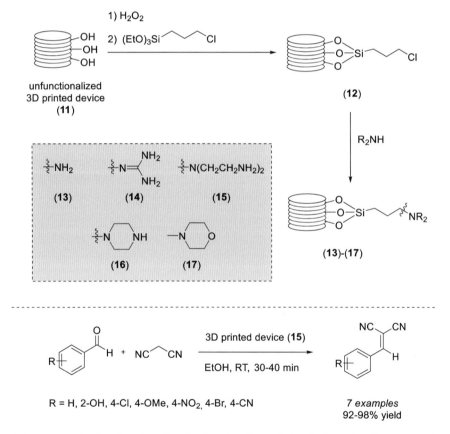

Scheme 7. Amino-functionalized 3D-printed catalytic devices for Knoevenagel reactions.[86]

composed of bisphenol A ethoxylate diacrylate **18**, isobornyl acrylate **19**, trimethylolpropane triacrylate **20**, and diphenyl(2,4,6-trimethylbenzoyl) phosphine oxide **21**. The resin was mixed with a catalyst chosen from among *p*-toluene sulfonic acid (p-TSA), DMAP, Zn(OTf)$_2$, Cu$_2$OTf$_2$, and Pd(PPh$_3$)$_4$ (Scheme 8). Embedded catalytic stirrer holders **22–26**) were then employed as catalysts in Mannich reactions, phenol acetylations, C–N bond formations, azide–alkyne couplings, and Suzuki couplings.

Another example of functionalized polymerizable resin was reported in 2017 by Slowing and coworkers: a UV-curable resin was mixed with

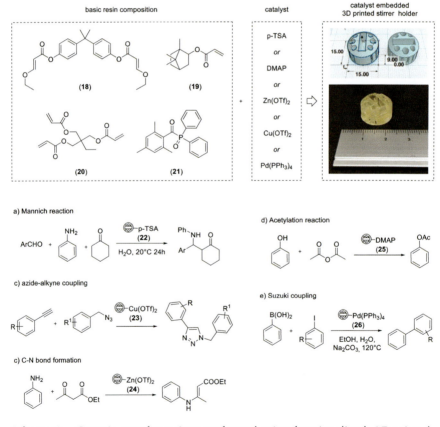

Scheme 8. Organic transformations performed using functionalized, 3D-printed, catalytic stirrer holders **22–26**. Adapted from Ref. 87 with permission from ChemRxiv.

acrylates bearing carboxylic acids, amines, or copper carboxylate functionalities for obtaining 3D-printable catalytic plastics to promote Mannich, aldol, and Huisgen cycloaddition reactions, respectively.[89] A 3D-printable acidic resin was prepared starting from a mixture of acrylic acid, PEG, and phenylbis(2,4,6-trimethylbenzoyl)phosphine oxide, and used to realize a device of particular shape (AL-COOH), which was employed as an acid catalyst for a three-component Mannich reaction (Scheme 9). Kinetic analysis performed on a mixture of benzaldehyde, aniline, and cyclohexanone, in the presence of a 3D-printed catalytic cuvette adaptor made in AL-COOH, showed a 2.5-fold enhancement in terms yield for the Mannich adduct **27**, in comparison to the noncatalyzed reaction. A bifunctionalized acid/base 3D-printed structure was also

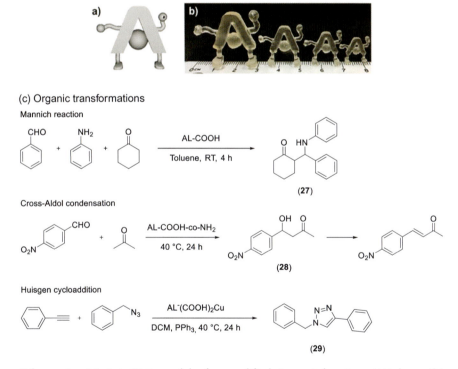

Scheme 9. (a) AutoCAD model of a modified Ames Laboratory (AL) logo. (b) 3D-printed structures of AL-COOH device with different sizes. (c) Organic reactions performed using AL-COOH/AL-COOH-co-NH and AL-(COOH)$_2$Cu devices. Adapted from Ref. 89 with permission © 2017 American Chemical Society.

realized using AL-COOH-co-NH resin, which was obtained by adding diallylamine to the previous resin mixture. AL-COOH-co-NH prints were tested as catalysts for the cross-aldol condensation between 4-nitrobenzaldehyde and acetone, leading to the formation of aldol product **28** with good results. An AL-(COOH)$_2$Cu device was instead synthetized by adding copper(II) acetate to the original AL-COOH mixture and used as catalyst for the Huisgen cycloaddition reaction between phenylacetylene and benzyl azide for the synthesis of triazole **29**. 3D-printed materials containing the AL-(COOH)$_2$Cu resin showed a homogenous distribution of copper. They were also employed to realize a fluidic device. Product formation was observed by UV spectroscopy, but no detailed information on yields was given. All these devices were successfully recycled without any indication of polymer degradation or leaching of the material.

The potential of catalytic 3D-printed structures was showed in another application of SLA technology focused on the fabrication of 3D-printed catalyst-impregnated devices to promote the organocatalyzed Friedel–Crafts alkylation of *N*-Me-indole with *trans-β*-nitrostyrene.[90] In this approach, devices **30–33** were designed and 3D printed using a photopolymerizable methacrylate–based resin, which was mixed with the Schreiner's thiourea organocatalyst **34** before the printing process. All devices were realized with a 10 wt% and 15 wt% loading of catalyst and present different shape: device **30** is a 6-mm diameter sphere with two circular channels (1.57 cm^2 as surface area); device **31** is a "gear wheel" of 6-mm diameter and 3-mm height (2.06 cm^2 surface area); device **32** is a cubic woodpile structure (6 × 6 × 6 mm, 3.78 cm^2 surface area); and device **33** is a cube with nine channels for each face (6 × 6 × 6 mm, 4.36 cm^2 surface area). These devices were employed as organocatalysts for the synthesis of the 3-substituted indole **35** and the desired compound was obtained in 79% yield when 15 wt% thiourea-embedded device **33** was employed in a 1:1 catalyst/substrate molar ratio at room temperature (Scheme 10). Other device shapes were less effective in promoting the synthesis of product **35**. Recycling investigation demonstrated that these 3D-printed devices can be reused at least one time without significant loss of catalytic performance.

In addition to RC and STL technology, also SLS was successfully employed in the production of catalytically active solid objects. In 2019,

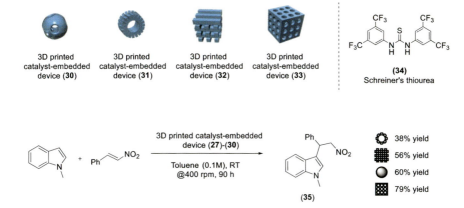

Scheme 10. STL. 3D-printed catalytically active devices for the Friedel–Crafts alkylation of N-Me-indole with trans-β-nitrostyrene. Adapted from Ref. 90 with permission from MDPI.

with the idea to further develop the catalytic stirrer proposed by Hilton, Haukka and coworkers reported the realization of a stirrer holder to perform heterogeneous hydrogenation of carbon–carbon double and triple bonds under hydrogen pressure.[91] Palladium on silica (Pd/SiO$_2$ with 5 wt% palladium) was mixed in a loading of 10 wt% with PP powder and the resulting mixture was subjected to SLS printing for the realization of stir bar sleeves **36** (Pd loading of the device = 0.5 wt%). Styrene, cyclohexene, and phenylacetylene were successfully reduced under 10 bar hydrogen pressure at room temperature for 2 h (Scheme 11(a)). Reactions performed in the presence of device **36** showed a slight decrease of activity compared to the reaction performed with pure Pd/SiO$_2$, but nearly identical conversions upon three consecutive recycle runs. Analysis of the palladium leaching revealed that only 0.57 wt% of the palladium present in a fresh 3D-printed stir bar sleeve was leached in the reaction mixture, a data that is comparable to that of the Pd/SiO$_2$ powder. With the same approach, Haukka reported also the realization of polyamide-12 SLS 3D-printed stir bar sleeves functionalized with gold nanoparticles **37** (gold was adsorbed as tetrachloroaurate), which were tested in the simple reduction of 4-nitrophenol to 4-aminophenol in the presence of NaBH$_4$ (Scheme 11(b)).[92] Reaction was completed in 2 h only, demonstrating that

a) Reductions of styrene, cyclohexene, and phenylacetylene catalyzed by Pd embedded 3D-Printed stir bar sleeves

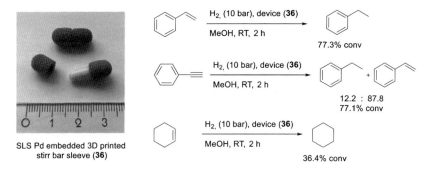

b) Reduction of nitrophenol catalyzed by Au embedded 3D-Printed stir bar sleeves

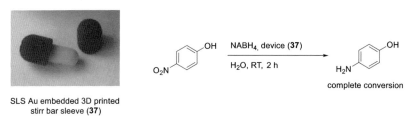

Scheme 11. (a) Reductions of styrene, cyclohexene, and phenylacetylene catalyzed by Pd-embedded 3D-printed stir bar sleeves. (b) Reduction of nitrophenol catalyzed by Au-embedded 3D-printed stir bar sleeves. Adapted from Refs. 91 and 92 with permission © 2019 American Chemical Society.

SLS process provides an alternative and effective way to produce highly active and easily reusable heterogeneous catalysts without significantly decreasing of the catalytic efficiency.

The architectural control offered by 3DP processes allows a straightforward production of devices endowed with different shapes and surface areas, and tunable properties as catalytic devices. More studies related to the characteristics of the materials used in the 3D printing of catalytic devices are needed; nevertheless, we believe that in the future, additive manufacturing might represent a great opportunity to fabricate catalytically active objects able to efficiently promote other organic reactions.

B. 3D-Printed Devices for In-flow Reactions

After initial skepticism, flow chemistry has been universally recognized by scientists as an important tool for the synthesis of chemicals, thanks to its advantages related to easy automation procedure, reproducibility, safety, and process reliability. The advantages include large surface to volume ratios for a better heat and mass transfer rate, small reagent volumes with consequent reduction or elimination of toxic or hazardous intermediates, possibility to safely operate under hard reaction conditions (e.g., above boiling points, high pressure), as well as a more efficient mixing, and control of reaction parameters.[93–102] Many applications have been reported both in academia and in the pharmaceutical industry,[103–108] giving witness to the fact that flow technology represent a valid — and in some cases also better — alternative to traditional batch production. Oversimplifying, we could say that continuous flow is the performance of chemical reactions in a pipe or tube, rather than in a traditional batch stirred vessel; in this contest, 3DP has introduced a way to create "*ad hoc*" fluidic devices in a cheap and easy way (3D-printed objects cost less than those sold by supply companies).[109] In 2012, Cronin and coworkers reported the first application of 3DP to the realization of fluidic devices for continuous-flow synthesis that came out with the name "3D-printed reactionware."[41,44] In this seminal work, a few simple reactors were printed in PP by the FDM approach and were successfully employed in reductive amination and alkylation reactions, polyoxometalate synthesis, gold nanoparticles and inorganic nanoclusters synthesis. After that, 3D-printed fluidic devices have been successfully employed for analytical and sensoristic applications,[39,110,111] and in further organic synthetic transformations.[20,65,112] According to their dimension, fluidic devices can be classified into micro-, meso-, and macro reactors.[113] Micro reactors present higher mixing efficiency (in the microfluidic scale, the flow regime is laminar), but are very sensible to solid occlusion and are subjected to strong limit in terms of productivity, which cannot be solved through the often proposed "numbering-up" approach.[114] On the other hand, meso- and macro reactors, thanks to their larger channel dimensions, enable higher throughputs, but are characterized by low mixing efficiency. However, in spite of appearances, fluidic devices are able to promote both heterogeneous and homogeneous reactions. A survey of the most significant examples will be discussed in this section.

1. Reactions under heterogeneous conditions

In 2017, Sans and coworkers reported the realization of a miniaturized continuous-flow, STL-based, 3D-printed oscillatory baffled reactor (a particular fluidic device that ensures the mixing of reagents by oscillating the reaction fluid through orifice plate baffles placed at controlled intervals in the flow path) for the synthesis of silver nanoparticles.[115] This device is characterized by the presence of three inlets necessary for the synthesis of silver nanoparticles, and two inlets for residence time distribution studies, with a total volume of 2.7 mL distributed along a tubular section of 2.5 mm of diameter. The fluidic device and apparatus setup are schematically described in Figure 1(a). The setup consists in two high-performance liquid chromatography (HPLC) pumps, the 3D-printed reactor, a UV-visible spectrometer with its light source, and a back-pressure regulator. An HPLC pump was employed to generate the oscillation (0.1 mm, 40 Hz), using an immiscible silicon oil to prevent back mixing of the reagents, whereas another pump was employed to feed the reactor with a 0.25 mM solution of $AgNO_3$ in CH_3CN and a 0.75 mM mixture of $NaBH_4$ and polyvinylpyrrolidone in CH_3CN. The synthesis of silver nanoparticles was run for 100 min at a combined flow rate of 2 mL min^{-1}. The entire system

a) Oscillatory baffled reactor

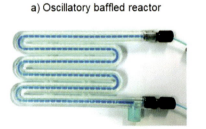

Figure 1. (a) 3D-Printed fluidic oscillatory baffled reactor. Adapted from Ref. 115 with permission from the Royal Society of Chemistry. (b) Composed FDM-based 3D-printed microfluidic device showing fouling along the channel. Adapted from Ref. 116 with permission from Elsevier.

was pressurized at 4 bar using the back-pressure regulator and the formation of product was monitored by UV-visible at the outlet of the reactor. It was found that these fluidic devices offer improved mixing conditions at a millimeter scale, when compared to tubular reactors, leading to the formation of nearly monodisperse particles (particle size of 5.0 nm ± 1.2 nm) with an excellent stability over time due to the reduced fouling obtained by a simple optimization of the oscillation conditions.

Following this example, Silva and coworkers reported the fabrication of a composed FDM-based, 3D-printed microfluidic device made by poly(lactic acid) and poly(methyl methacrylate) for the synthesis of silver and gold nanoparticles under continuous-flow conditions (Figure 1(b)).[116] The particularity of this device is the presence of a transparent window to perform electrochemical measurements and to visualize the flow inside the channels. In order to reduce fouling inside the microchannels, the device was optimized to use a segmented flow of mineral oil. Silver nanoparticles were synthesized at 20°C starting from $AgNO_3$ in the presence of sodium borohydride, whereas gold nanoparticles were synthesized at 90°C starting from a solution of gold(III) chloride in the presence of trisodium citrate as reducing agent. Nanoparticles were characterized by UV-visible spectroscopy and scanning and transmission electron microscopy, which allows to identify silver nanoparticles with sizes ranging from 5 ± 2 to 8 ± 3 nm and gold nanoparticles with sizes ranging from 20 ± 9 to 34 ± 12 nm.

3D-Printed fluidic devices with heat exchange ability have found applications in continuous-flow cooling crystallizations. In this field, Sans and coworkers designed and manufactured a modular jacketed flow device for the crystallization of paracetamol form II (Figure 2).[117] The device employed was realized using a SLA 3D printer (total volume of 2.95 mL, with 2.5 mm as channels diameter) and fed with an aqueous isopropyl alcohol solution of paracetamol form I, combined with metacetamol as a cocrystallizing agent. The recrystallization process was performed with a flow rate of 0.5 mL min^{-1} (residence time = 5.9 min) at constant temperature; the process achieved a steady state after 30 min, and a significantly larger amount of form II rather than form I was observed. Although the crystallization of paracetamol polymorph form I was not quantified, this process represents the first example in which polymorphs formation has been controlled under continuous-flow conditions.

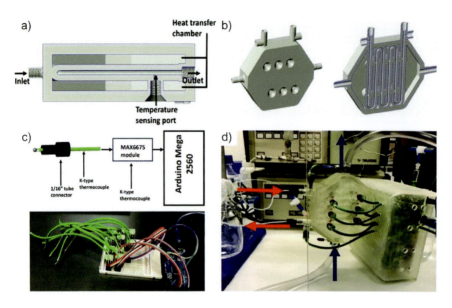

Figure 2. (a) Cross section of the device showing the heat exchange chamber, the crystallizer channel, and the temperature-sensing ports. (b) CAD model of the fluidic devices. (c) Schematic design of the temperature measuring devices and a picture of the actual setup. (d) Rig setup for heat transfer characterization in the additively manufactured crystallizer. The setup comprises a recirculating chiller, HPLC pump, and temperature sensors connected to the crystallizers. Reproduced from Ref. 117 with permission from the Royal Society of Chemistry.

Very recently, McDonough, Harvey, and coworkers reported the realization of 3D-printed flow helical tubes[118] and coil-in-coil flow reactor by a 3DP SLA approach that consists in a complex tube arrangement designed by adding a helical coil tube insert to a helical-coiled tube operating in a net flow with superimposed oscillation.[119] Even if a synthetic application was not reported, it was demonstrated that this device exhibits exceptional plug flow qualities, which can be implemented in crystallization, biofuel synthesis, and polymerization processes.

2. Reactions under homogeneous conditions

The first examples of organic reactions performed under continuous-flow conditions using 3D-printed fluidic devices consist in the synthesis of aromatic anilines *via* reduction of the corresponding imines.[44,74,120] A few basic reactors, bearing standardized inlet and outlet connections, were realized in PP by a FDM 3DP approach. Reductive amination was performed according to a tandem methodology in which two reactors were connected in sequence; imines formation occurs in the first reactor (circular channel shape, ID: 1.5 mm) where a solution of benzaldehyde and a solution of a substituted aniline dissolved in MeOH were fed by means of a syringe pump. The outlet of the first reactor was then connected to the inlet of the second 3D-printed device in which the resulting imine solution was combined with a solution of NaBH$_3$CN. The molar and volumetric hydride/imine ratios were kept constant (1:1), and the corresponding amines were obtained in good yields after 7-min residence time (Scheme 12(a)).

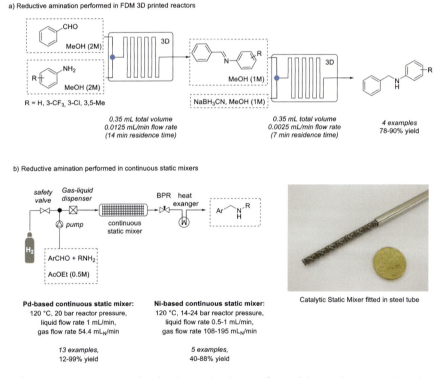

Scheme 12. Examples of reductive amination performed in continuous static mixers. Adapted from Ref. 121 with permission from Springer.

An evolution of this system was recently reported by Hornung and coworkers, who realized a 3D-printed catalytic static mixer to promote the reductive amination of aldehydes and ketones inside a continuous-flow reactor.[121] A tubular reactor geometry with 3D-printed and catalytically active tube insert, coated with a catalytically active layer of either Pd or Ni, was realized by a combination of additive manufacturing and metal deposition techniques (electroplating and meta cold spraying). The Pd^0 catalyst was supported on a 3D-printed metal scaffold, which was inserted in a steel column (inner diameter 6 mm, length 150 mm) where hydrogen was fluxed at 120°C and 16 bar of pressure (gas flow rate of 67.9 mL/min). After 60 min, the reactor was flushed with ethyl acetate, then a substrate solution containing benzaldehyde and aliphatic amine was introduced. Upon start-up of the reactor system, hydrogen gas was introduced (120°C, 20 bar pressure, liquid flow rate 1 mL/min, gas flow rate 4 mL/min) and the reaction was monitored by ^{1}H-NMR and GC. Analysis of the results revealed that the reductive aminations reached high conversions to the desired functional amines, with up to 99% yield using both Pd-based and Ni-based continuous-flow reactors, even if, in the latter case, further optimization of the reaction conditions was necessary (Scheme 12(b)).

These nickel(0)- and palladium(0)-coated catalytic static mixers are very versatile, since it was previously reported that they are able to promote other hydrogen-based reductions, such as the in-flow hydrogenation of oleic acid, vinyl acetate, cinnamaldehyde,[122] as well as hydrogenation of alkenes, alkynes, halides, carbonyls, and organic nitrogen compounds (Scheme 13(a)).[123] Catalytic static mixers were also used in the preparation of a key intermediate for the synthesis of the antimicrobial drug linezolid (Scheme 13(b)).[124]

A different catalytic static mixer was previously reported by Rudolf von Rohr and coworkers who described the use of a porous-structured reactor consisting in an aluminium oxide–zinc oxide–based layer, coated with palladium. This reactor was employed in a proof of concept experiment, in the reduction of 2-methyl-3-butyn-2-ol to the corresponding saturated alcohol, under solvent-free, continuous-flow conditions.[125] The device was realized by a SLS approach and it allowed to achieve higher performances in terms of selectivity, yield, and turnover frequency compared to the traditional batch process.

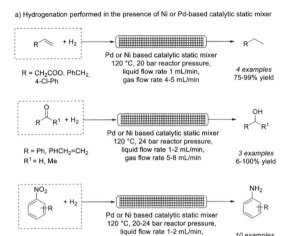

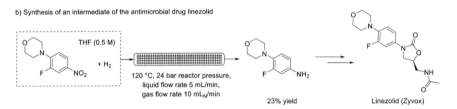

Scheme 13. Continuous-flow hydrogenations using catalytic static mixers inside a tubular reactor.[122–124]

The advantage associated to these tubular reactors is related to a better control of heat and mass transfer along the length of the device, avoiding issues related to the nonuniformity typically observed in batch, and to the possibility to operate under high pressure and to control the mixing performance by the realization of specific structures by 3DP.

3DP *via* SLM was employed by Kappe and coworkers for the realization of a continuous-flow reactor printed in stainless steel and specifically designed for the fast difluoromethylation of 2,2-diphenylacetonitrile **38**, with *n*-BuLi as base and gaseous fluoroform as atom-economic and inexpensive reagent.[126] The reactor presents four inlets (three inlets are dedicated to substrates, whereas the last one is reserved to the quench solution) and a cylindrical channel (internal diameter of 0.8 mm, outer diameter of

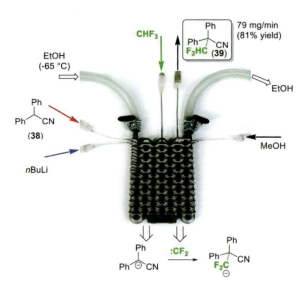

Figure 3. Continuous-flow α-difluoromethylation with fluoroform. Adapted from Ref. 126 with permission from the Royal Society of Chemistry.

2.4 mm), which is directly attached onto the cooling serpentine to reduce heat transport distances and to facilitate the cooling process. The substrate and n-BuLi were fed into the device at different flow rates (800 μL min^{-1} and 360 μL min^{-1}, respectively) and cooled to −65°C in a precooling zone, before to be combined with CHF$_3$ introduced by a mass flow controller (26.7 mL min^{-1} as flow rate). After a zigzag reaction channel, the fourth inlet guarantees the introduction of MeOH as a quench solution, at a flow rate of 1 mL min^{-1} (Figure 13). Difluoromethyldiphenylacetonitrile **39** was obtained in excellent purity and in 81% yields in less than 2 min.

In another example, 3D-printed stainless steel metal micro reactors, fabricated by high-resolution SLM, were employed to avoid the Fries rearrangement of phenolic carbamate **40** into the corresponding o-hydroxybenzamide. Using various chlorides (RCl) as reactants, the intermolecular trapping products **41–44** could be obtained, with high conversion rates and yields.[127] The microfluidic device of 25 nL total volume is composed by circular cross-sectional fluidic channels arranged in a simple T-shaped geometry, which was extremely important for high mixing efficiency and for the suppression of side reactions (Scheme 14). Computational flow

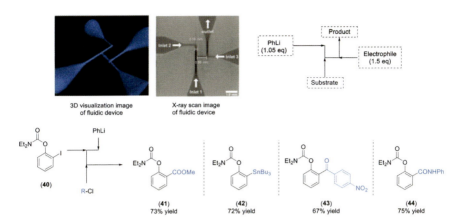

Scheme 14. Conceptual scheme and actual images of the fabricated microfluidic device (top). Control of the Fries rearrangement of compound **40** and intermolecular trapping with various electrophiles (bottom). Adapted from Ref. 127 with permission © Wiley-VCH GmbH, Weinheim.

dynamics simulations were essential for the design of the microfluidic device.

Another interesting application of 3DP technology in the field of flow chemistry was the creation of a polydimethoxysilane microfluidic systems for the acetylation of different amines, under microwave irradation.[128] At the contrary of traditional approaches, Hur *et al.* used DIW technology to print sacrificial microchannels using a water-soluble Pluronic F127 ink. The microchannels were encapsulated in a mixture of polydimethoxysilanes and the device was cured at room temperature for 24 h (Scheme 15). After that, channels were dissolved by washing with cold water and ethanol. The device so constructed was employed in the reaction between *N*-acetylbenzotriazole (**45**) and several amines under 100 W microwave irradiation, using tetrahydrofuran as solvent and with a reagents flow rate of 0.3 mL min^{-1}. The desired acetamides were obtained in high yields, showing a dramatically reduced reaction time compared with batch reactions performed under microwave or conventional heating.

In the same year, Pauer and coworkers reported the realization of a FDM 3D-printed tubular bended macro reactor made by PLA, to perform emulsion copolymerizations of styrene, butyl acrylate, and vinyl

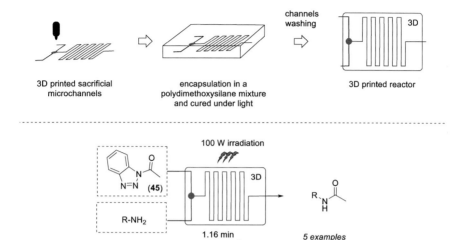

Scheme 15. Microwave-assisted *N*-acetylation of amines performed in a polydimethoxysilane microfluidic system.[128]

acetate.[129] The reactor is constituted by a circular channel four-fold serpentine with a total volume of 330 mL (18 mm inner diameter), with a total length of 1340 mm and with a watertight up to 8 bars of pressure. This fluidic device has been inserted in a modular reaction plant for the preparation of uniform copolymer particles at various solid contents. Continuous, redox-initiated emulsion copolymerizations of styrene — butyl acrylate and of vinyl acetate — neodecanoic acid vinyl esters were successfully performed with 20 and 40 wt% monomer in the feed (Scheme 16). The emulsion copolymerizations were initiated at room temperature using a redox initiator system (TBHP/AsAc/Fe-cat), with a mean residence time between 5 and 15 min, depending on the final mixture desired, allowing to observe conversions between 81% and 99%. The resulting emulsions were stable with Zeta potential values up to −132.0 mV and no product adhesion to the reactor wall was found. In addition, no fouling, clogging, or deformation of the 3D-printed reactor was observed.

In 2019, Ziegler and coworkers reported multistep sequential glycosylation reactions performed in 3D-printed FDM fluidic devices and continuous stirred tank reactors with the assistance of a 3D-printed syringe pump

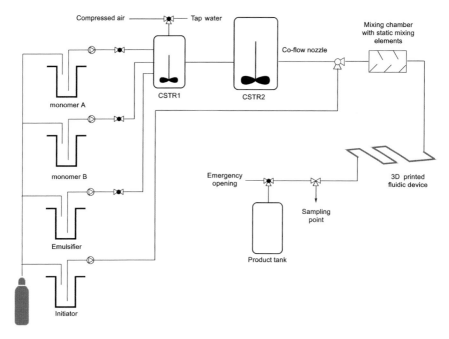

Scheme 16. Instrumentation diagram of the reactor setup for continuous emulsion copolymerizations.

controlled by open-source Arduino software.[130] In this work, Ziegler and coworkers describe the fabrication of many fluidic devices in PP, with different shapes and dimensions, for the synthesis of the glycosyl donor acetobromo-α-D-glucose **47**, its corresponding methyl glycoside **48**, and the imidate glycosyl donor **50**. Compound **47** was synthetized by reaction of pentaacetylglucose **46** with HBr in acetic acid (33%) using a fluidic device; reaction workup and isolation of the product were performed using two continuous stirred tank reactors (86% yield, Scheme 17(a)). After isolation, acetobromo-α-D-glucose (**47**) was subjected to Koenigs–Knorr reaction in another flow setup for the synthesis of compound **48** in 44% yield (Scheme 17(b)). In addition, compound **46**, after cleavage of the anomeric acetyl group using hydrazine acetate under traditional batch conditions, was converted into trichloroacetimidate **50** according to the schematic pathway reported in Scheme 17(c). In a further application, two-step glycosylation reactions were also performed under a tandem approach, leading to the formation of various glycosides in 43–69% yield.

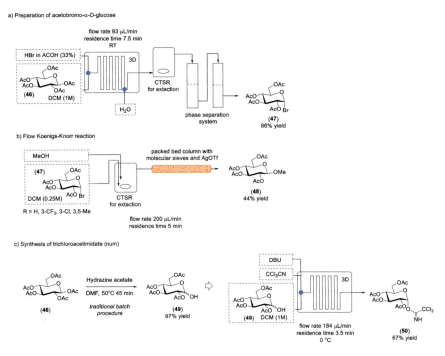

Scheme 17. Glycosylation reactions performed in multiple 3D-printed flow reactors.[130]

In the same year, Hilton and coworkers reported the development of a modular 3D-printed, compressed air-driven continuous-flow systems for chemical synthesis entirely printed using a FDM 3D printer.[131] This system results compatible with traditional stirrer hotplates equipped with DrySyn Multi-E adaptor. Interestingly, flow conditions are guaranteed by the use of a compressed air-based controller, instead of syringe pumps. The applicability of this system was demonstrated performing nucleophilic aromatic substitution (SNAr) reactions between 2-chloro-5-nitropyridine **51** and differently substituted phenols for the synthesis of corresponding ethers, with good results (Scheme 18). The system is composed by a solvent block and an injection block to control the disposal of reagents, a flow block with a compressed air-based controller to regulate the flow and a basement that hosts circular fluidic devices printed in PP (total internal volume 4.2 mL). A solution of **51** in acetonitrile was reacted

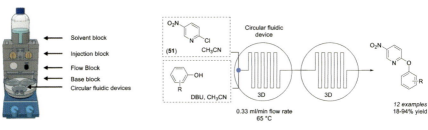

Scheme 18. Modular 3D-printed compressed air-driven continuous-flow systems for SNAr reactions. Adapted from Ref. 131 with permission © Wiley-VCH GmbH, Weinheim.

with a mixture of the desired phenol and 1,8-diazabiciclo[5.4.0]undec-7-ene (DBU) in acetonitrile at 65°C, with a flow rate of 0.33 mL/min. Two circular disk reactors connected in sequence were necessary to allow the synthesis of desired compounds with modest to good yields.

An advantage of this system is the possibility to perform the reaction under flow conditions using a low-cost equipment, without investing in expensive apparatus.

The creation of new 3D-printed fluidic devices and their application in organic synthesis is in continuous expansion: the fabrication of a serpentine micro channel reactor realized by thermally bonding stacked polyimide films, to promote the biohydration of acrylonitrile,[132] or 3D-printed stainless steel reactors with inline oxygen sensors for the aerobic oxidation of Grignard reagents under continuous-flow conditions[133] are just some of other examples.

Very recently it was shown that the synthesis of organic compounds can be performed using immobilized enzymes on 3D-printed supports, for applications in continuous-flow chemistry.[134] In 2018, Rabe and coworkers developed an inexpensive, agarose-based, thermoreversible, 3D-printed enzyme-integrated hydrogel that has been employed for the realization of flow reactor cartridges.[135] A thermostable mutant of ketoisovalerate decarboxylase (KIVD mutant) from *Lactococcus lactis* was encapsulated in 3D-printed grid-structured disks, which were used for modular, tunable, two-step biotransformations under continuous-flow conditions. The decarboxylation of ketoisovalerate **52** to isobutyraldehyde performed

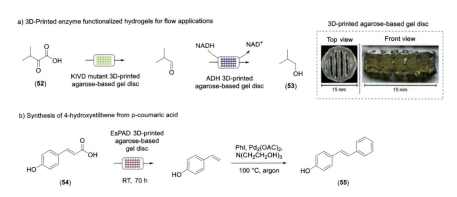

Scheme 19. Examples of enzyme-functionalized 3D-printed devices applied to various organic reactions under continuous-flow conditions. Adapted from Ref. 135 with permission © Wiley-VCH GmbH, Weinheim.

using KIVD-mutant disks and the reduction of isobutyraldehyde to isobutanol **53** using a alcohol dehydrogenase (ADH)-functionalized disk was successfully accomplished in a modular fashion, giving 40% conversion when operating with 25 mL min^{-1} as a flow rate (Scheme 19(a)). By an identical approach, the same authors reported also the realization of 3D-bioprinted disk-shaped agarose-based modules functionalized with phenacrylate decarboxylase, an enzyme derived from *Enterobacter* (EsPAD). These modules were placed in a fluidic device and used for the decarboxylation of *p*-coumaric acid **54**. This process, combined with a subsequent palladium(II) acetate-catalyzed Heck reaction led to the synthesis of 4-hydroxystilbene **55** starting from *p*-coumaric acid **54**, with a total yield of 14.7% on a milligram scale (Scheme 19(b)).[136]

The numerous examples of reactions performed under continuous-flow conditions reported in this section clearly demonstrate how 3D printing can play a crucial role in the development of flow technology–assisted organic reactions, including multistep continuous-flow process for the synthesis of pharmaceutically relevant products.

C. 3DP for Drugs and Pharmaceuticals Applications

3DP has started to acquire an important role in the development of drugs, thanks to its capability to create precise dosage formulation in various shapes, sizes, and textures (which are difficult to produce using traditional techniques), and thanks to the possibility of making available drug production in locations where the traditional distribution is not allowed (i.e., space expeditions or war zones).[137–141] In addition, 3DP technology opened the way to the creation of patient-based pharmaceuticals, which meet the specific pharmacogenomic, anatomical, and physiological requirements of the single patient.[142,143] Even if a large number of hurdles needs to be overcome to make this belief a reality, such as problems related to drug stability, sterilization, and satisfaction of all regulations in quality control,[144–147] many studies related to the application of 3DP technologies in pharmaceutical manufacturing have highlighted the numerous potential advantages (and disadvantages) of this approach.[148–151] Thus, a huge number of studies on new orally disintegrating tablets and new porous substrates for a better drug delivery were reported in the literature.[152–154] The two most important 3DP methodologies employed in drug development and fabrication are powder-based printing[155–159] and IJP,[160–162] but examples related to FDM,[163–166] hot melt extrusion,[167] SLS,[9,168] STL,[169–172] electrohydrodynamic printing,[173] and syringe/extrusion 3DP[174] are also known. Since the availability of a pharmaceutical compound in the body is correlated to a specific carrier, careful consideration must be taken in the selection of the appropriate 3D-printable "ink," especially when multidrug pills are realized with a compartmentalized polypill approach.[175,176] Actually, the application of 3DP technology to pharmaceuticals still presents problems related to the printable formulation, and the conventional methods to produce tablets will not be replaced soon by 3DP.[177] Nevertheless, in 2016, the Food and Drug Administration (FDA) approved the first 3D-printed tablet containing levetiracetam manufactured by 3DP (Spritam,® Aprecia Pharmaceuticals),[178] making closer the realization of tablets containing the exact drug dose to suit an individual patient need and creating a significant milestone in 3DP.

Although many examples in the literature are focused on the realization of printable formulations, the manufacturing processes of pharmaceutical products that involve 3DP technology are rather limited. An interesting approach in this field was reported by Cronin and coworkers, who developed an automated synthesis robot, which is able to fabricate PP reaction vessels, by FDM 3DP, and subsequently to employ these devices to synthesize ibuprofen **56** *via* a consecutive one-pot three-step reaction.[179] Starting from isobutylbenzene and propanoic acid, the synthesis of ibuprofen was achieved in six fully automatized operations and a simple tuning of the software parameters allows to scale up the reaction up to 60 mg of final product. The robot was programmed to print the reaction vessel and to control syringe pumps for dispensing the reagents inside the vessels. Reactions were conducted at room temperature in air, and the combined processes (printing and synthetizing) were completed in 24 h, affording compound **56** in 34% overall yield, without the assistance of an operator (Scheme 20). Another synthetic process entirely performed in 3DP reactionwares was already discussed in Scheme 1 and relies to the synthesis of baclofen, lamotrigine, and zolimidine.[75] In this case, a minimal human interaction is required, but it is limited to simple interventions at specific time periods (loading reagents, opening valves, etc.).

Recent advances of 3DP technologies in pharmaceutical applications are devoted to the development of personalized topical drug delivery systems. For example, in 2016, Gaisford and coworkers reported the realization of a facial mask containing an antiacne drug, based on a 3D reconstruction of the patient face.[180] 3D scanning of a volunteer subject was performed in order to capture a 3D image of his face that was used as a mold for the realization of a SLA 3D-printed mask, made with a mixture of photo cross-linkable resin and salicylic acid (Figure 4). A 4:6 mixture of polyethylene glycol diacrylate:PEG (PEGDA:PEG) was premixed with 2%wt of salicylic acid and used as feedstock material for 3DP. The drug completely dissolved in the photopolymer solution and no crystallization occurred during the photopolymerization process. Interestingly, drug diffusion values from the masks after 3 h (291 $\mu g/cm^2$) were higher compared to those obtained using skin as a membrane in Franz cells after 24 h (<60 $\mu g/cm^2$), demonstrating that the combination of 3D scanning and 3DP technologies has the potential to create personalized drug-loaded devices, adaptable in size and shape to individual patients.

The number of pharmaceuticals 3DP applications is continuously growing, and nowadays 3D-printed drug-functionalized orthodontic retainers,[181,182] intratumoral implants,[183] or wound dressings[184] have been realized.[185,186] 3DP has the right credentials to take a leading role in new formulations production, to move away from mass production and to

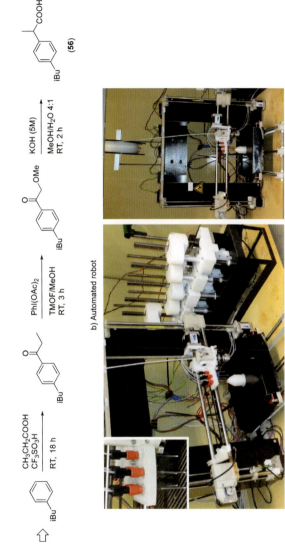

Scheme 20. (a) Synthesis of racemic ibuprofen using modified 3D printer. (b) 3D printer for the automated synthesis of ibuprofen: (left) full view of robotic platform (left inset), dispensing needle carriage for 3DP/liquid deposition; (right) front view of the 3DP section of the robotic setup with a 3D-printed reaction vessel showing the PP feedstock for reaction vessel printing. Reproduced from Ref. 179 with permission from Beilstein Journal of Organic Chemistry.

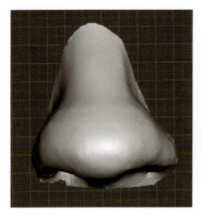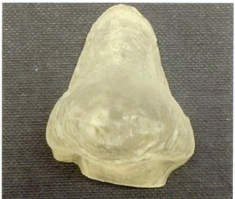

Figure 4. A model of a nose (left) and nose-shaped mask fabricated by SLA 3DP PEGDA/PEG (4:6)-salicylic acid (right). Reproduced from Ref. 180 with permission from Elsevier.

create highly flexible and customized devices for a patient-based medical health. Nonetheless, more investigations are required on its manufacturing aspects in terms cost effectiveness, reproducibility, and bioequivalence.

III. 3D-Printed Devices in Stereoselective Reactions

In the last few years, the application of 3DP technologies in organic synthesis led to the development of a huge variety of devices and reactors that were implemented in many synthetic methodologies, leading to more efficient transformations, compared to the traditional approach. Despite the growing number of applications in organic synthesis, only a few examples of stereoselective transformations for the synthesis of chiral molecules have been reported so far. The first pioneering example dates back to 2017, when products of pharmaceutical interest such as metaraminol **57**, norephedrine **58**, and methoxamine **59** were synthetized under continuous-flow conditions using a FDM 3D-printed meso reactor (Scheme 21).[187] The approach reported for the synthesis of these compounds is a two-step sequence that combines an enantioselective and catalytic Henry reaction with a hydrogenation process. In the first step, aromatic aldehydes were combined with nitroethane under the presence of a chiral copper organocatalyst to generate

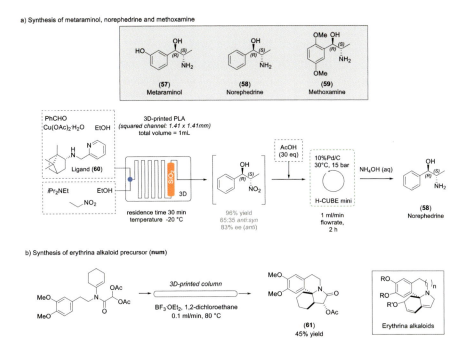

Scheme 21. Stereoselective catalytic synthesis of active pharmaceutical ingredients in homemade 3D-printed mesoreactors.[187]

the corresponding nitroalcohol, using 3D-printed flow reactors differing in channel shapes and dimensions and in building materials (PLA, HIPS, and nylon). The nitroalcohol obtained was then subjected to a short silica pad filtration before undergoing a Pd/C catalyzed in flow hydrogenation using an H-cube apparatus, which generates hydrogen gas from water under safe conditions (Scheme 21(a)). Thanks to the 3DP technology, the silica pad was directly inserted into the fluidic device, allowing to perform the filtration step under continuous-flow condition, with a consequent decrement of the operations required for the synthesis of the desired norephedrine **58**, notably. The chiral copper complex catalyst obtained by combining Cu(OAc)$_2$·2H$_2$O and chiral ligand **60**, derived from camphor, ensured good *anti* diastereoselectivity. In one of the best setup arrangements, norephedrine was synthetized in 90% overall yield, 65:35 *anti:syn* ratio and with an enantiomeric excess of 80% (*anti* isomer), starting from commercially available benzaldehyde and nitroethane.

Another example of 3D-printed fluidic device applied to stereoselective transformations was reported by Hilton and coworkers in 2019, where a 3D-printed PP continuous-flow column reactor was incorporated into an existing continuous-flow system and applied to the diastereoselective synthesis of racemic bicyclic and tetracyclic heterocycles.[188] The stereoselectivity of this transformation is not affected by the geometry of the fluidic device, thus compound **61**, a precursor of the erythrina alkaloids family, was isolated as a single diastereoisomer in 45% yield when operating at 80°C with a flow rate of 0.1 mL/min (Scheme 21(b)). In order to facilitate prolonged contact with the outside hot surface of the column block, the reactor was designed with an internal spiral flow path coil with a diameter of 2 mm and a theoretical internal volume of 1.6 mL. The thermal resistance was ensured using PP as material, and the device was printed using a FDM 3D printer.

The use of enzyme-functionalized devices under flow conditions is an alternative approach for the controlled introduction of stereogenic centers into a desired molecule, and examples of continuous biotransformations using immobilized enzymes are nowadays quite common.[134,189] In this sense, in 2017, Sans and coworkers developed a method to efficiently immobilize the ω-transaminase ATA117 enzyme on a 3D-printed continuous-flow device and its application in the kinetic resolution of α-methylbenzylamine (Scheme 22).[190] Interestingly, the suitable enzyme and immobilization conditions could be identified rapidly by parallel screening in 3D-printed multiplexed multiwell plates. The results of the screening could be transferred then readily into the continuous-flow micro-structured devices. The biofluidic reactor presents a channel diameter of 1.5 mm with a total volume of 0.5 mL. It was built with nylon Taulman 618 polymer using a FDM 3D printer and the surface of the device was sequentially modified to enable the adhesion of the enzyme. The selective conversion of (R)-α-methylbenzylamine into acetophenone was investigated under fluidic conditions (Scheme 22(a)). When operating with a flow rate of 10 μL min^{-1} (50 min as residence time) in the presence of pyruvic acid and pyridoxal phosphate (PLP), almost full conversion of the (R)-α-methylbenzylamine was observed (>49%) yielding the unreacted (S)-amine **62** with >99% enantiomeric excess. It was demonstrated that the reactor was stable for 100 h, but despite the good performances, the total enzyme activity decreased of 75% compared to the free enzyme.

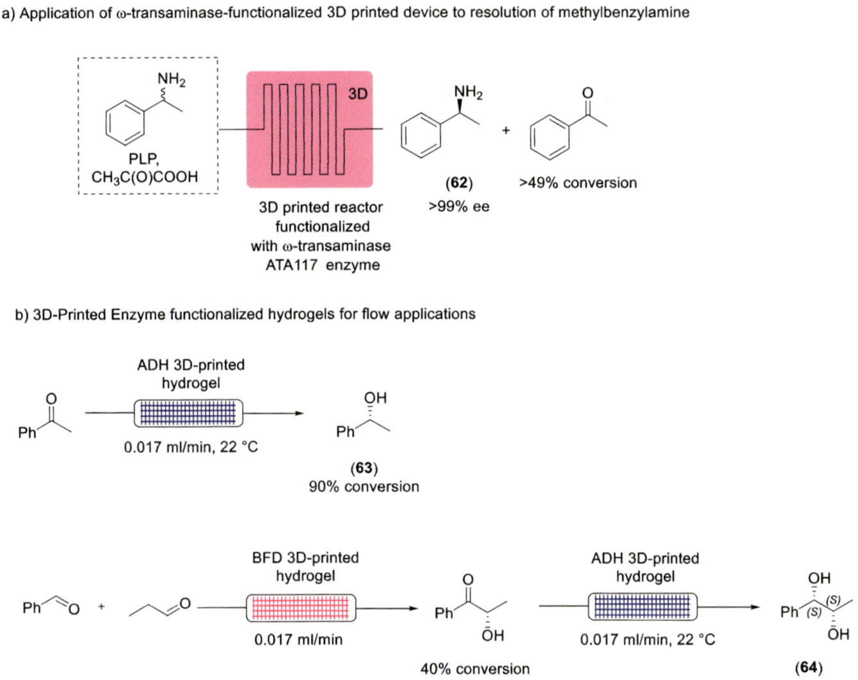

Scheme 22. Examples of enzyme-functionalized 3D-printed devices applied to stereoselective reactions under continuous-flow conditions.[190,191]

In a more recent approach, Franzreb and coworkers describe a direct immobilization of enzymes by physical entrapment into 3D-printed hydrogel lattices that were placed inside a fluidic device.[191] An ADH from *Lactobacillus brevis* (ADH) and a benzoylformate decarboxylase from *Pseudomonas putida* (BFD) were entrapped in a crosslinked poly(ethylene glycol) diacrylate-based hydrogel, realized using an extrusion-based bioprinter. The 3D-printed lattice is a woodpile structure (13 × 13 × 3 mm^3 with a strand distance of 1.5 mm) that was inserted into a 3D-printed reactor housing. The system was able to operate under continuous-flow conditions to perform different enzymatic reactions. The ADH-functionalized device was employed in the synthesis of (*R*)-phenylethanol **63** from acetophenone, whereas the combined use of BFD- and ADH-functionalized flow reactors allowed to obtain (1*S*,2*S*)-1-phenylpropane-1,2-diol **64** in good yield and with complete stereocontrol (Scheme 22(b)).

In a very recent approach, also penicillin G acylase, protease, glycosidase, and lipase enzymes were anchored to different 3D scaffolds, and employed as proof of concept in amoxicillin and lactosucrose synthesis.[192]

IV. Conclusion

Additive manufacturing is revolutionizing our world with advances in medicine, food, energy, and water technologies, since it allows one to fabricate structures with complex geometries that cannot be achieved through traditional methods. Over recent years, 3DP has moved toward printing functional components that can be used in a variety of applications including electronics, energy storage, thermal management, healthcare monitoring, sensors, and robotics.

Although, at present, 3DP revolution failed to replace manufacturing, mainly because it would require expensive upfront investment, especially in terms of creating designs for the desired parts, it excels at specialist innovations. As a cutting-edge technology today, 3DP is best used for targeted, specific functions and its applications are often innovative and unique, as demonstrated by its numerous medical applications. As recent developments show, the future of 3D printers lies not in mass production, but in specialized applications, such as crafting replacement organs or artificial limbs. In the last decade, 3DP has found many applications also in chemistry, and, in the last very few years, chemists have started to combine 3DP technology with organic compounds for the development of "3D-printed catalysts," which have been employed in organic synthesis.

Many challenges still remain for the future: 3D printers have to become more robust, more user-friendly, and more efficient. In addition, the development of novel materials with customized properties, finalized to the final application, is an open field for future investigations; these devices should also be adapted to use sustainable materials, such as upcycled waste plastic. Although several technical and economy-related issues need to be solved, 3D printing could change the way we will produce our catalysts in the future.

V. Acknowledgments

The authors thank PSR 2019 (Università degli Studi di Milano); AP thanks Transition grant 2018.

VI. References

1. Yap, Y. L.; Yeong, W. Y. *Virtual Phys. Prototyp.* **2014**, *9*, 195–201.
2. Kim, S.; Seong, H.; Her, Y.; Chun, J. *Fash Tex.* **2019**, *6*, 9. doi:10.1186/s40691-018-0162-0
3. Yan, Q.; Dong, H.; Su, J.; Han, J.; Song, B.; Wei, Q.; Shi, Y. *Engineering* **2018**, *4*, 729–742.
4. Dawood, A.; Marti-Marti, B.; Sauret-Jackson, V.; Darwood, A. *Br. Dent. J.* **2015**, *219*, 521–529.
5. Chaudhary, R.; Doggalli, N.; Chandrakant, H. V.; Patil, K. *Int. J. Forensic Odontol.* **2018**, *3*, 59–65.
6. Roberts, B. H. *The Third Industrial Revolution: Implications for Planning Cities and Regions.* URBAN FRONTIERS PTY LTD Brisbane, QLD, Australia, **2015**, p. 1.
7. Joshi, S. C.; Sheikh, A. A. *Virtual Phys. Prototyp.* **2015**, *10*, 175–185.
8. Froes, F.; Boye, R. Eds. *Additive Manufacturing for the Aerospace Industry*, 1st ed. Amsterdam, The Netherlands: Elsevier, **2019**.
9. Fina, F.; Goyanes, A.; Gaisford, S.; Basit, A. W. *Int. J. Pharm.* **2017**, *529*, 285–293.
10. Melchels, F. P.; Feijen, J.; Grijpma, D. W. *Biomaterials* **2010**, *31*, 6121–6130.
11. Sun, J.; Zhou, W.; Huang, D.; Fuh, J. Y. H.; Hong, G. S. *Food Bioprocess Technol.* **2015**, *8*, 1605–1615.
12. Guo, C.; Zhang, M.; Bhandari, B. *Compr. Rev. Food Sci. Food Saf.* **2019**, *18*, 1052–1069.
13. Bikas, H.; Stavropoulos, P.; Chryssolouris, G. *Int. J. Adv. Manuf. Technol.* **2015**, *83*, 389–405.
14. Dahotre, N. B.; Harimkar, S. *Laser Fabrication and Machining of Materials.* Springer Science + Business Media, LLC, 233 Spring Street, New York, NY 10013, USA, **2008**.
15. Hartings, M. R.; Ahmed, Z. *Nat. Rev. Chem.* **2019**, *3*, 305–314.
16. Maruo, S.; Nakamura, O.; Kawata, S. *Opt. Lett.* **1997**, *22*, 132–134.
17. Carlotti, M.; Mattoli, V. *Small* **2019**, *15*, e1902687.
18. Costa, L.; Vilar, R. *Rapid Prototyping J.* **2009**, *15*, 264–279.
19. Louvis, E.; Fox, P.; Sutcliffe, C. J. *J. Mater. Process. Technol.* **2011**, *211*, 275–284.
20. Parra-Cabrera, C.; Achille, C.; Kuhn, S.; Ameloot, R. *Chem. Soc. Rev.* **2018**, *47*, 209–230.
21. Singh, R.; Garg, H. K. Fused Deposition Modeling — A State of Art Review and Future Applications. In *Reference Module in Materials Science and Materials Engineering.* Elsevier, **2016**

22. Hurt, C.; Brandt, M.; Priya, S. S.; Bhatelia, T.; Patel, J.; Selvakannan, P. R.; Bhargava, S. *Catal. Sci. Technol.* **2017**, *7*, 3421–3439.
23. Ko, D.-H.; Gyak, K.-W.; Kim, D.-P. *J. Flow Chem.* **2017**, *7*, 72–81.
24. Vyavahare, S.; Teraiya, S.; Panghal, D.; Kumar, S. *Rapid Prototyping J.* **2020**, *26*, 176–201.
25. Feilden, E.; Blanca, E. G.-T.; Giuliani, F.; Saiz, E.; Vandeperre, L. *J. Eur. Ceram. Soc.* **2016**, *36*, 2525–2533.
26. Chen, Z.; Li, Z.; Li, J.; Liu, C.; Lao, C.; Fu, Y.; Liu, C.; Li, Y.; Wang, P.; He, Y. *J. Eur. Ceram. Soc.* **2019**, *39*, 661–687.
27. Cooper, K. G. *Rapid Prototyping Technology: Selection and Application.* CRC Press, **2019**.
28. Safari, A.; Danforth, S. C.; Allahverdi, M.; Venkataraman, N. Rapid prototyping. In *Encyclopedia of Materials: Science and Technology*; Buschow, K. H. J.; Cahn, R. W.; Flemings, M. C.; Ilschner, B.; Kramer, E. J.; Mahajan, S.; Veyssière, P., Eds. Elsevier: Oxford, **2001**, pp. 7991–8003.
29. Vayre, B.; Vignat, F.; Villeneuve, F. *Mech. Ind.* **2012**, *13*, 89–96.
30. Kloski, L. W.; Kloski, N. *Getting Started with 3D Printing: A Hands-on Guide to the Hardware, Software, and Services Behind the New Manufacturing Revolution.* Maker Media, Inc: San Francisco, **2016**.
31. Kotz, F.; Risch, P.; Helmer, D.; Rapp, B. E. *Adv. Mater.* **2019**, *31*, e1805982.
32. Erokhin, K. S.; Gordeev, E. G.; Ananikov, V. P. *Sci. Rep.* **2019**, *9*, 20177.
33. Bose, S.; Vahabzadeh, S.; Bandyopadhyay, A. *Mater. Today* **2013**, *16*, 496–504.
34. Butscher, A.; Bohner, M.; Hofmann, S.; Gauckler, L.; Müller, R. *Acta Biomater.* **2011**, *7*, 907–920.
35. Lee, M.; Wu, B. Recent Advances in 3D Printing of Tissue Engineering Scaffolds. In *Computer-Aided Tissue Engineering*; Liebschner, M. A. K., Ed. Springer Science+Business Media, LLC, 233 Spring Street, New York, NY 10013, USA, **2012**, Vol. 868, pp. 257–267.
36. Shirazi, S. F.; Gharehkhani, S.; Mehrali, M.; Yarmand, H.; Metselaar, H. S.; Adib Kadri, N.; Osman, N. A. *Sci. Technol. Adv. Mater.* **2015**, *16*, 033502.
37. Sun, W.; Starly, B.; Nam, J.; Darling, A. *Comput.-Aided Des.* **2005**, *37*, 1097–1114.
38. Zein, I.; Hutmacher, D. W.; Tan, K. C.; Teoh, S. H. *Biomaterials* **2002**, *23*, 1169–1185.
39. Gross, B.; Lockwood, S. Y.; Spence, D. M. *Anal. Chem.* **2017**, *89*, 57–70.
40. Salentijn, G. I.; Oomen, P. E.; Grajewski, M.; Verpoorte, E. *Anal. Chem.* **2017**, *89*, 7053–7061.
41. Symes, M. D.; Kitson, P. J.; Yan, J.; Richmond, C. J.; Cooper, G. J.; Bowman, R. W.; Vilbrandt, T.; Cronin, L. *Nat. Chem.* **2012**, *4*, 349–354.
42. Erkal, J. L.; Selimovic, A.; Gross, B. C.; Lockwood, S. Y.; Walton, E. L.; McNamara, S.; Martin, R. S.; Spence, D. M. *Lab Chip* **2014**, *14*, 2023–2032.
43. Ahn, B. Y.; Duoss, E. B.; Motala, M. J.; Guo, X.; Park, S. I.; Xiong, Y.; Yoon, J.; Nuzzo, R. G.; Rogers, J. A.; Lewis, J. A. *Science* **2009**, *323*, 1590–1593.

44. Kitson, P. J.; Rosnes, M. H.; Sans, V.; Dragone, V.; Cronin, L. *Lab Chip* **2012**, *12*, 3267–3271.
45. Ho, C. M.; Ng, S. H.; Li, K. H.; Yoon, Y. J. *Lab Chip* **2015**, *15*, 3627–3637.
46. Tsuda, S.; Jaffery, H.; Doran, D.; Hezwani, M.; Robbins, P. J.; Yoshida, M.; Cronin, L. *PLoS ONE* **2015**, *10*, e0141640.
47. Amin, R.; Knowlton, S.; Hart, A.; Yenilmez, B.; Ghaderinezhad, F.; Katebifar, S.; Messina, M.; Khademhosseini, A.; Tasoglu, S. *Biofabrication* **2016**, *8*, 022001.
48. Au, A. K.; Huynh, W.; Horowitz, L. F.; Folch, A. *Angew. Chem. Int. Ed.* **2016**, *55*, 3862–3881.
49. Bhattacharjee, N.; Urrios, A.; Kang, S.; Folch, A. *Lab Chip* **2016**, *16*, 1720–1742.
50. He, Y.; Wu, Y.; Fu, J.-z.; Gao, Q.; Qiu, J.-j. *Electroanalysis* **2016**, *28*, 1658–1678.
51. Monaghan, T.; Harding, M. J.; Harris, R. A.; Friel, R. J.; Christie, S. D. *Lab Chip* **2016**, *16*, 3362–3373.
52. Morgan, A. J.; Hidalgo San Jose, L.; Jamieson, W. D.; Wymant, J. M.; Song, B.; Stephens, P.; Barrow, D. A.; Castell, O. K. *PLoS ONE* **2016**, *11*, e0152023.
53. Zhang, N.; Liu, J.; Zhang, H.; Kent, N. J.; Diamond, D.; M, D. G. *Micromachines (Basel)* **2019**, *10*, 595.
54. Yuen, P. K. *Lab Chip* **2016**, *16*, 3700–3707.
55. Kalsoom, U.; Nesterenko, P. N.; Paull, B. *Trac-Trends Anal. Chem.* **2018**, *105*, 492–502.
56. Simon, U.; Dimartino, S. *J. Chromatogr. A* **2019**, *1587*, 119–128.
57. Mathew Thomas, K.; Lakerveld, R. *Ind. Eng. Chem. Res.* **2019**, *58*, 20381–20391.
58. Gordeev, E. G.; Degtyareva, E. S.; Ananikov, V. P. *Russ. Chem. Bull.* **2016**, *65*, 1637–1643.
59. Baden, T.; Chagas, A. M.; Gage, G. J.; Marzullo, T. C.; Prieto-Godino, L. L.; Euler, T. *PLoS Biol.* **2015**, *13*, e1002086.
60. Wijnen, B.; Hunt, E. J.; Anzalone, G. C.; Pearce, J. M. *PLoS ONE* **2014**, *9*, e107216.
61. Rossi, S.; Benaglia, M.; Brenna, D.; Porta, R.; Orlandi, M. *J. Chem. Educ.* **2015**, *92*, 1398–1401.
62. Rossi, S.; Dozzi, M. V.; Puglisi, A.; Pagani, M. *Chem. Teach. Int.* **2019**. doi:10.1515/cti-2019-0010
63. Pernaa, J.; Wiedmer, S. *Chem. Teach. Int.* **2019**. doi: 10.1515/cti-2019-0005
64. Rossi, S.; Rossi, C.; Accorigi, N. *J. Chem. Educ.* **2020**, *5*, 1391–1395.
65. Rossi, S.; Puglisi, A.; Benaglia, M. *ChemCatChem* **2018**, *10*, 1512–1525.
66. Saggiomo, V. 3D Printed Devices for Catalytic Systems. In *Catalyst Immobilization: Methods and Applications*; Benaglia, M.; Puglisi A. Eds. Wiley-VCH Verlag GmbH & Co. KGaA, Boschstr. 12, 69469, Weinheim, Germany, **2019**, pp. 369–408.
67. Zhou, X.; Liu, C.-J. *Adv. Funct. Mater.* **2017**, *27*, 1701134.
68. Johnson, R. D. *Nat. Chem.* **2012**, *4*, 338–339.
69. Kitson, P. J.; Symes, M. D.; Dragone, V.; Cronin, L. *Chem. Sci.* **2013**, *4*, 3099–3103.
70. Kitson, P. J.; Marshall, R. J.; Long, D.; Forgan, R. S.; Cronin, L. *Angew. Chem. Int. Ed.* **2014**, *53*, 12723–12728.

71. Williams, I. D. *Nat. Chem.* **2014**, *6*, 953–954.
72. Gordeev, E. G.; Galushko, A. S.; Ananikov, V. P. *PLoS ONE* **2018**, *13*, e0198370.
73. Lederle, F.; Meyer, F.; Kaldun, C.; Namyslo, J. C.; Hübner, E. G. *New J. Chem.* **2017**, *41*, 1925–1932.
74. Kitson, P. J.; Glatzel, S.; Chen, W.; Lin, C. G.; Song, Y. F.; Cronin, L. *Nat. Protoc.* **2016**, *11*, 920–936.
75. Kitson, P. J.; Marie, G.; Francoia, J. P.; Zalesskiy, S. S.; Sigerson, R. C.; Mathieson, J. S.; Cronin, L. *Science* **2018**, *359*, 314–319.
76. Middelkoop, V.; Slater, T.; Florea, M.; Nea u, F.; Danaci, S.; Onyenkeadi, V.; Boonen, K.; Saha, B.; Baragau, I.-A.; Kellici, S. *J. Clean Prod.* **2019**, *214*, 606–614.
77. Danaci, S.; Protasova, L.; Snijkers, F.; Bouwen, W.; Bengaouer, A.; Marty, P. *Chem. Eng. Process* **2018**, *127*, 168–177.
78. Lefevere, J.; Gysen, M.; Mullens, S.; Meynen, V.; Van Noyen, J. *Catal. Today* **2013**, *216*, 18–23.
79. Middelkoop, V.; Coenen, K.; Schalck, J.; Van Sint Annaland, M.; Gallucci, F. *Chem. Eng. J.* **2019**, *357*, 309–319.
80. Govender, S.; Friedrich, H. *Catalysts* **2017**, *7*, 62
81. Tubio, C. R.; Azuaje, J.; Escalante, L.; Coelho, A.; Guitian, F.; Sotelo, E.; Gil, A. *J. Catal.* **2016**, *334*, 110–115.
82. Azuaje, J.; Tubio, C. R.; Escalante, L.; Gomez, M.; Guitian, F.; Coelho, A.; Caamano, O.; Gil, A.; Sotelo, E. *Appl. Catal. A-Gen.* **2017**, *530*, 203–210.
83. Díaz-Marta, A. S.; Tubío, C. R.; Carbajales, C.; Fernández, C.; Escalante, L.; Sotelo, E.; Guitián, F.; Barrio, V. L.; Gil, A.; Coelho, A. *ACS Catal.* **2017**, *8*, 392–404.
84. Sanchez Diaz-Marta, A.; Yanez, S.; Tubio, C. R.; Barrio, V. L.; Pineiro, Y.; Pedrido, R.; Rivas, J.; Amorin, M.; Guitian, F.; Coelho, A. *ACS Appl. Mater. Interfaces* **2019**, *11*, 25283–25294.
85. Díaz-Marta, A. S.; Yañez, S.; Lasorsa, E.; Pacheco, P.; Tubío, C. R.; Rivas, J.; Piñeiro, Y.; Gómez, M. A. G.; Amorín, M.; Guitián, F.; Coelho, A. *ChemCatChem* **2020**, *12*, 1762–1771.
86. Jiang, Y.; Jiang, F.-Q.; Liao, X.; Lai, S.-L.; Wang, S.-B.; Xiong, X.-Q.; Zheng, J.; Liu, Y.-G. *J. Porous Mater.* **2020**, *27*, 779–788.
87. Hilton, S. T.; Penny, M. R.; Dos Santos, B. S.; Patel, B. Three-dimensional printing of impregnated plastics for chemical reactions, Patent Application WO2017158336A1.
88. Penny, M. R.; Hilton, S. T. *React. Chem. Eng.* **2020**, *5*, 853–858.
89. Manzano, J. S.; Weinstein, Z. B.; Sadow, A. D.; Slowing, I. I. *ACS Catal.* **2017**, *7*, 7567–7577.
90. Rossi, S.; Puglisi, A.; Raimondi, L. M.; Benaglia, M. *Catalysts* **2020**, *10*, 109.
91. Lahtinen, E.; Turunen, L.; Hanninen, M. M.; Kolari, K.; Tuononen, H. M.; Haukka, M. *ACS Omega* **2019**, *4*, 12012–12017.
92. Lahtinen, E.; Kukkonen, E.; Kinnunen, V.; Lahtinen, M.; Kinnunen, K.; Suvanto, S.; Vaisanen, A.; Haukka, M. *ACS Omega* **2019**, *4*, 16891–16898.

93. Plutschack, M. B.; Pieber, B.; Gilmore, K.; Seeberger, P. H. *Chem. Rev.* **2017**, *117*, 11796–11893.
94. Jas, G.; Kirschning, A. *Chem. Eur. J.* **2003**, 9, 5708–5723.
95. Jensen, K. F.; Reizman, B. J.; Newman, S. G. *Lab Chip* **2014**, 14, 3206–3212.
96. Darvas, F.; Dormán, G.; Hessel, V. Flow Chemistry — Fundamentals. In *Flow Chemistry. De Gruyter Textbook Series.* Walter de Gruyter GmbH, **2014**, Vol. 1.
97. Reschetilowski, W. Ed. *Microreactors in Preparative Chemistry: Practical Aspects in Bioprocessing, Nanotechnology, Catalysis and More.* Wiley-VCH Verlag GmbH & Co. KGaA, Boschstr. 12, 69469, Weinheim, Germany, **2013**, pp. 1–335.
98. Wirth, T. *Microreactors in Organic Synthesis and Catalysis*, 2nd ed. Wiley-VCH: Weinheim, **2013**.
99. Wegner, J.; Ceylan, S.; Kirschning, A. *Chem. Commun.* **2011**, *47*, 4583–4592.
100. Hartman, R. L.; McMullen, J. P.; Jensen, K. F. *Angew. Chem. Int. Ed.* **2011**, *50*, 7502–7519.
101. Wiles, C.; Watts, P. *Chem. Commun.* **2011**, *47*, 6512–6535.
102. Hessel, V. *Chem. Eng. Technol.* **2009**, *32*, 1655–1681.
103. Baumann, M.; Moody, T. S.; Smyth, M.; Wharry, S. *Org. Process Res. Dev.* **2020**, *24*, 1802–1813.
104. Badman, C.; Cooney, C. L.; Florence, A.; Konstantinov, K.; Krumme, M.; Mascia, S.; Nasr, M.; Trout, B. L. *J. Pharm. Sci.* **2019**, *108*, 3521–3523.
105. Martin, B.; Lehmann, H.; Yang, H.; Chen, L.; Tian, X.; Polenk, J.; Schenkel, B. *Curr. Op. Green Sus. Chem.* **2018**, *11*, 27–33.
106. Porta, R.; Benaglia, M.; Puglisi, A. *Org. Process Res. Dev.* **2015**, *20*, 2–25.
107. Baxendale, I. R.; Braatz, R. D.; Hodnett, B. K.; Jensen, K. F.; Johnson, M. D.; Sharratt, P.; Sherlock, J.-P.; Florence, A. J. *J. Pharm. Sci.* **2015**, *104*, 781–791.
108. Malet-Sanz, L.; Susanne, F. *J. Med. Chem.* **2012**, *55*, 4062–4098.
109. Capel, A. J.; Edmondson, S.; Christie, S. D.; Goodridge, R. D.; Bibb, R. J.; Thurstans, M. *Lab Chip* **2013**, *13*, 4583–4590.
110. Hampson, S. M.; Rowe, W.; Christie, S. D. R.; Platt, M. *Sens. Actuator B-Chem.* **2018**, *256*, 1030–1037.
111. Hampson, S. M.; Pollard, M.; Hauer, P.; Salway, H.; Christie, S. D. R.; Platt, M. *Anal. Chem.* **2019**, *91*, 2947–2954.
112. Stark, A. K. *AIChE J.* **2018**, *64*, 1162–1173.
113. Okafor, O.; Goodridge, R.; Sans, V. *Chem. Today* **2017**, *35*, 4–6.
114. Kuijpers, K. P. L.; van Dijk, M. A. H.; Rumeur, Q. G.; Hessel, V.; Su, Y.; Noël, T. *React. Chem. Eng.* **2017**, *2*, 109–115.
115. Okafor, O.; Weilhard, A.; Fernandes, J. A.; Karjalainen, E.; Goodridge, R.; Sans, V. *React. Chem. Eng.* **2017**, *2*, 129–136.
116. Bressan, L. P.; Robles-Najar, J.; Adamo, C. B.; Quero, R. F.; Costa, B. M. C.; de Jesus, D. P.; da Silva, J. A. F. *Microchem. J.* **2019**, *146*, 1083–1089.

117. Okafor, O.; Robertson, K.; Goodridge, R.; Sans, V. *React. Chem. Eng.* **2019**, *4*, 1682–1688.
118. McDonough, J. R.; Murta, S.; Law, R.; Harvey, A. P. *Chem. Eng. J.* **2019**, *358*, 643–657.
119. McDonough, J.; Armett, J.; Law, R.; Harvey, A. P. *Ind. Eng. Chem. Res.* **2019**, *58*, 21363–21371.
120. Dragone, V.; Sans, V.; Rosnes, M. H.; Kitson, P. J.; Cronin, L. *Beilstein J. Org. Chem.* **2013**, *9*, 951–959.
121. Genet, C.; Nguyen, X.; Bayatsarmadi, B.; Horne, M. D.; Gardiner, J.; Hornung, C. H. *J. Flow Chem.* **2018**, *8*, 81–88.
122. Avril, A.; Hornung, C. H.; Urban, A.; Fraser, D.; Horne, M.; Veder, J. P.; Tsanaktsidis, J.; Rodopoulos, T.; Henry, C.; Gunasegaram, D. R. *React. Chem. Eng.* **2017**, *2*, 180–188.
123. Hornung, C. H.; Nguyen, X.; Carafa, A.; Gardiner, J.; Urban, A.; Fraser, D.; Horne, M. D.; Gunasegaram, D. R.; Tsanaktsidis, J. *Org. Process Res. Dev.* **2017**, *21*, 1311–1319.
124. Gardiner, J.; Nguyen, X.; Genet, C.; Horne, M. D.; Hornung, C. H.; Tsanaktsidis, J. *Org. Process Res. Dev.* **2018**, *22*, 1448–1452.
125. Elias, Y.; Rudolf von Rohr, P.; Bonrath, W.; Medlock, J.; Buss, A. *Chem. Eng. Process.* **2015**, *95*, 175–185.
126. Gutmann, B.; Kockinger, M.; Glotz, G.; Ciaglia, T.; Slama, E.; Zadravec, M.; Pfanner, S.; Maier, M. C.; Gruber-Wolfler, H.; Kappe, C. O. *React. Chem. Eng.* **2017**, *2*, 919–927.
127. Lee, H. J.; Roberts, R. C.; Im, D. J.; Yim, S. J.; Kim, H.; Kim, J. T.; Kim, D. P. *Small* **2019**, *15*, e1905005.
128. Hur, D.; Say, M. G.; Diltemiz, S. E.; Duman, F.; Ersoz, A.; Say, R. *ChemPlusChem* **2018**, *83*, 42–46.
129. Bettermann, S.; Schroeter, B.; Moritz, H. U.; Pauer, W.; Fassbender, M.; Luinstra, G. A. *Chem. Eng. J.* **2018**, *338*, 311–322.
130. Neumaier, J. M.; Madani, A.; Klein, T.; Ziegler, T. *Beilstein J. Org. Chem.* **2019**, *15*, 558–566.
131. Penny, M. R.; Rao, Z. X.; Peniche, B. F.; Hilton, S. T. *Eur. J. Org. Chem.* **2019**, 3783–3787.
132. Guo, M.; Hu, X.; Yang, F.; Jiao, S.; Wang, Y.; Zhao, H.; Luo, G.; Yu, H. *Ind. Eng. Chem. Res.* **2019**, *58*, 13357–13365.
133. Maier, M. C.; Lebl, R.; Sulzer, P.; Lechner, J.; Mayr, T.; Zadravec, M.; Slama, E.; Pfanner, S.; Schmolzer, C.; Pochlauer, P.; Kappe, C. O.; Gruber-Woelfler, H. *React. Chem. Eng.* **2019**, *4*, 393–401.
134. Thompson, M. P.; Peñafiel, I.; Cosgrove, S. C.; Turner, N. J. *Org. Process Res. Dev.* **2018**, *23*, 9–18.
135. Maier, M.; Radtke, C. P.; Hubbuch, J.; Niemeyer, C. M.; Rabe, K. S. *Angew. Chem. Int. Ed.* **2018**, *57*, 5539–5543.

136. Peng, M.; Mittmann, E.; Wenger, L.; Hubbuch, J.; Engqvist, M. K. M.; Niemeyer, C. M.; Rabe, K. S. *Chem. Eur. J.* **2019**, 25, 15998–16001.
137. Capel, A. J.; Rimington, R. P.; Lewis, M. P.; Christie, S. D. R. *Nat. Rev. Chem.* **2018**, 2, 422–436.
138. Chandekar, A.; Mishra, D. K.; Sharma, S.; Saraogi, G. K.; Gupta, U.; Gupta, G. *Curr. Pharm. Des.* **2019**, 25, 937–945.
139. Narkevich, I. A.; Flisyuk, E. V.; Terent'eva, O. A.; Semin, A. A. *Pharm. Chem. J.* **2018**, 51, 1025–1029.
140. Kotta, S.; Nair, A.; Alsabeelah, N. *Curr. Pharm. Des.* **2018**, 24, 5039–5048.
141. Firth, J. Regulatory Perspectives on 3D Printing in Pharmaceuticals. In *3D Printing of Pharmaceuticals. AAPS Advances in the Pharmaceutical Sciences Series*; Basit, A.; Gaisford, S. Eds. Springer: Cham, **2018**, Vol. 31, pp. 153–162.
142. Zhang, J.; Vo, A. Q.; Feng, X.; Bandari, S.; Repka, M. A. *AAPS PharmSciTech* **2018**, 19, 3388–3402.
143. Trivedi, M.; Jee, J.; Silva, S.; Blomgren, C.; Pontinha, V. M.; Dixon, D. L.; Van Tassel, B.; Bortner, M. J.; Williams, C.; Gilmer, E.; Haring, A. P.; Halper, J.; Johnson, B. N.; Kong, Z.; Halquist, M. S.; Rocheleau, P. F.; Long, T. E.; Roper, T.; Wijesinghe, D. S. *Addit. Manuf.* **2018**, 23, 319–328.
144. Preis, M.; Oblom, H. *AAPS PharmSciTech* **2017**, 18, 303–308.
145. Zidan, A.; Alayoubi, A.; Asfari, S.; Coburn, J.; Ghammraoui, B.; Aqueel, S.; Cruz, C. N.; Ashraf, M. *Int. J. Pharm.* **2019**, 555, 109–123.
146. Rahman, Z.; Ali, S. F. B.; Ozkan, T.; Charoo, N. A.; Reddy, I. K.; Khan, M. A. *AAPS J.* **2018**, 20, 101.
147. Khairuzzaman, A. Regulatory Perspectives on 3D Printing in Pharmaceuticals. In *3D Printing of Pharmaceuticals. AAPS Advances in the Pharmaceutical Sciences Series*; Basit, A.; Gaisford, S. Eds. Springer: Cham, **2018**, Vol. 31, pp. 215–236.
148. Park, B. J.; Choi, H. J.; Moon, S. J.; Kim, S. J.; Bajracharya, R.; Min, J. Y.; Han, H.-K. *J. Pharm. Investig.* **2019**, 49, 575–585.
149. Warsi, M. H.; Yusuf, M.; Al Robaian, M.; Khan, M.; Muheem, A.; Khan, S. *Curr. Pharm. Des.* **2018**, 24, 4949–4956.
150. Algahtani, M. S.; Mohammed, A. A.; Ahmad, J. *Curr. Pharm. Design* **2018**, 24, 4991–5008.
151. Palo, M.; Hollander, J.; Suominen, J.; Yliruusi, J.; Sandler, N. *Expert Rev. Med. Devices* **2017**, 14, 685–696.
152. Ursan, I. D.; Chiu, L.; Pierce, A. *J. Am. Pharm. Assoc. (2003)* **2013**, 53, 136–144.
153. Khatri, P.; Shah, M. K.; Vora, N. *J. Drug Deliv. Sci. Technol.* **2018**, 46, 148–155.
154. Ameeduzzafar, A.; Alruwaili, N. K.; Rizwanullah, Md.; Bukhari, S. N. A.; Amir, M.; Ahmed, M. M.; Fazil, M. *Curr. Pharm. Des.* **2018**, 24, 5009–5018.
155. Yu, D. G.; Yang, X. L.; Huang, W. D.; Liu, J.; Wang, Y. G.; Xu, H. *J. Pharm. Sci.* **2007**, 96, 2446–2456.
156. Yu, D.-G.; Shen, X.-X.; Branford-White, C.; Zhu, L.-M.; White, K.; Yang, X. L. *J. Pharm. Pharmacol.* **2009**, 61, 323–329.

157. Katstra, W. E.; Palazzolo, R. D.; Rowe, C. W.; Giritlioglu, B.; Teung, P.; Cima, M. J. *J. Control. Release* **2000**, *66*, 1–9.
158. Infanger, S.; Haemmerli, A.; Iliev, S.; Baier, A.; Stoyanov, E.; Quodbach, J. *Int. J. Pharm.* **2019**, *555*, 198–206.
159. Tian, P.; Yang, F.; Xu, Y.; Lin, M. M.; Yu, L. P.; Lin, W.; Lin, Q. F.; Lv, Z. F.; Huang, S. Y.; Chen, Y. Z. *Drug Dev. Ind. Pharm.* **2018**, *44*, 1918–1923.
160. Trenfield, S. J.; Madla, C. M.; Basit, A. W.; Gaisford, S. Binder Jet Printing in Pharmaceutical Manufacturing. In *3D Printing of Pharmaceuticals. AAPS Advances in the Pharmaceutical Sciences Series*; Basit, A.; Gaisford, S. Eds. Springer: Cham, **2018**, Vol. 31, pp. 41–54.
161. Lee, B. K.; Yun, Y. H.; Choi, J. S.; Choi, Y. C.; Kim, J. D.; Cho, Y. W. *Int. J. Pharm.* **2012**, *427*, 305–310.
162. Kyobula, M.; Adedeji, A.; Alexander, M. R.; Saleh, E.; Wildman, R.; Ashcroft, I.; Gellert, P. R.; Roberts, C. J. *J Control. Release* **2017**, *261*, 207–215.
163. Alhijjaj, M.; Nasereddin, J.; Belton, P.; Qi, S. *Pharmaceutics* **2019**, *11*, 633.
164. Melocchi, A.; Uboldi, M.; Maroni, A.; Foppoli, A.; Palugan, L.; Zema, L.; Gazzaniga, A. *Int. J. Pharm.* **2020**, *579*, 119155.
165. Solanki, N. G.; Tahsin, M.; Shah, A. V.; Serajuddin, A. T. M. *J. Pharm. Sci.* **2018**, *107*, 390–401.
166. Goyanes, A.; Buanz, A. B.; Basit, A. W.; Gaisford, S. *Int. J. Pharm.* **2014**, *476*, 88–92.
167. Pietrzak, K.; Isreb, A.; Alhnan, M. A. *Eur. J. Pharm. Biopharm.* **2015**, *96*, 380–387.
168. Leong, K. F.; Phua, K. K.; Chua, C. K.; Du, Z. H.; Teo, K. O. *Proc. Inst. Mech. Eng. Part H — J. Eng. Med.* **2001**, *215*, 191–201.
169. Kadry, H.; Wadnap, S.; Xu, C.; Ahsan, F. *Eur. J. Pharm. Sci.* **2019**, *135*, 60–67.
170. Acosta-Velez, G. F.; Zhu, T. Z.; Linsley, C. S.; Wu, B. M. *Int. J. Pharm.* **2018**, *546*, 145–153.
171. Hollander, J.; Hakala, R.; Suominen, J.; Moritz, N.; Yliruusi, J.; Sandler, N. *Int. J. Pharm.* **2018**, *544*, 433–442.
172. Zhang, J.; Hu, Q.; Wang, S.; Tao, J.; Gou, M. *Int. J. Bioprint.* **2019**, *6*, 12–27.
173. Wu, S.; Li, J. S.; Mai, J.; Chang, M. W. *ACS Appl. Mater. Interfaces* **2018**, *10*, 24876–24885.
174. Rattanakit, P.; Moulton, S. E.; Santiago, K. S.; Liawruangrath, S.; Wallace, G. G. *Int. J. Pharm.* **2012**, *422*, 254–263.
175. Khaled, S. A.; Burley, J. C.; Alexander, M. R.; Yang, J.; Roberts, C. J. *J. Control. Release* **2015**, *217*, 308–314.
176. Khaled, S. A.; Burley, J. C.; Alexander, M. R.; Roberts, C. J. *Int. J. Pharm.* **2014**, *461*, 105–111.
177. Lamichhane, S.; Bashyal, S.; Keum, T.; Noh, G.; Seo, J. E.; Bastola, R.; Choi, J.; Sohn, D. H.; Lee, S. *Asian J. Pharm. Sci.* **2019**, *14*, 465–479.
178. Aprecia Pharmaceuticals. 3D Printing/ZipDose Technology. **2016**. Available at: https://www.aprecia.com/zipdose-platform/3d-printing.php

179. Kitson, P. J.; Glatzel, S.; Cronin, L. *Beilstein J. Org. Chem.* **2016**, *12*, 2776–2783.
180. Goyanes, A.; Det-Amornrat, U.; Wang, J.; Basit, A. W.; Gaisford, S. *J. Control. Release* **2016**, *234*, 41–48.
181. Jiang, H.; Fu, J.; Li, M.; Wang, S.; Zhuang, B.; Sun, H.; Ge, C.; Feng, B.; Jin, Y. *AAPS PharmSciTech* **2019**, *20*, 260.
182. Ou, Y. H.; Ou, Y. H.; Gu, J.; Kang, L. F. *Int. J. Bioprint.* **2019**, *5*, 15–21.
183. Yang, N.; Chen, H.; Han, H.; Shen, Y.; Gu, S.; He, Y.; Guo, S. *Int. J. Pharm.* **2018**, *552*, 91–98.
184. Muwaffak, Z.; Goyanes, A.; Clark, V.; Basit, A. W.; Hilton, S. T.; Gaisford, S. *Int. J. Pharm.* **2017**, *527*, 161–170.
185. Trenfield, S. J.; Madla, C. M.; Basit, A. W.; Gaisford, S. The Shape of Things to Come: Emerging Applications of 3D Printing in Healthcare. In *3D Printing of Pharmaceuticals. AAPS Advances in the Pharmaceutical Sciences Series;* Basit, A.; Gaisford, S. Eds. Springer: Cham, **2018**, Vol. 31, pp. 1–19.
186. Awad, A.; Trenfield, S. J.; Gaisford, S.; Basit, A. W. *Int. J. Pharm.* **2018**, *548*, 586–596.
187. Rossi, S.; Porta, R.; Brenna, D.; Puglisi, A.; Benaglia, M. *Angew. Chem. Int. Ed.* **2017**, *56*, 4290–4294.
188. Rao, Z. X.; Patel, B.; Monaco, A.; Cao, Z. J.; Barniol-Xicota, M.; Pichon, E.; Ladlow, M.; Hilton, S. T. *Eur. J. Org. Chem.* **2017**, *44*, 6499–6504.
189. Romero-Fernandez, M.; Paradisi, F. *Curr. Opin. Chem. Biol.* **2019**, *55*, 1–8.
190. Peris, E.; Okafor, O.; Kulcinskaja, E.; Goodridge, R.; Luis, S. V.; Garcia-Verdugo, E.; O'Reilly, E.; Sans, V. *Green Chem* **2017**, *19*, 5345–5349.
191. Schmieg, B.; Dobber, J.; Kirschhofer, F.; Pohl, M.; Franzreb, M. *Front. Bioeng. Biotechnol.* **2018**, *6*, 211.
192. Ye, J. J.; Chu, T. S.; Chu, J. L.; Gao, B. B.; He, B. F. *ACS Sustain. Chem. Eng.* **2019**, *7*, 18048–18054.

© 2022 World Scientific Publishing Company
https://doi.org/10.1142/9789811248436_0008

8. Catalytic Strategies for the Enantioselective Amination of C(sp^3)–H Bonds

Philippe Dauban*,[†] and Tanguy Saget*

*Université Paris-Saclay, CNRS, Institut de Chimie des Substances Naturelles, UPR 2301, avenue de la Terrasse, 91198 Gif-sur-Yvette, France
[†]philippe.dauban@cnrs.fr

Table of Contents

List of Abbreviations	256
I. Introduction	256
II. Catalytic Asymmetric C(sp^3)–H Amination with Azides	259
A. Metal-catalyzed Concerted Processes	259
B. Metal-catalyzed Radical Processes	261
C. Biocatalytic Amination with Engineered Enzymes	263
III. Catalytic Asymmetric C(sp^3)–H Amination with Dioxazolones	266
A. Introduction	266
B. Design of Chiral Hydrogen-bond-donor Catalysts	267

C. Design of Catalysts with Chiral Hydrophobic Pockets　268
　　　D. Design of an Achiral Cp*M(III)/Chiral Carboxylic
　　　　　Acid Hybrid Catalytic System　270
　IV. Catalytic Asymmetric C(sp^3)–H Amination
　　　with Iodine(III) Oxidants　271
　　　A. Introduction　271
　　　B. Design of a Chiral Rhodium(II)–Carboxamidate
　　　　　Complex　272
　　　C. Design of a Chiral Rhodium(II)–Carboxylate Complex　273
　　　D. Design of a Bisoxazoline Ligand with Quaternary
　　　　　Stereocenters　274
　　　E. Design of Supramolecular Metal Complexes　275
　V. Conclusion　278
　VI. Acknowledgments　278
　VII. References　278

List of Abbreviations

CMD	concerted metalation deprotonation
Cp*	pentamethylcyclopentadienyl
DPP-4	dipeptidyl peptidase 4
Esp	$\alpha,\alpha,\alpha',\alpha'$-tetramethyl-1,3-benzenedipropionic acid
FDA	Food and Drug Administration
Nap	N-Tosyl-3-aminopiperidinone
Ns	p-nitrobenzenesulfonyl
Pfbs	pentafluorobenzyloxysulfonyl
PMB	p-methoxybenzyl
Ses	2-(trimethylsilyl)ethanesulfonyl
Tces	2,2,2-Trichloroethoxysulfonyl
Tfptad	tetrakis[(S)-(N-tetrafluorophthaloyl)-1-adamantylglycine]
TON	turnover number

I. Introduction

Looking back at the history of organic chemistry, chiral amines have often been at the heart of breakthrough discoveries in a number of fields.

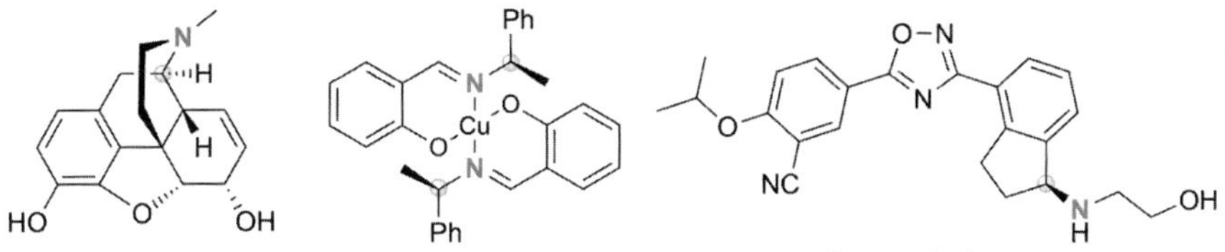

Figure 1. Relevant examples of chiral amines.

The early nineteenth century witnessed the birth of natural product chemistry with the isolation of morphine **1**[1] (Figure 1), followed by that of many other landmark alkaloids such as strychnine, quinine, or colchicine. In 1966, Nozaki and Noyori reported the use of the bis(chiral aldimine)-copper complex **2** (Figure 1) in catalytic carbene transfer reactions[2] thereby opening the Nobel Prize-winning domain of asymmetric catalysis. In 2020, chiral amines are still important motifs in life and material sciences. The sphingosine-1-phosphate receptor agonist Zeposia® **3**[3] stands as a relevant example, having been approved by the Food and Drug Administration (FDA) in 2020 for the treatment of relapsing multiple sclerosis.

The paramount importance of nitrogen in science has been a source of inspiration for synthetic organic chemists. Particularly, the ubiquity of chiral amines stimulated the discovery of several methods for the stereoselective formation of C–N bonds.[4,5] Along this line, the reductive amination of carbonyl compounds represents a pivotal transformation both in nature, for example, in the biosynthesis of amino acids, and in the pharmaceutical industry where it accounts for more than 20% of C–N bond-forming reactions.[6] The rise of asymmetric catalysis in the last 40 years has also deeply impacted the field and found applications for the manufacturing of pharmaceuticals. For example, the dipeptidyl peptidase 4 (DPP-4) inhibitor Januvia® **4**, a first-in-class drug approved in 2006 for the treatment of type 2 diabetes, has been produced in only four steps (Scheme 1). The strategy includes a Rh(I)-*t*Bu-JOSIPHOS-catalyzed enantioselective hydrogenation of an enamine generated from its ketone precursor.[7] Though it relies on an expensive platinum group metal as the catalyst, the low loading and the easy recovery of the Rh(I) complex make

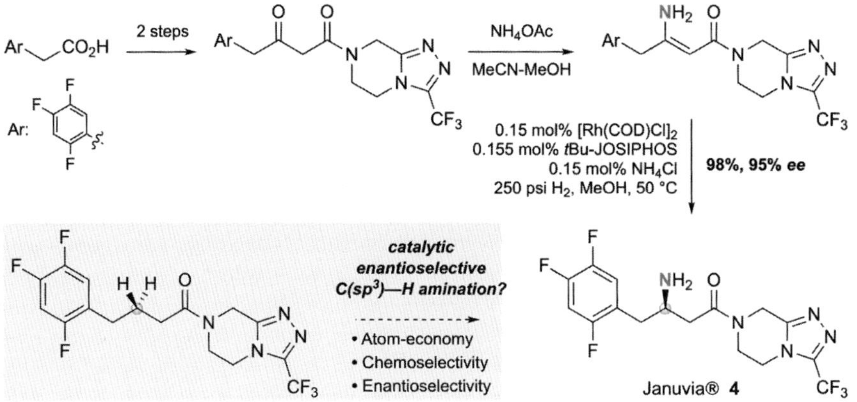

Scheme 1. Januvia®: second-generation industrial synthesis[7] and a hypothetical approach by C(sp³)-H amination.

the process cost-effective. Moreover, the total waste was reduced by 80% compared to the first-generation strategy.

The search for efficient catalytic C–H amination reactions has challenged the creativity of organic chemists over the last 20 years.[8–12] With the constant need to improve the sustainability of C–N bond-forming reactions, the introduction of a nitrogen group onto nonfunctionalized positions represents another step forward in terms of atom- and step economy. By comparison with the aforementioned reductive amination approach to Januvia® in Scheme 1, the direct conversion of a C–H to a C–N bond could in theory avoid the extra steps for the incorporation of the keto function in the substrate. Catalytic C–H functionalization reactions could also reduce the amount of waste arising from the elimination of the preinstalled functional group, a relevant problem particularly when the C–N bond is created by a catalytic coupling from a halide starting material.[13]

Although fully attractive with respect to sustainability issues, C–H functionalization reactions raise the challenging question of the chemoselectivity resulting from the ubiquity of C–H bonds in organic compounds.[14,15] Suitable conditions should be targeted for the selective conversion of either C(sp²)–H bonds or primary, secondary, and tertiary

C(sp^3)–H bonds that often display similar chemical reactivity. In this context, the structure of Januvia® makes a textbook case. The strategy should indeed favor the amination of a position known to be both less acidic than the C–H bonds α- to the carbonyl, and less reactive than the benzylic position of the substrate. Moreover, the stereoselectivity issue is equally difficult to tackle.[16] One must find a compromise between the high energy required for the cleavage of the rather inert C–H bonds and the mild reaction conditions enabling the differentiation of two enantiotopic sites.

The area of catalytic enantioselective C(sp^3)–H amination reaction was pioneered by Müller and coworkers at the end of the twentieth century. The use of chiral dirhodium(II) complexes in the presence of hypervalent iodine reagents allowed to perform the asymmetric C(sp^3)–H nitrene insertion on a benzylic site, albeit with low enantiomeric excesses (ees) in the 7–31% range.[17] Many developments inspired by the seminal study of Müller and coworkers were made since then.[18] However, new strategies have been explored more recently that led to significant achievements.[19] The purpose of this chapter is to highlight some of the most significant approaches in asymmetric C(sp^3)–H amination reactions. They are presented as a function of the nitrogen reagent and the concept developed for the enantiocontrol.

II. Catalytic Asymmetric C(sp^3)–H Amination with Azides

A. Metal-catalyzed Concerted Processes

Pioneering studies in catalytic C(sp^3)–H amination were reported with organic azides[20] that are versatile precursors to access metal-bound nitrenes with a range of transition metals. Notably, this process does not require oxidative conditions and is atom-economic since only one equivalent of N$_2$ is extruded. Metal-bound nitrenes are highly reactive intermediates, which are able to insert into aliphatic C–H bonds to directly introduce a nitrogen function on a saturated carbon skeleton *via* a concerted outer-sphere pathway (Scheme 2). When meso or prochiral substrates are reacted, the use of chiral complexes enables enantioselective C–N bond-forming reactions.

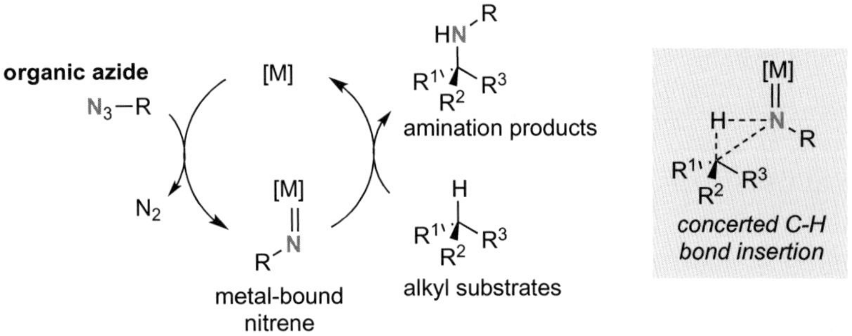

Scheme 2. General mechanism for the metal-catalyzed C(sp³)–H amination with azides.

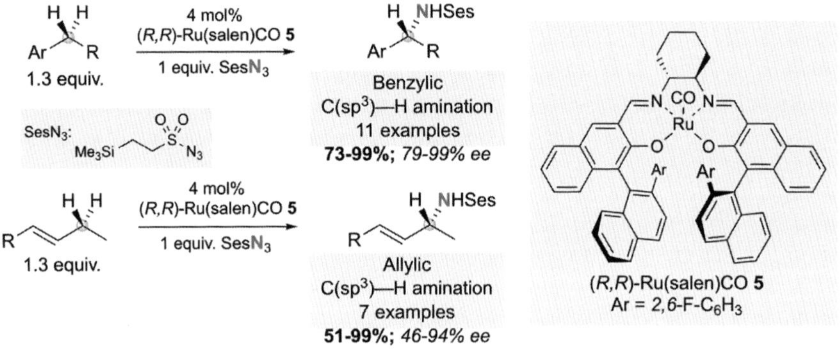

Scheme 3. Ruthenium-catalyzed asymmetric intermolecular C(sp³)–H amination with SesN₃.[21]

In 2013, Katsuki and coworkers reported an intermolecular enantioselective C–H amination reaction using the ruthenium complex **5**, bearing a chiral salen ligand, in the presence of 2-(trimethylsilyl)-ethanesulfonyl azide (SES-N₃) (Scheme 3).[21] After an extensive screening, a salen derived from 1,2-diaminocyclohexane combined with an axially chiral binaphthyl–derived aldehyde was identified as an efficient ligand for the regio- and enantioselective formation of C–N bonds at benzylic and allylic positions. The reaction at benzylic sites only occurs with cyclic substrates or ethylbenzene derivatives, as substrates with longer alkyl chains proved to be unreactive under these conditions. In addition, the present catalytic

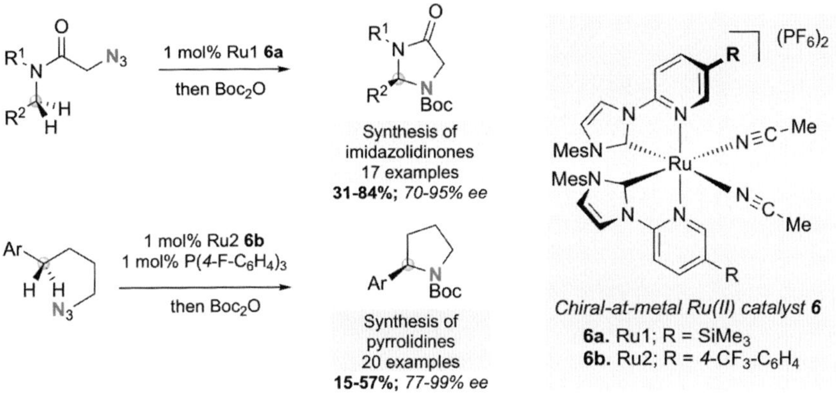

Scheme 4. Chiral-at-metal ruthenium complexes for asymmetric intramolecular C(sp^3)–H amination.[23,24]

system also enables the regioselective amination of (*E*)-alkenes displaying two different allylic positions. Importantly, a substrate/azide molar ratio of 1.3:1 is optimal to achieve good to excellent yields of the enantioenriched aminated products. Worth of mention is the removal of the SES protecting group under mild conditions using fluoride anion, which gives access to the free amines without epimerization.

Instead of the classical development of chiral ligands in asymmetric catalysis, Meggers *et al.* have recently started to investigate the use of chiral-at-metal complexes in enantioselective processes.[22] Thus, the chiral-at-metal octahedral complexes **6** proved to be versatile catalysts for the intramolecular C–H aminations of 2-azidoacetamides[23] and aliphatic azides[24] (Scheme 4). Complexes **6** display two achiral bidentate ligands that are coordinated in a C_2-symmetric fashion to form a stereogenic ruthenium center. This strategy enabled the straightforward synthesis of valuable enantioenriched imidazolidin-4-ones and pyrrolidines. In the case of less reactive aliphatic azides, a dual catalytic system incorporating a phosphine cocatalyst facilitates the metallonitrene formation *via* a Staudinger reaction.

B. Metal-catalyzed Radical Processes

Metalloradical catalysis has emerged as a powerful strategy to perform a range of C–H amination reactions.[25] In particular, Zhang and coworkers[26]

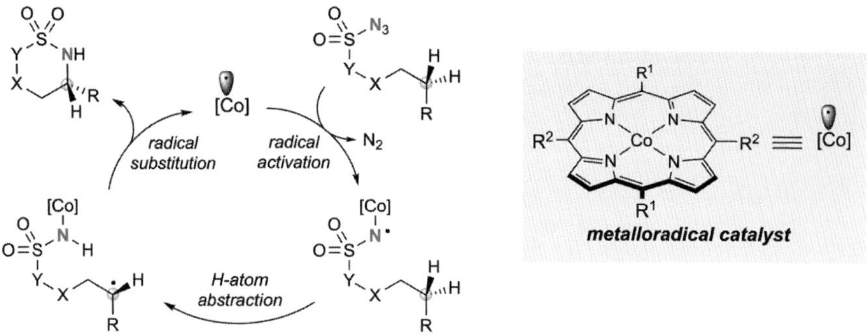

Scheme 5. General mechanism for the metal-catalyzed radical C(sp^3)–H amination.

discovered that Co(II)-porphyrin complexes, which are stable 15-electron metalloradicals, react with azides to form α-Co(III)-aminyl radicals. These species can undergo an intramolecular hydrogen atom transfer, leading to the formation of a carbon-based radical. Then, a radical substitution step affords the cyclic aminated products, while regenerating the Co(II) catalyst, according to the mechanism described in Scheme 5.

Importantly, the judicious design of D_2-symmetric phorphyrins enabled the development of enantioselective Co(II)-catalyzed amination reactions. In particular, the incorporation of chiral rigid amide motifs around the porphyrin backbone led to efficient catalysts, which can perform intramolecular aminations with good to excellent enantiocontrol. For example, catalyst [Co(P1)] **7a** is an effective catalyst for the synthesis of cyclic sulfamides through enantioselective 1,6-aminations of a variety of C(sp^3)–H bonds (Scheme 6).[26] Indeed, several benzylic, allylic, propargylic, and aliphatic positions could be aminated with good enantioselectivities. Moreover, the obtained sulfamides can be readily deprotected to access valuable chiral 1,3-diamines. This methodology was later extended to the construction of five-membered cyclic sulfonamides using a slightly modified catalyst [Co(P2)] **7b**.[27] In this case, only benzylic or allylic C–H bonds could be aminated with good selectivity (80–99% ee). Substrates with regular aliphatic chains led to mixtures of five-membered and six-membered sulfonamides, with much lower enantioselectivities (7–44% ee).

Zhang and coworkers also investigated the use of macrocyclic lactams as chiral elements in the design of amidoporphyrins. Notably, this

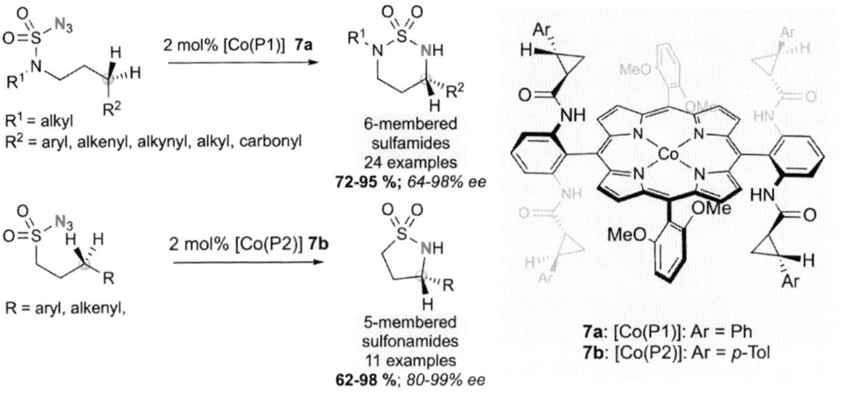

Scheme 6. Cobalt-catalyzed radical intramolecular C(sp³)–H amination.²⁶,²⁷

approach allows a fine tuning of the chiral pocket around the Co center by varying the nature of the tether between two amide units. A survey of several Co(II) complexes led to the identification of [Co(P3)] **8a** and [Co(P4)] **8b** as the most suitable catalysts. Both catalysts incorporate identical chiral cyclopropane units, bridged across the porphyrin plane by two tethers having different lengths (Scheme 7).²⁸ Interestingly, catalyst [Co(P3)] affords one enantiomer of the desired five-membered sulfamides with high ees, whereas [Co(P4)] leads to the other enantiomer of the products with similar selectivity. Mechanistic studies for this enantiodivergent process point toward an enantiodiscriminating H-atom abstraction, followed by a stereoretentive radical substitution. Moreover, when compared to unbridged complexes, these two catalysts lead to different regioselectivity in the intramolecular amination of sulfamoyl azides having two competitive amination sites. Indeed, [Co(P3)] and [Co(P4)] afford five-membered sulfamides, whereas [Co(P1)] and [Co(P2)] favor the formation of less-strained six-membered sulfamides.

C. Biocatalytic Amination with Engineered Enzymes

Enzymes that catalyze the oxidative amination of C–H bonds have not been identified in nature so far. However, directed evolution has recently enabled the discovery of engineered metalloenzymes able to perform such type of

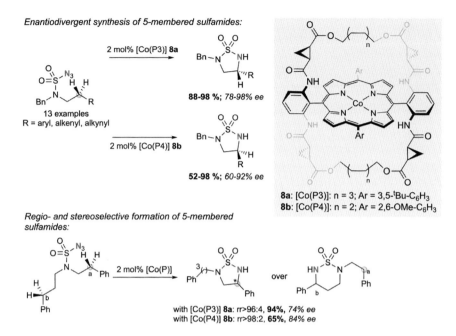

Scheme 7. Selectivity control in the cobalt-catalyzed radical C(sp³)–H amination.[28]

transformations.[29] Notably, the group of Arnold reported a series of enantioselective aminations catalyzed by engineered cytochrome P450 enzymes and operating through putative transient iron nitrenoids (Scheme 8).

A key aspect of these engineered enzymes is the presence of a serine axial ligand to the heme in place of the wild-type cysteine ligand. Moreover, all reactions are performed under anaerobic conditions with an iron at the ferrous state. Under such reaction conditions, Arnold et al. developed an intermolecular enantioselective amination of benzylic C–H bonds with tosyl azide as nitrene precursor and the alkane substrate as limiting reagent.[30] A range of cyclic and acyclic benzylic positions were aminated with excellent selectivities and turnover number (TON) in the 45–730 range depending on the substrates.

Following on these results, Arnold and coworkers later reported an intramolecular asymmetric amination of primary, secondary, and tertiary C–H bonds to access a variety of enantioenriched cyclic sulfamides (Scheme 9).[31] The engineered enzymes used in this study are extremely reactive and provide TON in the 770–5140 range. Remarkably, these

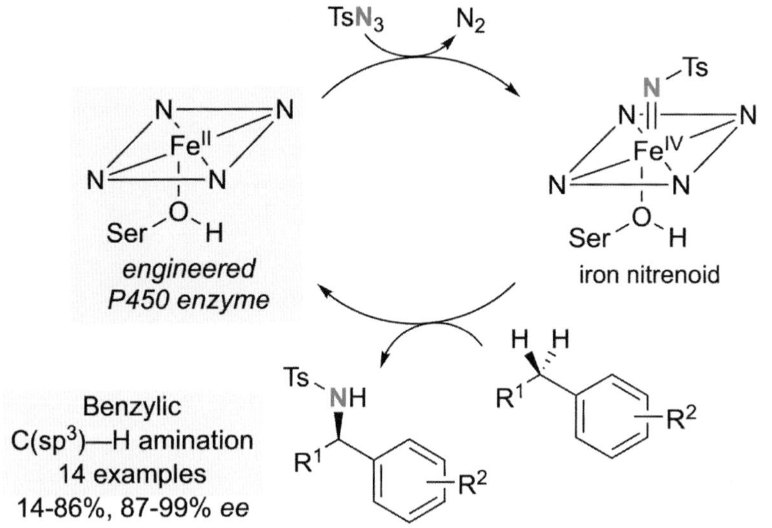

Scheme 8. Engineered enzyme-catalyzed intermolecular benzylic C(sp^3)–H amination.^{30}

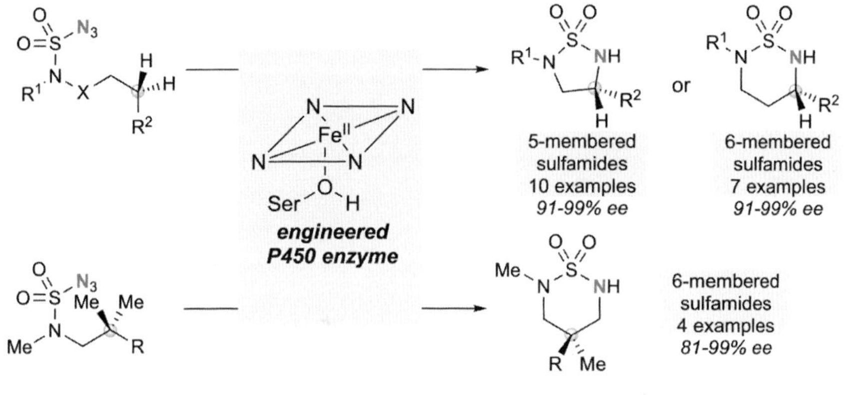

Enantioselective amination of primary and secondary C-H bonds:

Scheme 9. Engineered enzyme-catalyzed intramolecular C(sp^3)–H amination of 1° and 2° C(sp^3)–H bonds.^{31}

catalysts enable the enantioselective amination of both enantiotopic methylenic C–H bonds and enantiotopic methyl groups.

In addition, the same class of engineered enzymes also converts racemic substrates into enantioenriched six-membered sulfamides through an enantioconvergent amination of tertiary C–H bonds (Scheme 10).^{31} This

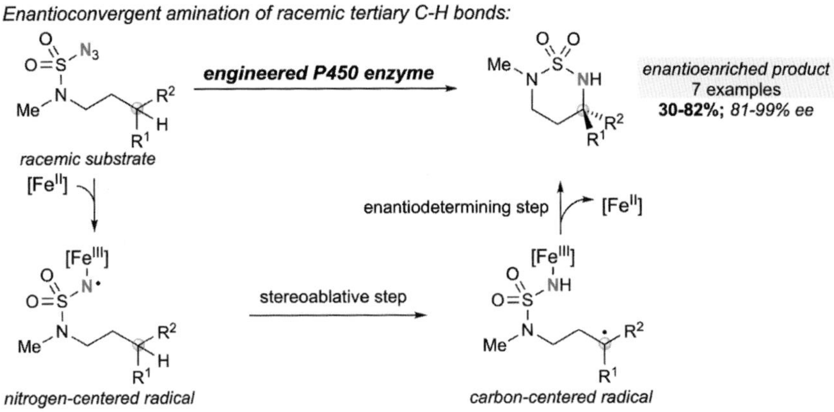

Enantioconvergent amination of racemic tertiary C-H bonds:

Scheme 10. Engineered enzyme-catalyzed intramolecular $C(sp^3)$–H amination of 3° $C(sp^3)$–H bonds.[31]

attractive feature originates from a radical mechanism similar to the one described in Scheme 5, where, however, the stereogenic information of the substrate is lost in a stereoablative hydrogen atom transfer. Then, an enantiodetermining intramolecular radical substitution step enables the access to enantioenriched sulfamides.

III. Catalytic Asymmetric $C(sp^3)$–H Amination with Dioxazolones

A. Introduction

1,4,2-Dioxazol-5-ones, or dioxazolones,[32] are readily available heterocycles that were first isolated in 1951 by reacting hydroxamic acids with phosgene.[33] Their use as nitrogen group donors in catalytic nitrene additions was reported in the mid-2010s with the pioneering studies of the Bolm group.[34] Since then, dioxazolones have appeared as safe, sustainable surrogates for acyl azides also in catalytic C–H amination reactions (Scheme 11).[35] They display a good thermal stability and their conversion to acyl nitrenes generates only CO_2 as a by-product. Moreover, whereas acyl azides–derived nitrenes have a high propensity to undergo Curtius-type rearrangements to isocyanates, the reactivity of dioxazolones in the presence of [Cp*M^{III}]-type complexes (M: Co, Rh, Ir) can be tamed to give the desired aminated products.

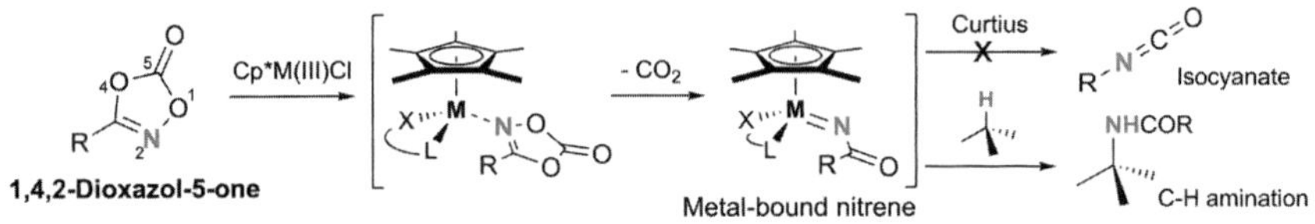

Scheme 11. Dioxazolones and their reactivity in the presence of [Cp*M^III] complexes.

B. Design of Chiral Hydrogen-bond-donor Catalysts

A major difficulty in performing C–H insertion reactions with acylnitrenes *via* an outer-sphere mechanism[15] is to avoid a competitive rearrangement of nitrenes to isocyanates. This pathway is highly favored when free nitrenes are generated under thermal or photochemical conditions; however, the careful design of organometallic catalysts has allowed to address this challenging issue. Thus, the synthesis of γ-lactams could be performed efficiently from dioxazolones by an intramolecular C(sp^3)–H amination. Key to the success was the seminal discovery made by the Chang group on the influence of the electronic density around the metal center on the energy barrier of the Curtius rearrangement.[36] The energy was estimated by density functional theory (DFT) calculations and found higher for metal complexes with strong electron-donating ligands, thus favoring the C–H insertion pathway. The iridium(III) complex **9** with a bidentate anionic 8-carbamoyl-4-methoxyquinoline ligand proved to be optimal to this end (Scheme 12).

The bidentate anionic ligand was a starting point to develop a catalytic asymmetric variant through a screening of various chiral *N,N* and *N,O* donor ligands. The identification and use of the chiral complex **10** enabled the synthesis of a wide range of γ-lactams, substituted by either aromatic groups (R = aryl) or aliphatic side chains (R = alkyl), in good yields and with optimal ees of up to 99% (Scheme 12).[37] DFT calculations allowed to rationalize these excellent results. The chirality of the ligand was first responsible for the diastereoselective coordination of the dioxazolone to the Ir complex. Then, after a facile decarboxylation, the resulting Ir-bound acylnitrene reacts through the stereodetermining half-chair transition state shown in Scheme 12, where the R group is located in pseudoequatorial position. Importantly, an intramolecular hydrogen bond between the primary amine of the ligand and the carbonyl of the acylnitrene is responsible

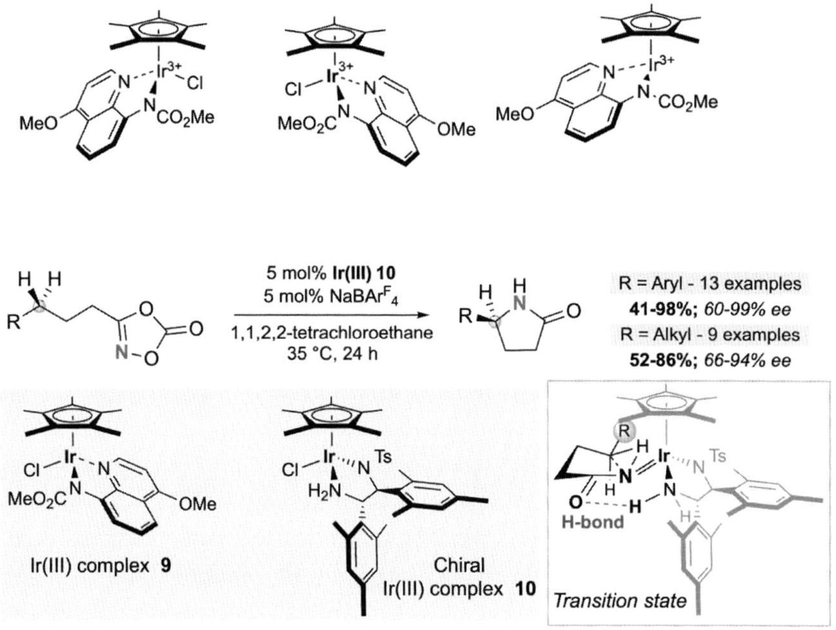

Scheme 12. Iridium-catalyzed enantioselective intramolecular C(sp^3)–H aminations via H-bonding intermediates.[37]

for the excellent levels of enantiocontrol. The hydrogen bonding interaction was corroborated by the X-ray structure of an Ir–amido complex, analogue of the Ir–acylnitrene intermediate. Its key role in the catalytic asymmetric C–H amination was evidenced also by the low ees observed with the *N*-methyl and *N*,*N*-dimethyl analogues of complex **10**. Concomitant to this study, a comparable approach based on ruthenium(II) complexes with Noyori-type chiral 1,2-diamine ligands proved as efficient for the enantioselective formation of γ-lactams from dioxazolones.[38]

C. Design of Catalysts with Chiral Hydrophobic Pockets

A conceptually different strategy, inspired by the same iridium complex **9**, was investigated then for the enantioselective synthesis of γ-lactams by intramolecular C(sp^3)–H amination. Although hydrogen bonds are the attractive interactions of choice in asymmetric catalysis, noncovalent π-interactions can provide opportunities for mimicking the hydro-

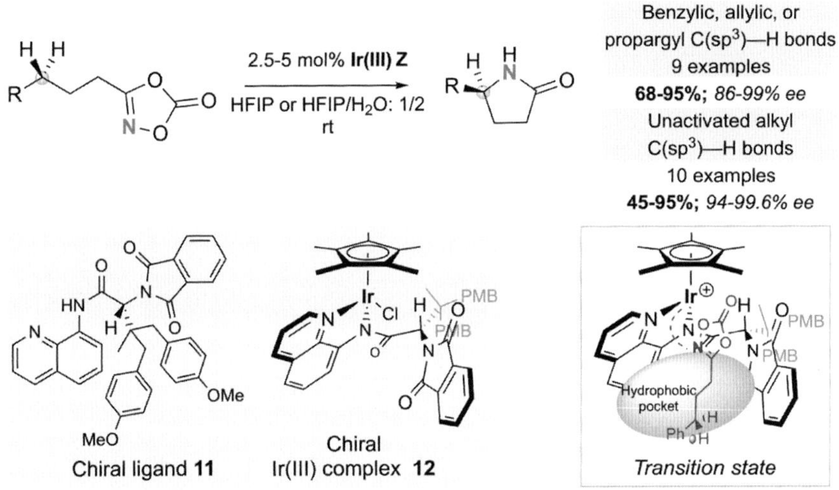

Scheme 13. Iridium-catalyzed enantioselective C(sp³)–H amination induced by an hydrophobic pocket.[40]

phobic pocket of enzymes and designing highly selective catalysts.[39] With this goal in mind, the new chiral ligand **11** was prepared by combining the 8-aminoquinoline scaffold with a substituted α-amino acid unit (Scheme 13). The resulting Ir complex **12** proved to be a highly efficient catalyst for the asymmetric intramolecular C(sp³)–H amination reaction of dioxazolones. Particularly, it expands the scope of the previous process to the amination of nonactivated alkyl C(sp³)–H bonds. Thus, γ-lactams with aliphatic substituents were isolated with very high ees, greater than 94%.[40]

A key feature of the transformation is the use of polar and/or aqueous solvents that facilitate the dissociation of the chloride from complex **12**, and favor the binding of apolar substrates. When bound to the metal, the ligand, indeed, forms a chiral hydrophobic pocket delineated by the Cp*, quinoline, and phthaloyl rings. The dioxazolone enters the pocket with the carbonyl group pointing toward the Cp* ring to limit the steric constraints, and the nitrogen atom binds then to the cationic iridium center. Once in the pocket, the substrate undergoes decarboxylation, followed by an enantiodetermining nitrene C(sp³)–H insertion. Formation of the (*S*)-product (when R is a phenyl group) is favored by the stronger attractive π–π and C(sp³)–H/π interactions between the substrate and the ligand in the corresponding transition state. On the other hand, the transition state leading to the (R) enantiomer is destabilized by repulsive interactions between the carbonyl groups of the substrate and the phthaloyl group of the ligand. Interestingly, this enantioselective C(sp^3)–H amination could be applied to the desymmetrization of meso compounds, such as cycloalkylmethyl-substituted dioxazolones.

D. Design of an Achiral Cp*M(III)/Chiral Carboxylic Acid Hybrid Catalytic System

Dioxazolones have also been applied to catalytic C–H amination reactions that operate through an inner-sphere C–H activation mechanism, in the presence of [Cp*M^{III}]-type complexes. According to this mechanism, activation of the C–H bond by the metal occurs first to afford an organometallic intermediate displaying a M–C bond. This intermediate then reacts with dioxazolones, which have a strong affinity for cationic M(III) complexes[41] and generate the metal-bound nitrene species after CO$_2$ extrusion. Final insertion of the nitrene into the M–C bond is favored due to a low activation barrier, thereby preventing the competitive rearrangement to isocyanates.

The C–H activation elementary step can proceed according to several pathways.[42] One of the most frequently observed mechanism relies on a concerted metalation deprotonation (CMD) mediated by a carboxylate ligand bound to the metal. Because the CMD is also the enantiodetermining step of the overall catalytic process, the use of a chiral carboxylic acid with an achiral metal complex can provide a relevant opportunity to perform catalytic asymmetric reactions. Such a strategy was applied to the enantioselective C(sp^3)–H amidation of thioamides (Scheme 14). The combination of the tert-leucine-derived acid 13 with the achiral Co(III) complex 14 allowed to discriminate between two enantiotopic methyl groups of a series of thioamides and create quaternary carbon centers with ees ranging from 80% to 88%.[43]

The same approach was then used for the enantioselective C(sp^3)–H amidation of 8-alkylquinolines.[44] The design of a sophisticated binaphthyl-derived chiral carboxylic acid, 15, combined with a Cp*Rh(III) complex led to the expected C–H–aminated products with ees between 82% and 88% (Scheme 15).

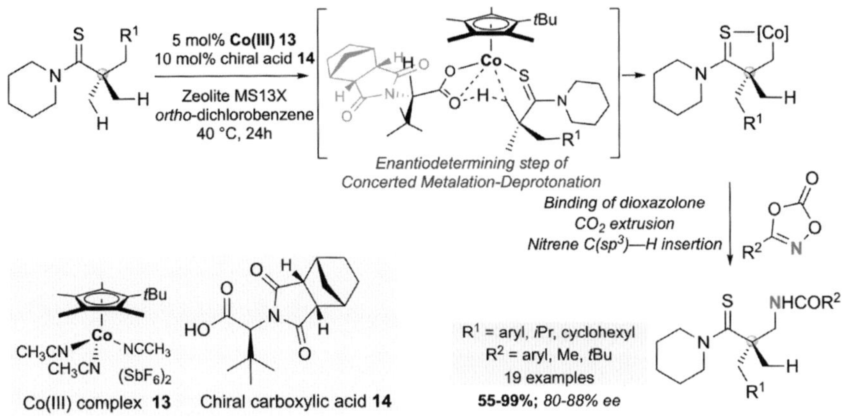

Scheme 14. Chiral carboxylic acid–mediated cobalt-catalyzed enantioselective C(sp^3)–H amination of thioamides.[43]

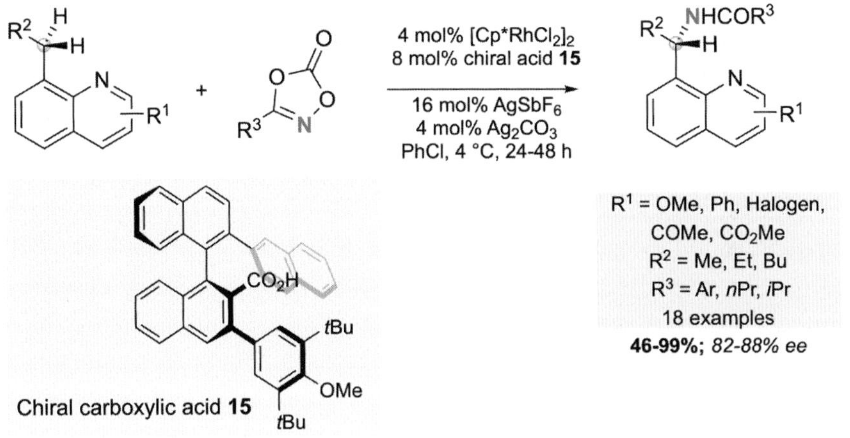

Scheme 15. Chiral carboxylic acid–mediated rhodium-catalyzed enantioselective C(sp^3)–H amination of 8-alkylquinolines.[44]

IV. Catalytic Asymmetric C(sp^3)–H Amination with Iodine(III) Oxidants

A. Introduction

The application of hypervalent iodine chemistry to catalytic C(sp^3)–H amination reactions was first reported by Breslow and Gellman in the

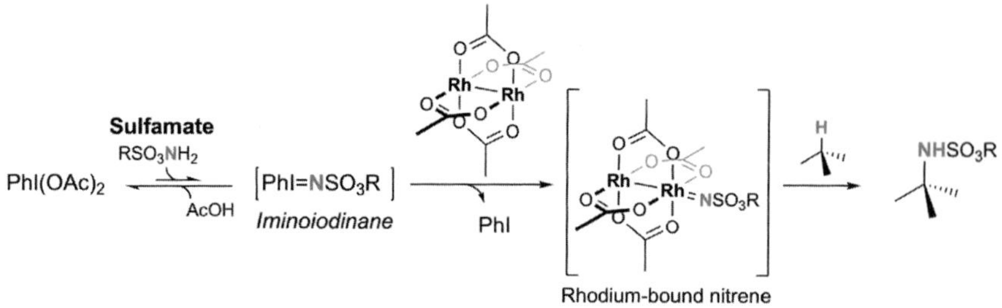

Scheme 16. Iminoiodanes, their generation from sulfamates and their reactivity in the presence of Rh(II) complexes.

early 1980s.[45] Their study particularly showcased the ability of dirhodium(II) complexes to catalyze the intramolecular C–H amination of N-(sulfonyl)-iminoiodinanes.[46] In 2001, the Du Bois group described practical and efficient conditions for this transformation that rely on the combination of the commercially available (iodosylbenzene)diacetate PhI(OAc)$_2$ with simple carbamates or sulfamates (Scheme 16).[47,48] The method allows for generating *in situ* the reacting oxidized nitrogen source, here the iminoiodinane, contrary to the reactions involving azides or dioxazolones that require preformed oxidized reagents. However, the use of hypervalent iodine reagents leads to the formation of stoichiometric amounts of iodobenzene as a side product. It is noteworthy that iodine(III) oxidant–mediated C(sp^3)–H amination reactions have been described from a variety of nitrogen functionalities, that is, sulfonamides, sulfamides, ureas, guanidines, or sulfonimidamides, among which carbamates and particularly sulfamates stand as the most efficient and versatile for synthetic applications.[49,50]

B. Design of a Chiral Rhodium(II)–Carboxamidate Complex

Following the seminal study of Müller and coworkers,[17] various chiral rhodium(II) complexes have been designed for catalytic asymmetric C(sp^3)–H amination.[18,19] Among these complexes, the chiral rhodium(II) carboxamidate Rh$_2$(*S*-nap)$_4$ **16** developed by Du Bois[51] stands out as an exceptional catalyst. The latter is indeed one of the rare rhodium(II) carboxamidates able to catalyze nitrene transfer under oxidizing conditions.

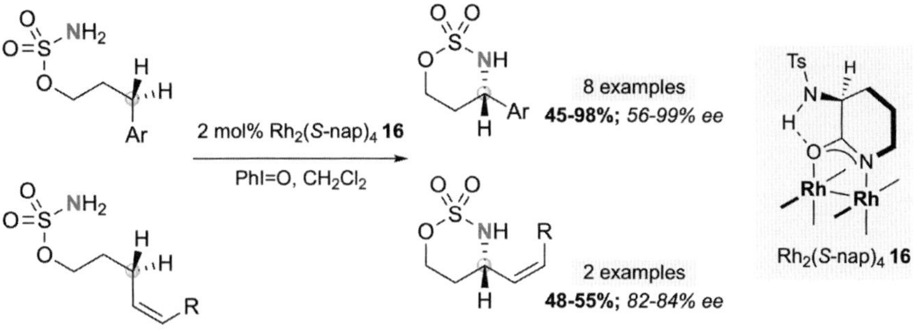

Scheme 17. Catalytic asymmetric intramolecular C(sp³)–H amination with a chiral Rh(II)–carboxamidate complex.[51]

Rhodium(II) carboxamidates are less electrophilic than their carboxylate analogs, so they are more prone to react with the iodine(III) reagent to give a deactivated Rh(II)/Rh(III) species. However, the higher redox potential of Rh$_2$(S-nap)$_4$ **16** prevents this side reaction to take place. This stable chiral rhodium(II) carboxamidate catalyzes the intramolecular benzylic and allylic C(sp³)–H aminations of sulfamates with excellent ees of up to 99% (Scheme 17). Importantly, in the case of allylic aminations, complete chemoselectivity is observed as no product arising from alkene aziridination is obtained. Moreover, the efficiency of the allylic amination depends on the olefin geometry, the best results being obtained with (Z) alkenes. It should be mentioned that complementary results in terms of scope were reported by Blakey and coworkers with a ruthenium(II)–pybox complex, which proved to be more efficient starting from (E)-alkenes.[52]

C. Design of a Chiral Rhodium(II)–Carboxylate Complex

Efficient chiral rhodium tetracarboxylate complexes have been designed by capitalizing on the diversity of α-amino acids. Following the initial study of Hashimoto and coworkers,[53] a wide range of catalysts were prepared by modifying either the amino acid side chain and/or the nitrogen-protecting group. A crucial feature of the resulting complexes is their low-energy "all-up" conformation in which the four adamantyl side chains are all positioned on one face of the dirhodium core.[54] Though the molecular structure displays an apparently achiral C_4 symmetry, it can induce enantioselective processes because the four phthalimido units

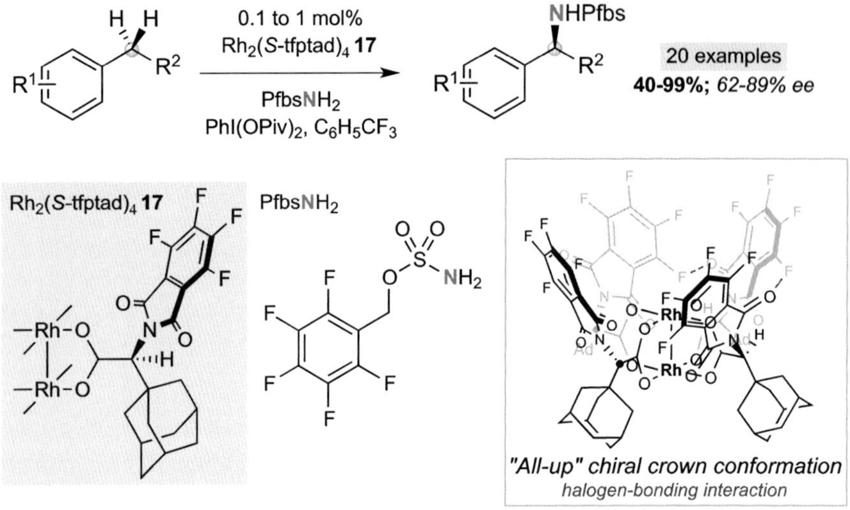

Scheme 18. Catalytic asymmetric intermolecular C(sp³)–H amination with a chiral Rh(II)–carboxylate complex.[55]

delineate a chiral hydrophobic pocket into which the nitrene binds and then reacts with the substrate. The complex Rh$_2$(S-tfptad)$_4$ **17** adopts such a chiral conformation that is stabilized by halogen bonding interactions between the fluorine and the carbonyls of the perfluorinated phthalimido groups (Scheme 18). This complex proved to be highly efficient, in combination with the perfluorinated benzylic sulfamate PfbsNH$_2$ (Pfbs = pentafluorobenzyloxysulfonyl), to mediate the intermolecular C(sp³)–H amination of a large range of benzylic substrates, with ees of up to 89%.[55] The reaction could be performed on a 50 mmol scale with a reduced catalyst loading of 0.1 mol%, and it was applied to the late-stage amination of complex products. In addition, the resulting C–H aminated products could be deprotected under mild conditions to provide the free amines in quantitative yield.[56]

D. Design of a Bisoxazoline Ligand with Quaternary Stereocenters

Bisoxazolines are privileged ligands in asymmetric catalysis.[57] Suffice to mention their efficient use in metal-catalyzed alkene aziridination with nitrenes, to highlight their importance.[58] Schomaker and coworkers have

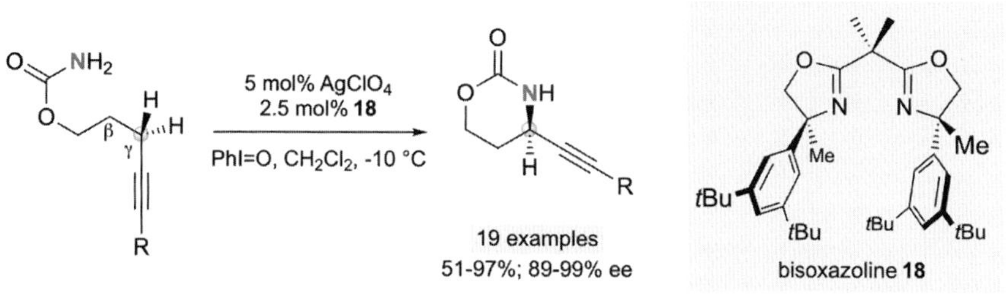

Scheme 19. Catalytic asymmetric intramolecular propargylic C(sp^3)–H amination with a chiral silver–bisoxazoline complex.[60]

recently revisited the chemistry of these C_2-symmetrical scaffolds with the design of new bisoxazolines displaying a fully substituted stereocenter α- to the nitrogen. They first demonstrated that this type of ligands enables the chemoselective intramolecular C(sp^3)–H amination of carbamates in γ-position, under silver catalysis. Of note, carbamates are well known to undergo amination at the β-position to give oxazolidinones.[47] However, the use of a α,α-dimethylbisoxazoline favors the formation of oxazinanones through the functionalization of the γ-C(sp^3)–H bond.[59] Then, the authors prepared the chiral bisoxazoline **18** that displays two quaternary stereocenters. The combination of silver(I) perchlorate with bisoxazoline **18** at −10°C proved to be optimal for the intramolecular γ-amination of carbamates in propargylic positions, giving good yields and excellent enantioselectivities (Scheme 19).[60] Instead of reducing the facial differentiation, the quaternary center imposes a steric crowding that restricts the rotation of the bulky aryl substituent, a conformational constrain that decreases the flexibility in the enantiodetermining transition state.

E. Design of Supramolecular Metal Complexes

The highly active bisdicarboxylate catalyst Rh$_2$(esp)$_2$ **19** (Scheme 20, esp = espinoate = α,α,α′,α′-tetramethyl-1,3-benzenedipropionate) has emerged as a complex of choice to perform intermolecular C(sp^3)–H amination reactions with the sulfamate TcesNH$_2$.[61] The chelate effect of the bidentate ligand allows to avoid any ligand exchange with the carboxylates released from the hypervalent iodine reagent PhI(OAc)$_2$ during the reaction. This feature secures notably the kinetic stability of the Rh$_2$(esp)$_2^{5+}$

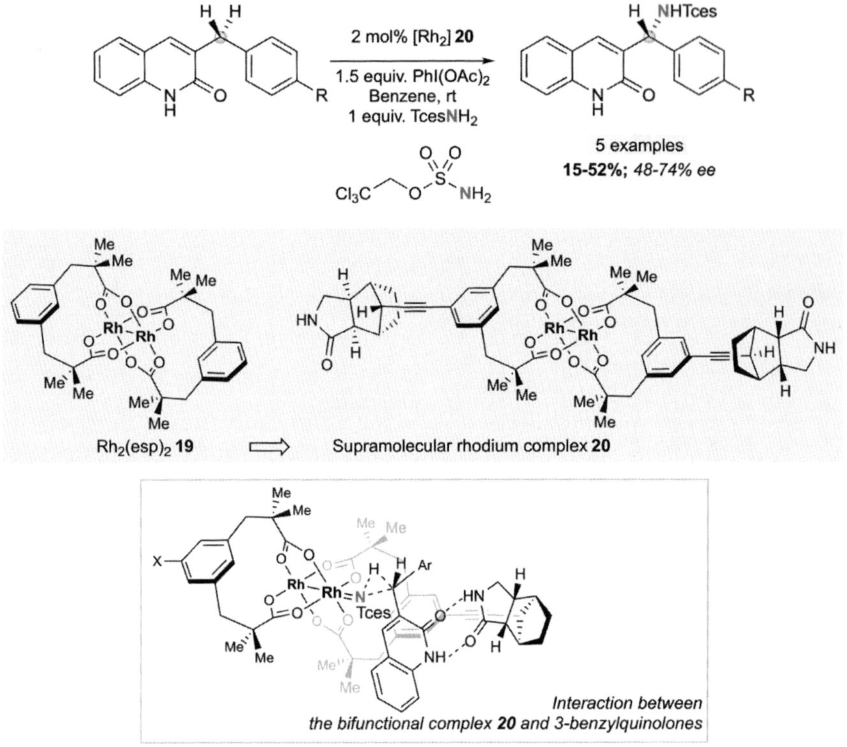

Scheme 20. Catalytic asymmetric intermolecular C(sp^3)–H amination with a supramolecular Rh(II) complex.[64]

amido species that are generated following the reaction between the dirhodium(II) complex and the sulfamates, through a proton-coupled electron transfer.[62,63] Accordingly, Rh$_2$(esp)$_2$ **19** is able to mediate catalytic C(sp^3)–H aminations in good yields and with low catalyst loadings. It thus appears as a relevant starting point to address the issue of the catalytic asymmetric intermolecular C–H amination that remains a challenge in terms of both reactivity and enantioselectivity.

It was with this goal in mind that Bach and coworkers designed the rhodium(II) complex **20** whose structure is inspired by the Rh$_2$(esp)$_2$ (Scheme 20).[64] By tethering chiral bicyclic lactams to the Rh$_2$(esp)$_2$ core, by means of an alkyne unit, the authors envisioned that the resulting bifunctional catalyst could interact with an appropriate substrate through an hydrogen bond network, and, thus, discriminate between two enantiotopic C(sp^3)–H bonds. 3-Benzylquinolones proved relevant to this end as they display hydrogen bond donor and acceptor units that are complementary to those of complex **20**. Formation of the two expected hydrogen bonds therefore led to enantioselective intermolecular benzylic C(sp^3)–H aminations, which proceeded with moderate to good selectivity and limited scope.

Better results were obtained later in the amination of analogous substrates under Ag(I) catalysis, by tethering the same chiral bicyclic lactams at the C-4 position of 1,10-phenanthroline. The resulting ligand **21** binds to a silver(I) salt to produce a complex which, in the presence of 1,10-phenanthroline, catalyzes the intermolecular C(sp^3)–H amination of quinolones and pyridones with ees ranging from 83% to 97% (Scheme 21).[65] Again, a hydrogen bond network is responsible for the selective removal of the pro-*S* hydrogen atom. The corresponding enantioenriched free amines are easily accessible by cleavage of the *p*-nitrobenzenesulfonyl (Ns) group with thiophenol under basic conditions.

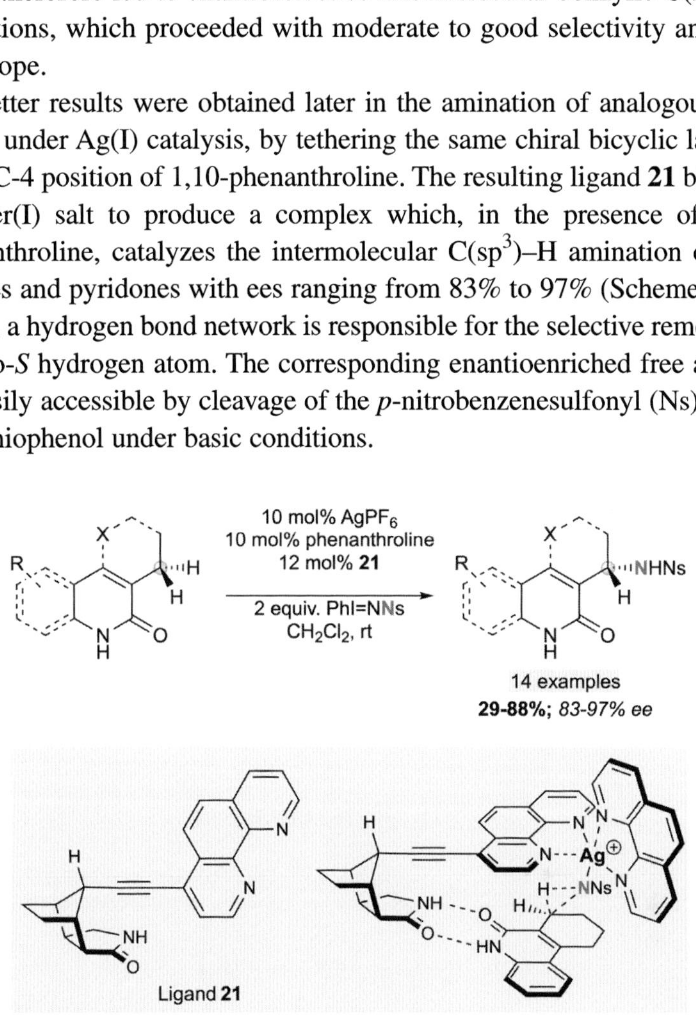

Scheme 21. Catalytic asymmetric intermolecular C(sp^3)–H amination with a supramolecular Ag(I) complex.[65]

V. Conclusion

Catalytic $C(sp^3)$–H amination reactions have long mostly relied on the application of hypervalent iodine chemistry in combination with transition metal complexes. However, the last decade has witnessed the discovery of other strategies involving azides or dioxazolones as starting materials, which have resulted in the design of various processes for the catalytic asymmetric amination of $C(sp^3)$–H bonds. The enantioselective methods described in this chapter are a perfect reflection of modern asymmetric catalysis as they rely on emerging concepts such as noncovalent interactions, hydrophobic pockets, multifunctional and/or supramolecular catalysis, or engineered biocatalysts. These strategies allow to discriminate efficiently various types of enantiotopic $C(sp^3)$–H bonds to efficiently access enantiopure 1,2- or 1,3-diamines or aminoalcohols, lactams, and benzylic amines. These processes also highlight the key role played by noble metals, as well as the emerging opportunities offered by first-row metals. Such new catalytic asymmetric methods should find application both in the total synthesis of nitrogen-containing natural products and in medicinal chemistry.

VI. Acknowledgments

The authors wish to thank the French National Research Agency (program no. ANR-11-IDEX-0003-02, CHARMMMAT ANR-11-LABX-0039 and ANR-19-CE07-0043-02), and the Institut de Chimie des Substances Naturelles for their support.

VII. References

1. Serturner, F. W.; *Annalen der Physik* **1817**, *55*, 56–89.
2. Nozaki, H.; Moriuti, S.; Takaya, H.; Noyori, R. *Tetrahedron Lett.* **1966**, *7*, 5239–5244.
3. Scott, F.L.; Clemons, B.; Brooks, J.; Brachmachary, E.; Powell, R.; Dedman, H.; Desale, H. G.; Timony, G. A.; Martinborough, E.; Rosen, H.; Roberts, E.; Boehm, M. F.; Peach, R. J. *Br. J. Pharmacol.* **2016**, *173*, 1778–1792.
4. Mailyan, A. K.; Eickhoff, J. A.; Minakova, A. S.; Gu, Z.; Lu, P.; Zakarian, A. *Chem. Rev.* **2016**, *116*, 4441–4557.
5. Nugent, T. C. *Chiral Amine Synthesis*. Wiley-VCH: Weinheim, **1998**.

6. Afanasyev, O. I.; Kuchuk, E.; Usanov, D. L.; Chusov, D. *Chem. Rev.* **2019**, *19*, 11857–11911.
7. Hansen, K. B.; Hsiao, Y.; Xu, F.; Rivera, N.; Clausen, A.; Kubryk, M.; Krska, S.; Rosner, T.; Simmons, B.; Balsells, J.; Ikamoto, N.; Sun, Y.; Spindler, F.; Malan, C.; Grabowski, E. J. J.; Armstrong III, J. D. *J. Am. Chem. Soc.* **2009**, *131*, 8798–8804.
8. Davies, H. M. L.; Manning, J. R.; *Nature* **2006**, *451*, 417–424.
9. Collet , F.; Dodd, R. H.; Dauban, P. *Chem. Commun.* **2009**, *45*, 5061–5074.
10. Zalatan, D. N.; Du Bois, J. *Top. Curr. Chem.* **2010**, *292*, 347–378.
11. Louillat, M. L.; Patureau, F. W. *Chem. Soc. Rev.* **2014**, *43*, 901–910.
12. Park, Y.; Kim, Y.; Chang, S. *Chem. Rev.* **2017**, *117*, 9247–9301.
13. Ruiz-Castillo, P.; Buchwald, S. L. *Chem. Rev.* **2016**, *116*, 12564–12649.
14. Godula, K.; Sames, D. *Science* **2006**, *312*, 67–72.
15. Dick, A. R.; Sanford, M. S. *Tetrahedron* **2006**, *62*, 2439–2463.
16. Saint-Denis, T. G.; Zhu, R.-Y.; Chen, G.; Wu, Q.-F.; Yu, J.-Q. *Science* **2018**, *359*, eaao4798.
17. Nägeli, I.; Baud, C.; Bernardinelli, G.; Jacquier, Y.; Moran, M.; Müller, P. *Helv. Chim. Acta* **1997**, *80*, 1087–1105.
18. Collet, F.; Lescot, C.; Dauban, P. *Chem. Soc. Rev.* **2011**, *40*, 1926–1936.
19. Hayashi, H.; Uchida, T. *Eur. J. Org. Chem.* **2020**, 909–916.
20. Kwart, H.; Kahn, A. A. *J. Am. Chem. Soc.* **1967**, *89*, 1950–1951.
21. Nishioka, Y.; Uchida, T.; Katsuki, T. *Angew. Chem. Int. Ed.* **2013**, *52*, 1739–1742.
22. Zhang, L.; Meggers, E. *Acc. Chem. Res.* **2013**, *50*, 320–330.
23. Zhou, L.; Chen, S.; Qin, J.; Nie, X.; Zheng, X.; Harms, K.; Riedel, R.; Houk, K. N.; Meggers, E. *Angew. Chem. Int. Ed.* **2019**, *58*, 1088–1093.
24. Qin, J.; Zhou, Z.; Cui, T.; Hemming, M.; Meggers, E. *Chem. Sci.* **2019**, *10*, 3202–3207.
25. Singh, R.; Mukherjee, A. *ACS Catal.* **2019**, *9*, 3604–3617.
26. Li, C.; Lang, K.; Lu, H.; Hu, Y.; Cui, X.; Wojtas, L.; Zhang, X. P. *Angew. Chem. Int. Ed.* **2018**, *57*, 16837–16841.
27. Hu, Y.; Lang, K.; Li, C.; Gill, J. B.; Kim, I.; Lu, H.; Fields, K. B.; Marshall, M.; Cheng, Q.; Cui, X.; Wojtas, L.; Zhang, X. P. *J. Am. Chem. Soc.* **2019**, *141*, 18160–18169.
28. Lang, K.; Torker, S.; Wojtas, L.; Zhang, X. P. *J. Am. Chem. Soc.* **2019**, *141*, 12388–12396.
29. Mahy, J.-P.; Cieselski, J.; Dauban, P. *Angew. Chem. Int. Ed.* **2014**, *53*, 6862–6864.
30. Prier, C. K.; Zhang, R. K.; Buller, A. R.; Brinkmann-Chen, S.; Arnold, F. H. *Nat. Chem.* **2017**, *9*, 629–634.
31. Yang, Y.; Cho, I.; Qi, X.; Liu, P.; Arnold, F. H. *Nat. Chem.* **2019**, *11*, 987–993.
32. Van Vliet, K. M.; De Bruin, B. *ACS Catal.* **2020**, *10,* 4751–4769.
33. Beck, G. *Chem. Ber.* **1951**, *84*, 688–689.
34. Bizet, V.; Buglioni, L.; Bolm, C. *Angew. Chem. Int. Ed.* **2014**, *53*, 5639–5642.
35. Park, Y.; Jee, S.; Kim, J. G.; Chang, S. *Org. Process Res. Dev.* **2015**, *19*, 1024–1029.
36. Hong, S. Y.; Park, Y.; Hwang, Y.; Kim, Y. B.; Baik, M.-H.; Chang, S. *Science* **2018**, *359*, 1016–1021.
37. Park, Y.; Chang, S. *Nat. Catal.* **2019**, *2*, 219–227.

38. Xing, Q.; Chan, C.-M.; Yeung, Y.-W.; Yu, W.-Y. *J. Am. Chem. Soc.* **2019**, *141*, 3849–3853.
39. Neel, A. J.; Hilton, M. J.; Sigman, M. S.; Toste, F. D. *Nature* **2017**, *543*, 637–646.
40. Wang, H.; Park, Y.; Bai, Z.; Chang, S.; He, G.; Chen, G. *J. Am. Chem. Soc.* **2019**, *141*, 7194–7201.
41. Park, Y.; Park, K. T.; Kim, J. G.; Chang, S. *J. Am. Chem. Soc.* **2015**, *137*, 4534–4542.
42. Roudesly, F.; Oble, J.; Poli, G. *J. Mol. Cat. A: Chem.* **2017**, *426*, 275–296.
43. Fukagawa, S.; Kato, Y.; Tanaka, R.; Kojima, M.; Yoshino, T.; Matsunaga, S. *Angew. Chem. Int. Ed.* **2019**, *58*, 1153–1157.
44. Fukagawa, S.; Kojima, M.; Yoshino, T.; Matsunaga, S. *Angew. Chem. Int. Ed.* **2019**, *58*, 18154–18158.
45. Breslow, R.; Gellman, S.H. *J. Am. Chem. Soc.* **1983**, *105*, 6728–6729.
46. Yamada , Y.; Yamamoto, T.; Okawara, M. *Chem. Lett.* **1975**, *4*, 361–362.
47. Espino, C. G.; Du Bois, J. *Angew. Chem. Int. Ed.* **2001**, *40*, 598–601.
48. Espino, C. G.; When, P. M.; Chow, J.; Du Bois, J. *J. Am. Chem. Soc.* **2001**, *123*, 6935–6936.
49. Roizen, J. L.; Harvey, M. E.; Du Bois, J. *Acc. Chem. Res.* **2012**, *45*, 911–922.
50. Darses, B.; Rodrigues, R.; Neuville, L.; Mazurais, M.; Dauban, P. *Chem. Commun.* **2017**, *53*, 493–508.
51. Zalatan, D. N.; Du Bois, J. *J. Am. Chem. Soc.* **2008**, *130*, 9220–9221.
52. Milczek, E.; Boudet, N.; Blakey, S. *Angew. Chem. Int. Ed.* **2008**, *47*, 6825–6828.
53. Yamawaki, M.; Tsutsui, H.; Kitagaki, S.; Anada, M.; Hashimoto, S. *Tetrahedron Lett.* **2002**, *43*, 9561–9564.
54. De Angelis, A.; Dmitrenko, O.; Yap, G. P. A.; Fox, J. M. *J. Am. Chem. Soc.* **2009**, *131*, 7230–7231.
55. Nasrallah, A.; Boquet, V.; Hecker, A.; Retailleau, P.; Darses, B.; Dauban, P. *Angew. Chem. Int. Ed.* **2019**, *58*, 8192–8196.
56. Nasrallah, A.; Lazib, Y.; Boquet, V.; Darses, B.; Dauban, P. *Org. Process. Res. Dev.* **2020**, *24*, 724–728.
57. Yoon, T. P.; Jacobsen, E. N. *Science* **2003**, *299*, 1691–1693.
58. Ju, M.; Weatherly, C. D.; Guzei, I. A.; Schomaker, J. M. *Angew. Chem. Int. Ed.* **2017**, *56*, 994–9948.
59. Ju, M.; Huang, M.; Vine, L. E.; Dehghany, M.; Roberts, J. M.; Schomaker, J. M. *Nat. Catal.* **2020**, *2*, 899–908.
60. Ju, M.; Zerull, E. E.; Roberts, J. M.; Huang, M.; Guzei, I. A.; Schomaker, J. M. *J. Am. Chem. Soc.* **2020**, *142*, 12930–12936.
61. Espino, C. G.; Fiori, K. W.; Kim. M.; Du Bois, J. *J. Am. Chem. Soc.* **2004**, *126*, 15378–15379.
62. Zalatan, D. N.; Du Bois, J. *J. Am. Chem. Soc.* **2001**, *130*, 7558–7559.
63. Kornecki, K. P.; Berry, J. F. *Chem. Eur. J.* **2011**, *17*, 5827–5832.
64. Höke, T.; Herdtweck, E.; Bach, T. *Chem. Commun.* **2013**, *49*, 8009–8011.
65. Annapureddy, R. R.; Jandl, C.; Bach, T. *J. Am. Chem. Soc.* **2020**, *142*, 7374–7378.

© 2022 World Scientific Publishing Company
https://doi.org/10.1142/9789811248436_0009

9 Bifunctional Homogeneous Catalysts Based on Ruthenium, Rhodium and Iridium in Asymmetric Hydrogenation

Christophe Michon[*,‡] and Francine Agbossou-Niedercorn[†,§]

[*]Université de Strasbourg, Université de Haute-Alsace, Ecole Européenne de Chimie, Polymères et Matériaux, CNRS, LIMA, UMR 7042, 25 rue Becquerel, 67087 Strasbourg, France
[†]Université de Lille, CNRS, Centrale Lille, Université d'Artois, UMR 8181 - UCCS - Unité de Catalyse et Chimie du Solide, F-59000 Lille, France
[‡]cmichon@unistra.fr
[§]francine.niedercorn@univ-lille.fr

Table of Contents

List of Abbreviations	283
I. Introduction	284
II. ATHs Using Bifunctional Catalysts	287

A. Ruthenium-Catalyzed Transfer Hydrogenations		287
1. Noyori-type catalysts		287
2. Shvo-type catalysts		305
B. Rhodium-Catalyzed Transfer Hydrogenations		307
1. Diamine-tethered η^5-cyclopentadienyl rhodium complexes with active NH functions		307
2. Amino-acid-derived amides and thioamides with active NH functions		309
C. Iridium-Catalyzed Transfer Hydrogenations		312
1. Amino- and amido iridium(III) catalysts with active NH functions		312
2. Octahedral chiral-at-metal iridium(III) catalysts with π-stacking and H-bonding ligands		316
D. Osmium-Catalyzed Transfer Hydrogenations		318
III. AHs Using Bifunctional Catalysts		319
A. Ruthenium-Catalyzed Hydrogenations		319
1. Shvo-type catalysts		319
2. Noyori-type catalysts		320
B. Rhodium-Catalyzed Hydrogenations		332
1. Catalysts implying hydrogen bondings between ligands and either substrates or halide counteranions		332
2. Catalysts implying ion pairs as noncovalent interactions between ligands and substrates		342
C. Iridium-Catalyzed Hydrogenations		345
1. Iridium catalysts involving active NH functions and related cases		346
2. Catalysts implying ion pair or hydrogen bonding interactions		349
3. Iridium phosphates and triflylphosphoramides as bifunctional catalysts		354
IV. Conclusion		356
V. Acknowledgments		357
VI. References		357

List of Abbreviations

AH	asymmetric hydrogenation
ATH	asymmetric transfer hydrogenation
BArF$_{24}$	tetrakis[(3,5-trifluoromethyl)phenyl]borate
BINAP	1,1'-binaphthalene-2,2'-diyl)bis(diphenylphosphine)
BINOL	1,1'-binaphthalene-2,2'-diol
cat	catalyst
CBA	chiral Brønsted acid
COD	cyclooctadiene
conv	conversion
Cp	cyclopentadienyl
Cp*	pentamethylcyclopentadienyl
de	diastereomeric excess
DENEB	N-[1,2-diphenyl-2-(2-(4-methylbenzyloxy)ethylamino)-ethyl]-(chloro)ruthenium(II)
DKR	dynamic kinetic resolution
DMSO	dimethylsulfoxide
DPEN	diphenylethylenediamine
dr	diastereomeric ratio
ee	enantiomeric excess
HFIP	hexafluoroisopropanol
Ms	methanesulfonyl
NBD	norbornadiene
NHC	N-heterocyclic carbene
NMR	nuclear magnetic resonance
S/C	substrate-to-catalyst ratio
SPO	secondary phosphine oxide
TBDPS	*tert*-butyldiphenylsilyl
TIPP	2,4,6-triisopropylphenyl
THF	tetrahydrofurane
TOF	turnover frequency
TON	turnover number
TRIP	3,3'-bis(2,4,6-triisopropylphenyl)-1,1'-binaphthyl-2,2'-diyl hydrogenphosphate
Ts	*p*-toluenesulfonyl

I. Introduction

Asymmetric hydrogenation (AH) is a fundamental and mature synthetic methodology providing high degree of stereocontrol for the preparation of optically pure fine chemicals like pharmaceuticals, agrochemicals, flavors, and fragrances. AHs of functionalized olefins and ketones have been realized using chiral homogeneous organometallic catalysts in the presence of hydrogen gas and several were applied to single enantiomer syntheses for advanced intermediates at industrial scales.[1–13] Most of the early methods employed catalysts based on rhodium(I) and required functional groups on the substrates to enable hydrogenation of the unsaturated bonds in highly efficient and selective manner.[14] In addition, such enantioselective hydrogenation catalysts remained almost inert toward simple ketones.[15]

In 1995, a catalyst based on the combination of a ruthenium precursor, BINAP, a chiral diamine and KOH, was reported by Noyori and co-workers to be able to hydrogenate simple aromatic ketones with high enantioselectivity in the presence of a hydrogen donor such as isopropanol.[16,17] Subsequently, the same authors had a major breakthrough by introducing η^6–arene–ruthenium catalysts with N-sulfonylated 1,2-diamines or amino alcohol as chiral ligands, for the highly efficient asymmetric transfer hydrogenation (ATH) of ketones.[18–20] For both the AH[15,21] and ATH[22–23] of ketones using these ligands, a metal–ligand bifunctional mechanism was established[21,24] by means of kinetic studies and isotopic labelling,[22,25,26] as well as via computational analyses.[26–28] It was demonstrated that the reaction goes through a six-membered pericyclic transition state involving a ruthenium hydride and one of the NH functions of the diamino (or amino alcohol) ligand (Scheme 1).

The N–H function of the ligand operates in a cooperative manner with the metal center during the hydrogenation reaction, which typifies a bifunctional transition metal-based molecular catalyst.[5,29,30] Since then, the concept of cooperative catalysis has been further applied in C–H, C–C, C–N, and CO formation.[31–38] In transfer hydrogenations, 2-propanol and formic acid are commonly used because these two cheap, nontoxic, and environmentally safe hydrogen sources do not require any specific equipment. A drawback of the use of 2-propanol is the thermodynamic reversibility of the transfer hydrogenation reaction leading to a limited

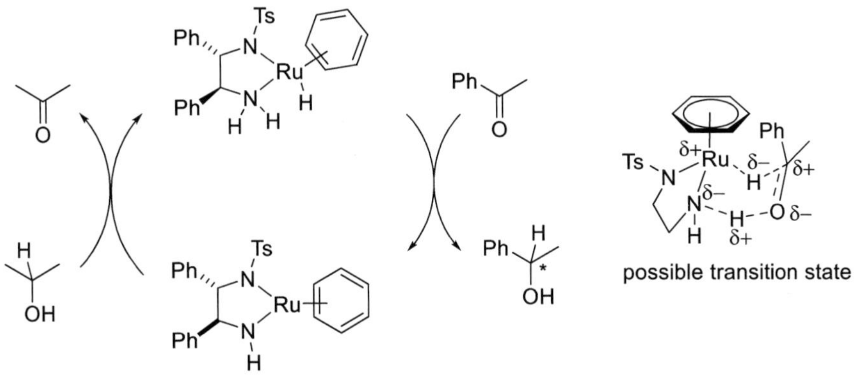

Scheme 1. Postulated mechanism for the asymmetric transfer hydrogenation of acetophenone in the presence of 2-propanol.[23]

conversion and to a gradual decrease of the enantiomeric purity of the targeted chiral alcohol. This can be overcome by the use of formic acid, since the CO_2 released can be removed to favor full conversions. Overall, the Noyori half-sandwich bifunctional catalysts in Scheme 1 are very effective for ATH using either 2-propanol or an azeotropic mixture of formic acid as hydrogen sources and offer a great potential in enantioselective synthesis.

These ruthenium-based bifunctional catalysts raised a huge interest among the chemistry community and, therefore, the initial studies were followed by numerous other reports focusing on the development of new catalysts based on late-transition metals and, more recently, on first row transition metals. In order to develop more robust catalysts for AH or ATH, half-sandwich η^6–arene–Ru complexes were modified notably by means of covalent tethers between the arene and the chiral diamine ligands. First introduced by Wills and co-workers, this new family of catalysts exhibited enhanced catalytic performances and stability (Scheme 2).[39–42]

Furthermore, the Noyori's ruthenium/diamine/diphosphine catalysts initially applied in AH have inspired the design of tetradentate analogues (Scheme 3, left) for the ATH of aryl,alkyl-ketones,[43] as well as many other variants such as the phosphinite complexes reported by Morris (Scheme 3, right).[44,45]

Since the 1990s and to date, bifunctional catalysts for ATH and AH have aroused high interest. This long-term interest for innovative catalyses

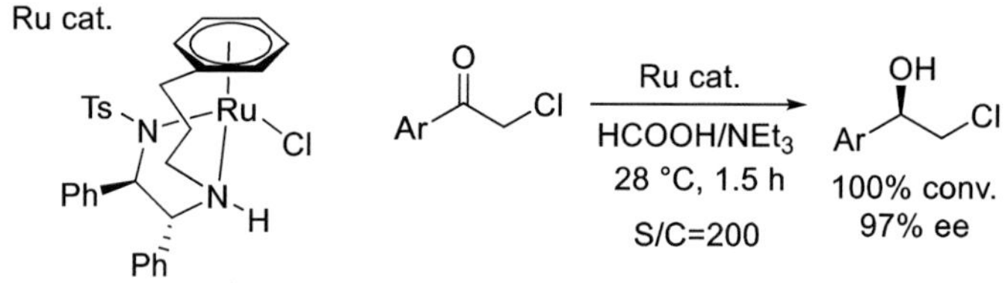

Scheme 2. ATH of α-chloroketones with a diamine-tethered arene–ruthenium catalyst.[39–42]

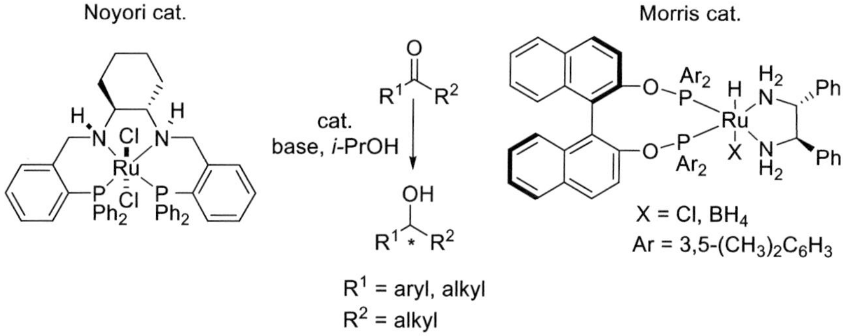

Scheme 3. Ruthenium/P2N2 ATH catalysts developed by Noyori and co-workers (left) and Morris and co-workers (right).[43–45]

is due to a lack of general solutions for the catalytic AHs and has led to the development of a number of metal catalysts exhibiting new functionalities.[46] These new generations of catalysts that rely on both hydrogen bonding, CH-π stacking and ion-pair interactions, have enabled bifunctional hydrogen transfer as in the Noyori's prototypical examples[47] but also activation of organic substrates through secondary interactions (cooperative catalysis)[48] or heterolytic activation of dihydrogen itself. Interestingly, the significant contributions to enantioselection of weak noncovalent attractive interactions between the substrate and the chiral ligand in hydrogenation reactions have been enlightened by many recent theoretical studies.[49–51]

For clarity sake, the readers will find the following three definitions of the term "bifunctional catalyst":

(1) According to T. Ikariya[52]: "bifunctional catalysts have two functionalities cooperating in substrate activation and transformation."

(2) According to D. B. Grotjahn[53]: "...organometallic catalysts resulting from including ligands capable of proton transfer or hydrogen bonding."
(3) According to S. J. Miller[54]: "Bifunctional catalysts usually possess a nucleophilic moiety appended to Lewis basic or Brønsted acidic moieties capable of hydrogen bonding or charge stabilization. Coupling of these functionalities typically results in cooperative activation of substrates and stabilization of transition states leading to heightened reactivity and higher levels of enantiocontrol in various transformations."

This book chapter will illustrate the different classes of bifunctional (cooperative) catalysts and the role of noncovalent interactions, by considering ruthenium, rhodium, and iridium homogeneous AH catalysts. It will present selected examples of metal complexes developed over the last 10 years, that is, along the 2010–mid-2020 period, with a special focus on the understanding of their reaction mechanisms. Therefore, new applications of known catalysts in organic syntheses will not be reviewed. Furthermore, neither supported and immobilized catalysts[55–59] nor methods for the AHs in water[60] will be considered herein. ATHs and AHs will be presented successively in Sections II and III, respectively.

II. ATHs Using Bifunctional Catalysts

A. Ruthenium-Catalyzed Transfer Hydrogenations

1. Noyori-type catalysts

On the basis of the pioneering research of Noyori and co-workers on ruthenium catalysts for the ATH, over the years many efforts have been devoted to the development of new catalysts for the reduction of C=O and C=N double bonds that operate through concerted H-transfer from both the metal and the functional ligands (see Scheme 1). Since ligands with amino alcohols, diamines, natural products-based amines, and other chiral structural units had been set by Noyori and co-workers, research has continued following the same directions during the last decade. Thus, diamines and, to a lesser extent, β-amino alcohols have been further investigated in the ATH of ketones and imines by using either η^6-arene–Ru or diphosphine–Ru complexes.

Diphosphine–diamine–ruthenium catalysts

The Ru^{II}(diphosphane)(diamine) complexes in which the ruthenium center is coordinated by a nonchiral diphosphine and a chiral diamine drew special attention because of the easy variation of their chiral ligands.[43–45] In 2015, Karabuga and co-workers reported on ruthenium complexes that combine quinazoline-based chiral diamines (Scheme 4a) and dppb (1,4-diphenylphosphinobutane). These complexes catalyzed the ATH of aryl-methyl ketones with good conversions and variable enantioselectivities (9%–91% ee).[61] Moreover, Xing and co-workers disclosed recently Ru(diphosphane)(diamine) ATH catalysts that feature a single stereogenic element, by combining a simple and readily available aminopyridine ligand (1-(pyridin-2-yl)methanamine) with an achiral diphosphine, 1,3-diphenylphosphinopropane.[62] The design of this inspiring catalytic system relied on the hypothesis that this combination would replicate the stereo induction of the recognized Noyori's catalyst. Indeed, the catalytic system allowed the practical and highly enantioselective ATH of a wide range of ketones including aryl,alkyl- and aryl,*N*-heteroaryl ketones (Scheme 4b).

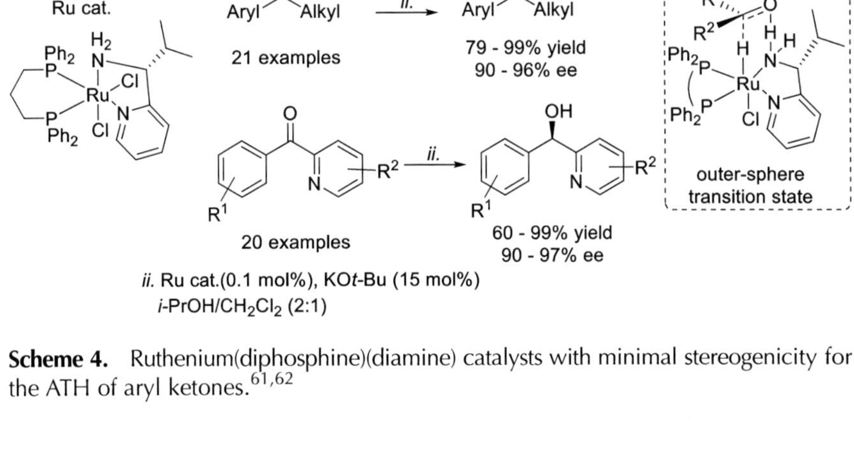

Scheme 4. Ruthenium(diphosphine)(diamine) catalysts with minimal stereogenicity for the ATH of aryl ketones.[61,62]

Amino-alcohol, diamine, and diamide-derived η^6-arene–ruthenium catalysts

Amino-alcohol-derived η^6-arene–ruthenium catalysts

As a complement to amino alcohols applied earlier in ATH, such as ephedrine,[63] 2-azanorbornyl-methanol derivatives,[64] and 1-aminoindane-2-ol,[65] Govender, Kruger, and co-workers reported in 2010 a series of C^1-substituted tetrahydroisoquinoline derivatives that were applied in the ATH of aromatic ketones. η^6-Arene–ruthenium complexes were generated *in situ* from [(*p*-cymene)RuCl$_2$]$_2$ and the chiral amino alcohol. The ATH reactions were carried out in the presence of KO*t*-Bu/*i*-PrOH and provided up to 94% ee in the hydrogenation of 4-methylacetophenone (Scheme 5).[66] The *trans*-relative orientation of the C^1 and C^3 substituents of the ligand proved essential for high activity and enantioselectivity, while substitutions on C^5 resulted in a decreased reactivity. The same ligands were also applied in rhodium-catalyzed ATH.

In 2011, Zheng and co-workers reported another series of amino alcohol ligands derived from 1-phenylethylamine (Scheme 5).[67] The less sterically demanding ligand of the series (R = H), associated to [RuCl$_2$(*p*-cymene)]$_2$ (0.5 mol%) in the presence of KOH (10 mol%) and *i*-PrOH allowed the reduction of aryl,alkyl-ketones in good yields (67%–95%) and enantioselectivities (67%–95% ee) at –10°C.

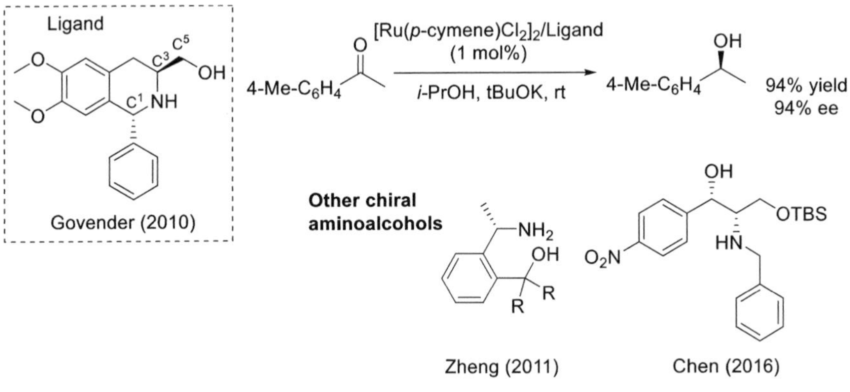

Scheme 5. Examples of recently developed amino alcohol and diamine ligands for ATH.[66–68]

Along the same lines, Chen and co-workers developed the novel chloramphenicol-based ligands in Scheme 5 for the ruthenium-catalyzed ATH of a N-Boc α-amino-β-ketoester. Under dynamic kinetic resolution (DKR) conditions in water, the reduced product was obtained in 85% yield, 92% de and 87% ee (conditions: [RuCl$_2$(p-cymene)]$_2$, chiral ligand 3 mol%, HCOONa/Tween 20, H$_2$O, 25°C).[68]

Aminoamide and diamine derived η^6-arene–ruthenium catalysts

Beside amino alcohols, both diamines and amino amides have been applied as ligands in the ATH of ketones. In 2011, following their continuous efforts to develop asymmetric catalysts for ATH, Wills and co-workers synthesized new ligands containing (R,R)- or (S,S)-DPEN and one or two proteinogenic (S)-proline groups.[69] The TsDPEN mono-proline ligands led to variable results in ruthenium-, iridium-, and rhodium-promoted ATH of ketones in water, with sodium formate as hydrogen donor. On the other hand, the DPEN/bis-proline ligands led to effective ruthenium catalysts in the ATH of cyclic ketones (Scheme 6). The ligand combining (S,S)-DPEN and (S)-proline units provided the most enantioselective catalyst. Though the ligand bearing two proline groups could coordinate up to two ruthenium atoms, control experiments showed that the active complex contains only one ruthenium per ligand, as well as that the selectivity is most probably controlled by steric factors.

Pericàs and co-workers prepared diamine ligands that display proline and triazole units, the latter being easily obtained through click chemistry.[70] It appeared that the compounds with an O-protected 4-hydroxyproline moiety was an efficient chiral ligand for the ruthenium-catalyzed ATH of

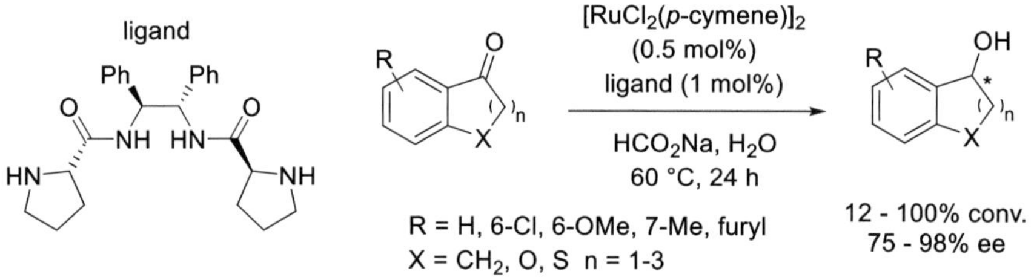

Scheme 6. DPEN/bis-proline ligands applied in ATH of cyclic ketones, by Wills and co-workers.[69]

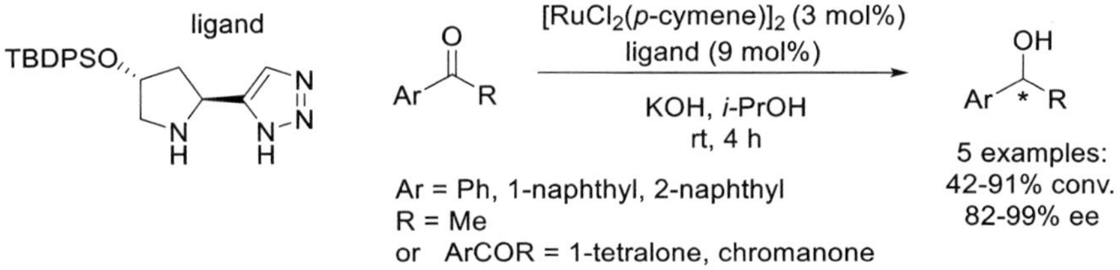

Scheme 7. Hydroxyproline/triazole ligands applied in the ATH of cyclic and acyclic ketones, by Pericas and co-workers.[70]

acetophenones and cyclic ketones, while nonsubstituted proline gave much lower ee (Scheme 7).

Amidoalcohol-derived η^6-arene–ruthenium catalysts

The easy availability and modular nature of α-amino acids makes them very attractive building blocks for the synthesis of chiral ligands. Thus, Adolfsson and co-workers have considered ruthenium and rhodium catalysts bearing pseudo-dipeptide-based ligands. The new ligands are N-Boc-protected α-aminoamides with β-amino alcohol functions. They were applied to ruthenium-catalyzed ATH of aryl,alkyl-ketones with excellent results; a variety of acetophenones could be hydrogenated with ee >80% (18 examples, Scheme 8a).[71–73] For such catalytic systems, the addition of LiCl was generally beneficial both in terms of enantioselectivity and conversion rate.[74] Thus, it was suggested that the reaction might involve the transfer of an hydride and lithium ion to the substrate and proceeds via the six-membered transition state shown in Scheme 8a. Interestingly, these pseudo-dipeptide ligands were also applied in the ruthenium-catalyzed ATH of propargylic ketones (Scheme 8b).[75] In this case, the addition of LiCl did not change the reaction outcome.

Afterwards, the authors applied further these ruthenium catalysts in the tandem alkylation/ATH of acetophenones with primary alcohols[76] and in the tandem-isomerization/ATH of allylic alcohols,[77] while ruthenium and rhodium complexes of this series were also widely tested in high-throughput screening experiments.[78]

Similarly, Diéguez, Adolfsson, and co-workers developed a library of 10 pseudo-dipeptide ligands in which the β-amino-alcohol moiety was obtained from readily available carbohydrates. The corresponding

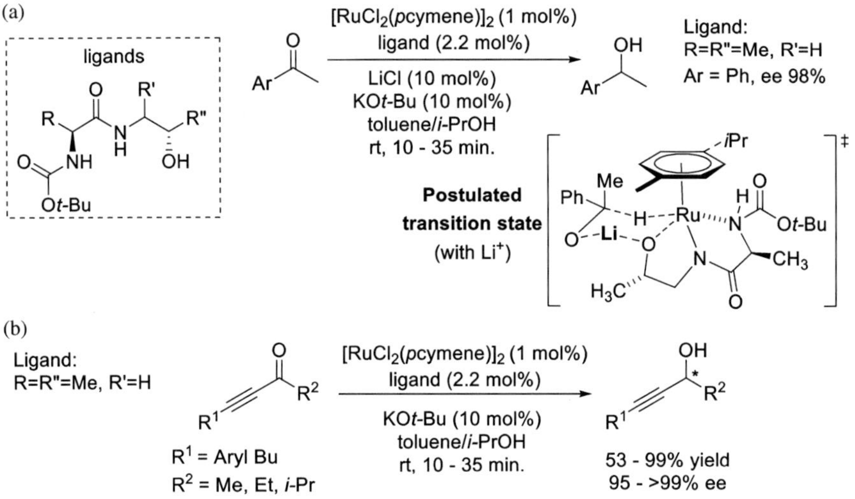

Scheme 8. Pseudo-dipeptide/ruthenium catalyzed ATH of acetophenones and propargylic ketones, by Adolfsson and co-workers.[71–75]

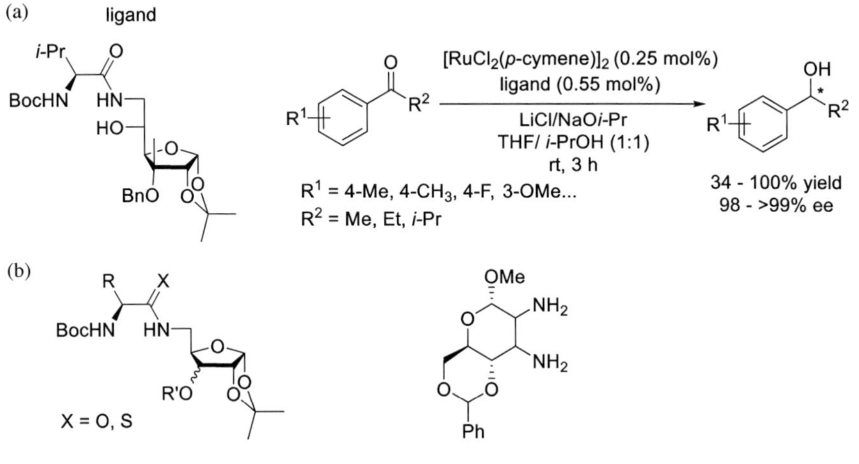

Scheme 9. Carbohydrate-derived pseudodipeptides for the ruthenium-catalyzed ATH of aryl,alkyl-ketones.[78–81]

ruthenium catalysts were highly efficient in ATH of aryl,alkyl-ketones (Scheme 9a).[79]

Furthermore, this approach was extended to other libraries of carbohydrate-based pseudo-dipeptides and thioamides, devoid of the CHOH function (Scheme 9b, left), that were used in ruthenium- and rhodium-catalyzed ATH of aryl,alkyl-ketones (ees up to >99%). The new ligands were readily prepared from inexpensive D-glucose and D-xylose. They proved to be highly tunable, as far as the substituents of the amide (or thioamide) moiety, the substituents at the C^5 and C^3 positions of the sugar, and the configuration of C^3 are easily adjustable.[80] In addition, hexapyranoside-derived diamines (Scheme 9b, right) were also used in the ATH of acetophenone, associated to ruthenium, rhodium, and iridium catalysts with variable results, the highest ee of 68% being obtained with a ruthenium catalyst.[81]

Diamine-derived η^6-arene–ruthenium catalysts

Despite the development of other diamines, TsDPENs remain ligands of high interest because of their highly enantio-discriminating properties, possibly tuned by the substituents on the primary amine bound to ruthenium(II).[82–83] If generally no loss of activity was observed by changing the amine substituent, in some instances beneficial effects were highlighted, such as higher activity in C=N reduction reactions[84–87] or increased water solubility.[88–90] Wills and co-workers reported a quite versatile series of TsDPENs in which the amine nitrogen atom was functionalized by heterocyclic residues.[91–93] These ligands were able to coordinate in either a bidentate or a tridentate mode, depending on the nature of the heterocycle. Good donors, such as pyrazoles, provided tridentate ligands and generated active catalysts from $Ru_3(CO)_{12}$ (Scheme 10a). Weak donors, such as furan or thiophene, provided bidentate ligands as typified by the (diamine)(benzene)RuCl complexes in Scheme 10b. The two families of complexes were good catalysts for the ATH of a large

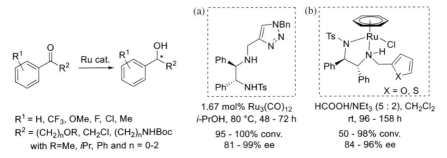

Scheme 10. ATH of ketones with ruthenium catalysts based on TsDPEN ligands bearing additional donor groups.[91–93]

range of functionalized ketones. In addition, this approach highlighted the subtle effects brought by changes of the nearby donor group, the two catalysts with appended thiophene and furan units being complementary in terms of substrate scope.

While investigating ammonia NH cleavage by metal–ligand cooperative catalysis, Kayaki, Dub, and co-workers[94] serendipitously discovered a conceptually novel DPEN-based Noyori-Ikariya-type precatalyst that displays a 2-amino-tetrafluorobenzenesulfonamide unit (Scheme 11). The authors underlined the proton-responsive nature of the aryl-amino group of the ligand. Indeed, the neutral Ru-complex ([Ru] in Scheme 11) could be protonated by various acids to give chiral cationic 18 electron complexes with tunable organic or inorganic counteranions (e.g., [Ru]H$^+$BF$_4^-$ in Scheme 11). Beside a full account detailing synthesis, characterization, and reactivity studies, the authors showed that both the ionic and nonionic ruthenium catalysts were efficient in the ATH of acetophenones and α-hydroxyacetophenone. The ionic complex [Ru]H$^+$BF$_4^-$ obtained by protonation with the strong acid HBF$_4$ and the nonionic complex led to similar results in the ATH of three different substrates.

Mohar and co-workers prepared γ-sultam-cored N,N-ligands. The *in situ* generated ruthenium catalysts were very efficient in ATH in particular for benzo-fused cyclic ketones. The DKR/ATH of a 1-indanone with a γ-carboxylic ester function provided the corresponding alcohol with an excellent stereoselectivity (Scheme 12).[95]

Ruthenium(II) NNN-type complexes containing a chiral pyrazolyl–pyridyl–oxazolinyl ligand were applied in the ATH of a series of aryl,alkyl-ketones (Scheme 13).[96] The best catalyst of the series was featuring a pyrazolyl NH functionality that is a regular structural unit of bifunctional ruthenium ATH catalysts.

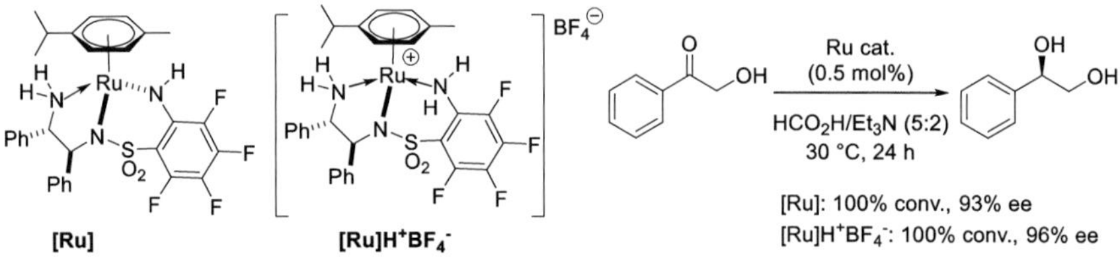

Scheme 11. Ruthenium catalysts developed by Kayaki, Dub and co-workers.[94]

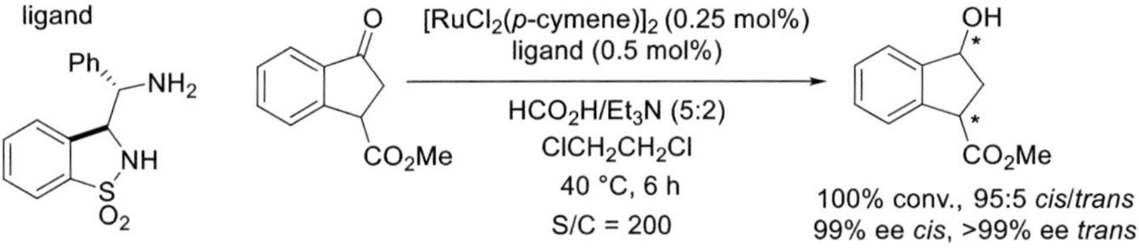

Scheme 12. DKR/ATH of a 1-indanone derivative, by Mohar and co-workers.[95]

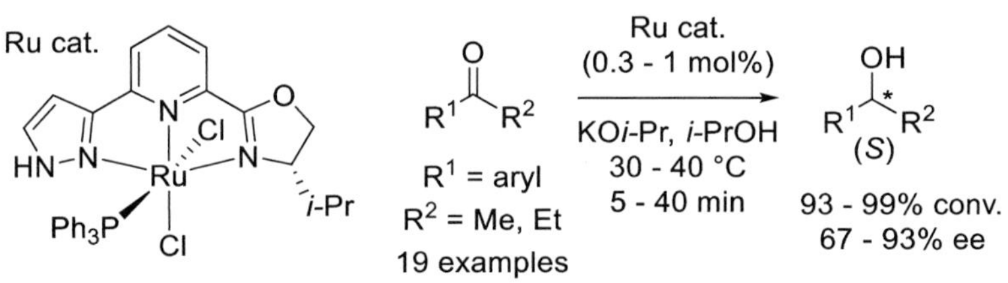

Scheme 13. Application of NNN-type chiral ligands in ATH of aryl,alkyl-ketones, by Yu and co-workers.[96]

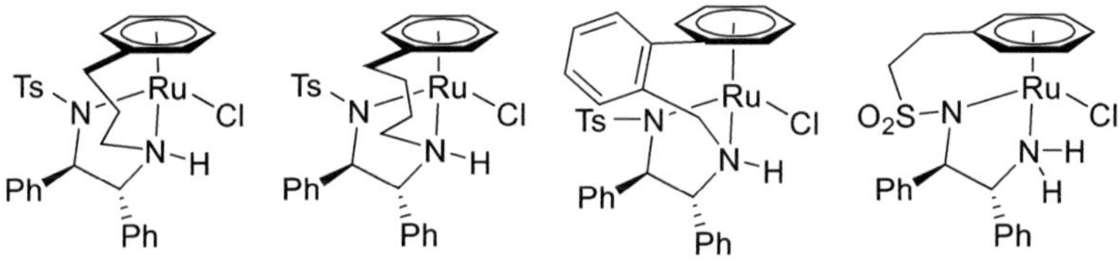

Scheme 14. Tethered η^6-arene–ruthenium pre-catalysts developed by Wills and co-workers.[98–103]

Diamine-tethered η^6-arene–ruthenium catalysts

As already mentioned in the general introduction, a structural change in half-sandwich arene/Ru/TsDPEN complexes reported by Wills and co-workers consisted in introducing a covalent linkage between the η^6-arene and the diamine units (Scheme 14). Such a modification was meant to prevent rotation of the arene ligand as well as to introduce electronic, steric, and other directing effects, by means of functional groups, in order to improve the catalyst efficiency, selectivity, and stability.[39–42] Hence, the strong chelation of sulfonamido-amine residues led to a persistent, forced coordination of the frequently labile η^6-arenes and lengthened therefore the life time of the catalytically active species. Moreover, the three-point coordination of the ligand resulted in a beneficial increase of the conformational rigidity of the complex and led to higher stereo-discrimination ability of the corresponding catalysts. These tethered ruthenium complexes were found to be highly active in the ATH of ketones in formic acid/triethylamine and proved more stable, over long reaction times, than the parent complex developed by Noyori and co-workers.[28–30] Using such tethered catalysts, the stereochemistry of the ATH of ketones was predominantly controlled by favorable edge-to-face electronic interactions between electron rich aryls in the substrate and the H-atoms of the electron deficient η^6-arene of the catalyst (CH–π interactions, Scheme 15).[40,41,97]

A first series of tethered catalysts, typified in Scheme 14, was prepared by connecting alkyl or aryl chains (2C to 5C alkyl chains or benzyl bridges) to the arenes (phenyl, 4-$CH_3C_6H_4$, OMe, 3,5-$(CH_3)_2C_6H_3$).[98–103] The Wills catalyst comprising a C3-tether exhibited high potential and possible industrial interest and therefore, Johnson Matthey developed a large-scale synthesis allowing its commercial availability.[102]

Thanks to the high activity and selectivity of the C3-tethered catalysts, hindered ketones (1,1-disubstituted tetralones, for example) could be reduced with high enantioselectivities (up to 99% ee) (Scheme 15).[104] According to the reaction scope, the directing effect of the CH/π interaction

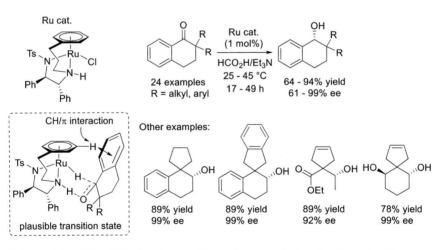

Scheme 15. ATH of hindered α- and β-tetralones and other functionalized ketones and plausible transition state.[104]

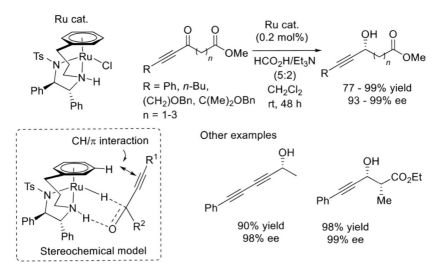

Scheme 16. ATH of acetylenic ketoesters and diynones and proposed stereochemical model by Wills and Fang.[105,106]

proved to be more effective for ketones displaying two aromatic rings or additional unsaturated groups, such as olefins or alkynes. It significantly contributed to the high levels of enantioselectivity.

The tethered catalysts also enabled the preparation of propargylic alcohols from functionalized ynones[105] or conjugated diynones[106] in excellent ees. Thus, alkynyl-substituted β, γ and δ-keto esters were reduced with high stereocontrol (Scheme 16). As in the case of ketones and tetralones seen earlier, the control of enantioselectivity takes place during the transfer of the two hydrogen atoms to the substrate. It was proposed that, at this step, the CH/π interactions between the alkyne and the catalyst have a major effect on stereocontrol (Scheme 16). In further studies, DKR could also be achieved on α-substituted β-ketoesters by using the same catalysts as reported by Ayad, Ratovelomanana-Vidal, and co-workers.[107]

Afterwards, Wills and co-workers applied the same strategy to successfully reduce aryl,alkynyl-ketones with C3-tethered catalysts displaying either a benzene or an anisole moiety coordinated to the ruthenium centre.[108] Worthy of note, the systematic screening of substrates with highly diverse substituents allowed to rationalize steric and electronic effects and consequently to define the best substrate/catalyst match.

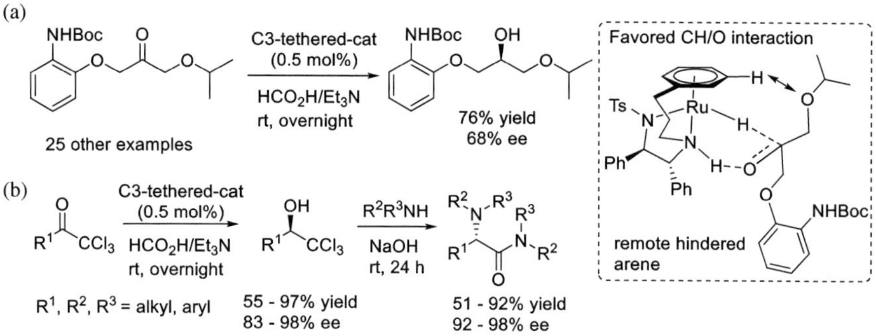

Scheme 17. ATH of a dissymmetric propanone and trichloromethyl ketones.[109,110]

In order to identify more precisely the discriminating factors, which underlie the high selectivity of tethered ruthenium catalysts, Wills and co-workers investigated the ATH of challenging 1,3-dialkyoxy/aryloxy 2-propanones presenting little difference between the two sides of the C=O functionality (Scheme 17a). They demonstrated that the C3-tethered ruthenium catalyst led to significant ees when both electronic and steric factors differentiate the substituents of the ketone, while tuning of electronic factors only (e.g., using substituted aryloxy groups) did not enable good enantioselectivity.[109]

The C3-tethered catalyst could also reduce efficiently trichloromethyl ketones, as reported by Fox and co-workers (7 substrates, up to 98% ee).[110] The corresponding optically enriched alcohols were converted then into amino-amides via stereospecific Jocic-type reactions (Scheme 17b).

In the series of alkyl-tethered ruthenium complexes, the variation of the size and electron-donor ability of the sulfonyl groups on the DPEN unit were also investigated. Thus, ruthenium catalysts were optimized by tuning both the tether length and the nature of the sulfonyl group. For example, in the reduction of α-hydroxyacetophenone high yields and enantioselectivities were obtained at low catalyst loading with the C3-tethered catalyst displaying a 2,4,6-triisopropylbenzenesulfonyl function (Scheme 18).[111]

The same family of C3-tethered ruthenium catalysts was also applied in the synthesis of chiral intermediates of interest, notably through ATH of imines,[112,113] β-azidocyclopropane carboxylates,[114] and α-azidoacrylates.[115]

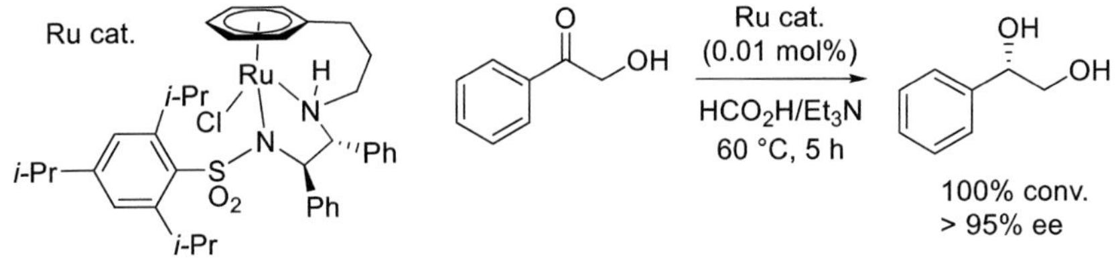

Scheme 18. ATH of α-hydroxyacetophenone catalyzed by a C3-tethered catalyst, by Wills and co-workers.[111]

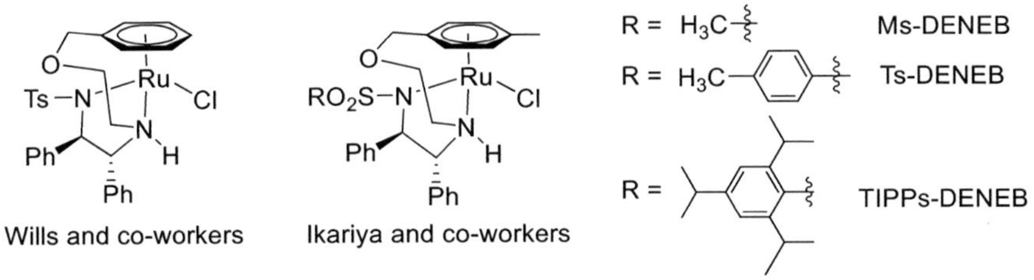

Scheme 19. Ether-tethered ruthenium catalysts developed by Wills and co-workers[119] and Ikariya and co-workers.[118]

In parallel, Wills and co-workers developed ruthenium catalysts containing a benzyl-type tether (see Scheme 14) that provided high enantioselectivity in the ATH of simple acetophenones and tetralones.[116,117] The first ligand of this series led also to active rhodium catalysts for the ATH of 2-substituted quinolines (up to 94% ee).[101]

The introduction of an oxygen atom in the catalyst-tethering chain was realized independently by Ikariya[118] and by Wills[119] and co-workers (Scheme 19). Thanks to the optimization of synthetic routes, the oxo-tethered catalyst Ts-DENEB® has been prepared on large scale by Takasago Int. Corp. and, like the C3-tethered Wills catalyst, it is commercially available. With respect to other tethered ruthenium catalysts described previously, these complexes exhibited either similar or complementary catalytic properties in the ATH of aryl,alkyl-ketones. Nevertheless, the Ikariya's oxo-tethered catalyst proved remarkable as far as it retained excellent catalytic performances under mild reaction conditions, with a substrate/catalyst ratio of 5000[118]

Ikariya, Kayaki, and co-workers applied the oxo-tethered catalyst to the ATH of benzophenones, in collaboration with Touge from the

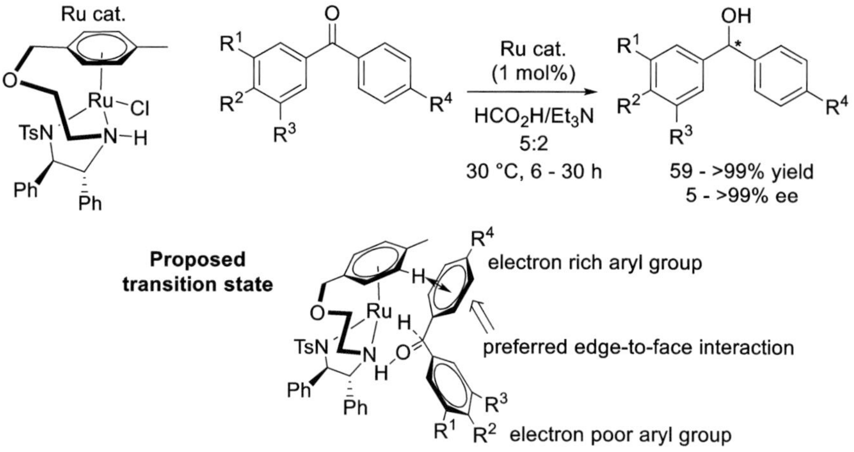

Scheme 20. ATH of benzophenones using an oxo-tethered catalyst and proposed transition state, by Ikariya, Kayaki and co-workers.[120]

Takasago Int. Corp.[120] The reductions proceeded smoothly at 30°C in a mixture of formic acid and triethylamine. For substrates possessing an *ortho*-substituted aryl group, the enantiodiscrimination was effective providing high enantiomeric excesses. Noteworthy, with non-*ortho*-substituted benzophenones, the catalyst was able to differentiate aromatic rings with different electron density, thus giving high levels of enantioselectivity (>99% ee) (Scheme 20). A plausible transition state was proposed implying an edge-to-face CH–π interaction between the catalyst and the more electron-rich aryl groups of the substrate.

Within these studies, Kayaki and co-workers also reported a concise synthesis of half-sandwich η^6-arene–Ru–diamine complexes bearing a perfluorinated phenylsulfonyl group in their tethering chains, that is, displaying N-SO$_2$-(C$_6$F$_4$)-o-O(CH$_2$)$_n$ (n = 1 or 2) tethers (see Scheme 14).[121]

In parallel, Zhou and co-workers applied with success O-tethered ruthenium catalysts in the synthesis of optically enriched aryl *N*-heteroaryl methanols, which are pharmaceutical intermediates of interest.[122] Aryl groups with opposite electronic and steric characters could be discriminated efficiently in these enantioselective ATH reactions. In this case also, the more electron-rich aryl fragments of the substrates approach preferentially the η^6-arene ligand.

Yuki, Ikariya, and co-workers applied successfully the O-tethered ruthenium catalysts in the ATH of α-functionalized ketones displaying halide, methoxy, nitro, dimethylamino, and ester groups. For example, reductions performed with HCO_2H and HCO_2K in a $EtOAc/H_2O$ solvent system allowed a highly selective access to halohydrins of high synthetic potential. The previous conditions suppressed the undesired formylation and other side-reactions often encountered under the classical reaction conditions using an azeotropic mixture of HCO_2H and Et_3N. The excellent activity and selectivity were preserved even with a substrate/catalyst ratio of 5000.[123] Interestingly, while optimizing an industrial flow process, Touge and co-workers applied this method to the ATH/DKR of an α-amido β-keto ester, as a suitable alternative to the previously used AH/hydroxyl inversion on the same substrate (Scheme 21). The oxo-tethered catalyst enabled the synthesis of the *anti* diastereomer of a key intermediate in the synthesis of the D-erythro-CER[NDS] ceramide at an 80 kg production scale (96% yield, 69% de, 97% ee).[124] It is worth to note that the robust catalytic process was developed in a pipes-in-series flow reactor. After crystallization, the enantiopure product was obtained at a >50 kg scale.

The mentioned synthetic methodology combining ATH and DKR and using O-tethered catalysts was recently applied by Kayaki, Touge, and co-workers in a one-step cascade synthesis of γ- and δ-lactones from racemic keto acids, with excellent control of multiple contiguous stereocenters (Scheme 22).[125] Wine lactone, a natural flavor and fragrance agent, was prepared with unprecedented level of diastereo- and enantioselectivity (95% yield, > 99:1 dr, 92% ee).

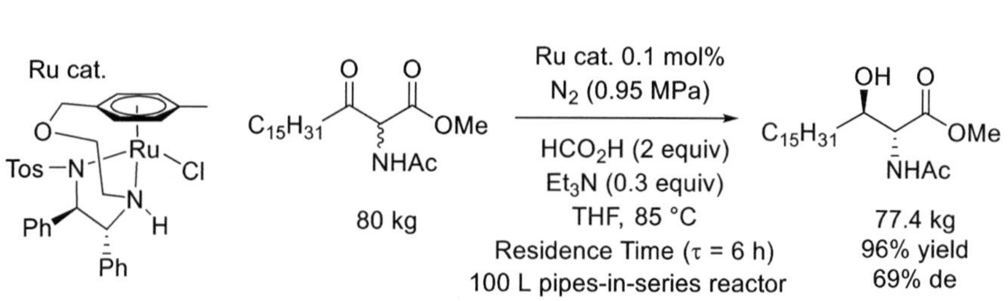

Scheme 21. Industrial DKR/ATH of an α-acetamido β-ketoester, by Touge and co-workers (Takasago Int. Corp.).[124]

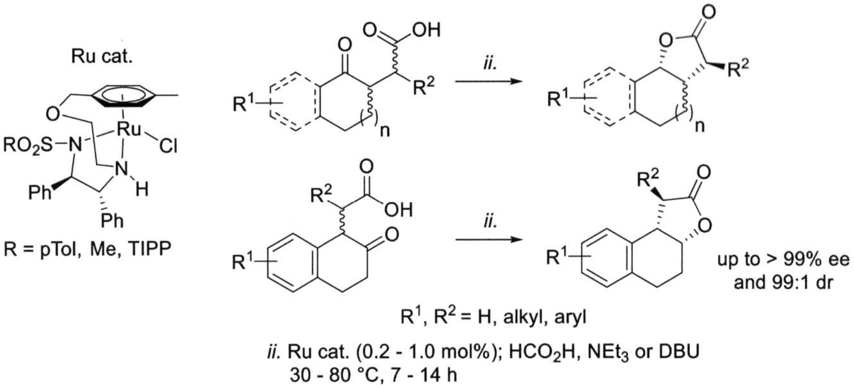

Scheme 22. DKR/ATH of racemic keto acids, by Kayaki, Touge, and co-workers.[125]

Another industrial application of the O-tethered ruthenium/DENEB catalyst was reported by Komiyama and co-workers at Teijin Pharma Limited who described the scalable synthesis of the key intermediate of a β2-adrenergic receptor agonist.[126] In this process, an α-NHCbz–acetophenone was reduced into the corresponding alcohol in 71% isolated yield and 99% ee at about 7-kg scale with an S/C ratio of 1000.

In parallel, Chung and co-workers at Merck reported on the development of an efficient process for the multikilogram scale synthesis of omarigliptin, a drug for the treatment of type 2 diabetes, applicable to a manufacturing plant.[127] The DKR/ATH of an α-NHBoc ketone provided the *anti* amino alcohol with 93% isolated yield, 24:1 dr, and >99% ee (0.365 mol of substrate, HCO_2H/DABCO, S/C close to 1000, 35°C, 20 h). Sparging with N_2 was critical to remove CO_2 and allow an efficient process.

By following a similar approach, Wang and co-workers reported on the application of DKR/ATH for the gram scale production of *syn*-β-hydroxy α-dibenzylamino esters, known intermediates in the synthesis of Doxidopa.[128]

Other tethered ruthenium catalysts were developed by Mohar and co-workers on the basis of an enantiopure N,C-(*N*-ethylene-*N*-methylsulfamoyl)-tethered DPEN–toluene hybrid ligand.[129] This new catalyst was rather efficient for the ATH of a large array of challenging 1-naphthyl ketones in the presence of HCO_2H/NEt_3 (Scheme 23). Similar to other

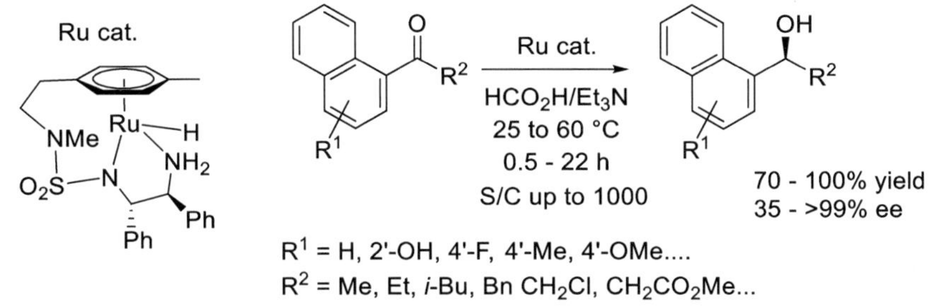

Scheme 23. ATH of 1-naphthyl ketones, by Mohar and co-workers.[129]

R^1 = H, 2'-OH, 4'-F, 4'-Me, 4'-OMe....
R^2 = Me, Et, i-Bu, Bn CH_2Cl, CH_2CO_2Me...

Ru cat. conditions: HCO_2H/Et_3N, 25 to 60 °C, 0.5 - 22 h, S/C up to 1000, 70 - 100% yield, 35 - >99% ee

tethered complexes, the high selectivity induced by this catalyst stemmed from dual synergistic stereoelectronic matches between the substrate and the catalyst.

Afterward, Mohar and co-workers explored the modulation of the same complexes by varying the length of the carbon tether (from 1 to 3 CH_2 units) and the nature of the *para* substituent (H, CH_3, *i*Pr, OMe) of the η^6-aryl. The new ruthenium complexes catalyzed the ATH of a broad scope of aryl,alkyl-ketones and α or β-diketones in HCO_2H/Et_3N, under mild conditions, with excellent selectivities (39 substrates, 94 to >99% ee).[130] Further modifications of the tethered ruthenium complexes led to new catalysts bearing R_2NSO_2DPEN-$(CH_2)_n(\eta^6$-aryl) type ligands[131] that performed well in the ATH of various classes of (hetero)aryl ketones using a mixture of formic acid and triethylamine. Benzo-fused cyclic ketones were hydrogenated with 98% to >99% ee at a low catalyst loading of 0.003 mol%.

Through further tuning of the same series of ruthenium catalysts, Mohar and co-workers achieved the selective single or double ATH of $CF_3C(O)$-substituted benzo-fused cyclic ketones, under DKR conditions (Scheme 24).[132] This efficient and practical methodology gave access to synthetic intermediates of interest such as CF_3-substituted 1,3-diols, in excellent enantiopurities and yields. To rationalize the selectivity, it was proposed that the stereocontrol was enforced by either CH/π or CH/F attractive electrostatic interactions. In the first hydrogenation step, the CH/F interaction overcomes the CH/π interaction, thus leading to the trifluoromethyl carbinol of (*S*) configuration as the mono-reduced product, on the way to the fully hydrogenated *anti* diol product.

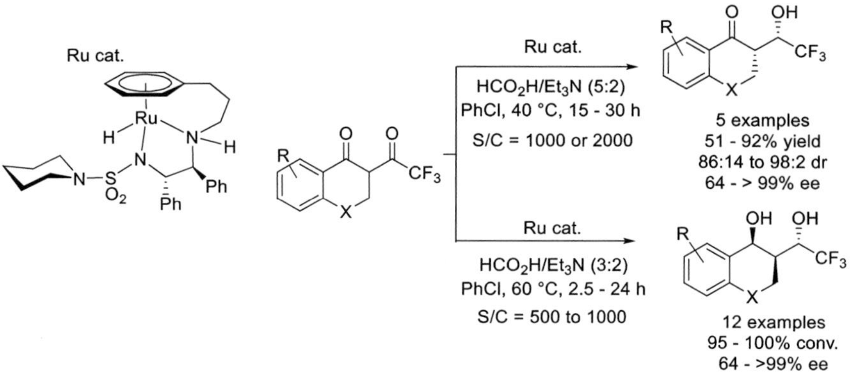

Scheme 24. ATH of benzo-fused β-diketones leading to either β-keto-alcohols or diols, by Mohar and co-workers.[132]

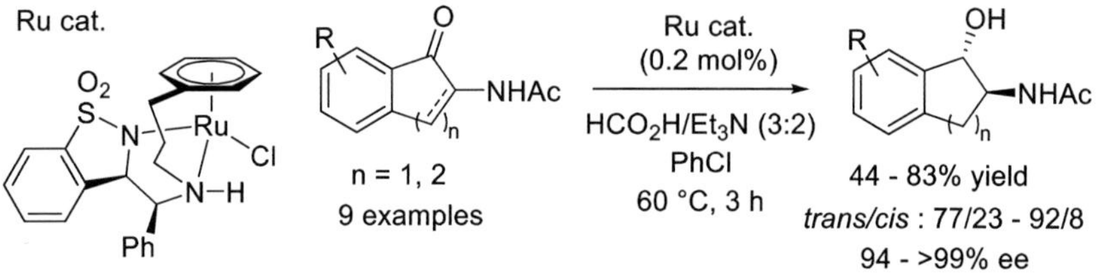

Scheme 25. ATH/DKR of α-acetamido benzocyclic ketones, by Mohar and co-workers.[135]

Analogous ruthenium catalysts were applied for the preparation of hydroxy benzosultams (S/C = 10,000, HCO_2H/Et_3N, up to >99.9% ee)[133] and 2,3-disubstituted-1-indanols through DKR and ATH.[134]

Recently, Mohar and co-workers reported another tethered ruthenium catalyst where the diamine ligand entails a cyclic sulfonamide moiety. This catalyst was efficient in the DKR/ATH of α-acetamido benzo-fused cyclic ketones (Scheme 25).[135] Calculations were carried out in order to rationalize the high *trans* selectivity of the transformation. Two attractive stabilizing effects were identified, the first being a "classical" CH/π interaction between the coordinated phenyl ring and the π system of the benzocyclic part of the substrate. The second interaction relied on a hydrogen bond between the SO_2 moiety of the catalyst and the acetamido group of the substrate.

2. Shvo-type catalysts

The Shvo catalyst is a cyclopentadienone-ligated dimeric ruthenium complex that was found to be one of the most active catalysts for the transfer hydrogenation of ketones and imines,[136–138] as well as for the chemoenzymatic DKR of secondary alcohols and amines (Scheme 26).[139] In solution, this binuclear complex is in equilibrium with the coordinatively unsaturated intermediate **A** and the catalytically active ruthenium hydride complex **AH$_2$**, both species interconverting each other through H$_2$ exchange. The OH function of **AH$_2$** participates, with the hydride ligand, in the H-transfer process. After hydrogen transfer to the substrate, the **AH$_2$** species is reformed through reaction with the hydrogen donor (HCO$_2$H or i-PrOH). Recently, chiral variants of these catalysts have been reported; they displayed, however, rather poor catalytic properties.

In a detailed study, Wills and co-workers addressed the preparation of a range of diastereopure cyclopentadienone–ruthenium–tricarbonyl complexes shown in Scheme 27 and their application to the ATH of

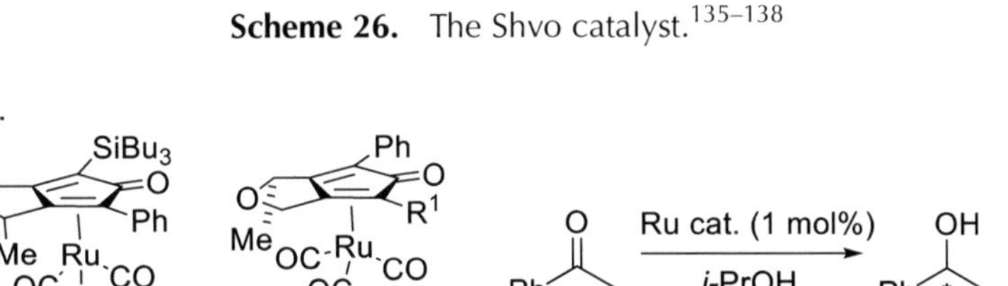

Scheme 26. The Shvo catalyst.[135–138]

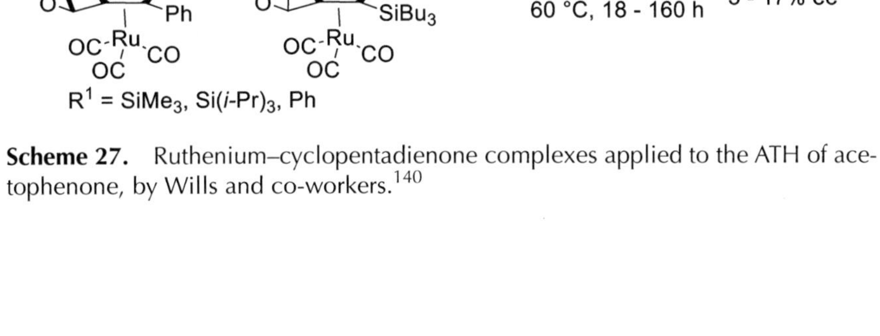

Scheme 27. Ruthenium–cyclopentadienone complexes applied to the ATH of acetophenone, by Wills and co-workers.[140]

acetophenone.[140] These species did not require removal of CO with Me$_3$NO to become catalytically active. The required hydrides were formed *in situ* prior to hydrogenation experiments, by reaction with NaOH in THF, followed by the addition of phosphoric acid. The catalytic activity was moderate, leading to low to good conversions (13%–61%) after 18–160 hours. The modest enantioselectivities of these hydrogenation reactions were regulated by both the stereogenic centers and the planar chirality of the catalysts. The most selective catalysts display a methyl substituent on the ligand, close to the Ru(CO)$_3$ moiety. After extended reaction times, racemization was observed.

In a novel contribution, Hayashi and Dou prepared planar chiral Shvo catalysts, where the chirality results solely from the nonsymmetric substitution patterns of the cyclopentadienone ligands, next to the CO function (Scheme 28).[141] The ATH of imines and ketones proceeded using a 0.5 mol% catalyst loading and afforded the corresponding reduction products in moderate to high yields with low to moderate enantioselectivities, with maximum ees of about 60%. Acetophenone was much less reactive than the other activated ketones, it required 20 hours to be hydrogenated, and the alcohol was obtained in low ee.

Overall, the known chiral Shvo-type ruthenium catalysts do not meet the current standards for ATH reactions, in terms of both catalytic activity and enantioselectivity. Nevertheless, the promising features of the parent

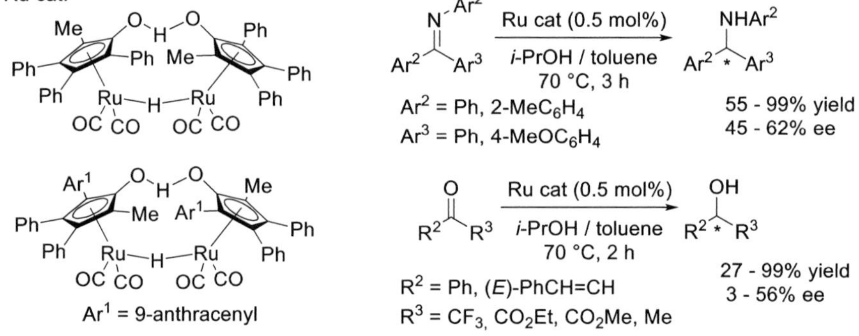

Scheme 28. Ruthenium–Shvo complexes applied to the ATH of ketones and imines, by Dou and Hayashi.[141]

achiral catalyst should encourage further efforts toward the design of chiral catalysts in the future.

B. Rhodium-Catalyzed Transfer Hydrogenations

Compared to Noyori-type ruthenium(II) catalysts, the corresponding isoelectronic bifunctional rhodium(III) complexes, namely Cp*Rh^{III}(N-Ts-diamine)-type complexes,[142] have been less investigated in ATHs. Notably, the rhodium catalysts exhibited good stability and provided variable conversions and enantioselectivies in the ATH of ketones.[8] In the time frame covered by this review, new catalytic applications of the known ligands have been developed and, moreover, two directions have been investigated intensively: the synthesis of ligands based on natural building blocks and the development of new tethered complexes. Significant advances have been made by using these ligands and complexes in the hydrogenation of a variety of ketones and functionalized olefins.

1. Diamine-tethered η^5-cyclopentadienyl rhodium complexes with active NH functions

Wills and co-workers[42,143,144] followed by Ratovelomanana-Vidal, Phansavath, and co-workers[145–152] developed rhodium catalysts displaying a benzyl tether between the cyclopentadienyl and the TsDPEN units, for applications in the ATH of ketones. The rhodium catalysts developed by Wills and co-workers proved to be very efficient for the reduction of a series of ketones, especially in aqueous solutions (Scheme 29).[42,143,144]

Ratovelomanana-Vidal and co-workers developed a series of new ligands closely related to the Wills ligands by introducing a substituent on the aromatic unit of the tether (Me, OMe, F, CH$_3$).[145–152] These ligands provided tethered rhodium(III) catalysts highly efficient in the ATH of alkyl,aryl-ketones[145–146] and in the DKR/ATH of α-amino-β-ketoesters, leading to functionalized alcohols with high syn/anti ratios and ee.[147] Furthermore, the DKR/ATH of α-methoxy-β-keto esters was carried out in water/cetyltrimethylammonium bromide by using a tethered rhodium(III) complex that displays

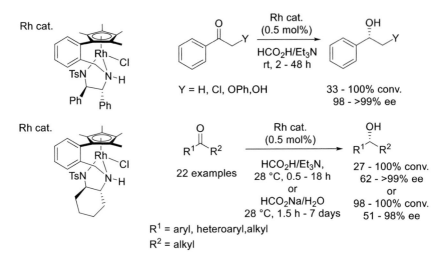

Scheme 29. ATH of ketones with tethered cyclopentadienyl-diamine rhodium catalysts, by Wills and co-workers. [42,143,144]

a N-$C_6F_5SO_2$-substituted DPEN ligand.[148] Ratovelomanana-Vidal and co-workers also prepared cis-3-(hydroxymethyl)chromanol derivatives with excellent diastereo- and enantioselectivity by applying the tethered rhodium(III) catalyst in the DKR/ATH of 3-substituted chromanones, under mild conditions and using HCO_2H/Et_3N as hydrogen source (Scheme 30).[149] The reaction proceeds via three separate hydrogenation steps. After reduction of the aldehyde and the C=C bond, the intermediate hydroxymethyl chromanone undergoes a highly enantioselective ATH reaction under DKR conditions.

Very recently, the same group described the first highly chemo- and stereoselective ATH of γ-keto-δ-acetal enamides in the presence of a tethered rhodium(III) catalyst (Scheme 31).[150] A high chemoselectivity in favor of the reduction of the carbonyl group over the C=C bond was observed. The reaction tolerated various aromatic substituents and demonstrated better enantioselectivities with substrates bearing a dioxane as the acetal moiety (up to 99% ee).

Furthermore, Ratovelomanana-Vidal and co-workers used the same rhodium(III) catalyst to develop the first ATH of 4-quinolone derivatives and obtained 1,2,3,4-tetrahydroquinolin-4-ol derivatives with excellent levels of enantioselectivity.[151] In collaboration with Zhang and co-workers

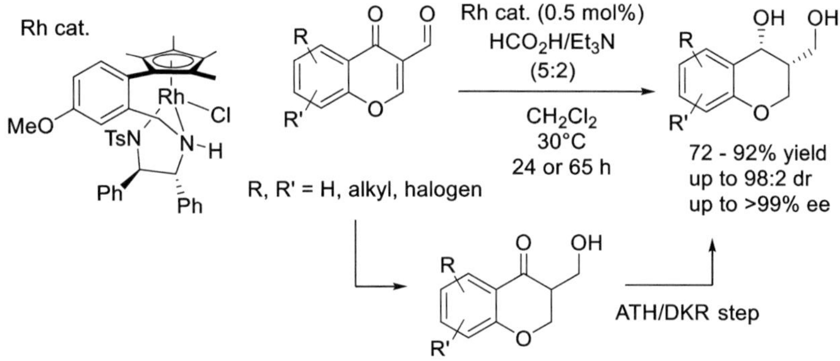

Scheme 30. DKR/ATH of 3-formyl chromanones, by Ratovelomanana-Vidal and co-workers.[149]

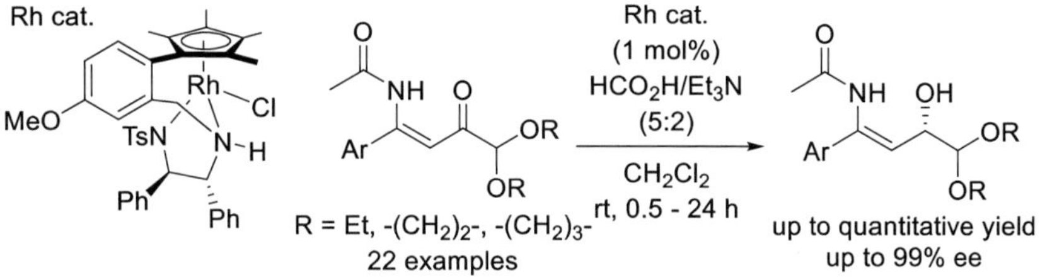

Scheme 31. Chemoselective ATH of γ-keto-δ-acetal enamides, by Ratovelomanana-Vidal and co-workers.[150]

the tethered rhodium(III) catalyst has been applied also in the diastereoelective transfer hydrogenation of optically pure α-aminoalkyl-α'-chloromethyl ketones. The method allowed to prepare the key intermediates for the synthesis of some HIV protease inhibitors (up to 99% yield, up to 99:1 dr).[152]

Overall, the previous results clearly demonstrate that tethering the Cp* and diamine ligands is a suitable strategy for the design of rhodium-based catalysts for enantioselective ATH reactions.

2. Amino-acid-derived amides and thioamides with active NH functions

Natural products have been also used as chiral building blocks for the development of original ligands. Essentially, ligands based on amino

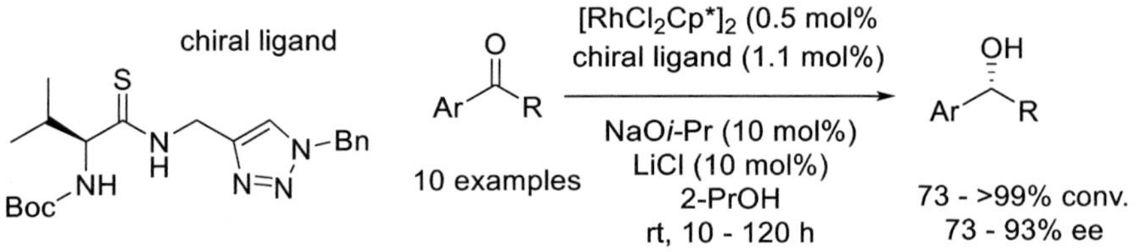

Scheme 32. Triazole containing pseudodipeptides in rhodium-catalyzed ATH of aryl,alkyl-ketones, by Adolfsson and Tinnis.[155]

acids and pseudopeptides, possibly combined with sugars, have been investigated as already highlighted in Section II.A.1. Thus, within their extensive studies on the use of amino acids as ligand precursors,[153,154] Adolfsson and co-workers synthesized a series of 1,2,3-triazole-functionalized, amino-acid-derived thioamide ligands. A ligand of this series, when combined with the dimeric Cp*-rhodium dichloride, provided a good catalyst for the ATH of aryl,alkyl-ketones in the presence of sodium *iso*-propoxide and lithium chloride in 2-propanol (Scheme 32).[155] The corresponding alcohols were obtained with good to high conversion rates, with enantioselectivities up to 93%. Himo, Adolfsson, and co-workers carried out then a theoretical study on the ATH of ketones catalyzed by these amino-acid-derived rhodium complexes.[156] Similar to the ruthenium complexes described previously (see for instance Schemes 1 and 13), the rhodium(III) complexes proved to operate as bifunctional catalysts and CH/π interactions between the substrate and the Cp* ligand were shown to significantly contribute to the stereoselectivity of these reductions.

The first generation of pseudodipeptide ligands modified by carbohydrates shown in Scheme 9 has been applied also successfully in rhodium-catalyzed ATH reactions.[79] Later on, the second and third generation ligands were prepared by Diéguez, Pàmies, and co-workers, as well as by Adolfsson and co-workers for applications in rhodium and ruthenium ATH of ketones. The second-generation ligands are thioamides obtained by combining D-glucose units and α-amino-acid derivatives. The resulting rhodium catalysts enabled ATH of aryl,alkyl- and heteroaromatic-ketones with excellent enantioselectivity.[79,157] The third-generation ligands are α-amino-acid-derived amides and thioamides combined with D-mannose units. The ruthenium complexes of β-hydroxy-amides of this series and

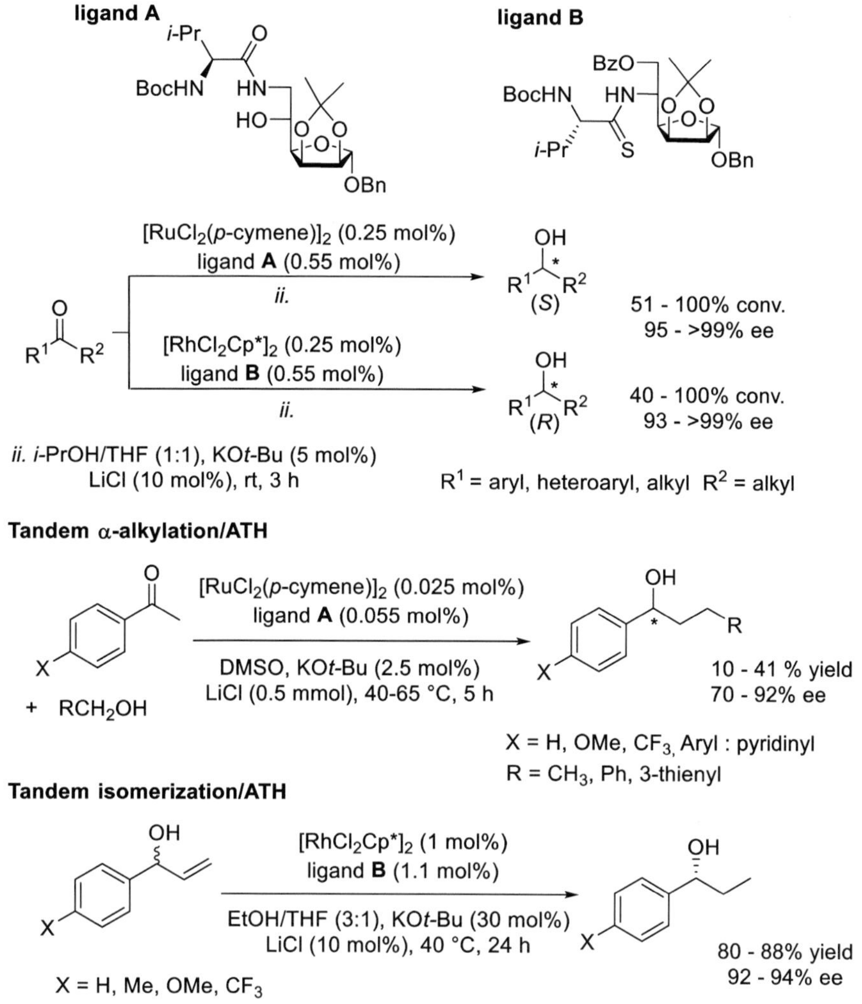

Scheme 33. Sugar-modified α-amino-acid derived hydroxyamide and thioamide ligands applied in the ATH and tandem reactions of ketones, by Diéguez, Pàmies, and co-workers.[158]

the rhodium complexes of thioamides acted as catalysts for the ATH of ketones in a complementary way and allowed to access both enantiomers of the desired alcohols (Scheme 33).[158] These catalysts were highly efficient in the ATH of a broad scope of ketones (41 examples, up to >99% ee), that is, aryl,alkyl-, alkyl,alkyl-, heteroaryl-, and unsaturated ketones but also in tandem isomerization/ATH reactions (rhodium and ruthenium catalysts, 11 examples, up to 99% ee) and in the tandem alkylation/ATH of acetophenones and acetyl pyridine (ruthenium catalyst, 8 examples, up to 92% ee).

These recent developments prove that chiral rhodium catalysts bearing either tethered Cp*/diamine ligands or amino acid, as well as pseudo-peptide-based ligands are suited for ATH of various ketones. Although they were less investigated than their ruthenium congeners, efforts to design new chiral ligands for rhodium-catalyzed ATH have led to significant achievements in the last decade.

C. Iridium-Catalyzed Transfer Hydrogenations

Only a few new bifunctional iridium catalysts have been developed in the last decade for applications in ATH. These catalysts may display either active NH functions, according to the Noyori's model, or remote H-bonding functions that participate in the stereochemical control by interacting with the substrates. Two innovative concepts have been introduced in terms of catalyst design, which is the use of iridium-based artificial metallo-enzymes and chiral-at-metal octahedral complexes.

1. Amino- and amido iridium(III) catalysts with active NH functions

Iridium catalysts displaying classical diamines, sulfonylated diamines, or tridentate phosphine-diamines as chiral ligands, were often investigated in parallel to ruthenium and rhodium catalysts and proved to be useful for the ATH of ketones and imines.[11,159] The common feature of these complexes is that an NH function of the ligand participates in the enantiodetermining step via H-transfer to the substrate.

Initially, Wills and co-workers applied proline-functionalized DPEN (see Section II.A.1) as chiral ligands in rhodium- and iridium-promoted ATH of simple ketones.[69] However, these catalysts were far less efficient than their ruthenium counterparts: combined with $[RhCl_2Cp^*]_2$, these ligands gave 3%–68% conversions and up to 85% ee; while with $[IrCl_2Cp^*]_2$, they gave 0%–33% conversion and 14%–47% ee (four examples). In parallel studies however, Carreira and co-workers designed new

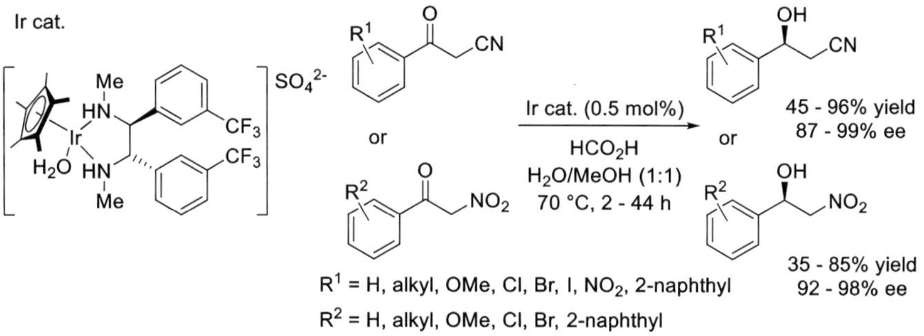

Scheme 34. ATH of α-cyano and α-nitro acetophenones with iridium aqua-diamine complexes, by Carreira and co-workers.[160]

aqua iridium(III) catalysts that proved successful for the challenging ATH of α-cyano and α-nitro acetophenones (Scheme 34).[160] In the ATH of 2-cyanoacetophenone with sodium formate in water/methanol, the stable aqua iridium(III)-diamine complexes led to the corresponding alcohols in 63%–90% yields and 47%–95% ees. The best catalyst proved to be a DPEN derivative with *m*-CF$_3$ substituents on the phenyl groups of the diamine. It proved highly active in the ATH of α-cyanoketones and little less active in the ATH of α-nitroketones, high enantioselectivities being reached in both cases. In particular, *ortho*-substituted acetophenones ketones were hydrogenated with excellent ees thanks to an "*ortho*-effect." Thus, these optically active 1,2-diamine ligands appeared as interesting alternatives to the commonly used monosulfonylated diamines.

Other catalytic processes were developed by Xie, Zhou, and co-workers: iridium(III) complexes of tridentate pyridine-aminophosphine ligands displaying a spiranic scaffold were applied to the ATH of simple ketones (Scheme 35).[161] These reactions proceeded in ethanol and afforded the corresponding chiral alcohols with excellent conversion rates and enantioselectivities. Authors demonstrated that ethanol provided the hydride required for the hydrogenation reactions, while it was converted into ethyl acetate. Zhou and co-workers subsequently investigated the ATH of alkynyl ketones using the same family of iridium complexes. Thanks to a combination of sodium formate and ethanol as hydrogen sources, the reaction did not require any additional base to proceed and provided a very efficient method for the ATH of alkynyl ketones (catalyst 0.5–1 mol%, 90°C, 4–48 h, 86–99% yield, and 84–98% ee).[162]

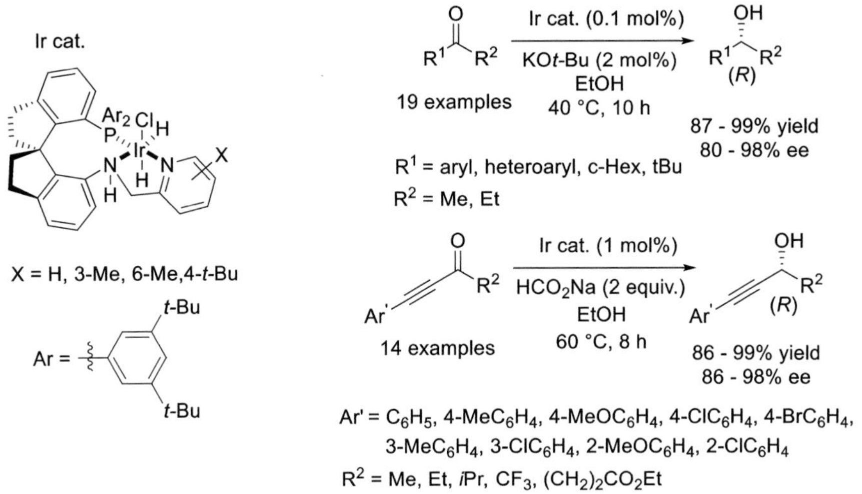

Scheme 35. P,N,N-tridentate ligands for the iridium-catalyzed ATH of aryl,alkyl-ketones and alkynyl ketones, by Xie, Zhou, and co-workers.[161,162]

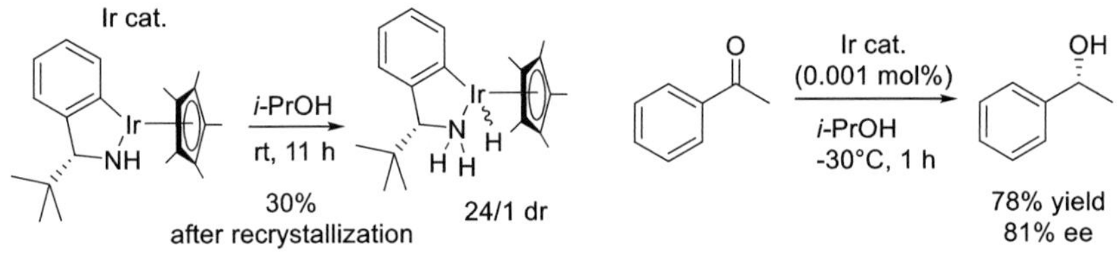

Scheme 36. Cyclometalated amidoiridium catalyst in the ATH of acetophenone, by Ikariya, Kayaki, and co-workers.[163]

Ikariya, Kayaki, and co-workers prepared C,N-chelated amidoiridium metallacycles displaying a stereogenic carbon center α to the amido moiety.[163] Monomeric and dimeric iridium complexes were characterized. Monomeric species exhibited outstanding catalytic activity in the ATH of acetophenone in i-PrOH, furnishing (S)-phenyl ethanol with high ee (S/C = 1000, 4 h at −30°C, 78% yield, 81% ee). It was demonstrated that the unsaturated amido iridium complex reacts with i-PrOH to give the corresponding amino-hydrido complex as a mixture of two diastereoisomers in a 24/1 ratio, with preferential formation of the R_CS_{Ir} isomer (Scheme 36). To account for the observed selectivity, it was postulated that the steric hindrance of the *t*-Bu group directs the addition of *i*-PrOH to the amido complex. In the ATH reaction, the iridium–hydride complex induces then hydride and proton transfers to the *Re*-face of the ketone, the reaction being controlled by the attractive CH/π interactions between the phenyl ring of the substrate and the Cp* ligand.

A highly innovative approach to enantioselective metal catalysis has been introduced by T. Ward who developed the so-called artificial metalloenzymes, by introducing an organometallic catalyst within a protein environment by means of noncovalent interactions. Thus, in 2008, Ward, Stenkamp, and co-workers disclosed η^6-arene–ruthenium complexes bearing a biotinylated aminosulfonamide ligand that, combined with suitable proteins, could be used in the ATH of ketones.[164] Later on, in 2011, Ward and co-workers described an analogous iridium complexes and its applications in ATH reactions.[165–168] A biotinylated Cp* iridium complex, combined with streptavidin (Sav), was highlighted as the most promising catalyst for the ATH of imines, while ruthenium and rhodium complexes were superior for the hydrogenation of ketones.[166] Iridium complexes were used notably for the preparation of Salsolidine through ATH of the isoquinoline precursor. After chemical and genetic optimization of the artificial transfer hydrogenase, the synthesis of Salsolidine was efficiently realized with up to 96% ee (Scheme 37). Based on X-Ray structural data,

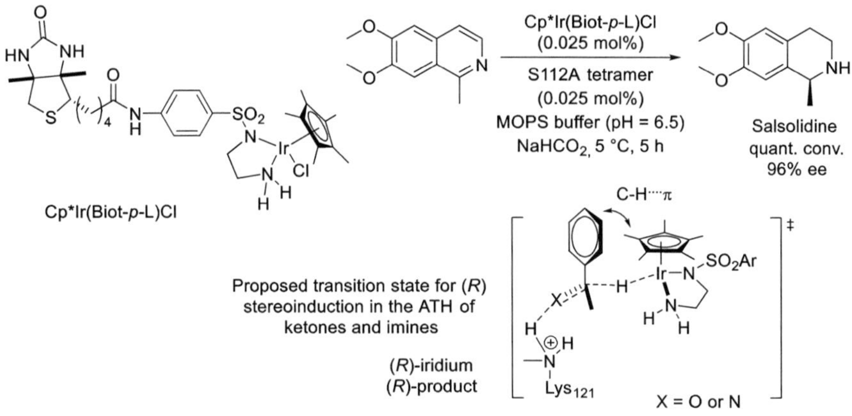

Scheme 37. Salsolidine synthesis by ATH catalyzed by an artificial enzyme and postulated transition state, by Ward and co-workers.[166]

the authors suggested that the reaction proceeds through a nonconcerted mechanism, that a CH–π interaction contribute to the stereochemical control and that the K121 residue of S112A Sav participates in the protonation step to afford the chiral amine (Scheme 37). This proton transfer from the ammonium group of the supramolecular biotine-Sav ligand to the imine seems essential for the ATH reaction to proceed.[167,168]

Thus, this work on iridium complexes provides new examples of reactions where a supramolecular ligand enables a catalytic process thanks to the involvement of a remote functional group.

2. Octahedral chiral-at-metal iridium(III) catalysts with π-stacking and H-bonding ligands

Within their extended studies on chiral-at-metal complexes, Meggers and co-workers developed bis-cyclometalated octahedral iridium(III) complexes in which the metal is the only stereogenic element.[169] These complexes contain two metallacyclic units made of C,N-bidentate benzoxazole-type ligands positioned in a propeller-type fashion. The chiral iridium center is also coordinated by a bidentate 5-trifluoroacetamido-3-(2-pyridyl)-1H-pyrazole ligand (Scheme 38). With these catalysts, the ATH of nitroalkenes with Hantzsch esters as hydrogen donors proceeded in high yields and enantioselectivities at low catalysts loadings. In these reactions, the trifluoroacetamido-pyrazole unit of the ligand should act as a double hydrogen-bond donor to the substrate. Concurrently, the hydroxymethyl

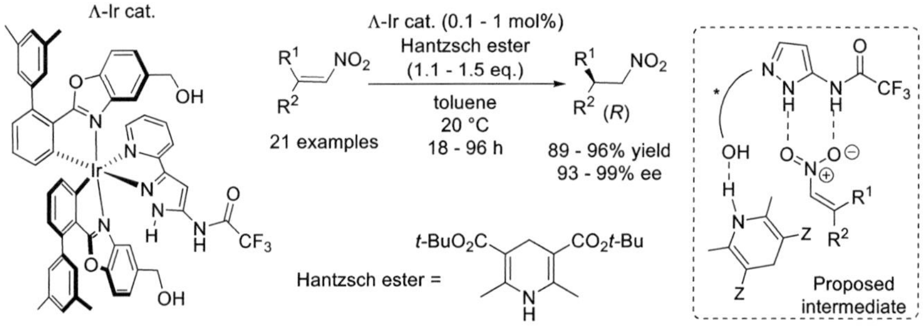

Scheme 38. Chiral-at-metal iridium catalysts for ATH of nitroalkenes and proposed reaction intermediate, by Meggers and co-workers.[169]

substituent of a benzoxazole ligand should serve as hydrogen-bond acceptor toward the Hantzsch ester. Thus, the excellent stereochemical control results from the conformational constraints generated by these cooperative interactions between the catalyst and the substrate.

In order to access catalysts for the ATH of ketones, Meggers and co-workers modified their iridium complexes by combining the bisphenylbenzothiazole metallacycles with monodentate ligands.[170] 3,5-dimethyl-pyrazole proved to be the best choice. It led to highly efficient and enantioselective iridium catalysts for ATHs using HCO_2NH_4 as hydrogen source in H_2O/THF as solvent mixture (Scheme 39). It was proposed that, after coordination of pyrazole to the Ir precursor by substitution of acetonitrile, the iridium complex reacts with ammonium formate to generate an active iridium hydride species, with the assistance of the pyrazole ligand. The complex behaves as a bifunctional catalyst, giving concerted transfer of both the hydride and the pyrazole NH proton to the carbonyl. The high level of asymmetric induction should result from steric constraints (minimal steric hindrance) as well as from additional π–π stacking between the aromatic ring of the substrate and one of the benzothiazole ligands (Scheme 39).

Overall, enantiomeric excesses of over 90% were attained in the ATH of a variety of aryl- or heteroaryl-methyl ketones, β-ketoesters,

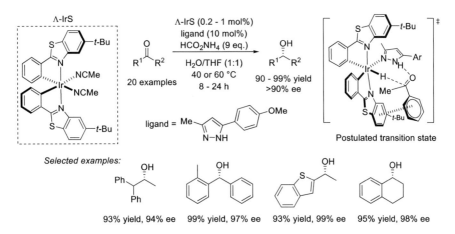

Scheme 39. Chiral–at-metal iridium catalysts for ATH of ketones and proposed transition state, by Meggers and co-workers.[170]

indanone, and a diaryl-ketone (20 examples). These excellent results have validated the Meggers's ambitious design of chiral-at-metal catalysts for ATH reactions.

D. Osmium-Catalyzed Transfer Hydrogenations

Considering the high efficiency of ruthenium catalysts in the ATH of ketones, it is not surprising that the potential of analogous osmium catalysts was similarly considered. Thus, piano stool osmium complexes were investigated that contain ligands like α-amino acids[171] and TsDPEN.[172] Carmona and co-workers disclosed monomeric and trimeric half-sandwich osmium complexes bearing amino-acid-type ligands.[171] The neutral monomeric and cationic trimeric compounds catalyzed the ATH of a few ketones, like acetophenone. With *i*-PrOH as the hydrogen source, ee up to 72% were obtained (Scheme 40a). It was assumed that these catalysts proceed through a Noyori-type bifunctional mechanism. Afterward, Sadler, Wills, and co-workers developed robust and effective (*p*-cymene) osmium catalysts containing the TsDPEN ligand coordinated in monodentate or bidentate modes. The ATH of acetophenone derivatives was

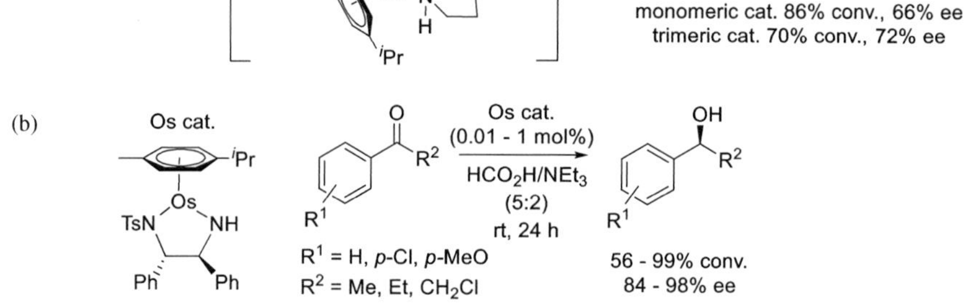

Scheme 40. Monomeric and trimeric arene–osmium catalyst for ATH of ketones, by Carmona and co-workers[171] (a) and by Sadler, Wills, and co-workers (b).[172]

successfully performed with moderate to high conversion rates and high ees (Scheme 40b).[172] The mechanism of ATH reactions with osmium catalysts was enlightened by theoretical studies for the specific case of pyruvic acid reduction.[173]

Overall, these new osmium–arene catalysts have demonstrated their efficiency in the ATH of prochiral ketones. They operate most probably following Noyori's bifunctional mechanism. The ease of preparation of the pre-catalysts and their high stability makes them attractive for applications in ATH reactions.

III. AHs Using Bifunctional Catalysts

This section relates to the catalytic hydrogenations of ketones, imines, and olefins, where molecular hydrogen itself is used as the reducing agent, under metal activation. As in the previous section, ruthenium, rhodium, and iridium-based catalysts that operate as bifunctional catalysts are considered. In terms of reaction pathways, the most frequent remains the cooperative H-transfer from the metal and from an amino group of the ligand. In addition, bifunctional catalysts giving H-bonds with the substrates have been reported.

A. Ruthenium-Catalyzed Hydrogenations

1. *Shvo-type catalysts*

Recently, a few attempts have been made to use chiral analogues of the Shvo catalyst in the AH of simple ketones. Despite the high catalytic activity, these attempts met with moderate success only in terms of enantioselectivity levels.

Thus, inspired by the Shvo catalyst (Scheme 26),[136–138] Yamamoto and co-workers reported the synthesis of the spirocyclic C-riboside-derived ruthenium complexes in Scheme 41.[174] Such chiral ruthenium catalysts efficiently promoted the AH of acetophenone, but the enantiomeric excesses were low. Their mode of action has been correlated to that of the bifunctional Shvo catalysts (see Section II.A.2), since the corresponding hydride complex that displays a ribose-derived hydroxycyclopentadienyl ligand could be prepared and used successfully.

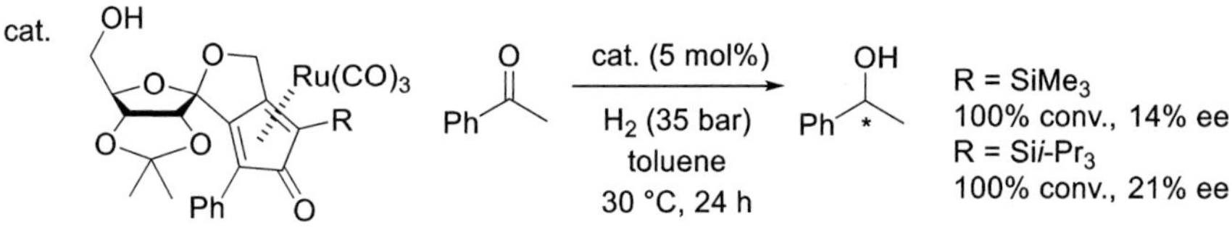

Scheme 41. AH of acetophenone using a ruthenium cyclopentadienone catalyst, by Yamamoto and co-workers.[174]

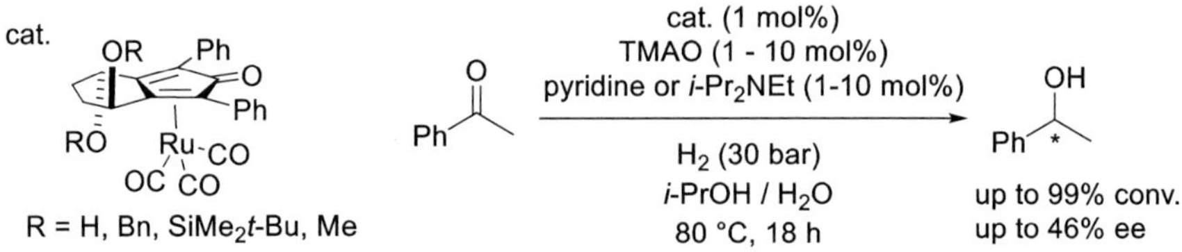

Scheme 42. AH of acetophenone using a ruthenium cyclopentadienone catalyst, by Wills and co-workers.[175]

In 2019, Wills and co-workers reported a series of four enantiomerically pure ruthenium tricarbonyl cyclopentadienone complexes derived from chiral bicyclic cyclopentenones (Scheme 42).[175] All these complexes were active in the model AH of acetophenone, but the enantiomeric excesses were low. Interestingly, the enantiomeric excess could be increased up to 46% by combining hydrogenation and transfer hydrogenation, that is to say, by running the reaction under hydrogen, in aqueous isopropanol with a catalytic amount of i-Pr$_2$NEt or pyridine.

2. Noyori-type catalysts

Parallel to the investigation of Noyori-type ruthenium catalysts for the ATH, research efforts have been directed toward the application of the same catalysts in AH in the presence of molecular hydrogen. During the last decade, several classes of chiral ligands such as diaminophosphines (P,N,N) and tetradentate amines (N,N,NH,NH) have been associated to ruthenium to produce hydrogenation catalysts. Also, η^6-arene–ruthenium complexes with phosphino-amine (P,N) and diamine (N,N) ligands have been investigated, together with diamines-tethered η^6-arene complexes.

Selected examples of applications of such catalysts are given in the following.

Tridentate phosphino-diamine ligands for ruthenium catalysts

Kitamura and co-workers developed a chiral ruthenium(II) complex that contains an axially chiral PN(H)N ligand, called binan-Py-PPh$_2$, and dimethylsulfoxides as additional ligands.[176–177] Once combined with a catalytic amount of base and an excess of dimethylsulfoxide, this complex catalyzed effectively the challenging AH of functionalized and unfunctionalized *tert*-alkyl ketones, the resulting alcohols being obtained in high yields, with good to excellent enantioselectivities (Scheme 43).[176,177]

Kitamura and co-workers investigated the mechanism of these catalyzed hydrogenations through structural analyses, NMR experiments including isotope labelings, kinetic and rate law studies, as well as by control experiments (Scheme 44).[177,178] They demonstrated that the linear and flexible tridentate ligand PN(H)N coordinates to ruthenium and forms selectively a *fac*-octahedral complex, *fac*-[Ru(PN(H)N)(dmso)$_3$](BF$_4$)$_2$. In the complex three DMSO ligands are coordinated to the soft ruthenium center through their soft sulfur atoms, in relative *cis*-positions. Afterward, the reaction of this pre-catalyst with two equivalents of

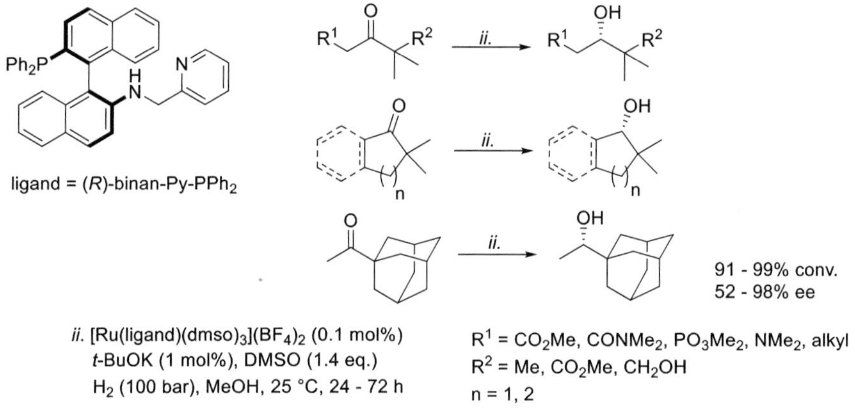

ligand = (*R*)-binan-Py-PPh$_2$

ii. [Ru(ligand)(dmso)$_3$](BF$_4$)$_2$ (0.1 mol%)
t-BuOK (1 mol%), DMSO (1.4 eq.)
H$_2$ (100 bar), MeOH, 25 °C, 24 - 72 h

R^1 = CO$_2$Me, CONMe$_2$, PO$_3$Me$_2$, NMe$_2$, alkyl
R^2 = Me, CO$_2$Me, CH$_2$OH
n = 1, 2

91 - 99% conv.
52 - 98% ee

Scheme 43. AH of sterically hindered ketones using a PN(H)N ligand-based ruthenium catalyst, by Kitamura and co-workers.[176,177]

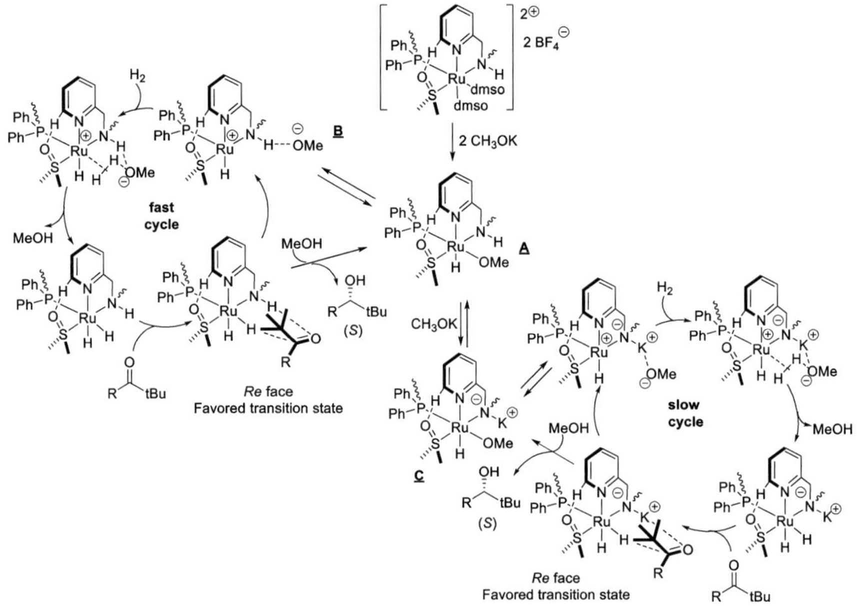

Scheme 44. Mechanism of the AH of sterically hindered ketones using a ruthenium catalyst based on a PN(H)N ligand, by Kitamura and co-workers.[177,178]

potassium methanolate, generated from *t*-BuOK in a 9:1 CH$_3$OH–DMSO mixture, led to the catalytically active ruthenium hydride methanolate complex (Complex **A** in Scheme 44). The hydrogenation of *tert*-alkyl ketones proceeded then via one of the two outer-sphere bifunctional mechanisms displayed in Scheme 44. Indeed, the hydride methanolate complex **A** is in equilibrium with its cationic form **B** or, alternatively, it will be deprotonated by MeOK to give the potassium amide complex **C**. Both complexes **B** and **C** proved able to split hydrogen and form active ruthenium dihydride complexes. The hydride transfer from these complexes takes place preferentially to the *Re*-face of the *tert*-alkyl ketone substrates. Finally, the (*S*)-configured alcohols are delivered by reaction with methanol. The catalytic cycle involving the potassium amide complex (Scheme 44, right) is slower and its contribution to the AH was found to depend on the concentration of the base and the pKa of the substrate.

Tetradentate N$_2$(NH)$_2$ ligands for ruthenium catalysts

Another study from Kitamura and co-workers focused on a ruthenium(II) catalyst displaying a chiral tetradentate amino-pyridine ligand, the Goodwin-Lions-type (R)-Ph-binan-H-Py ligand.[179] The AH of acetophenone was effectively performed in high enantiomeric excess and the reaction mechanism was investigated through structural analyses, NMR experiments and kinetic studies (Scheme 45). The authors showed that the reaction between the chiral N$_2$(NH)$_2$ ligand and the bis(2-methylallyl) cyclooctadiene ruthenium complex, under H$_2$ pressure, provides the expected *cis* ruthenium dihydride complex **A** in tiny amounts only, the starting materials remaining mostly unreacted. Nevertheless, this species

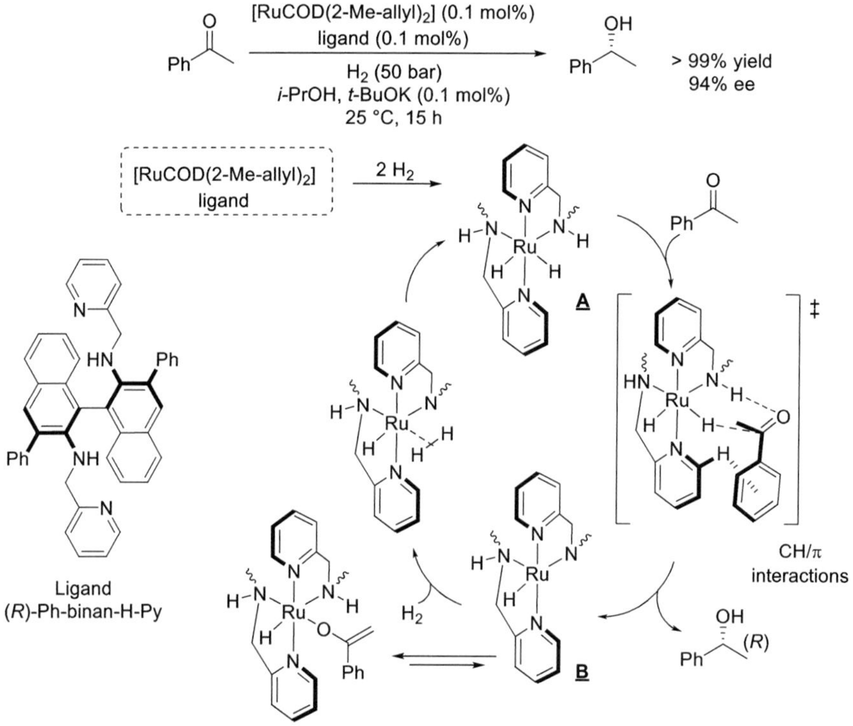

Scheme 45. Mechanism of the AH of ketones using a ruthenium catalyst based on an N$_2$(NH)$_2$ Goodwin–Lions-type ligand, by Kitamura and co-workers.[179]

proved able to catalyze the hydrogenation of acetophenone efficiently, through the usual outer sphere H-transfer bifunctional mechanism. The approach of acetophenone to the ruthenium complex through its *Si*-face is efficiently directed by CH/π and hydrogen bonding interactions. Then, the combined hydride and proton transfer affords (*R*)-phenylethanol and a ruthenium amide intermediate **B** that is reversibly trapped by acetophenone to give a ruthenium enolate species. Finally, the ruthenium amide complex **B** gives back the starting ruthenium dihydride by reacting with dihydrogen, the hydrogen-splitting step being rate-determining.

Phosphino amine derived η^6-arene–ruthenium catalysts.

In 2010, Ikaryia and co-workers reported AH reactions catalyzed by piano stool, Cp*–ruthenium(II) complexes comprising simple chiral amino-phosphine ligands. These complexes had been applied initially to ATH reactions. Thus, ruthenium(II) catalysts with chiral cyclic PN ligands were applied notably to the AH/desymetrization of cyclic and bicyclic imides (Scheme 46).[180] These reactions that involve cleavage of a C–N bond, led to secondary amides with remote hydroxymethyl groups, with high conversion rates and good to excellent enantioselectivities. Later on, analogous diamine-based catalysts have been applied to hydrogenative DKR of a lactone.[181]

Fan and co-workers investigated the use of η^6-arene–ruthenium(II)–MsDPEN (or other sulfonylated DPEN) catalysts in the AH of a broad

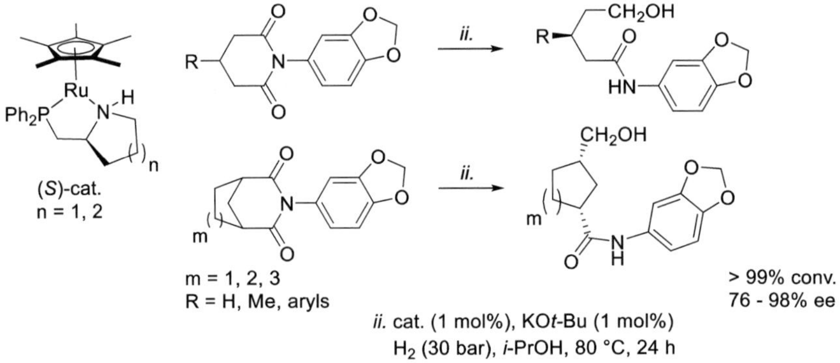

Scheme 46. AH of imides using a ruthenium catalyst based on a PN ligand, by Ikariya and co-workers.[180]

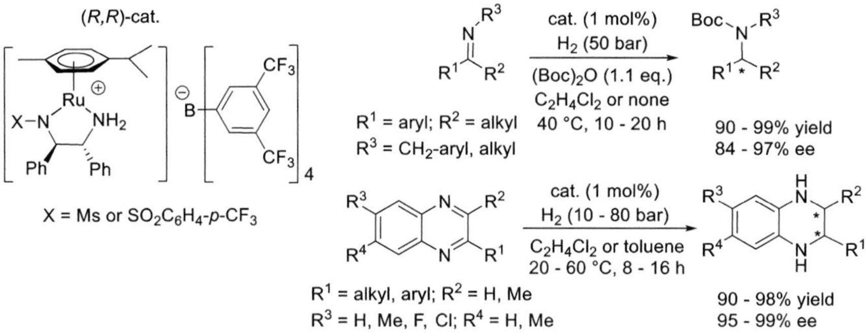

Scheme 47. AH of imines and quinoxalines using a cationic ruthenium amido catalyst, by Fan and co-workers.[182–184]

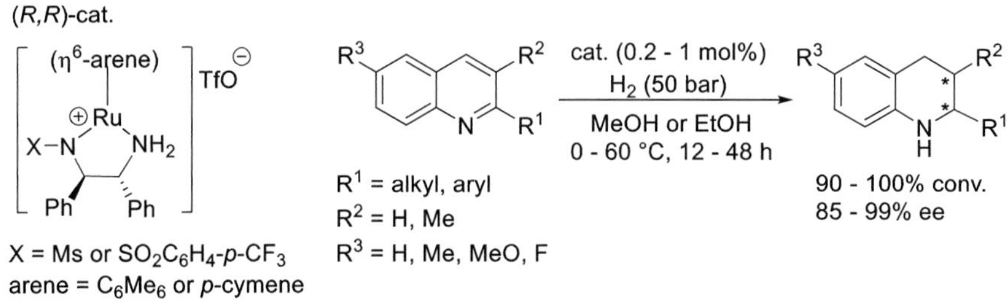

Scheme 48. AH of quinolines using a cationic ruthenium amido catalyst, by Chan and co-workers.[185]

range of acyclic and cyclic imines[182,183] including quinoxalines (Scheme 47).[184] They demonstrated that a cationic catalyst displaying the lipophilic, noncoordinating tetrakis[3,5-bis(trifluoromethyl)phenyl]borate (BArF$_{24}$) counterion enables the hydrogenation of these substrates in high yields and enantioselectivities.

In parallel studies, Chan, Fan, and co-workers applied analogous cationic η^6-arene–ruthenium(II)–MsDPEN or TsDPEN complexes with triflate counterions to the AH of a broad range of quinolines (Scheme 48).[185] The resulting 1,2,3,4-tetrahydroquinolines were obtained with full conversions and high enantioselectivities (ee up to 99%).

The reaction mechanism was studied through stoichiometric reactions, isolation of intermediates, isotope labelings and theoretical studies. These studies demonstrated a cationic cascade pathway including a

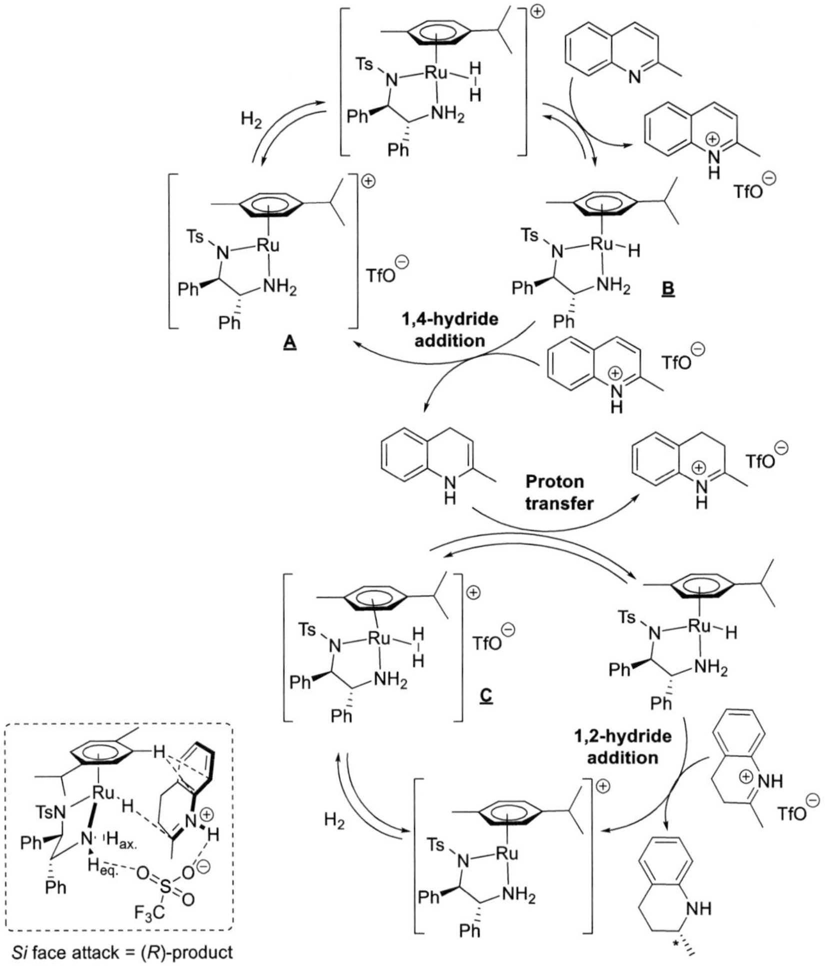

Scheme 49. Mechanism of the AH of quinolines using a cationic ruthenium amido catalyst, by Chan and co-workers.[185]

1,4-hydride addition, a protonation step and a 1,2-hydride addition (Scheme 49). At first, the cationic ruthenium complex **A** coordinates and splits dihydrogen with the assistance of the basic substrate. The resulting ruthenium hydride **B** reacts with the protonated quinoline through a 1,4-hydride transfer to release 1,4-dihydroquinoline and the starting complex **A**. The dihydroquinoline undergoes proton transfer from the ruthenium-H_2 complex **C** to generate the 3,4-dihydroquinolinium triflate, which serves then as a substrate for the second hydrogenation cycle. The enantiodetermining step consists in an irreversible 1,2-hydride transfer to the iminium salt, assisted by the triflate anion. According to theoretical studies, the transition state would involve CH/π interactions between the ruthenium arene ligand and the fused phenyl ring of the substrate, as well as hydrogen bonds between the triflate anion and the NH functions of both the substrate and the diamine ligand (Scheme 49). The hydride transfer would proceed on the *Si* face of the dihydroquinolinium, leading to an (*R*)-configured tetrahydroquinoline, as observed experimentally.

Fan and co-workers also applied cationic η^6-arene–ruthenium(II) Ms-DPEN or Tos-DPEN catalysts to the AH of 2,4-disubstituted 1,5-benzodiazepines (Scheme 50).[186] The reduced products were obtained in high yields, enantioselectivities and diastereoselectivities. In the hydrogenation of aryl-substituted benzodiazepines, the authors observed a reversal of the enantioselectivity upon changing the achiral counteranion from phosphate (PhO)$_2$PO$_2^{(-)}$ to BArF$_{24}^{(-)}$ In order to rationalize the enantioinversion, two

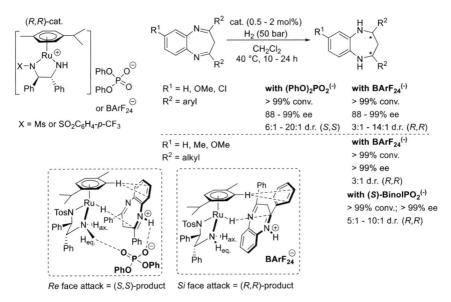

Scheme 50. AH of benzodiazepines using cationic η^6-arene–ruthenium amido catalysts, by Fan and co-workers.[186]

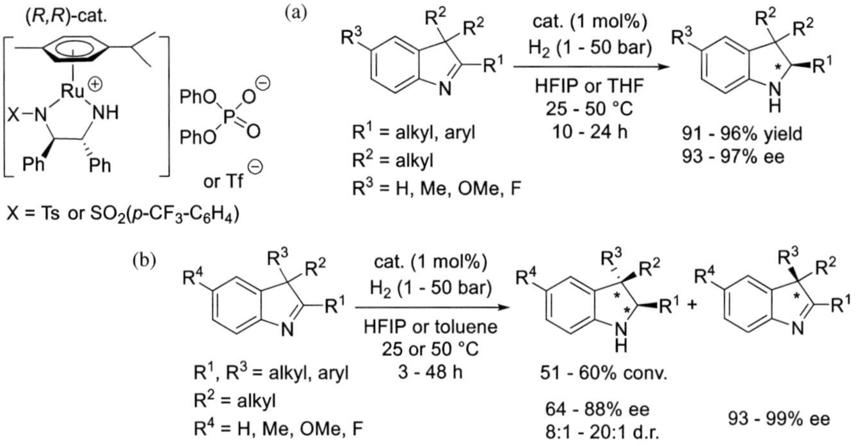

Scheme 51. AH of indoles using a η^6-arene amido–ruthenium catalyst, by Fan and co-workers.[187,188]

different transition states were proposed (Scheme 50). The first one would involve the (*R,R*)-ruthenium catalyst, the phosphate anion and the substrate. CH/π interactions are established between the η^6-arene ligand and the fused phenyl ring of the substrate, while hydrogen bonds are formed between the phosphate anion and the NH functions of both the substrate and the diamine ligand. This transition state would allow an hydride transfer to the *Re* face of the substrate and lead to an (*S,S*)-configured product, as observed experimentally. The second possible transition state involves the (*R,R*)-ruthenium$^+$ BArF$_{24}^-$ catalyst. It shows CH/π interactions between the arene ligand and a phenyl substituent of the substrate, while the BArF$_{24}$ anion does not interact with the substrate (Scheme 51). The hydride transfer takes place here to the *Si* face of the substrate and leads to the (*R,R*)-tetrahydrobenzodiazepine, as observed.

Further experimental data confirmed these models. Indeed, the enantioinversion was not observed with alkyl-substituted substrates. Moreover, the addition of increasing amounts of methanol, that is expected to suppress hydrogen bonds in the postulated transition state, led to a gradual decrease of the enantioselectivity and even to enantioinversion when phosphates are used as counterions.

Fan and co-workers applied further the same catalytic system to the AH of imines and *3H*-indoles (Scheme 51a),[187,188] as well as to the kinetic resolution of racemic 3,3-unsymmetrically disubstituted *3H*-indoles (Scheme 51b). These reactions proceeded in high yields and enantioselectivities

Diamine-tethered η^6-arene–ruthenium catalysts.

Following the successful development by Wills and co-workers of ruthenium catalysts with η^6-arenes tethered to diamine ligands, Ikariya and co-workers reported the synthesis and applications of oxo-tethered ruthenium(II) complexes for both ATH (see Scheme 18) and AH reactions (Scheme 52).[118] Thus, for instance, the bis-amido–ruthenium complex **A** in Scheme 52 and its HCl adduct **B** could be used in AH reactions. The bis-amido complex was usually more active than the ruthenium chloride analogue. By comparison to nontethered catalysts, the activity of these species was much higher, ketones being hydrogenated in high yields and enantioselectivities in shorter reaction times, at lower catalyst loadings. Results underlined that a fine-tuning of the electronic features of the metal and the nitrogen ligands is required to optimize the catalyst performances.

Wills and co-workers studied arene-diamine-tethered ruthenium(II) catalysts in the AH of several simple ketones (Scheme 53).[102] At the

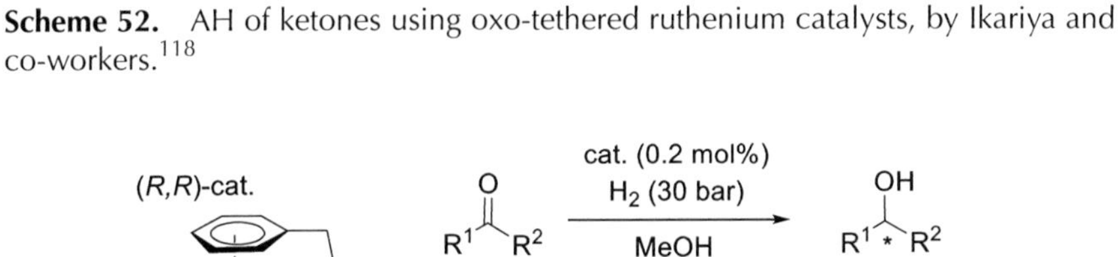

Scheme 52. AH of ketones using oxo-tethered ruthenium catalysts, by Ikariya and co-workers.[118]

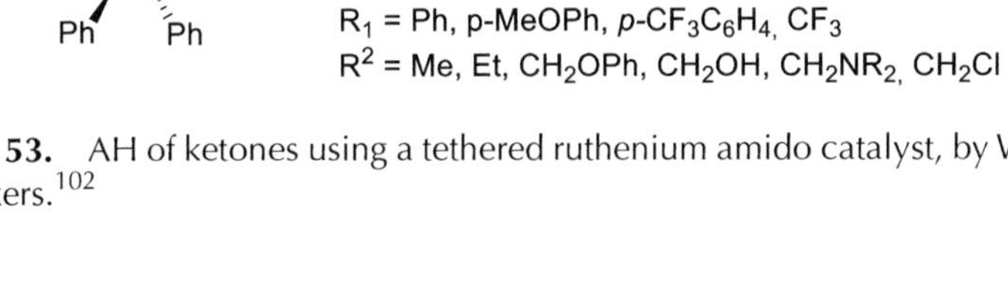

Scheme 53. AH of ketones using a tethered ruthenium amido catalyst, by Wills and co-workers.[102]

exception of fluorinated, sterically hindered and α,β-conjugated ketones, all substrates were reduced with high conversions and enantioselectivities using the ruthenium chloride pre-catalysts, without any additional base.

Outer sphere hydrogenation catalysts with active NH functions (Noyori's type)

Glorius and co-workers reported the synthesis and applications of some unusual ruthenium(II) complexes displaying chiral N-heterocyclic carbenes (NHC) and DPEN-type diamine ligands (Scheme 54).[189,190] These complexes feature a C,N-metallacyclic structure involving the phenyl substituent of the DPEN ligand. After activation with tBuOK, the complexes catalyze the AH of various isocoumarins in high yields and enantioselectivities, affording key building blocks for the total synthesis of compounds of biological interest like O-methylmellein, mellein, and ochratoxin A.[189]

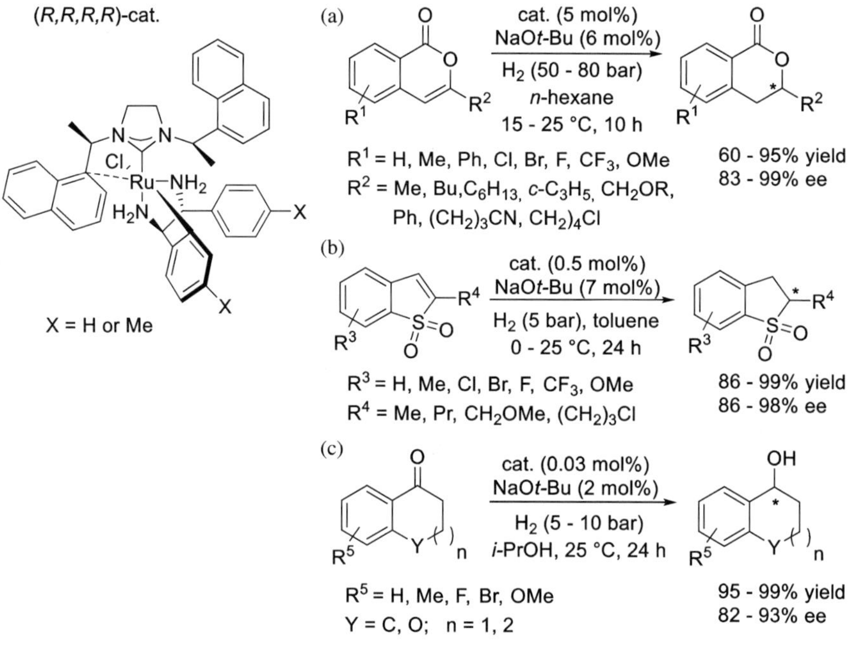

Scheme 54. AH of isocoumarins, benzothiophene 1,1-dioxide, and cyclic ketones using a ruthenium NHC catalyst, by Glorius and co-workers.[189,190]

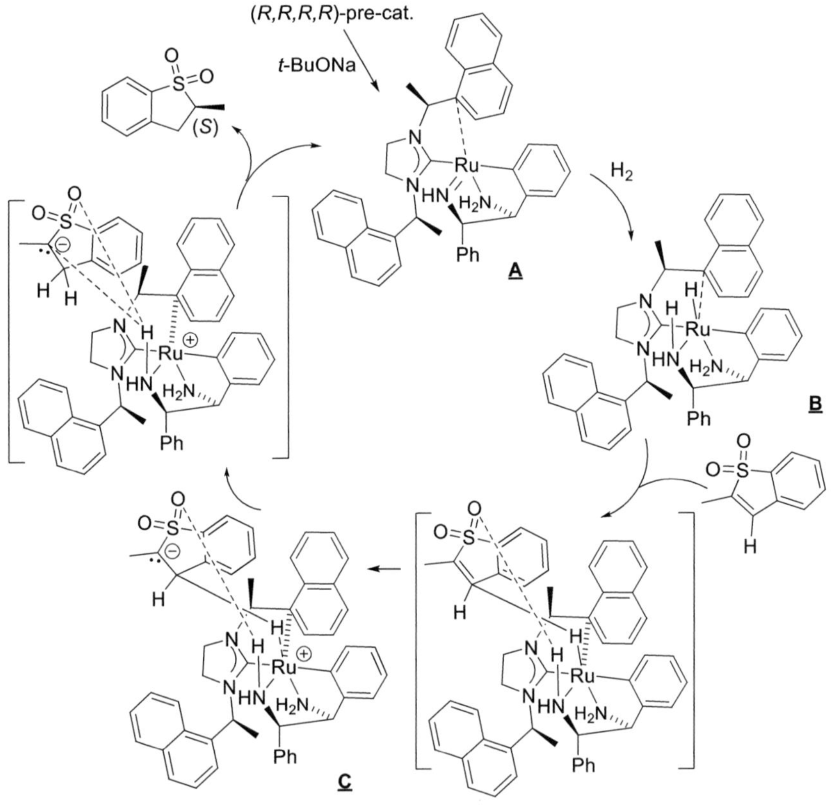

Scheme 55. Mechanism of the AH of benzothiophene 1,1-dioxide using a ruthenium NHC-diamine catalyst, by Glorius and co-workers.[190]

The catalyst was also effective in the AH of a range of benzothiophene 1,1-dioxides (Scheme 53b), and some cyclic ketones (Scheme 54c).[190] Studies based on stoichiometric reactions, NMR experiments and calculations led to a possible reaction mechanism for the hydrogenation of benzothiophene dioxides (Scheme 55). At first, the ruthenium pre-catalyst reacts with the base to give an amido complex **A**, which is able to split dihydrogen heterolytically, leading to the ruthenium hydride species **B**. Afterward, the benzothiophene 1,1-dioxide interacts with intermediate **B** through hydrogen bonding between the oxygen atom of the sulfone and the amine NH proton. The hydride and proton transfers proceed then in a stepwise manner, as known for other outer-sphere bifunctional hydrogenation processes. First, the nucleophilic ruthenium hydride adds to the β-position of the sulfone to give the stabilized intermediate **C**. Follows a proton transfer from the amine NH to the α-position of the sulfone, which takes place preferentially at the *Re* face of the substrate, leading to the (*S*)-configured reduced product. The ruthenium amido complex **A** is released as the catalyst resting state.

B. Rhodium-Catalyzed Hydrogenations

The strategy implying a rhodium catalyst assisted by ligands with functional groups that participate in the activation process, has been investigated also in AHs with molecular hydrogen. Noncovalent interactions such as hydrogen bonding and ion-pairing effects have been smartly investigated. The selected recent examples given below will highlight the value of such approaches.

1. *Catalysts implying hydrogen bondings between ligands and either substrates or halide counteranions*

Several rhodium-based catalysts for AH reactions have been developed recently by taking advantage of weak, noncovalent interactions between suitably functionalized ligands and substrates to improve the stereochemical control. If classical H-bonding with substrates is still prevailing in this field, the possible role of bondings between H-donors and halide counterions (anion binding catalysis) has been also highlighted and successfully exploited recently. A few examples of these two activation pathways are shown hereafter.

For more than a decade, Reek and co-workers have been working on the application of supramolecular ligands in homogeneous catalysis.[191,192] Special attention was paid to ligands able to give H-bondings to other ligands and, therefore, these studies also led to highlight H-bonding phenomena involving ligands and substrates. A first generation of supramolecular ligands had implied hydrogen bonding between the NH moiety of a phosphoramidite and the carbonyl of a urea-functionalized phosphine.[191] A second generation associated phosphoramidites and diphosphine

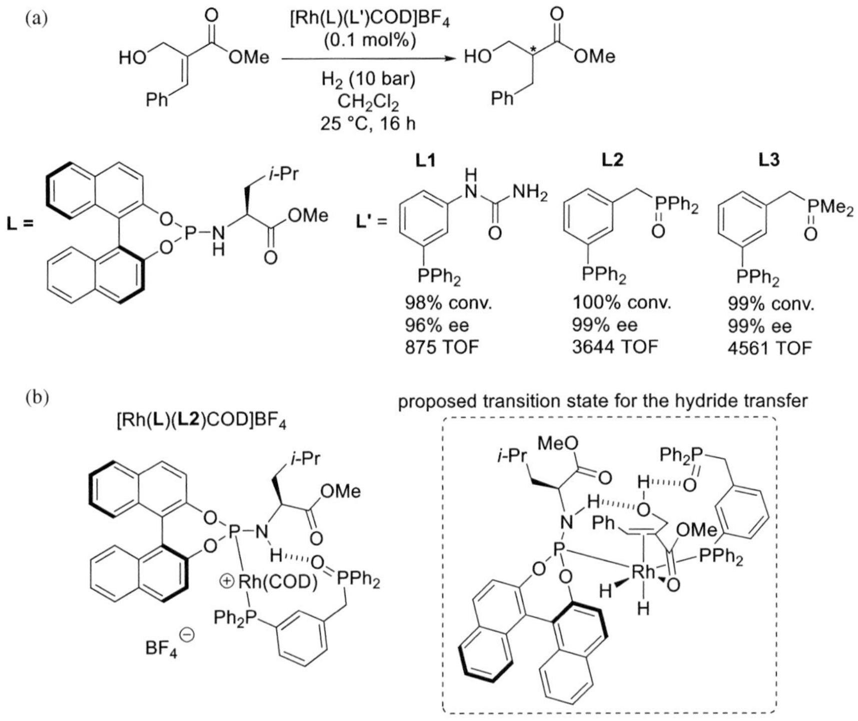

Scheme 56. AH of α,β-conjugated ester using supramolecular rhodium catalysts, by Reek and co-workers.[191–193]

monoxides.[192,193] While investigating these second-generation catalysts in the rhodium-catalyzed AH of α,β-unsaturated esters, the authors noticed especially high activities and enantioselectivities for substrates bearing a hydroxyl group (Scheme 56). According to NMR, kinetic, and theoretical studies, the catalyst and the α,β-conjugated ester interact here through coordination of the C=C and C=O double bonds to rhodium and, additionally, by two hydrogen bonds connecting the hydroxyl group to both the phosphine oxide and the phosphoramidite NH functions. These supramolecular interactions were shown to be strong enough to stabilize the transition state in the rate-determining hydride transfer step (see Scheme 56b), thus leading to higher catalytic activities.

In 2013, Zhang, Wang, and co-workers introduced a novel chiral ferrocene-based bisphosphine–thiourea ligand named ZhaoPhos, in order to

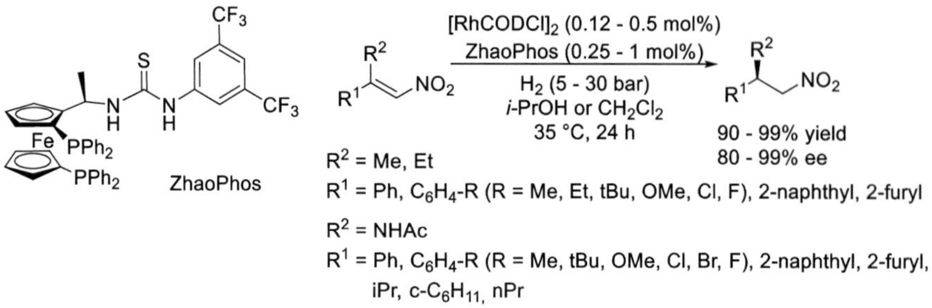

Scheme 57. AH of β,β-disubstituted nitroalkenes using a rhodium catalyst based on a bisphosphine–thiourea ligand, by Zhang, Wang, Dong, and co-workers.[194,195]

take advantage from thiourea as an activating and directing group (Scheme 57).[194] This ligand design proved especially successful for applications in the rhodium-promoted AHs of functionalized, cyclic, and acyclic olefins and iminium salts. A few recent examples are shown hereafter.

A rhodium(I)–ZhaoPhos complex was applied to the AH of some challenging β,β-disubstituted nitroalkenes and the hydrogenated products were obtained in high yields and enantioselectivities. According to control experiments and by comparison with other diphosphine ligands, it was demonstrated that the activation of the substrate was effectively directed by the thiourea moiety of the ligand. The catalytic system was further applied to the AH of β-amido nitroolefins, the resulting chiral β-amino nitroalkanes being obtained in excellent yields and enantioselectivities.[195]

Zhang and co-workers also reported the AH of α,β-unsaturated carbonyl compounds promoted by the rhodium–ZhaoPhos complexes (Scheme 58).[196] At the exception of sterically hindered substrates, the resulting chiral amides and esters were obtained in high yields and enantioselectivities. A reaction under deuterium gas led to an α- and β-deuterated product and therefore indicated an inner-sphere mechanism, like in most common rhodium-catalyzed hydrogenations. Authors proposed a transition state implying coordination of the C=C double bond to rhodium and H-bonding of the carbonyl to the thiourea moiety. Such catalytic system was then effectively applied to the AH of ethyl (Z)- and (E)-3-substituted-3-thio-acrylates[197] and α,β-unsaturated sulfones,[198] the latter requiring a specific solvent mixture in order to avoid decomposition of the catalyst (Scheme 58).

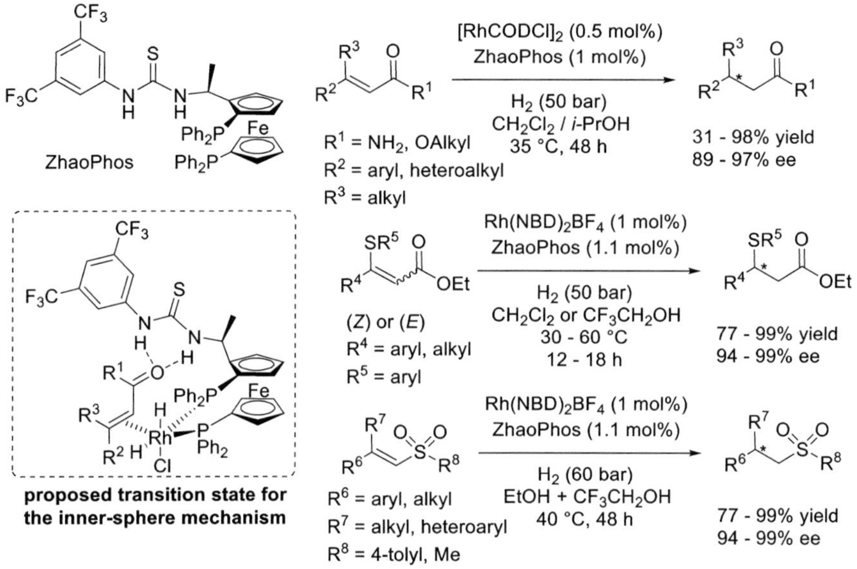

Scheme 58. AH of α,β-unsaturated esters and amides, 3-substituted-3-thioacrylates and α,β-unsaturated sulfones using a ZhaoPhos rhodium catalyst, by Zhang and co-workers.[196–198]

The same catalytic system was further applied to the AH of cyclic electron-poor olefins like 3-substituted maleic anhydrides and maleimides (Scheme 59a).[199,200] The corresponding 3-substituted succinic anhydrides and succinimides were obtained in high yields and enantioselectivities at low catalyst loadings. The method was successfully applied to the synthesis of bioactive compounds such as the hypoglycemic drug mitiglinide and a potent α-2-adrenoceptor antagonist.

In addition, β-substituted α,β-unsaturated lactams were reduced to the corresponding chiral γ-lactams in high yields and enantioselectivities, irrespective of the substituents on the lactam nitrogen (Scheme 59b).[201] This methodology allowed the synthesis of chiral bioactive compounds such as (R)-rolipram, pyrrolidines and γ-amino acids like (R)-baclofen, a GABA receptor agonist. 1,5-benzothiazepines were also prepared stereoselectively, without any catalyst inhibition by the sulfur atom of the substrate (Scheme 59c). The method allowed a direct access to the antidepressant drug (R)-(−)-thiazesim.[202] Finally, this catalytic AH could be applied to the reduction of exocyclic double bonds of α,β-unsaturated cyclic carbonyl compounds. The

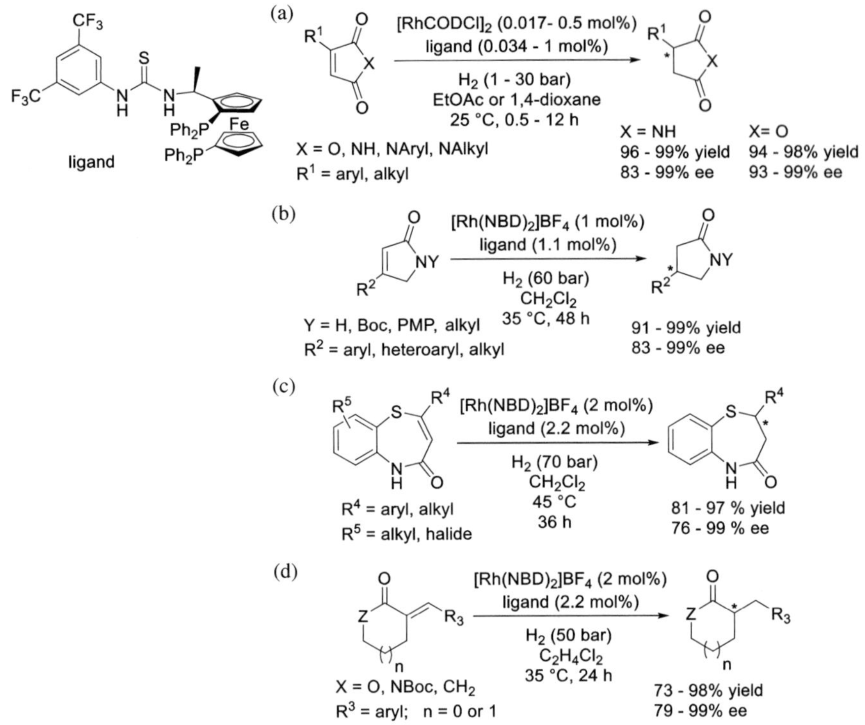

Scheme 59. AH of β-substituted α,β-unsaturated lactams, α,β-unsaturated carbonyl compounds, 3-substituted maleic anhydrides, and maleimides using a ZhaoPhos rhodium catalyst, by Zhang, Dong, Yin, Wen, Lv, and co-workers.[199–203]

corresponding lactones, lactams, and cyclic ketones were obtained in high yields and enantioselectivities (Scheme 59d).[203]

While studying the AH of β-cyanocinnamic esters, Zhang and co-workers demonstrated that the simple *N*-methylation of one NH group of the ZhaoPhos ligand improved significantly the enantioselectivity of the hydrogenation reaction, thus enabling an efficient synthesis of chiral γ-aminobutyric acid (GABA) derivatives (Scheme 60a).[204] Experiments and calculations were performed to investigate the noncovalent interactions between the substrate and the possible catalytically active species. The results indicated that the *N*-methylated ligand, a single hydrogen bond donor, affords a stronger hydrogen bonding than ZhaoPhos itself which is a potential double hydrogen bond donor. Finally, this catalytic

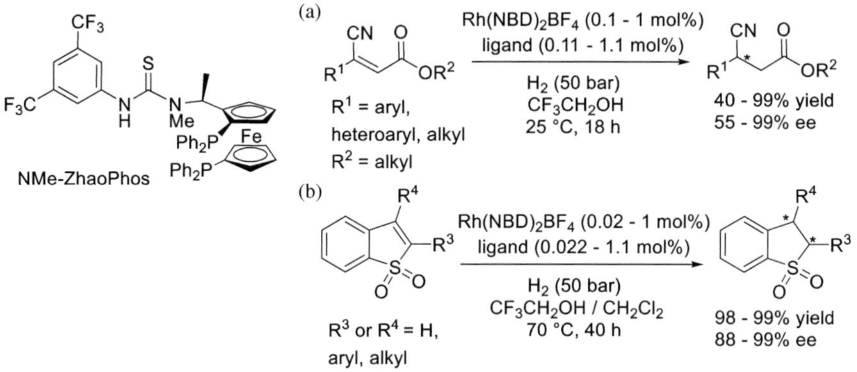

Scheme 60. AH of β-cyanocinnamic esters and benzo[b]thiophene 1,1-dioxides using a N-methylated ZhaoPhos-rhodium catalyst, by Zhang and co-workers.[204,205]

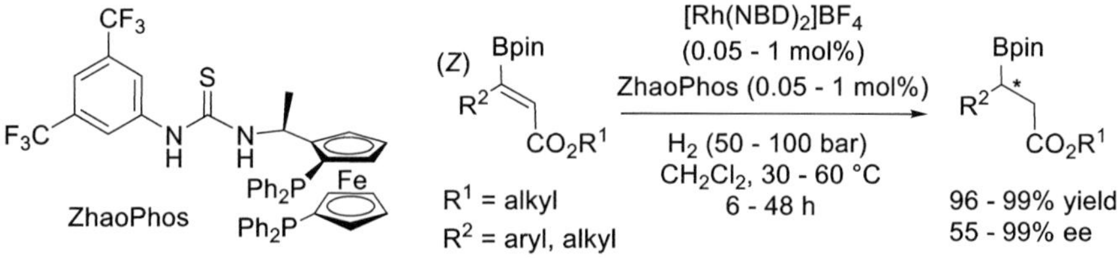

Scheme 61. AH of (Z)-β-substituted-β-boryl-α,β-unsaturated esters using a rhodium–ZhaoPhos catalyst, by Zhang and co-workers.[206]

AH was effectively applied to the synthesis of a number of compounds of interest like (S)-pregabalin, (R)-phenibut, (R)-baclofen, chiral δ-amino alcohols, chiral γ-lactam, and chiral pyrrolidines.

It is worth to note that a series of substituted benzo[b]thiophene 1,1-dioxides were also hydrogenated in high yields and enantioselectivities at low catalyst loadings (Scheme 60b).[205]

In a recent report, Zhang, Dong, and co-workers studied the AH of (Z)-β-substituted-β-boryl-α,β-unsaturated esters in the presence of the rhodium–ZhaoPhos catalyst (Scheme 61).[206] At the exception of alkyl derivatives (R^2 = alkyl), the saturated β-boryl esters were obtained in high yields and enantioselectivities, some of them being subsequently converted to other versatile synthetic intermediates, like methyl (S)-3-hydroxy-3-phenylpropanoate and methyl (S)-3-(furan-2-yl)-3-phenylpropanoate.

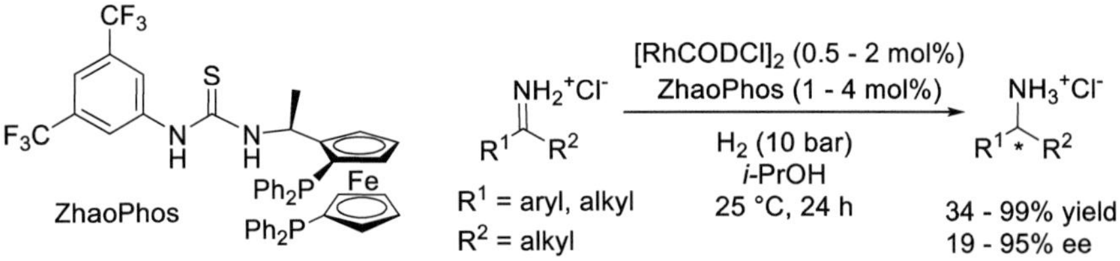

Scheme 62. AH of unprotected imines using a ZhaoPhos–rhodium catalyst, by Zhang, Anslyn, Wong, and co-workers.[207]

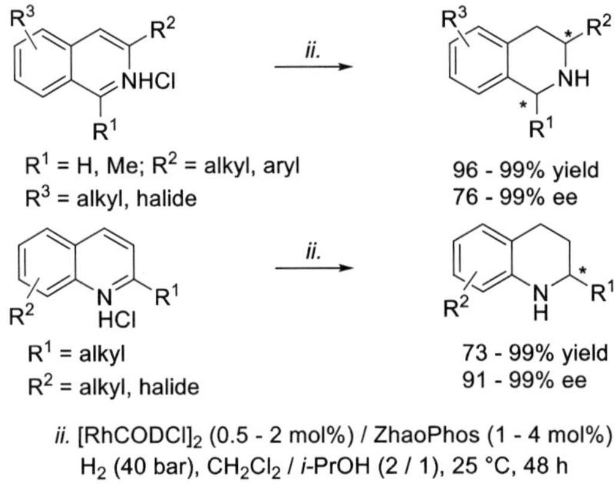

Scheme 63. AH of isoquinolines and quinolines using the ZhaoPhos–rhodium catalyst, by Zhang, Zhao, and co-workers.[208]

The application of the same bisphosphine–thiourea ligand to hydrogenations involving anion-binding type catalysis has been reported. This is typified by the AH of unprotected NH imines shown in Scheme 62. The catalyst was generated *in situ* from the ligand and [(COD)RhCl]$_2$ and the desired amines were obtained in high yields and enantioselectivities, at the exception of dialkylated substrates.[207] While control experiments highlighted that chloride was the best anion for an effective and enantioselective catalytic reaction, an NMR study confirmed that, in the key hydrogenation step, a dual NH–halogen bonding interaction took place between the thiourea unit of the ligand and the chloride counterion, while rhodium was bound to the C=N double bond of the imine substrate.

Zhang, Zhao, and co-workers further reported another example of anion binding catalysis by applying the ZhaoPhos ligand to rhodium-catalyzed AH of isoquinolines and quinolines activated by HCl (Scheme 63).[208]

A broad range of substrates were hydrogenated in high yields and enantioselectivities.

NMR studies and ligand screening evidenced anion binding between the thiourea moiety of the ligand and the chloride counterion of the heterocyclic substrate. On the basis of previous reports on the hydrogenation of protonated heterocycles, coordination of the substrate to the rhodium catalyst seemed unlikely. Therefore an inner-sphere reaction mechanism was discarded and an outer-sphere mechanism implying hydrogen transfer was proposed (Scheme 64). The starting rhodium(I) complex would undergo oxidative addition of dihydrogen leading to the rhodium(III) dihydride **A**. The subsequent reaction with isoquinolinium chloride provides an anionic rhodium dichloro dihydride complex **B**. Afterward, a hydride transfer shall proceed between the equatorial hydride and the isoquinolinium chloride. In this key step, the isoquinolinium chloride is activated by the interaction

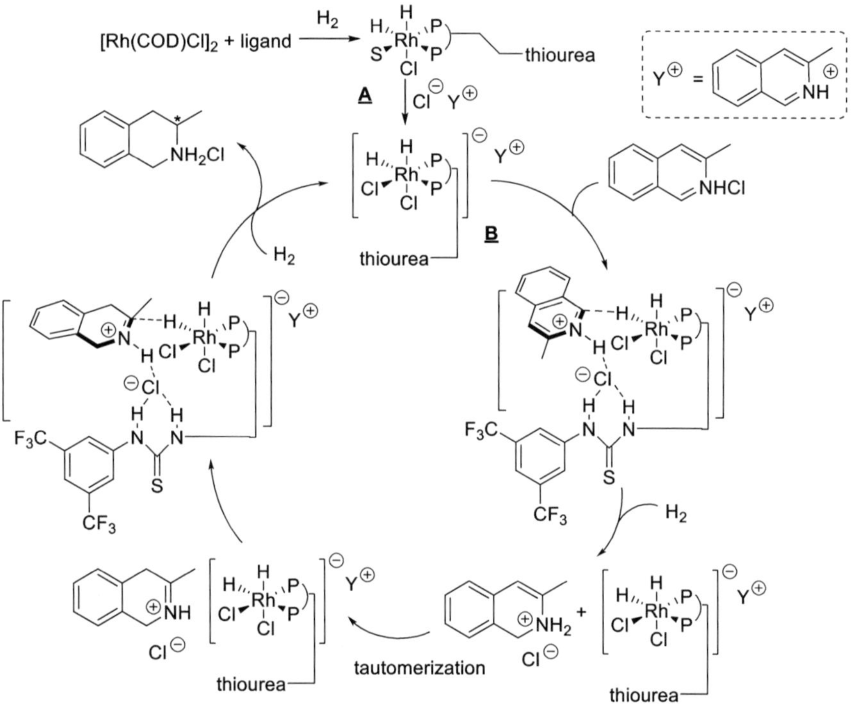

Scheme 64. Mechanism of the AH of isoquinolines using the ZhaoPhos–rhodium catalyst, by Zhang, Zhao, and co-workers.[208]

between the chloride and the thiourea moiety of the ligand. The 1,2-addition process is achieved then by reaction with H_2. The resulting dihydroisoquinolinium will undergo tautomerization into the corresponding iminium, as confirmed by deuterium labeling experiments. The second hydrogenation step proceeds in a similar way, the enantio-determining hydride transfer being directed by binding of the chloride anion to the thiourea moiety. Finally, reaction with dihydrogen would release the hydrogenated product and the rhodium dichloro dihydride catalyst.

A similar approach has been used for the rhodium-catalyzed AH of unprotected indoles in the presence of acids (Scheme 65).[209] At the exception of 2-aryl derivatives that were poorly reactive, all substrates led to chiral indolines in high yields and enantioselectivities. Once again, NMR and mechanistic studies evidenced an anion binding between the thiourea moiety of the catalyst and the chloride counterion of the heterocyclic substrate. An outer-sphere reaction mechanism was proposed according to deuterium labelling experiments and calculations (Scheme 66), although a mechanism implying an electrostatic interaction between an anionic rhodium complex and the cationic indolinium could not be excluded.

At first, the starting rhodium(I) complex undergoes an oxidative addition of hydrogen to generate the active five-coordinated rhodium(III) dihydride species **A**. Further reaction with the indole hydrochloride affords a catalyst–substrate adduct wherein the hydride transfer occurs at the *Re*-face of the protonated indole in position 2, following an outer-sphere process. The enantioselection is directed by the chloride bridge between the protonated substrate and the thiourea moiety. After release of

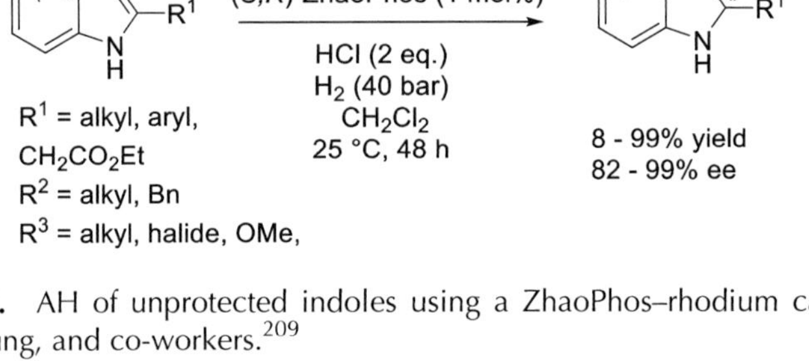

Scheme 65. AH of unprotected indoles using a ZhaoPhos–rhodium catalyst, by Zhang, Chung, and co-workers.[209]

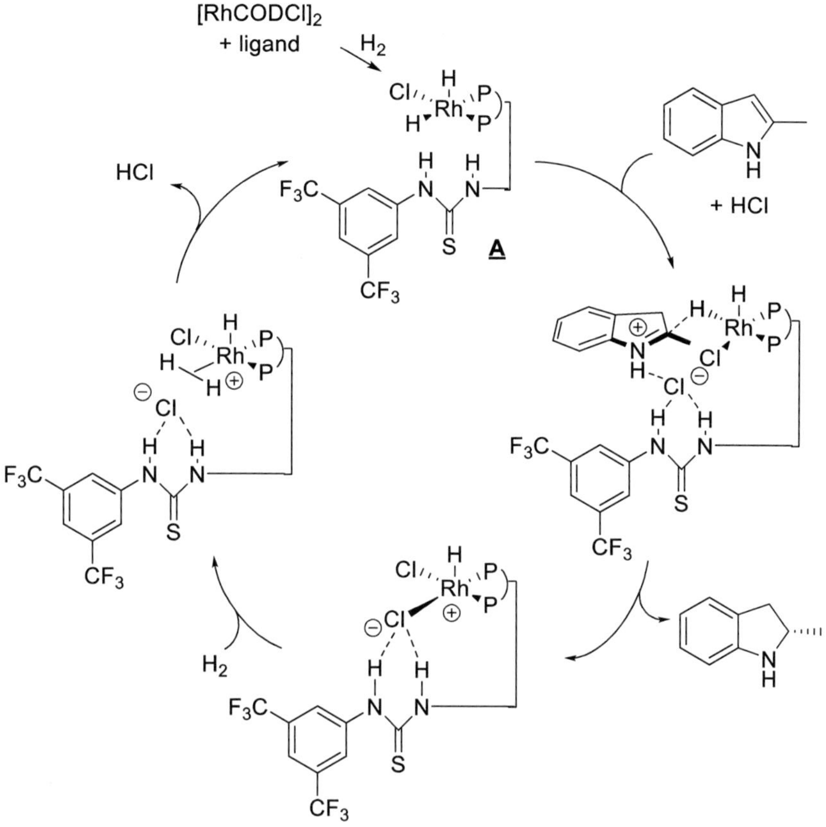

Scheme 66. Mechanism of the hydrogenation of unprotected indoles using a ZhaoPhos–rhodium catalyst, by Zhang, Chung, and co-workers.[209]

the indoline, the catalyst is activated by heterolytic cleavage of dihydrogen assisted by the chloride anion. This rate-determining step gives HCl and the active rhodium dihydride species.

Overall, the results summarized in Schemes 57–66 demonstrate the exceptional efficiency of the ferrocene-based thiourea diphosphine ZhaoPhos developed by Zhang and Dong, in a variety of AH reactions. This powerful synthetic method combines low catalyst loading, full conversion, and excellent enantioselectivity levels. The successful application of the same phosphine in iridium-promoted AH reactions will be discussed in Section III.C.2.

2. Catalysts implying ion pairs as noncovalent interactions between ligands and substrates

This section illustrates the use of chiral ligands with amino functions that, after protonation, may give ion pairs with anionic functions of the substrates. These noncovalent interactions alone, or combined with other H-bonding type interactions, enable excellent stereocontrol in the AH of selected olefinic substrates, under rhodium catalysis.

In 2016, Zhang, Dong, and co-workers reported rhodium-catalyzed AH of 2-substituted acrylic acids by applying the concept of ion-pairing as noncovalent directing interactions (Scheme 67a).[210] They used a bidentate ligand called Wudaphos, a P-stereogenic diphosphine that contains the planar chiral moiety of the Ugi amine. The AH of various acrylic acids with α-alkyl or aryl substituents proceeded with high conversion rates and enantioselectivities in base-free and mild reaction conditions, with an S/C ratio up to 20,000. Interestingly, the enantioselectivity dropped dramatically for olefins displaying carbon spacers between the C=C bond and the acid function (7% ee for the olefin with a CH_2 spacer). Thus, the attractive, noncovalent ion pair interaction between the carboxylate and the

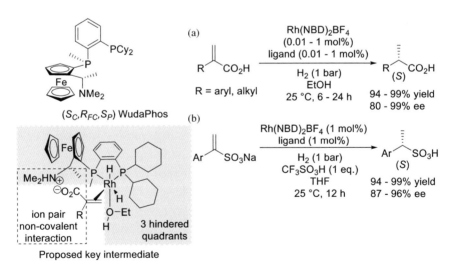

Scheme 67. AH of 2-substituted acrylic acids using a rhodium–WudaPhos catalyst, by Zhang, Dong, and co-workers.[210,211]

protonated dimethylamine moiety of the ligand was postulated to play a key role in the stereochemical control. This role was further confirmed by control experiments: esters were unreactive and the reaction in the presence of a base led to a large drop of enantioselectivity. On the basis of the X-ray structure of the ligand and previous calculations, authors rationalized the enantioselection by applying the quadrants model (Scheme 67). The substituted acrylic acid is assumed to coordinate to the rhodium catalyst in the free quadrant and a favorable ion pair interaction enables then discrimination between the two faces of the substrate, leading to an (S)-configured hydrogenation product. Afterward, the rhodium/Wudaphos combination was also applied to the AH of sodium α-arylethenylsulfonates (Scheme 67b).[211] The reactions proceeded only in the presence of one equivalent of triflic acid and afforded the α-arylethylsulfonic acids in high yields and enantioselectivities.

Zhang, Dong, Chung, and co-workers further improved their catalyst design by combining noncovalent ion pair interactions with hydrogen bondings. The WudaPhos ligand was modified by introducing a secondary phosphine oxide (SPO) moiety instead of the P-stereogenic tertiary phosphine function (Scheme 68).[212] It was expected indeed that

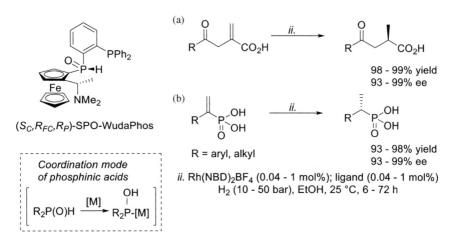

Scheme 68. Base-free AH of α-methylene-γ-keto carboxylic acids and α-substituted ethenylphosphonic acids using the SPO–WudaPhos rhodium catalyst, by Zhang, Dong, Chung, and co-workers.[212,213]

the SPO ligand would coordinate the rhodium center via the P-atom of its phosphinic acid form ($R_2P(OH)$, see Chapter 1.2 of this book). The free hydroxy group on phosphorus will enable then hydrogen bondings with the substrate to hydrogenate. The base-free AH of various α-methylene-γ-keto carboxylic acids afforded the corresponding γ-keto acids in high yields and enantioselectivities (Scheme 68a), the carbonyl group of the substrate behaving here as the H-bonding acceptor. Interestingly, a further AH of the resulting chiral γ-keto acids, in the presence of chiral iridium or ruthenium catalysts and a base, led, respectively, to the corresponding *syn-* and *anti-*lactones, which are interesting building blocks for the synthesis of bioactive compounds. As an example, the potent antimitotic antitumor agent cryptophycin has been prepared.

In these experiments, the use of the SPO–Wudaphos ligand proved to be critical to attain high enantioselection, as demonstrated by comparison with the WudaPhos ligand itself. The role of the dual ion pair and hydrogen bonding noncovalent interactions was further confirmed by reactions conducted in the presence of a base, a large drop of activity and enantioselectivity being observed. A similar drop was observed while hydrogenating substrates such as unsaturated carboxylic esters or carboxylic acids devoid of keto-functions. Theoretical studies also confirmed the key role of the strong ion pair between the dimethylamino moiety and the acidic group, combined with H-bonding between the hydroxy group of the SPO ligand and the ketone moiety. Both contribute to the extended interaction between the substrate and the catalyst and give selectively to the (*R*)-configured hydrogenation product.

The SPO–WudaPhos ligand was further combined with rhodium for the AH of α-substituted ethenylphosphonic acids (Scheme 67b).[213] The corresponding saturated phosphonic acids were obtained in high yields and enantioselectivities using low catalyst loading. The role of ion-pairing and hydrogen bonding interactions was confirmed by the large drop of enantioselectivity observed in the presence of a base. Similarly, no reaction was observed with 1-phenylethenylphosphonate. In this case, the H-bonding of the P-OH group is not mentioned explicitly as a regulating effect.

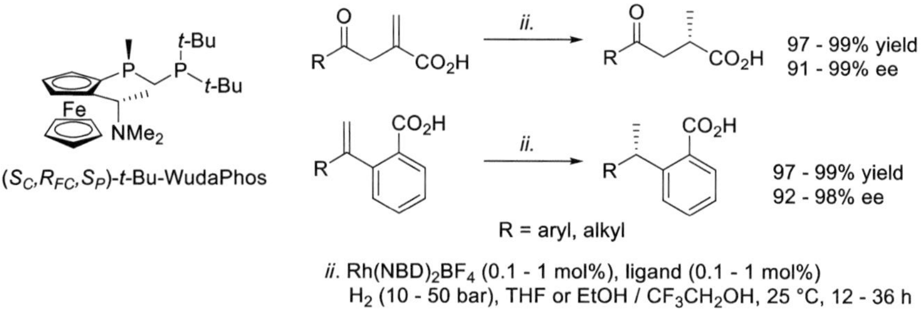

Scheme 69. AH of α-methylene-γ-keto carboxylic acids and disubstituted terminal olefins using a tBu-WudaPhos–rhodium catalyst, by Zhang, Dong, and co-workers.[214,215]

Pursuing their efforts in the development of catalysts involving ion pair interactions, Zhang, Dong, and co-workers reported a modified WudaPhos ligand comprising a bis-tertiobutylphosphine moiety (Scheme 69). The combination of this ligand with rhodium allowed the AH of α-methylene-γ-keto carboxylic acids[214] as well as the AH of disubstituted terminal olefins displaying a carboxylic acid function on an aryl substituent.[215] In both cases, the hydrogenated products were obtained in high yields and enantioselectivities. As in the case of the WudaPhos based catalysts in Scheme 67, the quadrants model, combined with the favorable ion pair directing interaction, accounts for the sense of chiral induction and the almost perfect enantioselectivity.

Overall, the few examples recalled in this section demonstrate the high potential of ion-pairing effects as tools for the stereochemical control in rhodium promoted AHs. Remarkably, in the mentioned examples, ion pairing effects increase significantly the discriminating ability of structurally simple and easily available chiral phosphorus ligands.

C. Iridium-Catalyzed Hydrogenations

This section deals first with iridium complexes of bidentate ligands that operate through monohydrido intermediates and outer sphere H-transfer to the substrate from both the metal and a coordinated NH moiety. In the second part of this section, examples of catalysts implying ion pair or hydrogen bonding interactions will be presented.

1. Iridium catalysts involving active NH functions and related cases

In line with their work on η^6-arene–ruthenium(II) catalysts with tethered amine–amido ligands (see Scheme 19 and following), Ikariya and co-workers developed η^5-cyclopentadienyl–iridium(III) complexes displaying a triflylamido ligand tethered to the cyclopentadienyl ligand and a chiral N-mesylated 1,2-diphenylethylendiamine (Scheme 70).[216] Such complexes were able to activate hydrogen and effectively catalyzed the AH of acetophenone. However, the length of the tethering carbon chain had a significant effect on the catalyst activity and stereoselectivity, complexes with 2- and 3-carbon chains leading to inferior performances compared to the complex with a C4-chain. According to the authors, the tether effect might result from a hydrogen bond between the NH_2 group of the diamine ligand and one of the triflylamide oxygens.

Ikariya and co-workers demonstrated that a diamido iridium(III)-Cp* complex derived from a chiral 1,2-diphenylethylendiamine, was able to activate hydrogen by forming a monohydrido-amine complex that catalyzed the AH of acetophenone in high yield and enantioselectivity (Scheme 71).[217] Deuterium labelling showed that, for reactions carried

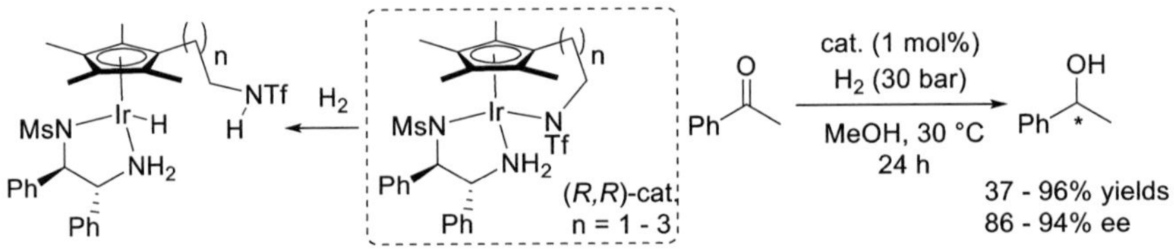

Scheme 70. AH of acetophenone using an iridium triflylamide tethered catalyst, by Ikariya and co-workers.[216]

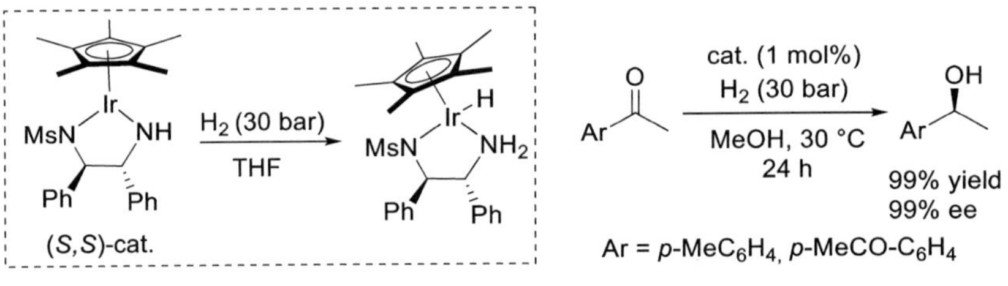

Scheme 71. Hydrogenation of acetophenone using an iridium amido pre-catalyst, by Ikariya and co-workers.[217]

out in methanol, the hydrogenation reaction competes with hydrogen transfer from the solvent and increases the enantioselectivity by preventing the reversible hydrogen transfer reaction to take place.

Kempe and co-workers also observed the favorable effect of combining hydrogenation and transfer hydrogenation conditions.[218] An iridium(I) complex based on an aminopyridinato ligand was applied to the AH of various ketones (Scheme 72). Better activities and enantioselectivities were obtained at low catalyst loading upon addition of large amounts of base and 2 equivalents of acetone, but conversion rates and enantioselectivities proved highly dependent on the substrate.

In 2011, Zhou, Xie, and co-workers developed iridium(III) dihydride catalysts based on P(NH)N tridentate ligands, called Spiro-PAP. The ligands display a spiranic backbone with a phosphine and a secondary amine functions, the latter having an appended pyridyl fragment (Scheme 73).[219] These catalysts proved to be highly effective and enantioselective for the AH of an impressive variety of ketones, enones, β-ketolactams, α-keto acids (Scheme 73a), β- or δ-ketoesters (Scheme 73b), and others.[220–224] Moreover, these iridium(III) catalysts were involved in several DKR processes and enabled the asymmetric total synthesis of compounds of biological interest like crinine-type alkaloids (Scheme 73c), gracilamine alkaloïds, (–)-goniomitine, mulinane diterpenoids, (–)-hamigeran B, (–)-mesembrine and others.[225–232] A recent application of these catalysts is the AH/DKR of racemic α-arylamino-γ-lactones and α-arylamino-δ-lactones leading to the corresponding chiral 2-amino diols with high yields and enantioselectivities (Scheme 73d).[233,234] According to theoretical studies, the hydride and proton transfer step proceeds via a cyclic 6-membered transition state. The calculated transition state of lowest energy leads to the (S)-2-amino-diol, as observed experimentally.

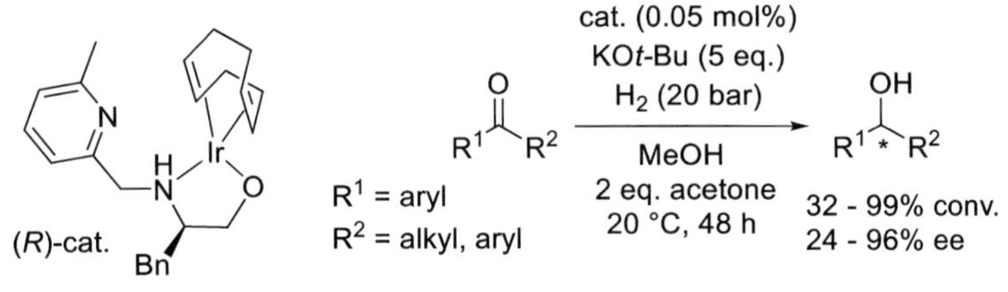

Scheme 72. Hydrogenation of ketones using an iridium catalyst based on an aminopyridinato ligand, by Kempe and co-workers.[218]

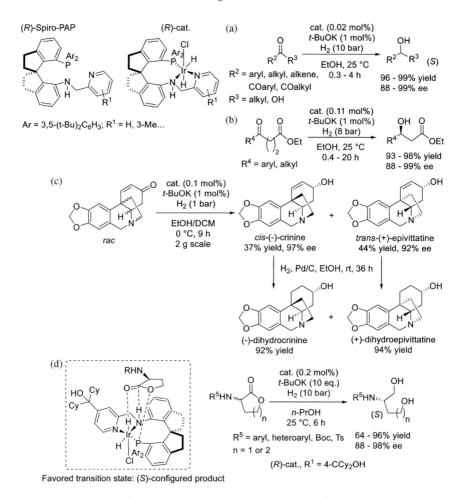

Scheme 73. AH of ketones, lactones, and β-ketoesters using an iridium catalyst based on a P(NH)N spiro ligand by Zhou, Xie, and co-workers.[219–234]

This method provides one of the most impressive recent applications of Ir-bifunctional catalysts in AH, in terms of both catalytic efficiency, enantioselectivity, substrate scope and synthetic potential.

It is worth to note that Zhang, Yin, and co-workers also reported the synthesis of rather similar oxa-spirocyclic chiral ligands and applied them to the iridium-catalyzed AH of Bringmann's lactones (Scheme 74a).[235] Moreover, in another report, Zhang, Dong, Lan, and co-workers developed

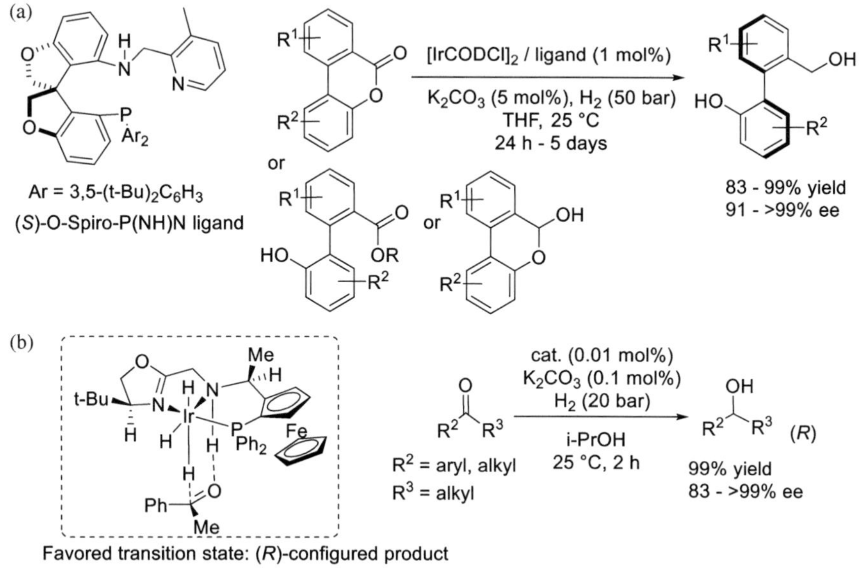

Scheme 74. AH of Bringmann's lactones and ketones with iridium catalysts based on P(NH)N tridentate ligands, by Zhang and co-workers.[235,236]

another P(NH)N tridentate ligand, based on a ferrocene-derived aminophosphine with a pendant oxazoline unit, which provided effective iridium catalysts for the hydrogenation of ketones, giving very high enantioselectivities and catalytic activities (TON and TOF) (Scheme 74b).[236] A theoretical study enlightened the hydrogenation mechanism and the origins of enantioselectivity, underlying the key role of the bifunctional amido-hydrido-iridium(III) species in the activation of hydrogen.

2. Catalysts implying ion pair or hydrogen bonding interactions

In 2017, Zhang, Dong, Chung, and co-workers reported the synthesis of chiral tridentate ferrocene ligands (f-Ampha) comprising an aminophosphine and a carboxylic acid functions (Scheme 75).[237] The catalyst formed by combining such ligands with [Ir(COD)Cl]$_2$ in basic media promoted the AH of various aryl,alkyl-ketones with high enantioselectivities. The catalytic activity was excellent, with catalyst loadings as low as 0.0002 mol%. Theoretical studies have suggested a catalytic cycle implying the

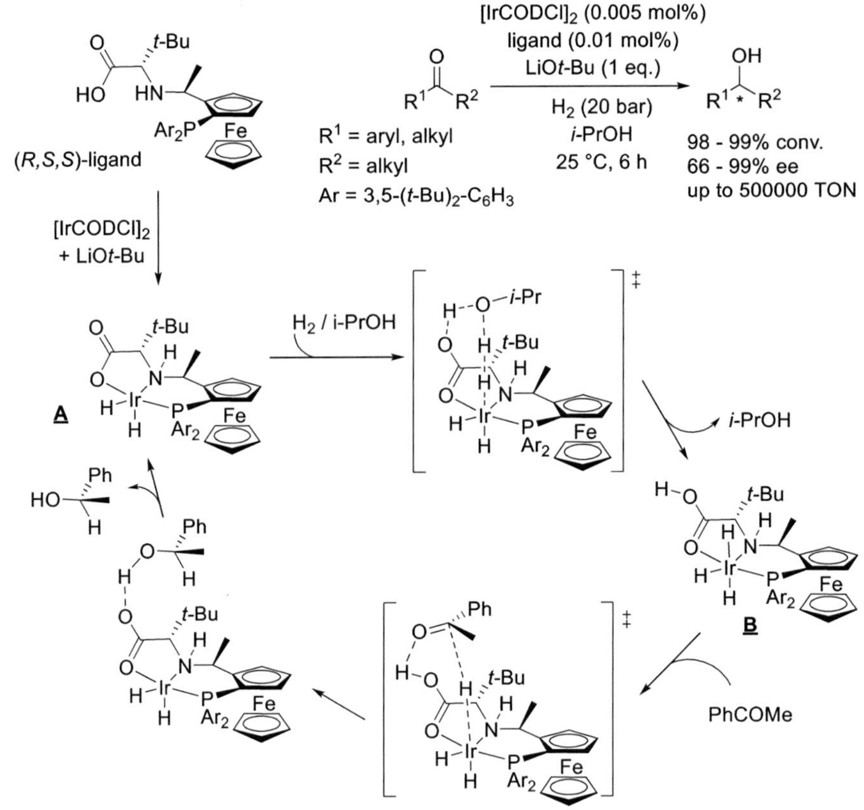

Scheme 75. AH of ketones using an iridium catalyst based on an amino-phosphine acid ligand (f-Ampha) by Zhang, Dong, Chung, and co-workers.[237]

iridium(III) dihydride complex **A**, in which the ligand behaves as a tridentate PNO species as the catalyst resting state (Scheme 75). This complex results from the reaction of the chiral ligand with [Ir(COD)Cl]$_2$ in the presence of a base, followed by oxidative addition of dihydrogen and cyclooctadiene decoordination. The iridium(III) dihydride complex **A** reacts first with dihydrogen which undergoes heterolytic splitting, assisted by both the 2-propanol used as the solvent and the carboxylate moiety of the ligand. The resulting iridium(III) trihydride **B** is involved then in the hydride transfer to the carbonyl group of the substrate. In this key, rate and enantiodetermining step, the substrate is activated by hydrogen bonding to the carboxylic acid moiety of the ligand, which finally behaves as the proton source. Overall, the stepwise transfer of a hydride from the metal and a proton from the ligand takes place through an outer-sphere mechanism, while steric repulsions between the substrate and the catalyst favor the (*R*) enantiomer. The final release of the alcohol product generates the iridium(III) dihydride complex **A** as the catalyst resting species.

Zhang, Dong, and co-workers applied a chiral ferrocene-based bisphosphine–thiourea ligand of the ZhaoPhos series (see Scheme 56 and following) to the iridium-catalyzed AH of benzoxazinones and quinoxalinones (Scheme 76).[238] The reactions proceeded at low catalyst loadings and afforded the reduced products in high yields and enantioselectivities, by combining the *N*-methylated ZhaoPhos ligand with [Ir(COD)Cl]$_2$, dihydrogen, and HCl. By considering the strong effect of the acid on the catalytic activity and stereoselectivity, the authors suggested that, after complexation, the imine function of the substrate will be protonated by

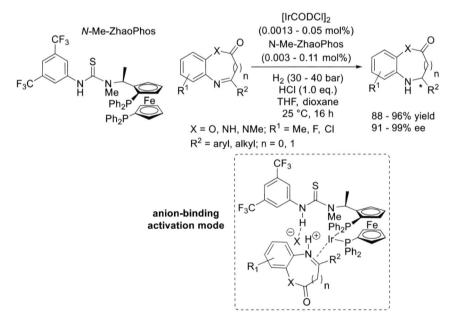

Scheme 76. AH of benzoxazinones and analogues using the *N*-Me-ZhaoPhos–iridium catalyst, by Zhang, Dong, and co-workers.[238]

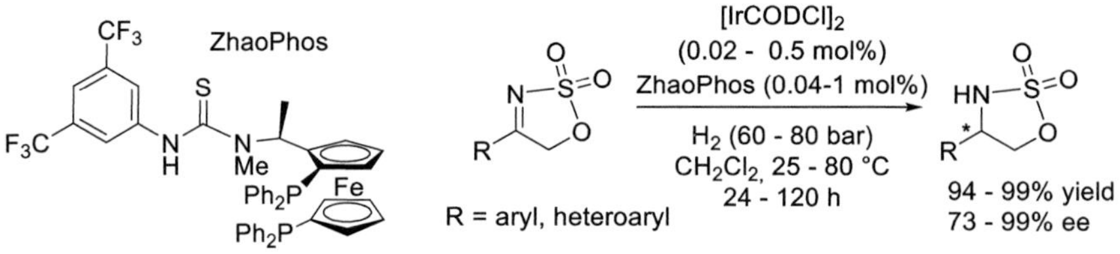

Scheme 77. AH of cyclic sulfamidate imines using N-Me-ZhaoPhos–iridium catalysts, by Zhang, Dong, and co-workers.[239]

HCl to give the corresponding iminium chloride and that a hydrogen–halide bond will be established between the chloride and the thiourea moiety of the ligand. No nonlinear effects were observed along these AH. This method was successfully applied in the synthesis of a modulator of the IgE/IgG receptor.

Zhang, Dong, and co-workers also reported the use of the methylated ZhaoPhos ligand in the iridium-catalyzed AH of cyclic sulfamidate imines (Scheme 77).[239] Compared to some of the privileged ligands like (S)-BINAP, (S)-SegPhos, and (R,S)-DuanPhos, the bisphosphine–thiourea produced the cyclic sulfamidates in higher yields and enantioselectivities at a gram scale, with rather low catalyst loadings.

In a recent work, Zhang, Wen, Lin, and co-workers applied the ZhaoPhos–iridium catalyst in the AH of isochroman acetals (Scheme 78).[240] The reactions were carried out in the presence of one equivalent of HCl and the corresponding isochromans were recovered in high yields, chemo- and enantioselectivities. In the presence of HCl, isochroman acetals are known to be in equilibrium with the corresponding α-chloroisochromans, from which chloride dissociation will generate reactive oxocarbeniums. Such oxocarbenium ions are postulated to be key intermediates in the iridium-promoted hydrogenations.

The reaction mechanism was investigated through experiments and calculations. While kinetic studies suggested a first-order dependence on the substrate and a half-order dependence on the catalyst, a positive nonlinear effect was observed and therefore dimerization of the dichloro precatalyst was suggested. According to calculations, the starting iridium(I) complex reacts with HCl through oxidative addition to generate a dichloro-hydrido-iridium(III) complex (Scheme 78). A subsequent isomerization

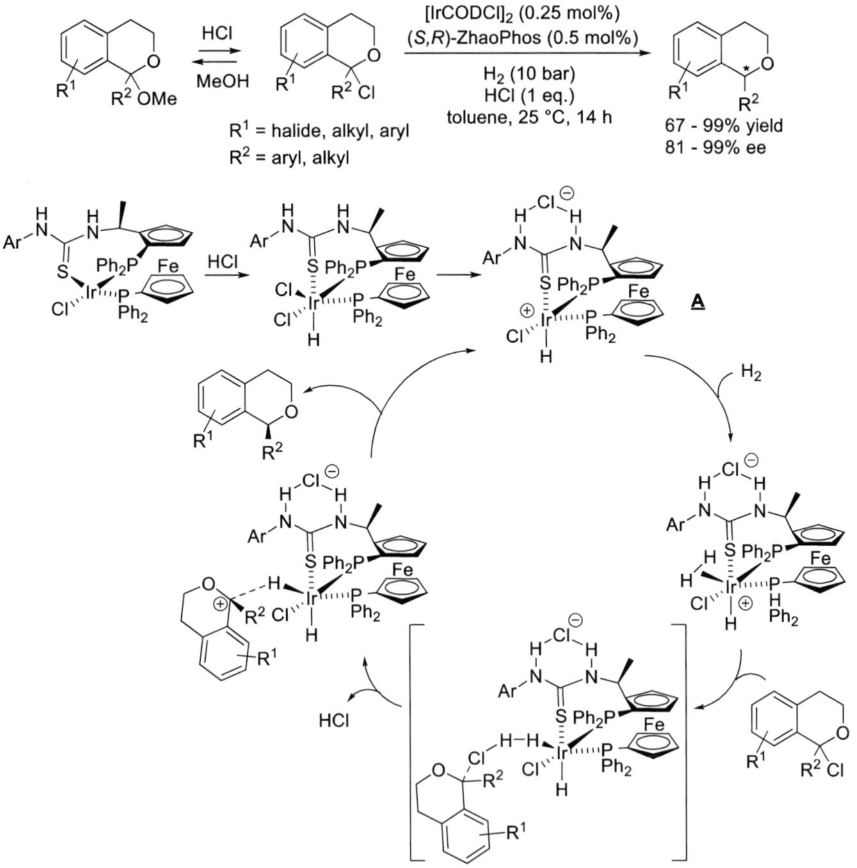

Scheme 78. AH of isochroman acetals using a ZhaoPhos–iridium catalyst, by Zhang, Wen, Lin, and co-workers.[240]

generates the cationic iridium(III) hydride **A**, where the thiourea fragment of the ligand binds the chloride counterion. This species is considered to be the catalyst resting state. After coordination of H_2, the heterolytic split of hydrogen proceeds concomitantly with the cleavage of the C–Cl bond of the substrate, resulting in the release of HCl and formation of the reactive oxocarbenium ion (ionic hydrogenation pathway). The carbenium binds to the ZhaoPhos–iridium dihydride complex through both ionic and Van der Waals interactions. In the enantiodetermining step, it undergoes hydride transfer to form the preferred (*S*)-configured isochroman product.

3. Iridium phosphates and triflylphosphoramides as bifunctional catalysts

The last examples hereafter deal with another class of bifunctional catalysts, that is, cationic amido–iridium complexes whose counteranion participates in the catalytic process via noncovalent interactions. In this field, initial disclosures from Rueping and Koenigs have been extended and enlightened later by Xiao and co-workers.

Rueping and Koenigs reported that N-sufonylated DPEN–iridium(III) complexes combined with chiral (R)-configured N-triflylphosphoramide Brønsted acids, provide active and effective catalysts for the AH of quinolines (Scheme 79).[241] Both racemic and optically pure DPEN–Ir precatalysts gave highly enantioenriched tetrahydroquinolines; however, the (R,R)-DPEN-iridium complex afforded the hydrogenation products with higher enantioselectivities with respect to the racemic complex. These results suggested the implication of chiral metal-phosphoramidite ion pairs and demonstrated the positive matching effect between the (R,R)-configured complex and its R-configured counterion in these hydrogenation reactions.

An analogous behavior has been highlighted in the iridium-promoted hydrogenation of imines in the presence of BINOL-derived phosphoric acids. Xiao and co-workers have demonstrated the combined use of (R)-TRIP phosphoric acid and an achiral pentamethylcyclopentadienyl iridium complex with a sulfonylated 1,2-ethylenediamine ligand in the AH of ketimines (Scheme 80).[242,243] High yields and enantioselectivities were obtained despite the rather long reaction times due to the steric hindrance imposed by the chiral phosphoric acid. Combined NMR and theo-

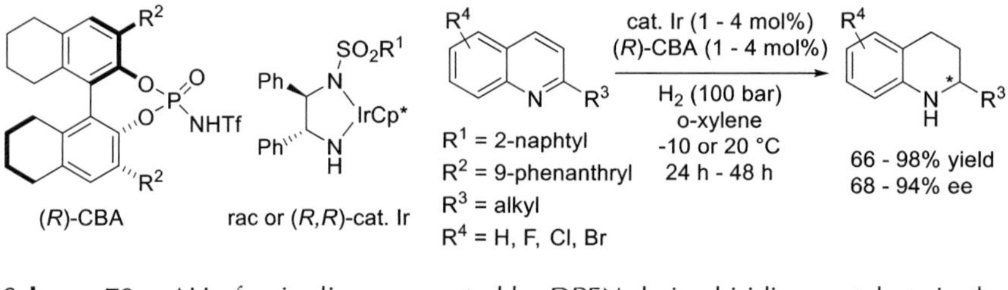

Scheme 79. AH of quinolines promoted by DPEN-derived iridium catalysts in the presence of N-triflylphosphoramide Brønsted acids, by Rueping and co-workers.[241]

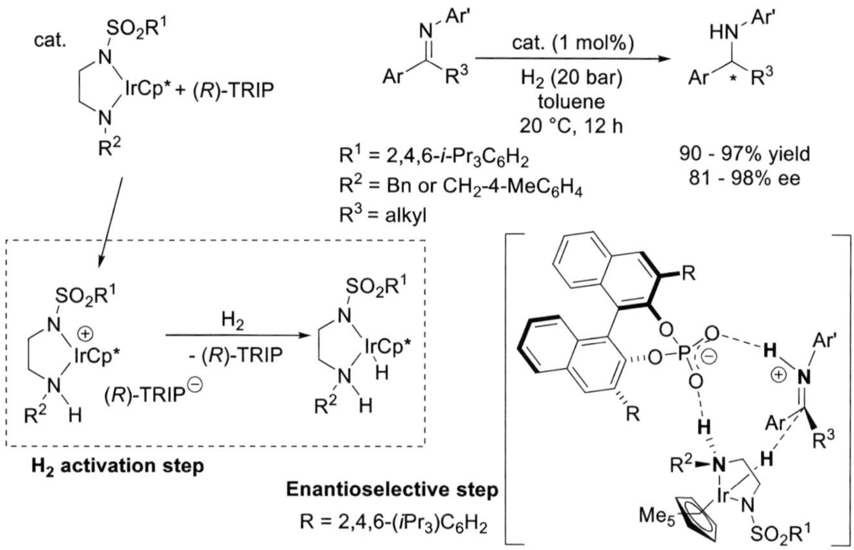

Scheme 80. AH of imines using achiral iridium–amido catalysts combined with a chiral phosphoric acid, by Xiao and co-workers.[242,243]

retical studies have shown that the Ir complex reacts with the phosphoric acid to give several isomeric cationic iridium hydrides, all bearing secondary amine ligands and phosphate counterions. In the catalytic cycle, the phosphate counterion participates in the heterolytic split of H_2 that generates an Ir-hydride intermediate. The resulting phosphoric acid acts as a proton relay for the protonation of the imine substrate. The enantioselective hydride transfer step is likely to proceed via a supramolecular ternary complex involving the iridium hydride complex, the chiral phosphate anion, and the protonated imine. Hydrogen bonding and CH/π interactions may contribute to the stereochemical control in this key step.

Afterward, Zhao and co-workers developed a highly selective ATH of ketimines using a similar catalytic system comprising a *meso* diol as hydrogen donor and a combination of the chiral TRIP phosphoric acid and an iridium(III) complex of a chiral sulfonylated diphenylethylendiamine ligand.[244] The nature of the alcohol hydrogen donor had a strong influence on the efficiency and selectivity of the hydrogenation, *meso* diols leading to good to high yields and enantioselectivities. Experimental and theoretical studies supported a reaction pathway involving an iridium alkoxide as the reducing species and proceeding through a concerted proton and hydride transfer. The enantioselectivity was apparently controlled by both the chiral iridium(III) complex and the chiral Brønsted acid, noncovalent interactions being observed between the acid, the *meso* diol and the substrate.

IV. Conclusion

We have seen that the applications of bifunctional catalysts to AHs are still blooming. New, highly efficient bifunctional catalysts implying ruthenium, rhodium, and iridium have been disclosed during the last decade, thanks to innovative ligand design. Beyond the classical bifunctional catalysts that rely on H-transfer from the ligand itself (Noyori approach), the development of cooperative approaches allowed noncovalent interactions to be established between the substrates and the ligands or counterions of specifically designed metal complexes. These non-covalent interactions operate as both activating tools (increased catalytic activity) and stereocontrol devices (increased enantioselectivity levels). Major advances in this field have been achieved thanks to a better knowledge of the catalysts features and behavior, by combining the development of synthetic methods with structural, kinetic, and theoretical studies.

Nowadays, AH and ATH are catalyzed by well-known series of privileged organometallic species which have complementary activities and stereo-, chemo-, and regioselectivities. Though these catalysts hydrogenate with high selectivities a great diversity of organic substrates with carbonyl, imine or olefin functions,[46] other substrates like alkyl ketones and imines,[245] carbocyclic arenes,[246,247] some nitrogen-based heterocycles[248] and tri-/tetrasubstituted alkenes[249,250] are still challenging and the development of chemoselective catalytic processes is also highly desirable.[251]

While new effective hydrogenation catalysts based on precious metals are still eagerly sought,[46] there's a growing interest in hydrogenation catalysts based on abundant metals, due to economic, environmental, and societal issues.[251–253] Several examples of bifunctional catalysts have already been reported in this field and others will surely appear in the future. Finally, the development of catalysts giving higher turnover numbers, as well as applications in flow chemistry,[124,254–256] shall certainly be privileged when considering industrial perspectives.[257,258]

V. Acknowledgments

The CNRS, the Chevreul Institute (FR 2638), the Ministère de l'Enseignement Supérieur et de la Recherche, the Région Hauts-de-France, and the FEDER program are acknowledged for supporting and funding partially this work. Sanofi, Oril Industrie, and Adisseo France SAS are acknowledged for past collaborations on AH catalysis. Dr. Francine Agbossou-Niedercorn and Dr. Christophe Michon thank the editorial team for the invitation to contribute to this book.

VI. References

1. Noyori, R. In *Asymmetric Catalysis in Organic Synthesis*. Wiley: New York, **1994**, Chapter 2.
2. Ohkuma, T.; Kitamura, M.; Noyori, R. In *Catalytic Asymmetric Synthesis 2nd Edition*; Ojima, I., Ed.; Wiley-VCH, New York, **2000**, p 1.
3. Blaser, H.-U.; Federsel, H.-J. Eds., *Asymmetric- Catalysis on Industrial Scale*. Wiley-VCH, **2010**.
4. Knowles, W. S. *Angew. Chem. Int. Ed.* **2002**, *41*, 1998–2007.
5. Noyori, R. *Angew. Chem. Int. Ed.* **2002**, 41, 2008–2022.
6. Noyori, R. *Adv. Synth. Catal.* **2003**, *345*, 15–32.
7. Blaser, H. U.; Pugin, B.; Spindler, F.; Saudan L. A.; Hydrogenation. *In Applied Homogeneous Catalysis with Organometallic Compounds 3rd Edition*, Cornils, B.; Herrmann, W. A.; Beller, M.; Paciello, R., Eds; Wiley-VCH, Weinheim, **2018**, 2, p 621–690.
8. Foubelo, F.; Najera, C.; Yus, M. *Tetrahedron: Asymmetry* **2015**, *26*, 769–790.
9. Nedden, H. G.; Zanotti-Gerosa, A.; Wills, M. *Chem. Record* **2016**, *16*, 2619–2639.
10. Stefane, B.; Pozgan, F. *Topics in Current Chem.* **2016**, *374*, 1–67.
11. Wang, D.; Astruc, D. *Chem. Rev.* **2015**, *115*, 6621–6686.
12. Matsunami, A.; Kayaki, Y. *Tetrahedron Lett.* **2018**, *59*, 504–513.
13. Baráth, E. *Catalysts* **2018**, *8*, 671–696.
14. Etayo, P.; Vidal-Ferran, A. *Chem. Soc. Rev.* **2013**, *42*, 728–754.
15. Ohkuma, T.; Noyori, R.; *Angew. Chem. Int. Ed.* **2001**, *40*, 40–73.
16. Ohkuma, T.; Ooka, H.; Hashiguchi, S.; Ikariya, T.; Noyori, R. *J. Am. Chem. Soc.* **1995**, *117*, 2675–2676.

17. Noyori, R.; Hashiguchi, S. *Acc. Chem. Res.* **1997**, *30*, 97–102.
18. Hashiguchi, S.; Fujii, A.; Takehara, J.; Ikariya, T.; Noyori, R. *J. Am. Chem. Soc.* **1995**, *117*, 7562–7563
19. Doucet, H.; Ohkuma, T.; Murata, K.; Yokozawa, T.; Kozawa, M.; Katayama, E.; England, A. F.; Ikariya, T.; Noyori, R. *Angew. Chem. Int. Ed.* **1998**, *37*, 1703–1707.
20. Ohkuma, T.; Koizumi, M.; Muñiz, K.; Hilt, G.; Kabuto, C. *J. Am. Chem. Soc.* **2002**, *124*, 6508–6509.
21. Sandoval, C. A.; Ohkuma, T.; Muñiz, K.; Noyori, R. *J. Am. Chem. Soc.* **2003**, *125*, 13490–13503.
22. Haack, K.-J.; Hashiguchi, S.; Fujii, A.; Ikariya, T.; Noyori, R. *Angew. Chem. Int. Ed.* **1997**, *36*, 285–288.
23. Palmer, M. J.; Wills, M. *Tetrahedron: Asymmetry* **1999**, *10*, 2045–2061.
24. Clamham, S.; Hadzovic, A.; Morris, R. H. *Coord. Chem. Rev.* **2004**, *248*, 2201–2237.
25. Casey, C. P.; Johnson, J. B. *J. Org. Chem.* **2003**, *68*, 1998–2001.
26. Alonso, D. A.; Brandt, P.; Nordin, S. J. M.; Andersson, P. G. *J. Am. Chem. Soc.* **1999**, *121*, 41, 9580–9588.
27. Yamakawa, M.; Ito, H.; Noyori, R. *J. Am. Chem. Soc.* **2000**, *122*, 1466–1478.
28. Joseph, S. M.; Samec, J. S.; Bäckvall, J.-E.; Andersson, P. G. *Chem. Soc. Rev.* **2006**, *35*, 237–248.
29. Ikariya, T.; Murata, K.; Noyori, R. *Org. Biomol. Chem.* **2006**, *4*, 393–406.
30. Ikariya, T.; Blacker, A. J. *Acc. Chem. Res.* **2007**, *40*, 1300–1308.
31. Sawamura, M.; Ito, Y. *Chem. Rev.* **1992**, *92*, 857–871.
32. Steinhagen, H.; Helmchen, G. *Angew. Chem. Int. Ed.* **1996**, *35*, 2339–2342.
33. Shibasaki, M.; Sasai, H.; Arai, T. *Angew. Chem. Int. Ed.* **1997**, *36*, 1236–1256.
34. Van den Beuken, E. K.; Feringa, B. L. *Tetrahedron* **1998**, *54*, 12985–13011.
35. Rowlands, G. J. *Tetrahedron* **2001**, *57*, 1865–1882.
36. Gröger, H. *Chem.–Eur. J.* **2001**, *7*, 5246–5251.
37. Shibasaki, M.; Yoshikawa, N. *Chem. Rev.* **2002**, *102*, 2187–2209.
38. Ma, J.-A.; Cahard, D. *Angew. Chem. Int. Ed.* **2004**, *43*, 4566–4583.
39. Hannedouche, J.; Clarkson, G. J.; Wills, M. *J. Am. Chem. Soc.* **2004**, *126*, 986–987.
40. Hayes, A. M.; Morris, D. J.; Clarkson, G. J.; Wills, M. *J. Am. Chem. Soc.* **2005**, *127*, 7318–7319.
41. Morris, D. J.; Hayes, A. M.; Wills, M. *J. Org. Chem.* **2006**, *71*, 7035–7044.
42. Matharu, D. S.; Morris, D. J.; Kawamoto, A. M.; Clarkson, G. J.; Wills, M. *Org. Lett.* **2005**, *7*, 5489–5491.
43. Gao, J.-X.; Ikariya, T.; Noyori, R. *Organometallics* **1996**, *15*, 1087–1089.
44. Guo, R.; Elpelt, C.; Chen, X.; Song, D.; Morris, R. H. *Chem. Commun.* **2005**, 3050–3052.
45. Guo, R.; Chen, X.; Elpelt, C.; Song, D.; Morris, R. H. *Org. Lett.* **2005**, *7*, 1757–1759.
46. Seo, C. S. G.; Morris, R. H. *Organometallics* **2019**, *38*, 47–65.

47. Ikariya, T.; Kayaki, Y. *Pure and Applied Chem.* **2014**, *86*, 933–943.
48. Dong, X.-Q.; Zhao, Q.; Li, P.; Chen, C.; Zhang, X. *Org. Chem. Front.* **2015**, *2*, 1425–1431.
49. Gridnev, I. V. *ChemCatChem* **2016**, *8*, 3463–3465.
50. Zhang, J.; Jia, J.; Zeng, X.; Wang, Y.; Zhang, Z.; Gridnev, I. D.; Zhang, W. *Angew. Chem. Int. Ed.* **2019**, *58*, 11505–11512.
51. Chen, J.; Gridnev, I. D. *iScience* **2020**, 23, 100960–100972.
52. Kuwata, S.; Ikariya, T. *Dalton Trans.* **2010**, *39*, 2984–2992.
53. Grotjahn, D. B. *Pure Applied Chem.* **2010**, *82*, 635–647.
54. Mbofana, C. T.; Miller, S. J. Bifunctional Catalysis with Lewis Base and X-H Sites That Facilitate Proton Transfer or Hydrogen Bonding (n→π*) in *Lewis Base Catalysis in Organic Synthesis*; Vedejs, E.; Denmark, S. E. Ed.; Wiley-VCH, Weinheim, **2016**, pp. 1259–1288.
55. Barath, E. *Synthesis* **2020**, *52*, 504–520.
56. Cheng, T.; Zhao, Q.; Zhang, D.; Liu, G. *Green Chem.* **2015**, *17*, 2100–2122.
57. Wang, Q. *Chem. Rev.* **2020**, *120*, 1438–1511.
58. Altava, B.; Burguete, M. I.; García-Verdugo, E.; Luis, S. V. *Chem. Soc. Rev.* **2018**, *47*, 2722–2771.
59. Baig, R. B. N.; Nadagouda, M. N.; Varma, R. S. *Coord. Chem. Rev.* **2015**, *287*, 137–156.
60. Wei, Y.; Wu, X.; Wang, C.; Xiao, J. *Catal. Today* **2015**, *247*, 104–116.
61. Kucukturkmen, C.; Agac, A.; Eren, A.; Karakaya, I.; Aslantas, M.; Celik, O.; Ulukanli, S.; Karabuga, S. *Catal. Commun.* **2016**, *74*, 122–125.
62. Chen, F.; He, D.; Chen, L.; Chang, X.; Wang, D. Z.; Xu, C.; Xing, X. *ACS Catal.* **2019**, *9*, 5562–5566.
63. Takehara, J.; Hashiguchi, S.; Fujii, A.; Inoue, S.; Ikariya, T.; Noyori, R. *J. Chem. Soc. Chem. Commun.* **1996**, 233–234.
64. Alonso, D. A.; Guijarro, D.; Pinho, P.; Temme, O.; Andersson, P. G. *J. Org. Chem.* **1998**, *63*, 2749–2751.
65. Palmer, M.; Walsgrove, T.; Wills, M. *J. Org. Chem.* **1997**, *62*, 5226–5228.
66. Chakka, S. K.; Andersson, P. G.; Maguire, G. E. M.; Kruger, H. G.; Govender, T. *Eur. J. Org. Chem.* **2010**, 972–980.
67. Han, M.-L.; Hu, X.-P.; Huang, J.-D.; Chen, L.-G.; Zheng, Z. *Tetrahedron: Asymmetry* **2011**, *22*, 222–225.
68. Wang, X.; Xu, L.; Yan, L.; Wang, H.; Han, S.; Wu, Y.; Chen, F. *Tetrahedron* **2016**, *72*, 1787–1793.
69. Manville, C. V.; Docherty, G.; Padda, R.; Wills, M. *Eur. J. Org. Chem.* **2011**, 6893–9901.
70. Cambeiro, X. C.; Pericàs, M. A. *Adv. Synth. Catal.* **2011**, *353*, 113–124.
71. Pastor, I. M.; Västilä, P.; Adolfsson, H. *Chem. Commun.* **2002**, *39*, 2046–2047.
72. Pastor, I. M.; Västilä, P.; Adolfsson, H. *Chem. Eur. J.* **2003**, *9*, 4031–4045.

73. Lundberg, H.; Hans Adolfsson, H. *Tetrahedron Lett.* **2011**, *52*, 2754–2758.
74. Shatskiy, A.; Kivijärvi, T.; Lundberg, H.; Tinnis, F.; Adolfsson, H. *ChemCatChem* **2015**, *7,* 3818–3821.
75. Västilä, P.; Zaitsev, A. B.; Wettergren, J.; Privalov, T.; Adolfsson, H. *Chem. Eur. J.* **2006**, *12*, 3218–3225.
76. Kovalenko, O. O.; Lundberg, H.; Hübner, D.; Adolfsson, H. *Eur. J. Org. Chem.* **2014**, 6639–6642.
77. Slagbrand, T.; Lundberg, H.; Adolfsson, H. *Chem. Eur. J.* **2014**, *20*, 16102–16106.
78. Buitrago, E.; Lundberg, H.; Andersson, H.; Ryberg, P.; Adolfsson, H. *ChemCatChem* **2012**, *4*, 2082–2089.
79. Coll, M.; Pàmies, O.; Adolfsson, H.; Diéguez, M. *Chem. Commun.* **2011**, *47*, 12188–12190.
80. Coll, M.; Ahlford, K.; Pàmies, O. Adolfsson, H.; Diéguez, M. *Adv. Synth. Catal.* **2012**, *354*, 415–427.
81. Boege, M.; Heck, J. *J. Mol. Catal. A: Chem* **2015**, *408*, 107–122.
82. Soni, R.; Cheung, F. K.; Clarkson, G. C.; Martins, J. E. D.; Graham, M. A.; Wills, M. *Org. Biomol. Chem.* **2011**, *9*, 3290–3294.
83. Barrios-Rivera, J.; Xu, Y.; Wills, M. *Org. Biomol. Chem.* **2019**, *17*, 1301–1321.
84. Martins, J. E. D.; Clarkson, G. J.; Wills, M. *Org. Lett.* **2009**, *11*, 847–850.
85. Martins, J. E. D.; Contreras Redondo, M. A.; Wills, M. *Tetrahedron: Asymmetry* **2010**, *21*, 2258–2264.
86. Soni, R.; Hall, T. H.; Morris, D. J.; Clarkson, G. J.; Owen, M. R.; Wills, M. *Tetrahedron Lett.* **2015**, *56*, 6397–6401.
87. Ma, W.; Zhang, J.; Xu, C.; Chen, F.; He, Y.-M.; Fan, Q.-H. *Angew. Chem. Int. Ed.* **2016**, *55*, 12891–12894.
88. Shan, W.; Meng, F.; Wu, Y.; Mao, F.; Li, X. J. *J. Organomet. Chem.* **2011**, *696*, 1687–1690.
89. Zimbron, J. M.; Dauphinais, M.; Charette, A. B. *Green Chem.* **2015**, *17*, 3255–3259.
90. Wang, R.; Wan, J.; Ma, X.; Xu X.; Liu L. *Dalton Trans.* **2013**, *42*, 6513–6522.
91. Johnson, T. C.; Totty, W. G.; Wills, M. *Org. Lett.* **2012**, *14*, 5230–5233.
92. Darwish, M. O.; Wallace, A.; Clarkson, G. J.; Wills, M. *Tetrahedron Lett.* **2013**, *54*, 4250–4253.
93. Barrios-Rivera, J.; Xu, Y.; Wills, M. *Org. Lett.* **2019**, *21*, 7223–7227.
94. Dub, P. A.; Matsunami, A.; Kuwata, S.; Kayaki, Y. *J. Am. Chem. Soc.* **2019**, *141*, 2661–2677.
95. Rast, S.; Modec, B.; Stephan, M.; Mohar, B. *Org. Biomol. Chem.* **2016**, *14*, 2112–2120.
96. Ye, W.; Zhao, M.; Yu, Z. *Chem. Eur. J.* **2012**, *18*, 10843–10846.
97. Nedden, H. G.; Zanotti-Gerosa, A.; Wills, M. *Chem. Rec.* **2016**, *16*, 2623–2643.
98. Cheung, F. K.; Lin, C.; Minissi, F.; Crivillé, A. L.; Graham, M. A.; Fox, D. J.; Wills; W. *Org. Lett.* **2007**, *9*, 4659–4662.

99. Martins, J. E. D.; Morris, D. J.; Tripathi, B.; Wills, M. *J. Organomet. Chem.* **2008**, *693*, 3527–3532.
100. Cheung, F. K.; Clarke, A. J.; Clarkson, G. J.; Fox, D. J.; Graham, M. A.; Lin, C.; Criville, A. L.; Wills, M. *Dalton trans.* **2010**, *39*, 1395–1402.
101. Parekh, V.; Ramsden, J. A.; Wills, M. *Tetrahedron Asym.* **2010**, *21*, 1549–1556.
102. Jolley, K. E.; Zanotti-Gerosa, A.; Hancock, F.; Dyke, A.; Grainger, D. M.; Medlock, J. A.; Nedden, H. G.; Le Paih, J. J. M.; Roseblade, S. J.; Seger, A.; Sivakumar, V.; Prokes, I.; Morris, D. J.; Wills, M. *Adv. Synth. Catal.* **2012**, *354*, 2545–2555.
103. Soni, R.; Jolley, K. E.; Clarkson, G. J.; Wills, M. *Org. Lett.* **2013**, *15*, 5110–5113.
104. Soni, R.; Collinson, J.-M.; Clarkson, G. C.; Wills, M. *Org. Lett.* **2011**, *13*, 4304–4307.
105. Fang, Z.; Wills, M. *J. Org. Chem.* **2013**, *78*, 8594–8605.
106. Fang Z, Wills M. *Org Lett.* **2014**, *16,* 374–377.
107. Monnereau, L; Cartigny, D.; Scalone, M.; Ayad, T.; Ratovelomanana-Vidal, V. *Chem. Eur. J.* **2015**, *21*, 11799–11806.
108. Vyas, V. K.; Knighton, R. C.; Bhanage, B. M.; Wills, M. *Org. Lett.* **2018**, *20*, 975–978.
109. Forshaw, S.; Matthews, A. J.; Brown, T. J.; Diorazio, L. J.; Williams, L. Wills, M. *Org. Lett.* **2017**, *19*, 2789–2792.
110. Perryman, M. S.; Harris, M. E.; Foster, J. L.; Joshi, A.; Clarkson, G. J.; Fox, D. J. *Chem. Commun.* **2013**, *49*, 10022–10024.
111. Hodgkinson, R.; Jurčík, V.; Zanotti-Gerosa, A.; Nedden, H. G.; Blackaby, A.; Clarkson, G. J.; Wills, M. *Organometallics* **2014**, *33*, 5517–5524.
112. Vyas, V. K.; Bhanage, B. M. *Org. Lett.* **2016**, *18*, 6436–6439.
113. Mangion, I. K.; Chen, C.-y.; Li, H.; Maligres, P.; Chen, Y.; Christensen, M.; Cohen, R.; Jeon, I.; Klapars, A.; Krska, S.; Nguyen, H.; Reamer, R. A.; Sherry, B. D.; Zavialov, I. *Org. Lett.* **2014**, *16*, 2310–2313.
114. Su, Y.; Tu, Y.-Q.; Gu, P. *Org. Lett.* **2014**, *16*, 4204–4207.
115. Ji, Y.; Xue, P.; Ma, D. D.; Li, X. Q.; Gu, P.; Li, R. *Tetrahedron Lett.* **2015**, *56*, 192–194.
116. Soni, R.; Jolley, K. E.; Gosiewska, S.; Clarkson, G. J.; Fang, Z.; Hall, T. H.; Treloar, B. N.; Knighton, R. C.; Wills, M. *Organometallics* **2018**, *37*, 48–64.
117. Knighton, R. C.; Vyas, V. K.; Mailey, L. H.; Bhanage, B. M.; Wills, M. *J. Organomet. Chem.* **2018**, *875*, 72–79.
118. Touge, T.; Hakamata, T.; Nara, H.; Kobayashi, T.; Sayo, N.; Saito, T.; Kayaki, Y.; Ikariya, T. *J. Am. Chem. Soc.* **2011**, *133*, 14960–14963.
119. Parekh, V.; Ramsden, J. A.; Wills, M. *Catal. Sci. Technol.* **2012**, *2*, 406–414.
120. Touge, T.; Nara, H.; Fujiwhara, M.; Kayaki, Y.; Ikariya, T. *J. Am. Chem. Soc.* **2016**, *138*, 10084–10087.
121. Matsunami, A.; Ikeda, M.; Nakamura, H.; Yoshida, M.; Kuwata, S. Kayaki, Y. *Org. Lett.* **2018**, *20*, 5213–5218.

122. Wang, B.; Zhou, H.; Lu, G.; Liu, Q.; Jiang, X. *Org. Lett.* **2017**, *19*, 2094–2097.
123. Yuki, Y.; Touge, T.; Nara, H.; Matsumura, K.; Fujiwhara, M.; Kayaki, Y.; Ikariya, T. *Adv. Synth. Catal.* **2018**, *360*, 568–574.
124. Touge, T.; Kuwana, M.; Komatsuki, K.; Tanaka, S.; Matsumu, K.; Sayo, N.; Kashibuchi, Y.; Saito, T. *Org. Process Res. Dev.* **2019**, *23*, 452–461.
125. Touge, T.; Sakaguchi, K.; Tamaki, N.; Hideki Nara, Yokozawa, T.; Matsumura, K.; Kayaki, Y. *J. Am. Chem. Soc.* **2019**, *141*, 16354–16361.
126. Komiyama, M.; Itoh, T.; Takeyasu, T. *Org. Process Res. Dev.* **2015**, *19*, 315–319.
127. Chung, J. Y. L.; Scott, J. P.; Andersson, C.; Bishop, B.; Bremeyer, N.; Cao, Y.; Chen, Q.; Dunn, R.; Kassim, A.; Lieberman, D.; Moment, A. J.; Sheen, F. Zacuto, M. *Org. Process Res. Dev.* **2015**, *19*, 1760–1768.
128. Sun, G.; Zhou, Z.; Luo, Z.; Wang, H.; Chen, L.; Xu, Y.; Li, S.; Jian, W.; Zeng, J.; Hu, B.; Han, X.; Lin, Y.; Wang, Z. *Org. Lett.* **2017**, *19*, 4339–4342.
129. Kisic, A.; Stephan, M.; Mohar, B. *Org. Lett.* **2013**, *15*, 1614–1617.
130. Kisic, A.; Stephan, M.; Mohar, B. *Adv. Synth. Catal.* **2014**, *356*, 3193–3198.
131. Kisic, A.; Stephan, M.; Mohar, B. *Adv. Synth. Catal.* **2015**, 357, 2540–2546.
132. Cotman, A. E.; Cahard, D.; Mohar, B. *Angew. Chem. Int. Ed.* **2016**, *55*, 5294–5298.
133. Jeran, M.; Cotman, A. E.; Stephan, M.; Mohar, B. *Org. Lett.* **2017**, *19*, 2042–2045.
134. Cotman, A. E.; Modec, B.; Mohar, B. *Org. Lett.* **2018**, *20*, 2921–2924.
135. Cotman, A. E.; Lozinsek, M.; Wang, B.; Stephan, M. Mohar, B. *Org. Lett.* **2019**, *21*, 3644–3648.
136. Shvo, Y.; Czarkie, D.; Rahamim, Y. *J. Am. Chem. Soc.* **1986**, *108*, 7400–7402.
137. Conley, B. L.; Pennington-Boggio, M. K. Boz, E.; Williams, T. J. *Chem. Rev.* **2010**, *110*, 2294–2312.
138. Warner, M. C.; Casey, C. P.; Bäckvall, J.-E. *Top. Organomet. Chem.* **2011**, *37*, 85–125.
139. Verho, O.; Bäckvall, J.-E. *J. Am. Chem. Soc.* **2015**, *137*, 3996–4009.
140. Hopewell, J. P.; Martins, J. E. D.; Johnson, T. C.; Godfrey, J.; Wills, M. *Org. Biomol. Chem.* **2012**, *10*, 134–145.
141. Dou, X.; Hayashi, T. *Adv. Synth. Catal.* **2016**, *358*, 1054–1058.
142. Nova, A.; Taylor, D. J.; Blacker, A. J.; Duckett, S. B.; Perutz, R. N.; Eisenstein, O. *Organometallics* **2014**, *33*, 3433–3442.
143. Matharu, D. S.; Morris, D.J.; Clarkson, G. J.; Wills, M.; *Chem Commun.* **2006**, *52*, 3232–3234.
144. Matharu, D. S.; Martins, J. E. D.; Wills, M.; *Chem. Asian J.* **2008**, *3*, 1374–1383.
145. Echeverria, P.-G.; Ferard, C.; Phansavath, P.; Ratovelomanana-Vidal, V. *Catal. Commun.* **2015**, *62*, 95–99.
146. Zheng, L.-S.; Llopis, Q.; Echeverria, P.-G.; Ferard, C.; Guillamot, G.; Phansavath, P.; Ratovelomanana-Vidal, V. *J. Org. Chem.* **2017**, *82*, 5607–5615.
147. Llopis, Q.; Ferard, C.; Guillamot, G.; Phansavath, P.; Ratovelomanana-Vidal, V. *Synthesis* **2016**, *48*, 3357–3363.
148. He, B.; Zheng, L.-S.; Phansavath, P.; Ratovelomanana-Vidal, V. *ChemSusChem* **2019**, *12*, 3032–3036.

149. He, B.; Phansavath, P.; Ratovelomanana-Vidal, V. *Org. Lett.* **2019**, *21*, 3276–3280.
150. Westermeyer, A.; Guillamot, G.; Phansavath, P.; Ratovelomanana-Vidal, V. *Org. Lett.* **2020**, *22*, 3911–3914.
151. He, B.; Phansavath, P.; Ratovelomanana-Vidal, V. *Org. Chem. Front.* **2020**, *7*, 975–979.
152. Wang, F.; Zheng, L.-S.; Lang, Q.-W.; Yin, C.; Wu, T.; Phansavath, P.; Chen, G.-Q.; Ratovelomanana-Vidal, V. *Chem. Commun.* **2020**, *56*, 3119–3122.
153. Ahlford, K. *Catal. Commun.* **2011**, *12*, 1118–1121.
154. Ahlford, K.; Ekström, J.; Zaitsev, A. B.; Ryberg, P.; Eriksson, L.; Adolfsson, H. *Chem. Eur. J.* **2009**, *15*, 11197–11209.
155. Tinnis, F.; Adolfsson, H. *Org. Biomol. Chem.* **2010**, *8*, 4536–4539.
156. Nordin, M.; Liao, R.-Z.; Ahlford, K.; Adolfsson, H.; Himo, F. *ChemCatChem* **2012**, *48*, 1095–1104.
157. Coll, M.; Pàmies, O.; Adolfsson, H.; Diéguez, M. *ChemCatChem* **2013**, *5*, 3821–3828.
158. Margalef, J.; Slagbrand, T.; Tinnis, F.; Adolfsson, H.; Diéguez, M.; Pàmies, O. *Adv. Synth. Catal.* **2016**, *358*, 4006–4018.
159. Bartoszewicz, A.; Ahlsten, N.; Martín-Matute, B. *Chem. Eur. J.* **2013**, *19*, 7274–7302.
160. Vázquez-Villa, H.; Reber, S.; Ariger, M. A.; Carreira, E. M. *Angew. Chem. Int. Ed.* **2011**, *50*, 8979–8981.
161. Liu, W.-P.; Yuan, M.-L.; Yang, X.-H.; Li, K.; Xie, J.-H.; Zhou, Q.-L. *Chem. Commun.* **2015**, *51*, 6123–6125.
162. Zhang, Y.-M.; Yuan, M.-L.; Liu, W.-P.; Xie, J.-H.; Zhou, Q.-L. *Org. Lett.* **2018**, *20*, 4486–4489.
163. Sato, Y.; Kayaki Y.; Ikariya, T. *Chem. Asian J.* **2016**, *11*, 2924–2931.
164. Creus, M.; Pordea, A.; Rossel, T.; Sardo, A.; Letondor, C.; Ivanova, A.; LeTrong, I.; Stenkamp, R. E.; Ward, T. R. *Angew. Chem. Int. Ed.* **2008**, *47*, 1400–1404.
165. Dürrenberger, M.; Heinisch, T.; Wilson, Y. M.; Rossel, T.; Nogueira, E.; Knörr, L.; Mutschler, A.; Kersten, K.; Zimbron, Malcolm, M.; Pierron, J.; Schnirmer, T.; Ward, T. R. *Angew. Chem. Int. Ed.* **2011**, *50*, 3026–3029.
166. Letondor, C.; Pordea, A.; Humbert, N.; Ivanova, A.; Mazurek, S.; Novic, M.; Ward, T. R. *J. Am. Chem. Soc.* **2006**, *128*, 8320–8328.
167. Blackmond, D. G.; Ropic, M. *Org. Process Res. Dev.* **2006**, *10*, 457–463.
168. Åberg, J. B.; Samec, J. S. M.; Bäckvall, J.-E. *Chem. Commun.* **2006**, 2771–2773.
169. Chen, L.-A.; Xu, W.; Huang, B.; Ma, J.; Wang, L.; Xi, J.; Harms, K.; Gong, L.; Meggers, E. *J. Am. Chem. Soc.* **2013**, *135*, 10598–10601.
170. Tian, C.; Gong, L.; Meggers, E. *Chem. Commun.* **2016**, 4207–4210.
171. Carmona, D.; Lahoz, F. J.; García-Orduña, P.; Oro, L. A.; Lamata, M. P.; Viguri, F. *Organometallics* **2012**, *31*, 3333–3345.
172. Coverdale, J. P. C.; Sanchez-Cano, C.; Clarkson, G. J.; Soni, R.; Wills, M.; Sadler, P. J. *Chem. Eur. J.* **2015**, *21*, 8043–8046.
173. Wang, W.; Xinzheng Yang, X. *Chem. Commun,* **2019**, *55*, 9633–9636.

174. Yamamoto, Y.; Yamashita, K.; Nakamura, M. *Organometallics* **2010**, *29*, 1472–1478.
175. Del Grosso, A.; Clarkson, G. J.; Wills, M. *Inorg. Chim. Acta* **2019**, *496*, 119043–119049.
176. Yamamura, T.; Nakatsuka, H.; Tanaka, S.; Kitamura, M. *Angew. Chem. Int. Ed.* **2013**, *52*, 9313–9315.
177. Yamamura, T.; Nakane, S.; Nomura, Y.; Tanaka, S.; Kitamura, M. *Tetrahedron* **2016**, *72*, 3781–3789.
178. Nakane, S.; Yamamura, T. M.; Sudipta K.; Tanaka, S.; Kitamura, M. *ACS Catal.* **2018**, *8*, 11059–11075.
179. Nakatsuka H.; Yamamura T.; Shuto Y.; Tanaka S.; Kitamura M.; Yoshimura M. *J. Am. Chem. Soc.* **2015**, *137*, 8138–8149.
180. Ito, M.; Kobayashi, C.; Himizu, A.; Ikariya, T. *J. Am. Chem. Soc.* **2010**, *132*, 11414–11415.
181. Ito, M.; Ootsuka, T.; Watari, R.; Shiibashi, A.; Himizu, A.; Ikariya, T. *J. Am. Chem. Soc.* **2011**, *133*, 4240–4242.
182. Chen, F.; Wang, T.; He, Y.; Ding, Z.; Li, Z.; Xu, L.; Fan, Q.-H. *Chem. Eur. J.* **2011**, *17*, 1109–1113.
183. Chen, F.; Ding, Z.; Qin, J.; Wang, T.; He, Y.; Fan, Q.-H. *Org. Lett.* **2011**, *13*, 4348–4351.
184. Qin, J.; Chen, F.; Ding, Z.; He, Y.-M.; Xu, L.; Fan, Q.-H. *Org. Lett.* **2011**, *13*, 6568–6571.
185. Wang, T.; Zhuo, L.-G.; Li, Z.; Chen, F.; Ding, Z.; He, Yanmei; Fan, Q.-H.; Xiang, J.; Yu, Z.-X.; Chan, A. S. C. *J. Am. Chem. Soc.* **2011**, *133*, 9878–9891.
186. Ding, Z.-Y.; Chen, F.; Qin, J.; He, Y.-M.; Fan, Q.-H. *Angew. Chem. Int. Ed.* **2012**, *51*, 5706–5710.
187. Chen, F.; Ding, Z.; He, Y.; Qin, J.; Wang, T.; Fan, Q.-H. *Tetrahedron* **2012**, *68*, 5248–5257.
188. Yang, Z.; Chen, F.; He, Y.; Yang, N.; Fan, Q.-H. *Angew. Chem. Int. Ed.* **2016**, *55*, 13863–13866.
189. Li, W.; Wiesenfeldt, M. P.; Glorius, F. *J. Am. Chem. Soc.* **2017**, *139*, 2585–2588.
190. Li, W.; Wagener, T.; Hellmann, L.; Daniliuc, C. G.; Mück-Lichtenfeld, C.; Neugebauer, J.; Glorius, F. *J. Am. Chem. Soc.* **2020**, *142*, 7100–7107.
191. Breuil, P.-A. R.; Patureau, F. W.; Reek, J. N. H. *Angew. Chem., Int. Ed.* **2009**, *48*, 2162–2165.
192. Daubignard, J.; Detz, R. J.; Jans, A. C. H.; de Bruin, B.; Reek, J. N. H. *Angew. Chem. Int. Ed.* **2017**, *56*, 13056–13060.
193. Daubignard, J.; Detz, R. J.; de Bruin, B.; Reek, J. N. H. *Organometallics* **2019**, *38*, 3961–3969.
194. Zhao, Q.; Li, S.; Huang, K.; Wang, R.; Zhang, X. *Org. Lett.* **2013**, *15*, 4014–4017.
195. Li, P.; Zhou, M.; Zhao, Q.; Wu, W.; Hu, X.; Dong, X.-Q.; Zhang, X. *Org. Lett.* **2016**, *18*, 40–43.
196. Wen, J.; Jiang, J.; Zhang, X. *Org. Lett.* **2016**, *18*, 4451–4453.

197. Liu, G.; Han, Z.; Dong, X.-Q.; Zhang, X. *Org. Lett.* **2018**, *20*, 5636–5639.
198. Sun, Y.; Jiang, J.; Guo, X.; Wen, J.; Zhang, X. *Org. Chem. Front.* **2019**, *6*, 1438–1441.
199. Han, Z.; Li, P.; Zhang, Z.; Chen, C.; Wang, Q.; Dong, X.-Q.; Zhang, X. *ACS Catal.* **2016**, *6*, 6214–6218.
200. Han, Z.; Wang, R.; Gu, G.; Dong, X.-Q.; Zhang, X. *Chem. Commun.* **2017**, *53*, 4226–4229.
201. Lang, Q.; Gu, G.; Cheng, Y.; Yin, Q.; Zhang, X. *ACS Catal.* **2018**, *8*, 4824–4828.
202. Yin, C.; Yang, T.; Pan, Y.; Wen, J.; Zhang, X. *Org. Lett.* **2020**, *22*, 920–923.
203. Yang, J.; Li, X.; You, C.; Li, S.; Guan, Y.-Q.; Lv, H.; Zhang, X. *Org. Biomol. Chem.* **2020**, *18*, 856–859.
204. Li, X.; You, C.; Yang, Y.; Yang, Y.; Li, P.; Gu, G.; Chung, L. W.; Lv, H.; Zhang, X. *Chem. Sci.* **2018**, *9*, 1919–1924.
205. Liu, G.; Zhang, H.; Huang, Y.; Han, Z.; Liu, G.; Liu, Y.; Dong, X-Q.; Zhang, X. *Chem. Sci.* **2019**, *10*, 2507–2512.
206. Liu, G.; Li, A.; Qin, X.; Han, Z.; Dong, X.-Q.; Zhang, X. *Adv. Synth. Catal.* **2019**, *361*, 2844–2848.
207. Zhao, Q.; Wen, J.; Tan, R.; Huang, K.; Metola, P.; Wang, R.; Anslyn, E. V.; Zhang, X. *Angew. Chem. Int. Ed.* **2014**, *53*, 8467–8470.
208. Wen, J.; Tan, R.; Liu, S.; Zhao, Q.; Zhang, X. *Chem. Sci.* **2016**, *7*, 3047–3051.
209. Wen, J.; Fan, X.; Tan, R.; Chien, H.-C.; Zhou, Q.; Chung, L. W.; Zhang, X. *Org. Lett.* **2018**, *20*, 2143–2147.
210. Chen, C.; Wang, H.; Zhang, Z.; Jin, S.; Wen, S.; Ji, J.; Chung, L. W.; Dong, X.-Q.; Zhang, X. *Chem. Sci.* **2016**, *7*, 6669–6673.
211. Yin, X.; Chen, C.; Dong, X.-Q.; Zhang, X. *Org. Lett.* **2017**, *19*, 2678–2681.
212. Chen, C.; Zhang, Z.; Jin, S.; Fan, X.; Geng, M.; Zhou, Y.; Wen, S.; Wang, X.; Chung, L. W.; Dong, X.-Q.; Zhang, X. *Angew. Chem. Int. Ed.* **2017**, *56*, 6808–6812.
213. Yin, X.; Chen, C.; Li, X.; Dong, X.-Q.; Zhang, X. *Org. Lett.* **2017**, *19*, 4375–4378.
214. Chen, C.; Wen, S.; Geng, M.; Jin, S.; Zhang, Z.; Dong, X.-Q.; Zhang, X. *Chem. Commun.* **2017**, *53*, 9785–9788.
215. Wen, S.; Chen, C.; Du, S.; Zhang, Z.; Huang, Y.; Han, Z.; Dong, X.-Q.; Zhang, X. *Org. Lett.* **2017**, *19*, 6474–6477.
216. Ito, M.; Endo, Y.; Tejima, N.; Ikariya, T. *Organometallics* **2010**, *29*, 2397–2399.
217. Moritani, J.; Kayaki, Y.; Ikariya, T. *RSC Adv.* **2014**, *4*, 61001–61004.
218. Kumar, P.; Irrgang, T.; Kostakis, G. E.; Kempe, R. *RSC Adv.* **2016**, *6*, 39335–39342.
219. Xie, J.-H.; Liu, X.-Y.; Xie, J.-B.; Wang, L.-X.; Zhou, Q.-L. *Angew. Chem., Int. Ed.* **2011**, *50*, 7329–7332.
220. Xie, J.-H.; Liu, X.-Y.; Yang, X.-H.; Xie, J.-B.; Wang, L.-X.; Zhou, Q.-L. *Angew. Chem. Int. Ed.* **2012**, *51*, 201–203.
221. Bao, D.-H.; Wu, H.-L.; Liu, C.-L.; Xie, J.-H.; Zhou, Q.-L. *Angew. Chem. Int. Ed.* **2015**, *54*, 8791–8794.

222. Zhang, Q.-Q.; Xie, J.-H.; Yang, X.-H.; Xie, Ji.-B.; Zhou, Q.-L. *Org. Lett.* **2012**, *14*, 6158–6161.
223. Yang, X.-H.; Xie, J.-H.; Liu, W.-P.; Zhou, Q.-L. *Angew. Chem. Int. Ed.* **2013**, *52*, 7833–7836.
224. Yan, P.-C.; Xie, J.-H.; Zhang, X.-D.; Chen, K.; Li, Y.-Q.; Zhou, Q.-L.; Che, D.-Q. *Chem. Commun.* **2014**, *50*, 15987–15990.
225. Yang, X.-H.; Wang, K.; Zhu, S.-F.; Xie, J.-H.; Zhou, Q.-L. *J. Am. Chem. Soc.* **2014**, *136*, 17426–17429.
226. Lin, H.; Xiao, L.-J.; Zhou, M.-J.; Yu, H.-M.; Xie, J.-H.; Zhou, Q.-L. *Org. Lett.* **2016**, *18*, 1434–1437.
227. Bao, D.-H.; Gu, X.-S.; Xie, J.-H.; Zhou, Q.-L. *Org. Lett.* **2017**, *19*, 118–121.
228. Liu, Y.-T.; Chen, J.-Q.; Li, L.-P.; Shao, X.-Y.; Xie, J.-H.; Zhou, Q.-L. *Org. Lett.* **2017**, *19*, 3231–3234.
229. Zuo, X.-D.; Guo, S.-M.; Yang, R.; Xie, J.-H.; Zhou, Q.-L. *Chem. Sci.* **2017**, *8*, 6202–6206.
230. Liu, Y.-T.; Li, L.-P.; Xie, J.-H.; Zhou, Q.-L. *Angew. Chem. Int. Ed.* **2017**, *56*, 12708–12711.
231. Zuo, X.-D.; Guo, S.-M.; Yang, R.; Xie, J.-H.; Zhou, Q.-L. *Org. Lett.* **2017**, *19*, 5240–5243.
232. Bin, H.-Y.; Wang, K.; Yang, D.; Yang, X.-H.; Xie, J.-H.; Zhou, Q.-L. *Angew. Chem. Int. Ed.* **2019**, *58*, 1174–1177.
233. Gu, X.-S.; Yu, N.; Yang, X.-H.; Zhu, A.-T.; Xie, J.-H.; Zhou, Q. L. *Org. Lett.* **2019**, *21*, 4111–4115.
234. Hua, Y.-Y.; Bin, H.-Y.; Wei, T.; Cheng, H.-A.; Lin, Z.-P.; Fu, X.-F.; Li, Y.-Q.; Xie, J.-H.; Yan, P.-C.; Zhou, Q.-L. *Org. Lett.* **2020**, *22*, 818–822.
235. Chen, G.-Q.; Lin, B.-J.; Huang, J.-M.; Zhao, L.-Y.; Chen, Q.-S.; Jia, S.-P.; Yin, Q.; Zhang, X. *J. Am. Chem. Soc.* **2018**, *140*, 8064–8068.
236. Wu, W.; Liu, S.; Duan, M.; Tan, X.; Chen, C.; Xie, Y.; Lan, Y.; Dong, X.-Q.; Zhang, X. *Org. Lett.* **2016**, *18*, 2938–2941
237. Yu, J.; Long, J.; Yang, Y.; Wu, W.; Xue, P.; Chung, L. W.; Dong, X.-Q.; Zhang, X. *Org. Lett.* **2017**, *19*, 690–693.
238. Han, Z.; Liu, G.; Wang, R.; Dong, X.-Q.; Zhang, X. *Chem. Sci.* **2019**, *10*, 4328–4333.
239. Liu, Y.; Huang, Y.; Yi, Z.; Liu, G.; Dong, X.-Q.; Zhang, X. *Adv. Synth. Catal.* **2019**, *361*, 1582–1586.
240. Yang, T.; Sun, Y.; Wang, H.; Lin, Z.; Wen, J.; Zhang, X. *Angew. Chem. Int. Ed.* **2020**, *59*, 6108–6114.
241. Rueping, M.; Koenigs, R. M. *Chem. Commun.* **2011**, *47*, 304–306.
242. Tang, W.; Johnston, S.; Iggo, J. A.; Berry, N. G.; Phelan, M.; Lian, L.; Bacsa, J.; Xiao, J. *Angew. Chem. Int. Ed.* **2013**, *52*, 1668–1672.
243. Tang, W.; Johnston, S.; Li, C.; Iggo, J. A.; Bacsa, J.; Xiao, J. *Chem. Eur. J.* **2013**, *19*, 14187–14193.

244. Pan, H.-J.; Zhang, Y.; Shan, C.; Yu, Z.; Lan, Y.; Zhao, Y. *Angew. Chem. Int. Ed.* **2016**, *55*, 9615–9619.
245. Ohkuma, T.; Sandoval, C. A.; Srinivasan, R.; Lin, Q.; Wei, Y.; Muniz, K.; Noyori, R. *J. Am. Chem. Soc.* **2005**, *127*, 8288–8289.
246. Kuwano, R.; Morioka, R.; Kashiwabara, M.; Kameyama, N. *Angew. Chem. Int. Ed.* **2012**, *51*, 4136–4139.
247. Wang, D.-S.; Chen, Q.-A.; Lu, S.-M.; Zhou, Y.-G. *Chem. Rev.* **2012**, *112*, 2557–2590.
248. Margarita, C.; Andersson, P. G. *J. Am. Chem. Soc.* **2017**, *139*, 1346–1356.
249. Kraft, S.; Ryan, K.; Kargbo, R. B. *J. Am. Chem. Soc.* **2017**, *139*, 11630–11641.
250. Gensch, T.; Teders, M.; Glorius, F. *J. Org. Chem.* **2017**, *82*, 9154–9159.
251. Bullock, R. M. *Science* **2013**, *342*, 1054–1055.
252. Ludwig, J. R.; Schindler, C. S. *Chem* **2017**, *2*, 313–316.
253. Agbossou-Niedercorn, F.; Michon, C. *Coord. Chem. Rev.* **2020**, *425*, 213523–213559.
254. Duque, R.; Pogorzelec, P. J.; Cole-Hamilton, D. J. *Angew. Chem. Int. Ed.* **2013**, *52*, 9805–9807.
255. O'Neal, E. J.; Lee, C. H.; Brathwaite, J.; Jensen, K. F. *ACS Catalysis* **2015**, *5*, 2615–2622.
256. Geier, D.; Schmitz, P.; Walkowiak, J.; Leitner, W.; Francio, G. *ACS Catalysis* **2018**, *8*, 3297–3303.
257. Gladysz, J. A. *Pure Appl. Chem.* **2001**, *73*, 1319–1324.
258. Hübner, S.; de Vries, J. G.; Farina, V. *Adv. Synth. Catal.* **2016**, *358*, 3–25.

© 2022 World Scientific Publishing Company
https://doi.org/10.1142/9789811248436_0010

10 Organocatalytic Enantioselective Reactions Involving Cyclic N-Acyliminium Ions

Milane Saidah* and Laurent Commeiras*,†

*Aix Marseille University, CNRS, Centrale Marseille, Institut des Sciences Moléculaires de Marseille, UMR 7313, 52 Avenue Escadrille Normandie Niemen, 13013 Marseille, France
†laurent.commeiras@univ-amu.fr

Table of Contents

List of Abbreviations	370
I. Introduction	371
II. Enantioselective Intramolecular Additions to NAI Intermediates	373
A. Ion-pairing Catalysis with Chiral Phosphoric Acids	374
1. Nucleophilic additions of indole derivatives	375
2. Nucleophilic additions of electron-rich arenes	378

B. Ion-paring Catalysis with Chiral Thioureas-based
   Catalysts   380
   1. Nucleophilic additions of indole and pyrrole derivatives   380
   2. Nucleophilic additions of olefins   383
C. Enamine Catalysis   384
III. Enantioselective Intermolecular Additions to NAI
   Intermediates   386
A. Ion-pairing Catalysis with CPAs and Analogs   387
   1. Nucleophilic additions of indole derivatives   387
   2. Nucleophilic additions of hydroxystyrenes and
      hydrazones   395
   3. Nucleophilic additions of sulfur, phosphorus,
      and hydride nucleophiles   396
B. Ion-paring Catalysis with Chiral Thioureas-based
   Catalysts   400
C. Enamine Catalysis   401
IV. Conclusion   403
V. Acknowledgments   404
VI. References   404

List of Abbreviations

BEMP	2-*tert*-butylimino-2-diethylamino-1,3-dimethylperhydro-1,3,2-diazaphosphorine
BINOL	1,1′-Bi-2-naphthol
Cbz	carboxybenzyl
CPA	chiral phosphoric acid
dr	diastereomeric ratio
ee	enantiomeric excess
GABA	gamma-aminobutyric acid
MVK	methyl vinyl ketone
NAI	*N*-Acyliminium ion
PS-BEMP	polymer-supported BEMP
SPINOL	1,1′-spirobiindane-7,7′-diol
TBME	*tert*-butyl methyl ether

TIPS   triisopropylsilyl
Ts   *p*-Toluenesulfonyl
TRIP   3,3′-Bis(2,4,6-triisopropylphenyl)-1,1′-binaphthyl-2,
                2′-diyl phosphate
VAPOL   2,2′-Diphenyl-(4-biphenanthrol)
rt   room temperature

I. Introduction

Iminium ions are important intermediates and substrates in organic synthesis, widely used for the creation of carbon–carbon or carbon–heteroatom bonds in the synthesis of heterocyclic compounds. Notably, two well-known organic reactions, that is, the Mannich reaction[1] and its intramolecular version, the Pictet–Spengler reaction,[2] involve iminium intermediates in α-aminoalkylation processes. Iminium ions have been discovered at the beginning of twentieth century and pioneering synthetic applications have led notably to elegant methods for the synthesis of alkaloids with indole and isoquinoline scaffolds.[3] Introducing an acyl group on the nitrogen atom generates even more reactive species, namely *N*-acyliminium ions (NAIs). The acyl group makes the iminium carbon more electron deficient, therefore much more reactive as an electrophile than simple *N*-alkyliminium ions. In addition, iminium ions being quite often prochiral species, their reactivity offers the possibility to construct new stereogenic centers *via* inter- or intramolecular α-aminoalkylation processes. In this context, the discovery of enantioselective methods based on the use of chiral catalysts has emerged since the 2000s. The purpose of this chapter is to present representative organocatalytic enantioselective transformations involving cyclic NAIs, in both intra- and intermolecular reactions, which enable the preparation of enantioenriched nitrogen-containing ring systems (Scheme 1(a)). These reactions will be organized according to the complementary strategies followed for the stereochemical control. The first strategy is the so-called *asymmetric counterion-directed catalysis* (Scheme 1(b)) in which a transient NAI/chiral catalyst ion pair is formed. In the enantioselective step, the chiral ion pair will react with nucleophiles to form the corresponding α-substi-

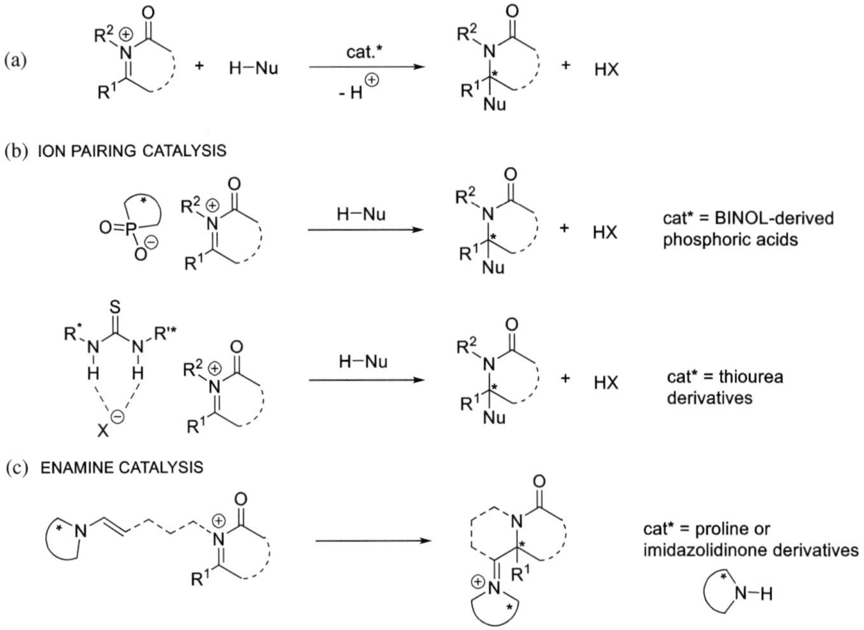

Scheme 1. Strategies for the enantioselective α-aminoalkylations involving NAIs.

tuted amides or lactams with good stereocontrol. For instance, chiral 1,1′-bi-2-naphthol (BINOL)-derived phosphoric acids are acidic enough to generate the *N*-acyliminium intermediates from suitable precursors, for example, from α-hydroxyamides by dehydratation, and to operate then as chiral counterions in tight ion pairs. Chiral ion pairs can be formed also in thioureas-promoted reactions in which thioureas bind the chloride counterion of *N*-acyliminium salts. The second mode of stereochemical control involves enamine catalysis (Scheme 1(c)). Here, an achiral NAI is generated and reacts then with chiral enamine-type nucleophiles that are formed mainly from carbonyl compounds and chiral imidazolidinone or proline derivatives.

Overall, this chapter intends to recall and emphasize the synthetic usefulness of enantioselective reactions involving cyclic NAIs, by showing a few representative examples only, focused on the construction of polycyclic scaffolds. Recent review articles are available for an exhaustive overview of the chemistry of NAIs and their synthetic applications.[4] Also, reviews have been specifically devoted to enantioselective reactions involving NAIs.[5]

II. Enantioselective Intramolecular Additions to NAI Intermediates

This section illustrates the synthetic usefulness of reactions involving cyclic NAIs and the intramolecular additions of carbon nucleophiles to these electrophilic substrates. In most cases, cyclic NAIs are generated by dehydratation of α-oxo-*N*-heterocycles, such as pyrrolidinones and piperidinones displaying hydroxy functions in their α-to nitrogen positions (see e.g. Schemes 2, 6, and 8). They are trapped then with indoles, pyrroles, electron-rich olefins, and, less commonly, electron-rich aryls. Alternative methods for the generation of cyclic NAIs include notably the protonation

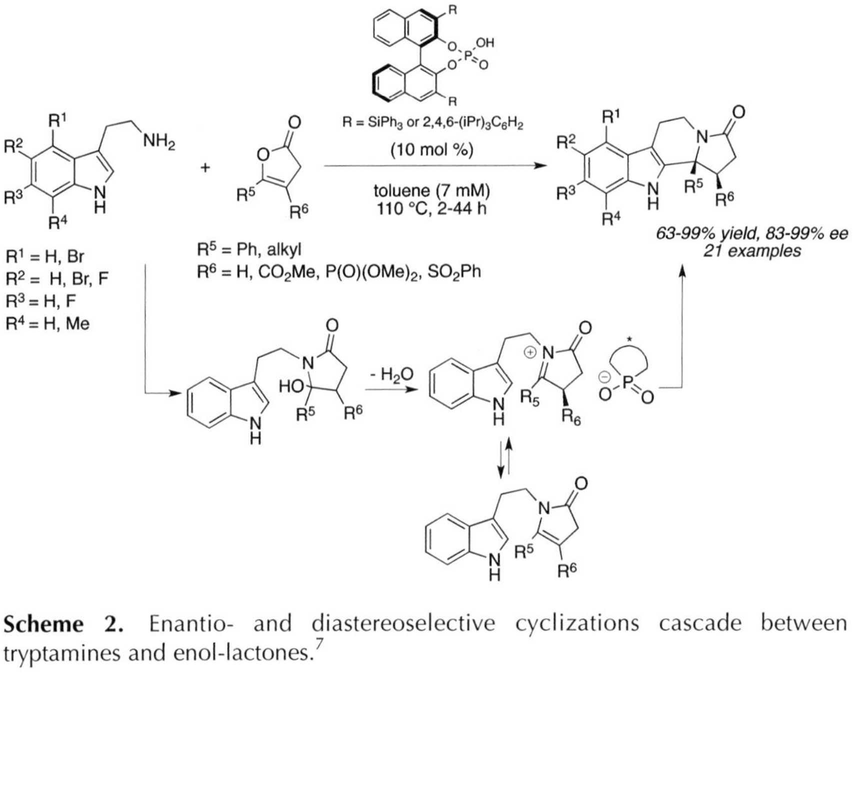

Scheme 2. Enantio- and diastereoselective cyclizations cascade between tryptamines and enol-lactones.[7]

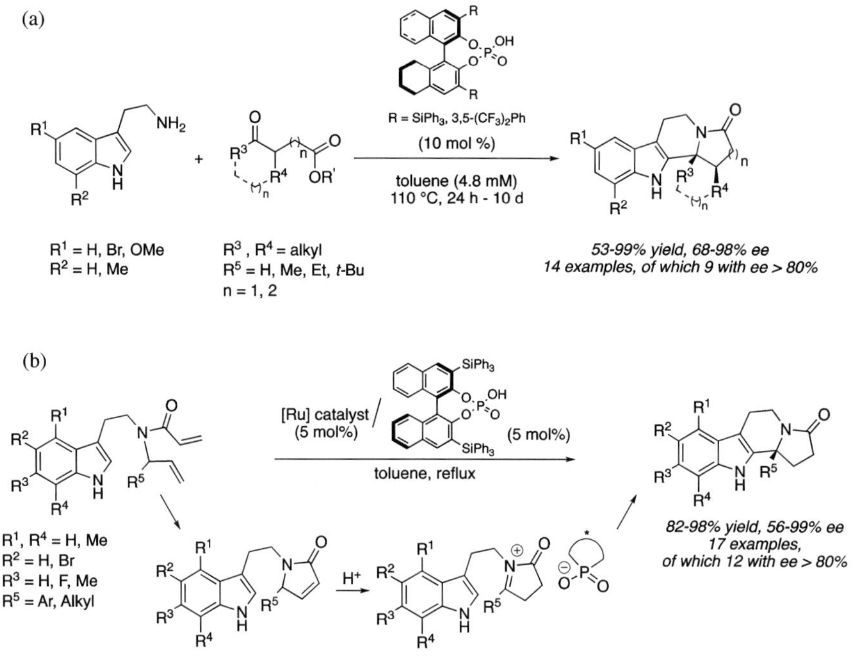

Scheme 3. (a) Enantioselective condensation/Pictet–Spengler cyclizations between tryptamines and keto esters; (b) sequential olefin metathesis/isomerization/Pictet–Spengler cyclizations on tryptamine derivatives.[9,10]

of dihydropyrrolones, as typified in Scheme 3(b). Cyclic NAIs generated by acylation of heteroaromatic substrates are not considered in this chapter.

In the following paragraphs, reactions involving chiral Brønsted acid catalysts for the generation of NAIs will be considered first. Then the use of chiral amines to generate nucleophilic reaction partners will be presented.

A. Ion-pairing Catalysis with Chiral Phosphoric Acids

Building of tight ion pairs with cationic substrates and intermediates is one of the multiple roles of chiral phosphoric acids (CPAs) in enantioselective catalysis. Although recognized only recently, this mode of action, the asymmetric counterion–directed catalysis, can lead to high stereocontrol in a number of organocatalytic and organometallic processes.[6]

The NAI chemistry has provided a particularly well-suited application of this concept. Thus, the following sections will show examples in which BINOL-derived phosphoric acids behave as chiral counterions for cyclic NAIs and enable the stereochemical control of the subsequent intramolecular addition of nucleophilic units.

1. *Nucleophilic additions of indole derivatives*

The intramolecular addition of π-nucleophiles to *N*-acyl iminium ions is a widespread method for the construction of nitrogen-containing ring systems. Among others, indoles proved to be well-suited nucleophiles in these reactions that are usually referred to as Pictet–Spengler type reactions.

Most often, this intramolecular addition step is incorporated into a cyclization cascade by which the cyclic NAI itself is formed. The cascade provides a powerful method for the one-pot production of architecturally complex polycyclic compounds. By following this approach, Dixon and coworkers reported in 2009 an enantioselective cyclization cascade leading to β-carbolines.[7] Compounds of this series are known to be biorelevant species, prevalent in many plants and animals and demonstrating various bioactivities with potential therapeutic applications.[8] The Dixon synthetic method involves enol lactones and tryptamines as starting materials. In the presence of BINOL-derived phosphoric acids, the reaction generates first a cyclic NAI by dehydrative condensation of the amine with the lactone (Scheme 2). Then, indole acts as a nucleophile toward the iminium ion, via its C2-carbon. The use of 10 mol% of chiral catalysts in refluxing toluene, under high-dilution conditions (7 mM), was found to be the best condition so far. A variety of β-carbolines were obtained with good to excellent yield (63–99%) and excellent enantioselectivity (83–99%) when using phosphoric acids derived from 3,3′-triphenylsilyl or aryl-substituted naphthols as Brønsted acid catalysts. Noteworthy, starting from 2,3-disubstituted lactones ($R^6 \neq H$), high levels of diastereoselectivity were observed, as only one isomer of the final product was isolated (up to 91% yield). Mechanistic studies suggested that the cyclic NAI intermediates are in equilibrium with the corresponding achiral enamines. Consequently, the final ring closure step is supposed to be the rate-determining and stereodetermining step, under catalyst control.

Despite the successful work mentioned earlier, the starting lactone was found to be occasionally nontrivial to prepare, depending on its substitution pattern. It was either isolated from separate experiments or generated *in situ* via gold-catalyzed cycloisomerization of alkynoic acids. To overcome this issue, the same group reported, one year later, an analogous method starting from commercially available, cyclic or acyclic keto-esters or keto-acids, instead of enol lactones (Scheme 3(a)). In this case, the desired polycyclic β-carboline moieties were obtained diastereoselectively (>98:2 diastereomeric ratio [dr]), in moderate to very good yields (53–99%) and high enantioselectivities (72–99% enantiomeric excess [ee]).[9]

The 3,3′-bis(triphenylsilyl)-substituted phosphoric acid was used also by You and coworkers in the Pictet–Spengler–type cyclizations of analogous cyclic NAI intermediates. The NAIs are formed here via a ruthenium-promoted ring-closing metathesis, followed by an isomerization/protonation sequence (Scheme 3(b)). In these sequential tandem reactions, excellent enantioselectivities were attained, with ee >80% in several instances.[10]

In other studies on BINOL-phosphoric acid-mediated enantioselective reactions, Dixon and coworkers have considered cyclization cascades that require both acid and base catalysts. The aim was to determine if acid/base catalysis could be performed in the same pot, without mutual inhibition, thanks to site isolation.[11] The authors turned their attention to the use of the polymer-supported 2-*tert*-butylimino-2-diethylamino-1,3-dimethylperhydro-1,3,2-diazaphosphorine (PS-BEMP), a commercially available microporous 1–2% cross-linked polystyrene-supported base. After titration tests of *(R)*-3,3′-bis(triphenylsilyl)-BINOL phosphoric acid against various amounts of PS-BEMP over time, the results pleasingly showed that both compounds could coexist without total annihilation. Eighty percent of the acid remained in solution after 17 h, and about 60% of PS-BEMP was still unquenched even after four days. This size-exclusion phenomenon enabled an acid/base–promoted cyclization cascade to be performed. A base-catalyzed Michael addition starting from methyl vinyl ketone (MVK) and malonate type nucleophiles (substrates displaying a malonamate and an indole unit), followed by an acid-catalyzed enatioselective NAI cyclization cascade represented an ideal test reaction, since it involves distinctive base- and acid-catalyzed steps

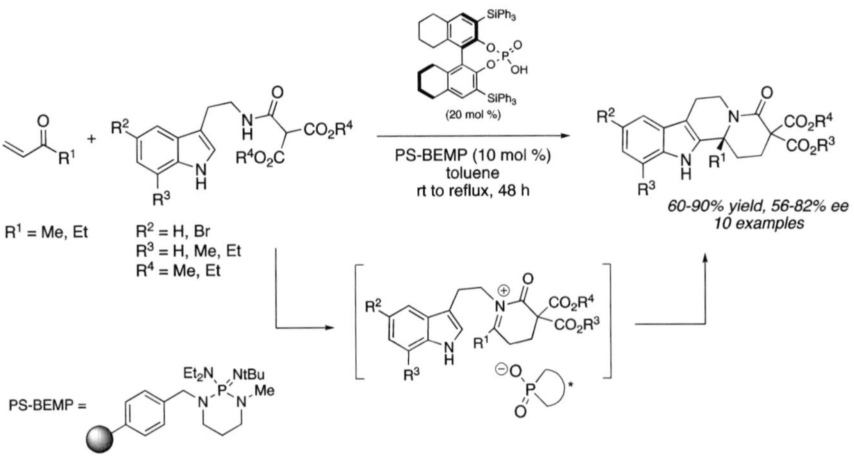

Scheme 4. Synthesis of polycyclic β-carbolines via a Michael addition/enantioselective N-acyl iminium cyclization cascade.[11b]

(Scheme 4). Michael addition to the vinyl ketone induces formation of the iminium ion intermediate that undergoes intramolecular enantioselective addition of the indole. After optimization, 10 mol% of PS-BEMP and 20 mol% of (R)-3,3′-bis(triphenylsilyl)-H$_8$-BINOL phosphoric acid were employed. The Michael additions of the malonamate derivative with 3 equivalents of vinyl ketones took place at room temperature (rt) for 24 h, then the enantioselective NAI cyclization was carried out in refluxing toluene for 24 h. These conditions were found to give the best enantiocontrol and reactivity (60–90% yield).[11b]

All things considered, the authors took advantage of a size-exclusion phenomenon between the polymer and sterically bulky BINOL-derived phosphoric acids to demonstrate a one-pot, base-catalyzed Michael addition/acid-catalyzed enantioselective NAI cyclization cascade that creates a large library of complex β-carbolines, with moderate to good enantioselectivity (56–82%). More importantly, they developed a viable and efficient alternative to the use of two different polymer-supported catalysts for reaction where site isolation is mandatory.

Quinolizidine scaffolds, closely related to those in Scheme 4, have been prepared with very high enantioselectivity via cascade reactions between enals and three functionalized indoles that proceed through

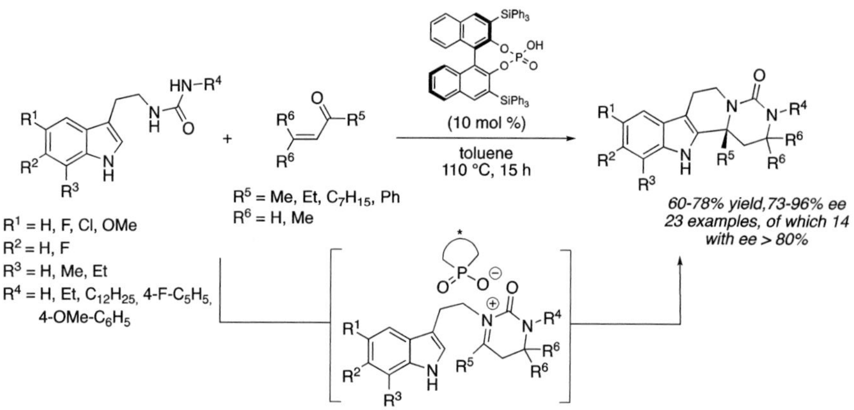

Scheme 5. Synthesis of tetracyclic β-carbolines via a cyclization cascade from tryptamine-derived ureas.[13]

analogous NAIs intermediates and Pictet–Spengler type cyclizations.[12] In these enantioselective catalytic reactions, however, the NAI intermediates are not involved in the enantiodetermining step. These processes and analogous multicomponent cascades fall therefore outside the scope of this review.

Additional examples of phosphoric acid catalyzed cyclization cascades have been reported by Dixon and coworkers in 2013 (Scheme 5).[13] Starting from tryptamine-derived ureas and enones, a Michael addition and the subsequent acid-catalyzed condensation of the ketone generate the cyclic iminium ion that undergoes then nucleophilic cyclization with the indole nucleus. A large library of complex tetra-heterocyclic compounds is produced with high enantioselectivity (up to 96% ee) when using the 3,3′-bis(triphenylsilyl)-substituted BINOL-derived phosphoric acid as the catalyst.

2. Nucleophilic additions of electron-rich arenes

Electron-rich aromatic substrates provide internal π-nucleophiles for the intramolecular addition to NAIs. This is typified by the reaction disclosed by Lete and coworkers who reported the 3,3′-Bis(2,4,6-triisopropylphenyl)-1,1′-binaphthyl-2,2′-diyl phosphate (TRIP)-mediated enantioselective

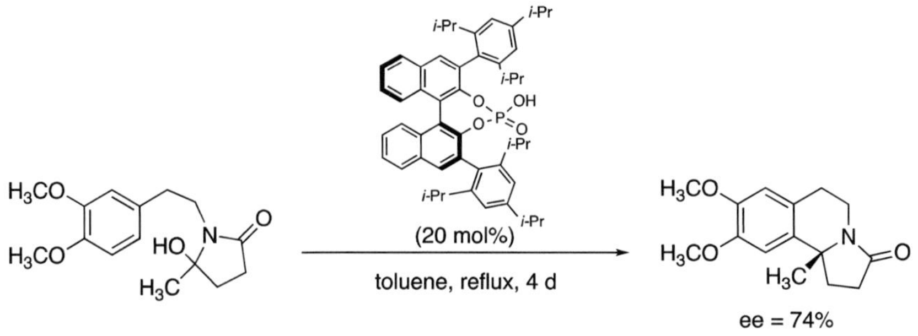

Scheme 6. Brønsted-acid-catalyzed intramolecular α-amidoalkylation of electron-rich arenes via NAI intermediates.[14a]

intramolecular α-amidoalkylation of a dimethoxybenzene derivative. The reaction enabled the synthesis of pyrrolo[2,1-*a*]isoquinolines (Scheme 6).[8a] The phosphoric acid generates the intermediate NAI by dehydration of the 5-hydroxypyrrolidinone, leading to a tight NAI/chiral phosphate ion pair. The bulky BINOL-derived phosphoric acid TRIP was required to have good enantiocontrol. Indeed, (*R*)-3,3′ bis(triphenylsilyl)-H8-BINOL phosphoric acid gave almost no enantioselectivity in these experiments, whereas the 2,4,6-triisopropylphenyl-substituted derivative gave enantioenriched isoquinolines with good ee (up to 74%). The authors also demonstrated the influence of the solvent on the reactivity and enantioselectivity. CH_3CN allowed to obtain pyrrolo[2,1-*a*]isoquinoline in good yield (80%) but with a low enantioselectivity (16%), while the reaction performed in toluene allowed a better enantioselectivity (74% ee) with a lower yield (23%). High catalyst loading, up to 30 mol%, is required in these reactions.

The method has been extended later to intramolecular amidoalkylations of electron-rich aryls (phenols and ethers) with phthalimide-derived NAIs.[14b] Overall, these intramolecular additions of electron-rich aryls to cyclic NAIs have the potential of being synthetically useful processes, but they did not attain so far satisfying levels of enantioselectivity and high yields concomitantly. Diastereoselective reactions involving nonchiral catalysts can give access to the same pyrroloisoquinoline scaffolds with higher efficiency.[15]

B. Ion-paring Catalysis with Chiral Thioureas-based Catalysts

In the following sections, examples are given in which, according to the postulated mechanisms, chiral thioureas coordinate the anionic counterions of cyclic NAI cations through their NH functions. The stereochemical features of the tight ion pair thus formed enable the stereochemical control of the subsequent nucleophilic additions.[16] The use of indoles, pyrroles, and olefin-type nucleophiles in intramolecular processes will be considered hereafter.

1. Nucleophilic additions of indole and pyrrole derivatives

In addition to employing chiral BINOL-phosphoric acid catalysts, N-acyl iminium intermediates can also be generated by hydrogen-bond donor catalysts such as chiral thioureas. Hence, Jacobsen and coworkers reported in 2007 the first enantioselective Pictet–Spengler-type cyclizations of tryptamine-derived hydroxylactams using chiral thiourea catalysts (Scheme 7).[17] After extensive screening of catalyst and activators, it was

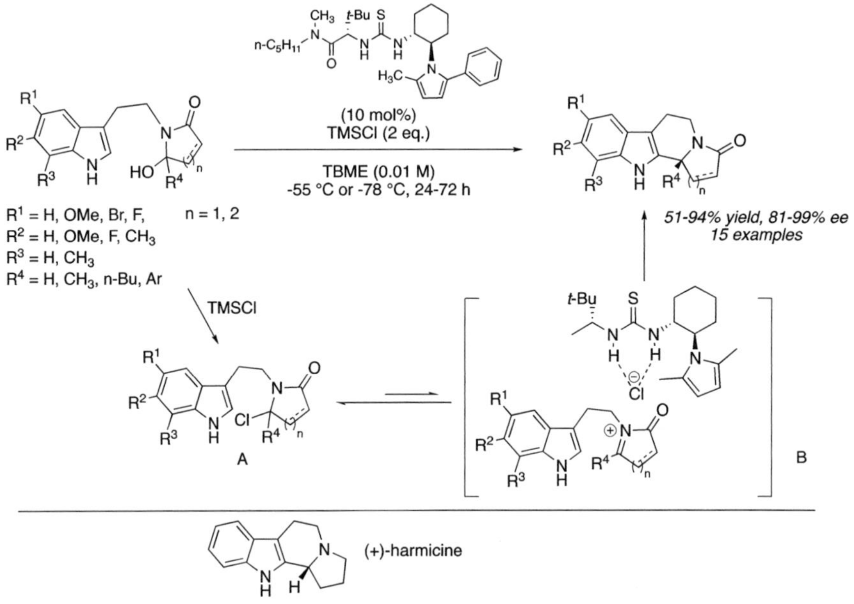

Scheme 7. Enantioselective Pictet–Spengler-type cyclizations of γ-hydroxylactams promoted by chiral thioureas.[17]

found that the combination of trimethylchlorosilane and a chiral pyrrole-thiourea provides high level of conversion and enantioselectivity. The catalyst entails several stereogenic centers as it combines an (R,R)-1,2-cyclohexandiamine and an (S)-configured t-butyl-substituted α-aminoamide unit. The cyclization of the β-indolylethyl hydroxylactams, previously formed from their succinimide and glutarimide precursors, was performed under optimized conditions (10 mol% of catalyst, in tert-butyl methyl ether (TBME) (0.01 M), TMSCl (2 equivalents), at −55°C for R^4 = H or −78°C for R^4 = alkyl, reaction time = 8–72 h) and afforded a large library of enantioenriched indolizidinones and quinolizidinones in good to excellent yield (52–94%) and excellent enantioselectivity (81–99% ee). Spectroscopic studies demonstrated that chlorolactam **A** was formed rapidly and irreversibly in the presence of TMSCl. The high enantioselectivity might be induced by a chiral N-acyliminium chloride–thiourea ion pair **B** that is formed after dissociation of the chloride as a counterion. Indeed, both the solvent and halide counterion effects have a significant impact on the enantioinduction, suggesting a SN_1-type mechanism for the final nucleophilic addition step.

It is worth noting that this new methodology was applied to the total synthesis of a natural product (+)-harmicine (Scheme 7), in four steps starting from tryptamine, with an excellent yield (90%) and an excellent enantioselectivity (ee = 99%).

In 2008, the intramolecular thiourea-catalyzed cyclization of γ-hydroxylactams has been extended by Jacobsen and coworkers to pyrroles as internal π-nucleophiles, affording a regio- and enantioselective approach to the corresponding pyrroloindolizidinone and pyrroloquinolizidinone derivatives (Scheme 8).[18] A series of hydroxylactames were prepared from β-pyrrolo-ethylsuccinimide by either reduction or addition of organometallic reagents. The N-acyl-Pictet–Spengler cyclization was investigated then. Under optimized conditions very similar to those developed previously, using the same thiourea catalyst, it allowed regioselective cyclizations at the C2 position of pyrrole and thus formation of pyrroloindolizidinones (n = 1) in good to excellent yields (51–86%). Very good enantioselectivities (up to 93%) were obtained with 5-alkyl-substituted hydroxylactames (R = alkyle), however, the reductively prepared hydroxylactam (R = H), as well as the phenyl-substituted substrate gave only

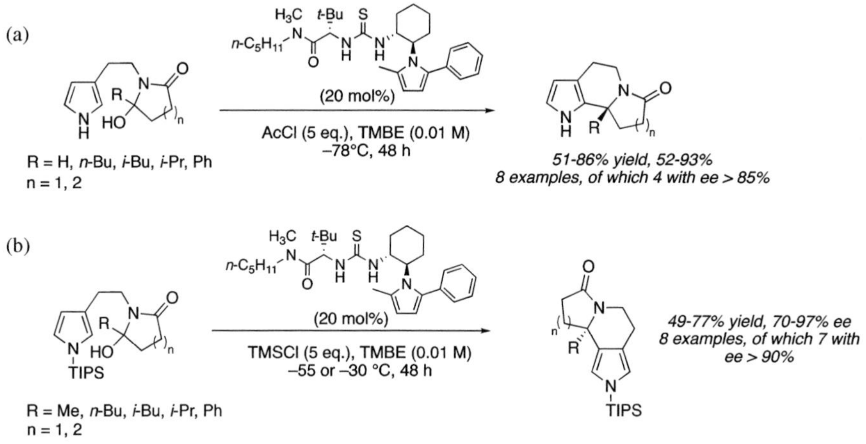

Scheme 8. Intramolecular addition of pyrroles to cyclic N-acyl iminium promoted by a chiral thiourea.[18]

modest ee (65% and 60%, respectively) under the same conditions. Pyrroloquinolizidinones ($n = 2$) were obtained via analogous procedures, with moderate enantioselectivity (up to 65%) though.

Alternatively, to change the regioselectivity of the cyclization, pyrroles substituted by the bulky N-triisopropylsilyl group on the nitrogen atom were considered. Herein, the intramolecular cyclizations occurred selectively at the C4-position and took place with high enantioselectivity (from 92% to 96%), both with succinimide- and glutarimide-derived hydroxylactams. Once again, cyclization of the nonsubstituted hydroxylactam (R = H, $n = 1$) displayed a significant drop of the enantioselectivity (ee = 70%). Comparative experiments have shown that, although cyclizations can be achieved with and without catalyst, the catalyst not only increases the reactivity, but also enhances the regioselectivity, affording C4 addition with a regioisomeric ratio (rr) > 50.1. These results highlight the dual role of the thiourea catalyst in the control of both regioselectivity and enantioselectivity. Finally, the increased reactivity and enantiocontrol obtained with alkyl-hydroxylactams suggest a S_N1-type mechanism, which is in line with the hypothesis of an anion-binding catalysis mode (see Scheme 7).

2. Nucleophilic additions of olefins

Cyclic NAIs can undergo intramolecular addition of olefins, with high stereocontrol, in the presence of especially designed chiral thiourea catalysts. An impressive example has been reported by the Jacobsen group in 2010, which relates to a cationic cyclization cascade (Scheme 9).[19] Starting from γ-hydroxylactams with an homoallyl substituent on the nitrogen atom, the cationic cascade is initiated by addition of the olefin function to the NAI. The subsequent cyclization steps involve aryl moieties and produce aza-steroids backbones. The final polycyclic products are obtained in good yields (51–77%) and high enantioselectivities (89–94%) when using a chiral thiourea catalyst that displays a pyrenyl substituent. The high degree of enantioselectivity was explained by a stabilizing cation–π interaction between the NAI intermediate and the pyrenyl substituent of the thiourea.

A second example of nucleophilic addition of olefins has been described by Jacobsen and coworkers, the intramolecular aza-Sakurai type cyclization of hydroxylactams with allylsilanes (Scheme 10).[20] The method makes use of a chiral-functionalized thiourea displaying a

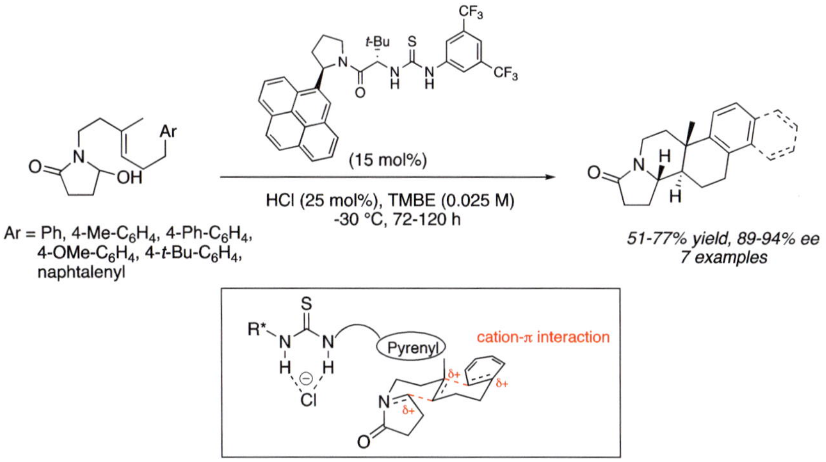

Scheme 9. Cationic polycyclizations involving γ-hydrolactams, promoted by a chiral thiourea.[19]

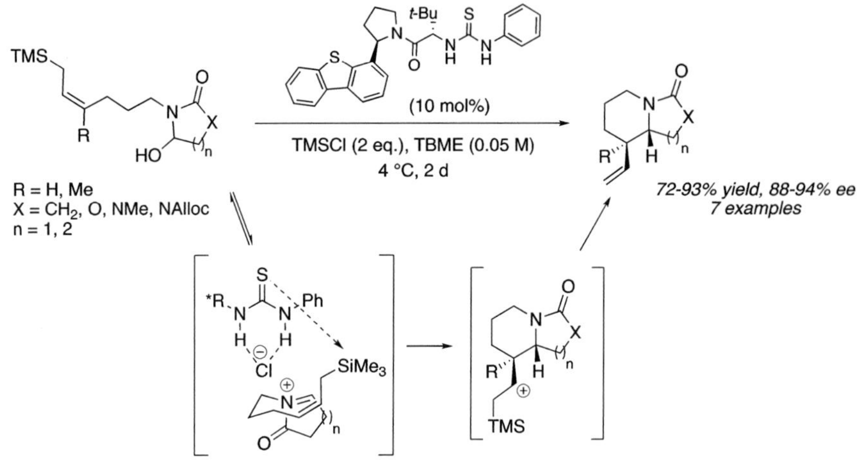

Scheme 10. Enantioselective intramolecular aza-Sakurai cyclizations.[20]

dibenzothiophenyl-pyrrolidino unit. It allows to prepare indolizidine, quinolizidines and related heterocycles with good to excellent yield (72–93%) and excellent enantioselectivities (88–94% ee). Mechanistic studies have been conducted with substrates containing different silyl groups, and using either thiourea or the corresponding urea. They suggested that the thiourea should play a dual role: it is not only engaged as a H-bond donor in anion-binding (halogen abstraction), but it is also involved in the Lewis base activation of the allylsilane nucleophile via its thiocarbonyl function. The dual activation taking place in the enantiodetermining step, it is assumed to account for the high enantioselectivity levels attained in these reactions.

C. Enamine Catalysis

For the intramolecular enantioselective additions of NAIs with π-nucleophiles, a suitable alternative to ion-pairing catalysis has been built based on the so-called enamine catalysis. In this approach, the electron-rich π-nucleophile is an enamine that is generated *in situ* by combining a carbonyl function of the substrate and a suitable chiral amine used as the catalytic auxiliary. This approach is illustrated hereafter.

Koley *et al.* reported in 2014 an asymmetric intramolecular Mannich cyclization of hydroxylactams with acetals for the synthesis of bicyclic

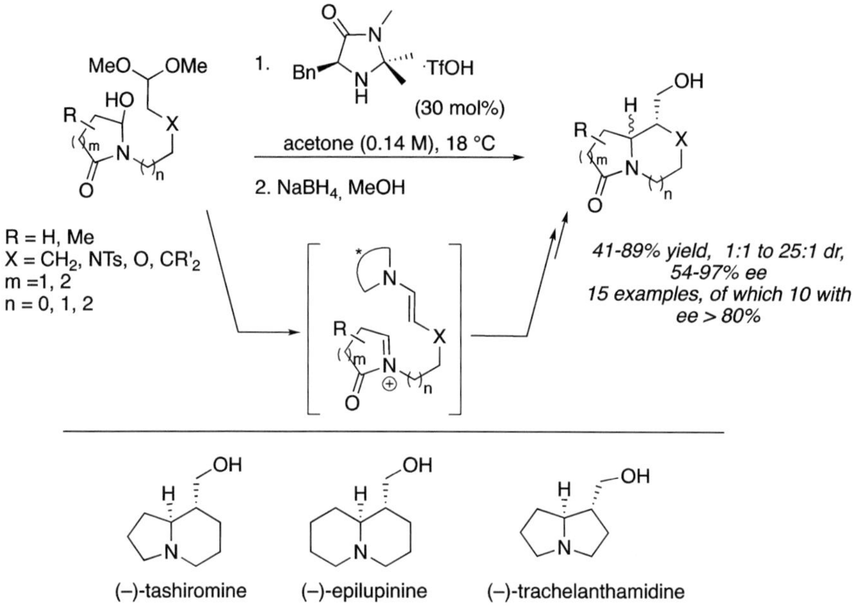

Scheme 11. Asymmetric organocatalytic cyclizations of hydroxylactam-acetals via Mannich reactions.[21]

indolizidine scaffolds (Scheme 11).[21] Although chiral Brønsted acids failed to provide good stereochemical control, the MacMillan imidazolidinone proved to be a suitable catalyst. The optimization of the reaction conditions demonstrated that the use of the triflic acid adduct of the imidazolinone in acetone can give high enantioselectivity in these cyclizations, leading to fused bicyclic alkaloid scaffolds with ee up to 97%. This one-pot process includes catalytic acetal removal, enamine formation, and NAI formation, followed by the nucleophilic addition of enamine to NAI. It is also worth to note that a drop of reactivity and enantioselectivity was observed in the presence of water, which hampers the formation of the enamine. However, only traces of product were formed in the presence of molecular sieves. The aldehydes produced in the cyclization step were reduced *in situ* with NaBH$_4$. Five-, six- and seven-membered rings were generally obtained in good yield (65–89% yields, 14 examples). The diastereo- and enantioselectivity of these cyclizations depends on the length and nature of the tethers between N atom and the acetal function: the highest ees are obtained for $n = 1$, that is, when six-membered rings are

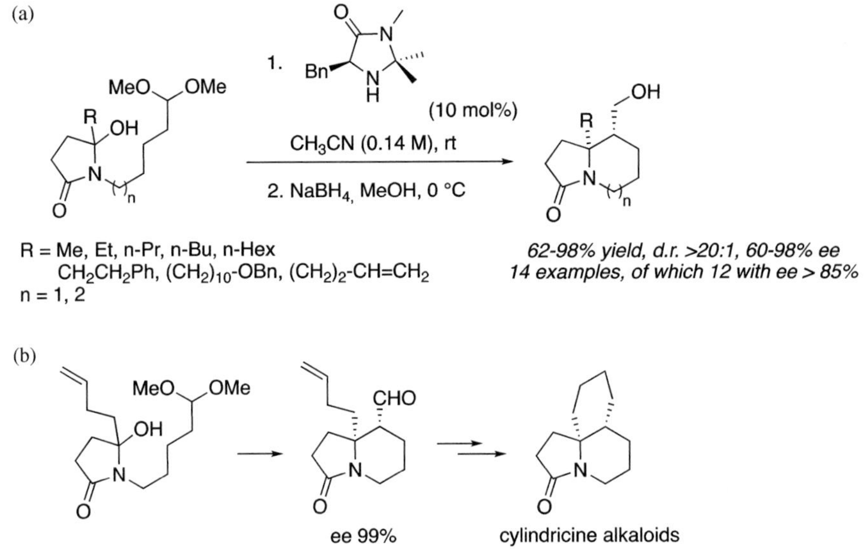
(a)
MeO
OMe
R
OH
N
n
O
1.
Bn
N
N
H
O
(10 mol%)
CH3CN (0.14 M), rt
2. NaBH4, MeOH, 0 °C
OH
R
N
n
O
R = Me, Et, n-Pr, n-Bu, n-Hex
CH2CH2Ph, (CH2)10-OBn, (CH2)2-CH=CH2
n = 1, 2
62-98% yield, d.r. >20:1, 60-98% ee
14 examples, of which 12 with ee > 85%
(b)
MeO
OMe
OH
N
O
CHO
N
O
ee 99%
N
O
cylindricine alkaloids

Scheme 12. Asymmetric organocatalytic cyclization of hydroxylactam-acetals leading to bicyclic scaffolds with bridgehead aza-quaternary centers.[22]

formed. For $n = 0$, moderate ees were obtained (54–74% ee). To illustrate the synthetic potential of this methodology, the total syntheses of three natural product, (–)-epilupinine, (–)-tashiromine, and (–)-trachelanthamidine, have been achieved.

A few years later, the same procedure has been applied to 5-substituted-5-hydroxypyrrolidinones affording the desired izidinone alkaloids that display an aza-quaternary stereogenic center (R ≠ H, Scheme 12(a)). These reactions proceeded in good yields, with excellent diastereocontrol (>20:1 drs for $n = 1$ or $n = 2$) and excellent enantioselectivities (up to 98% ee).[22] The method could be applied to the synthesis of the tricyclic skeleton of cylindricina alkaloids (Scheme 12(b)).

III. Enantioselective Intermolecular Additions to NAI Intermediates

The enantioselective methods for the intramolecular nucleophilic additions on cyclic NAI ions, which have been presented in the previous section, have been extended, as far as possible, to intermolecular reactions. In this case also, CPAs, thioureas, and chiral cyclic amines are privileged catalysts. The mechanisms involved are either ion-pairing catalysis or enamine catalysis.

A. Ion-pairing Catalysis with CPAs and Analogs

CPAs of the BINOL series have been largely employed as Brønsted acid catalysts to generate cyclic NAIs first, and to operate then as chiral counterions for the stereochemical control of their reactions with external nucleophiles. These processes are illustrated hereafter by the additions of carbon nucleophiles (indoles, hydrazones, electron-rich olefins), as well as by the addition of nitrogen, sulfide, and hydride nucleophiles to a variety of cyclic NAIs. Generally speaking, optimization of these reactions in terms of enantioselectivity requires notably fine tuning of the chiral scaffold of the phosphoric acid and its substitution pattern. 3,3′-disubstituted BINOL-derived acids are the most popular, however, in some instances 1,1′-spirobiindane-7,7′-diol (SPINOL)-derived acids have been used successfully.

1. *Nucleophilic additions of indole derivatives*

Indoles are the most common nucleophiles for the enantioselective additions to cyclic NAIs under acid catalysis. In most instances, they add to NAIs selectively through their electron-rich C3 position, however examples of N-alkylation have been reported also. For these reactions, cyclic NAIs have been generated from hydroxylated pyrrolidones and isoindolinones by dehydratation, as well as from *1H*-pyrrolones by protonation-induced migration of the double bond. These cyclic precursors led to the corresponding indolyl-substituted derivatives that often display interesting bioactivities.

Thus, for instance, 5-indolylpyrrolidones are interesting targets since not only these five-membered heterocycles are found in natural products and synthetic compounds displaying relevant biological properties, but also they could be precursors of gamma-aminobutyric acid (GABA) analogs via a hydrolysis reaction. For the synthesis of this class of compounds,

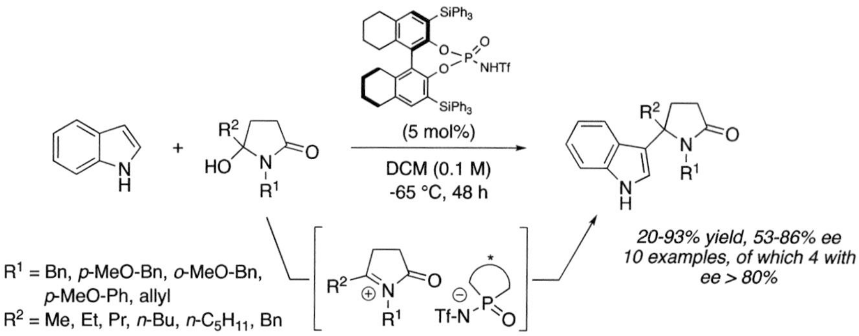

Scheme 13. Enantioselective Friedel–Crafts alkylation of indoles by γ-hydroxy-γ-lactames, using N-triflyl-phosphoramides as Brønsted acid catalysts.[23]

Rueping and Nachtsheim developed in 2010 an intermolecular enantioselective Brønsted acid–catalyzed nucleophilic substitution reaction of N-protected γ-substituted γ-hydroxy-γ-lactams with indoles (aza-Friedel–Crafts–type reactions), leading to enantioenriched γ,γ'-disubstituted γ-lactams (Scheme 13).[23] In contrast to the intramolecular versions described in Scheme 2 and following, the BINOL-derived phosphoric acids are not efficient enough to generate the NAI intermediates as chiral ion pairs. To overcome this lack of reactivity, more acidic Brønsted acids such as N-triflylphosphoramides were used. Under the optimal reaction conditions (5 mol% of catalyst, in DCM at –65 °C), several 3-indolyl-substituted γ-lactams were prepared with good yields (up to 93%) and good enantioselectivities (53–86% ee). Surprisingly, the addition of molecular sieves to trap water was prejudicial to the reaction in terms of both reactivity and enantioselectivity.

Shortly after, Masson and coworkers reported that BINOL-derived phosphoric acids are able to catalyze the enantioselective Friedel–Crafts alkylation of indoles using γ-hydroxy-γ-lactams as NAI precursors, provided that non-γ-substituted γ-lactams are employed (Scheme 14).[24] Indeed, N-aryl-γ-hydroxy-γ-lactams reacted with indoles substituted at the 2-, 5-, 6- and 7-positions, to furnish the desired 5-(3-indolyl)pyrrolidones in very good yields (60–99%) and excellent enantioselectivities (83–99% ee). The CPA displays a BINOL core with 9-phenanthryl substituents at the 3,3' positions. It operates as a stereodirecting counterion in the NAI

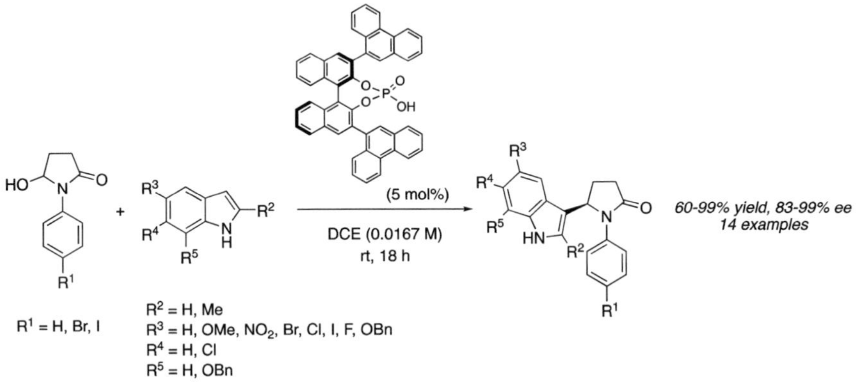

Scheme 14. Enantioselective aza-Friedel–Crafts alkylations of indoles with N-aryl-γ-hydroxy-γ-lactams.[24]

ion intermediate. The authors have shown that the enantioselectivity dropped markedly when the *N*-aryl group on the γ-hydroxy-γ-lactams is replaced by a *N*-benzyl moiety (ee = 64%), when the C2 position of the indole is substituted with a phenyl group (ee = 10%) and also by using an *N*-benzyl indole (ee = 19%). This last observation supports the hypothesis that a hydrogen bonding between the N–H site of the indole and the Lewis basic phosphoryl oxygen of the catalyst operates at the enantiodetermining step. When the reaction was performed with C3 substituted indoles, *N*-alkylation was observed instead of C3 alkylation, but the reaction gave low enantioselectivity (ee = 20%).

In addition to 5-hydroxy-γ-lactames, the benzo-fused analogs, 3-hydroxyisoindolin-1-ones, have been reported to react with indoles to access enantioenriched 3-substitued isoindolin-1-ones, a recurrent moiety present in several natural products and synthetic bioactive molecules. In 2011, Wang, Zhou, and coworkers reported the first examples of enantioselective Friedel–Craft alkylations of this class (Scheme 15).[25] A 5 mol% of a 3,3′-bis(2-naphtyl)-H_8-BINOL derived phosphoric acid catalyzed these reactions in $CHCl_3$ (0.2 M) at rt, to give 3-(3-indolyl)isoindolin-1-ones in good yields (90–99%) with moderate to good enantioselectivities (32–83% ee). Up to 99% ee were obtained after crystallization. A moderate 42% ee was obtained with the N-Me-substituted indole. As Rueping described previously,[23] the addition of molecular sieves is

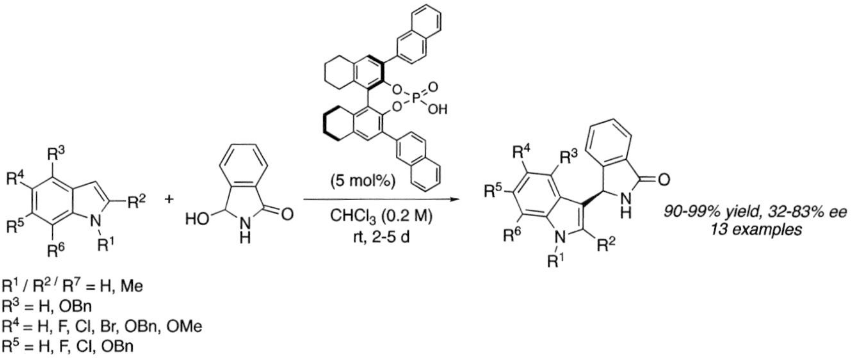

Scheme 15. Enantioselective aza-Friedel–Crafts alkylation of indoles with 3-hydroxyisoindolin-1-ones.[25]

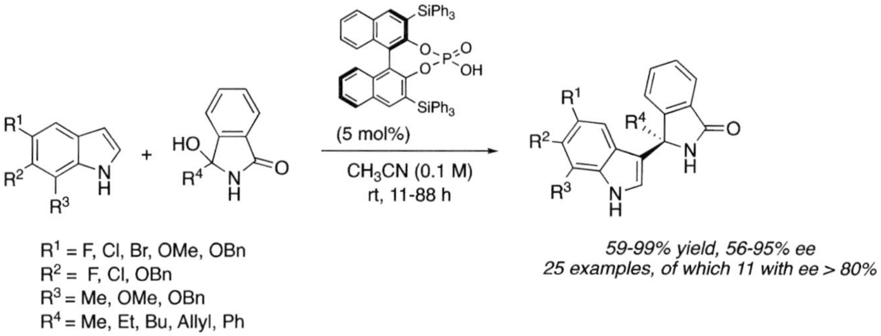

Scheme 16. Aza-Friedel–Crafts alkylations between indoles and 3-substituted 3-hydroxyisoindolin-1-ones.[26]

detrimental to the reaction in terms of both reactivity and enantioselectivity. The use of an N-methyl-substituted 3-hydroxyisoindolin-1-one led to the formation of the desired 3-indolylisoindolin-1-one (63%) but without inducing any enantioselectivity (ee = 2%).

Although this catalytic system worked well with 3-hydroxyisoindolin-1-ones, it failed to give 3, 3-disubstituted isoindolin-1-ones with a high degree of enantioselectivity, starting from 3-substituted isoindolinones. The use of 3,3′-bis(triphenylsilyl)-H8-BINOL-derived phosphoric acid in CH_3CN (0.1 M) circumvented this limitation (Scheme 16).[26] Whatever the nature of the substituents borne by both indoles and isoindolin-1-ones, several 3-substituted-3-indolylisoindolin-1-ones were obtained in excellent yield (59–99%) and good enantioselectivities (56–95% ee). The highest ees (80–95%) were attained with electron-rich indoles (R^1-R^3 = MeO or BnO, or R^3 = Me).

Even though the scope of the enantioselective phosphoric acid–catalyzed aza-Friedel–Crafts reactions above is rather broad, the addition of indoles to isoindolinones with an aryl substituent at the C3-position (R^4 = Ar), with a high level of enantioselectivity, was still unsatisfying. As a result, looking for more suitable catalysts, Gredićak, You, and coworkers have found that SPINOL-derived phosphoric acids are appropriate in terms of both reactivity and stereoselectivity for such transformation (Scheme 17).[27] The reactions of indole with various 3-aryl-3-hydroxy-isoindolinones were investigated revealing that, under these catalytic conditions, the desired 3-aryl-3-indolylisoindolinones are obtained in

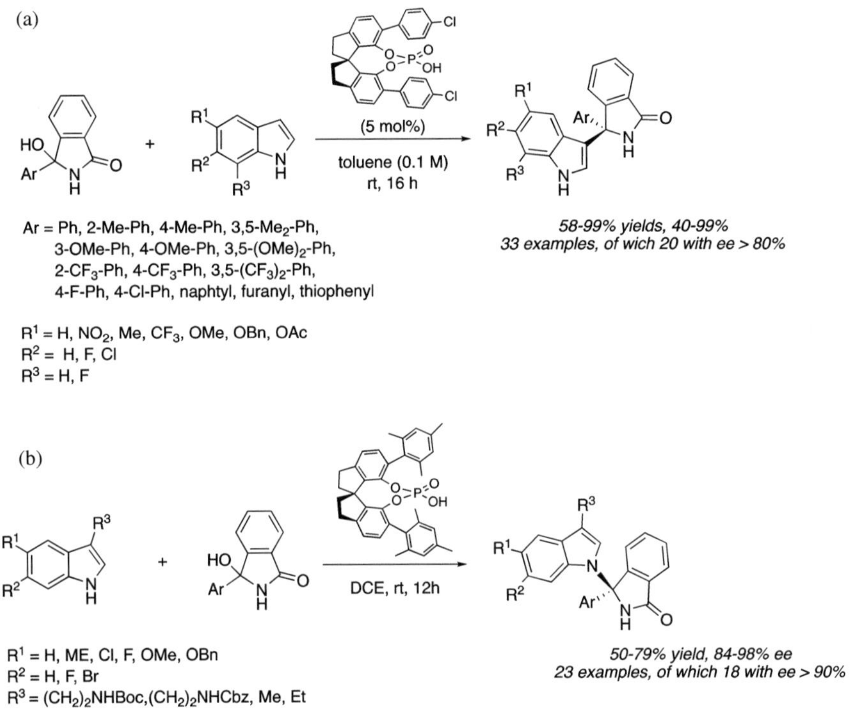
(a)
(5 mol%)
toluene (0.1 M)
rt, 16 h
Ar = Ph, 2-Me-Ph, 4-Me-Ph, 3,5-Me2-Ph,
3-OMe-Ph, 4-OMe-Ph, 3,5-(OMe)2-Ph,
2-CF3-Ph, 4-CF3-Ph, 3,5-(CF3)2-Ph,
4-F-Ph, 4-Cl-Ph, naphtyl, furanyl, thiophenyl
R1 = H, NO2, Me, CF3, OMe, OBn, OAc
R2 = H, F, Cl
R3 = H, F
58-99% yields, 40-99%
33 examples, of wich 20 with ee > 80%
(b)
DCE, rt, 12h
R1 = H, ME, Cl, F, OMe, OBn
R2 = H, F, Br
R3 = (CH2)2NHBoc,(CH2)2NHCbz, Me, Et
50-79% yield, 84-98% ee
23 examples, of which 18 with ee > 90%

Scheme 17. Enantioselective aza-Friedel–Crafts alkylation and N-alkylation of indoles with 3-aryl-3-hydroxyisoindolin-1-ones.[27,28]

good yields (up to 98%) and excellent enantioselectivities (82–99% ee). The reaction proved less efficient (prolonged reaction time and/or lower ee) when the aryl group is *ortho*-methyl or *ortho*-trifluoromethyl substituted (ee = 40–60%) or when Ar = 3-furanyl or 3-thiophenyl (ee = 60–64%). In addition, with substituted indoles, the nature and the position of the indole substituents have significant effects, although no general trends could be established. Very high ees were obtained notably in their reactions with isoindolinones bearing a 3,5-(MeO)$_2$-C$_6$H$_3$ substituent (6 examples, 96–99% ee).

Starting from 3-substituted indoles, the reaction with 3-aryl-3-hydroxy-isoindolin-1-ones under acidic conditions gives the *N*-alkylation products. Highly enantioselective *N*-alkylation has been reported by Zeng, Zhong, and coworkers using a SPINOL-derived phosphoric acid as the catalyst. Ees > 90% were obtained for a range of substrates (Scheme 17(b)).[28]

More sophisticated polycyclic compounds such as isoindoloisoquinolinones could be reacted also with indoles in Brønsted acid–catalyzed enantioselective α-amidoalkylations of the indole substrates. Thus, Lete and coworkers elegantly elaborated a sequence including a Parham cyclization and an enantioselective intermolecular α-amidoalkylation process to offer a suitable approach for the synthesis of 12b-isoindolo [1,2-*a*] isoquinolinones with stereogenic quaternary carbons (Scheme 18).[29]

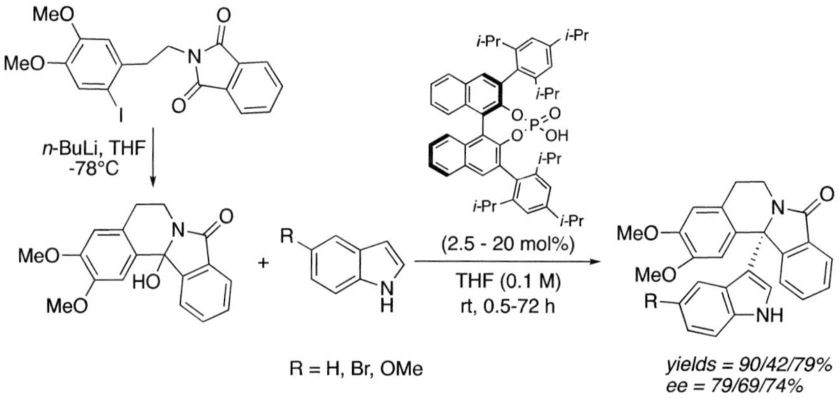

Scheme 18. Organocatalytic Friedel–Crafts alkylation of indoles with hydroxy-isoindolo-isoquinolinones.[29]

These interesting scaffolds were obtained by treatment of 12b-hydroxy-isoindoloisoquinolinones (prepared by Parham cyclization) with indole, 5-bromo-, or 5-MeO-indole, with up to 20 mol% of TRIP as the phosphoric acid catalyst. The reactivity and the enantioselectivity were strongly dependent on the solvents, temperature, additives, and the indole substituents. More specifically, no reactivity was observed when R = NO_2, while indole, 5-Br-indole, and 5-MeO-indole gave good yields and 79%, 69%, and 74% ee, respectively. The ee decreased to 37% with N-substituted indoles invoking a hydrogen bonding between the N–H site of indole and the Lewis basic phosphoryl oxygen of the catalyst in the enantiodetermining step. It is worth noting that the formation of a chiral ion pair was supported by ESI-MS and ESI-MS/MS experiments. These are the first examples of enantioselective reactions involving bicyclic N-acyliminium intermediates.

On the same line, Li, Shi, and coworkers reported the 13b-stereoselective functionalization of 13b-hydroxy-isoindolo-β-carbolines with indole nucleophiles, promoted by a 1 mol% amount of BINOL-derived phosphoric acid (Scheme 19).[30] In 1,4-dioxane (0,0125 M) at rt, the Friedel–Crafts reaction was found to be quite general, as it tolerates both electron-donating and electron-withdrawing groups on both the indoles and isoindolo-β-carbolines. The desired 13b-indolyl-isoindolo-β-carbolines were obtained in generally good yields (32–99%), with a high

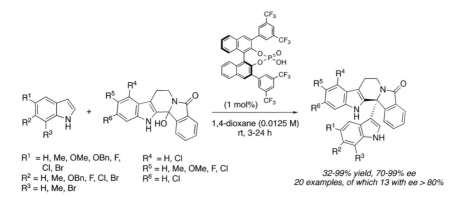

Scheme 19. Enantioselective Friedel-Crafts alkylation of indoles with hydroxy-isoindolo-β-carbolines.[30]

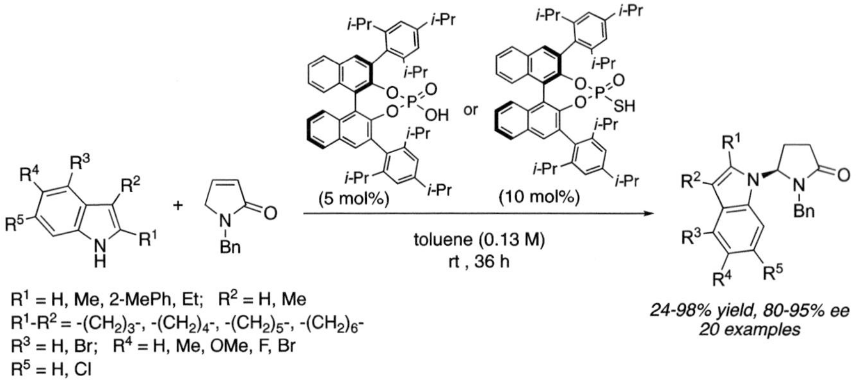
i-Pr
i-Pr
i-Pr
O
P
O
O
OH
or
i-Pr
i-Pr
i-Pr
(5 mol%)
O
P
O
O
SH
(10 mol%)
R4
R3
R2
R5
N
H
R1
+
N
Bn
O
toluene (0.13 M)
rt , 36 h
R1
R2
N
N
Bn
O
R3
R4
R5
R1 = H, Me, 2-MePh, Et; R2 = H, Me
R1-R2 = -(CH2)3-, -(CH2)4-, -(CH2)5-, -(CH2)6-
R3 = H, Br; R4 = H, Me, OMe, F, Br
R5 = H, Cl
24-98% yield, 80-95% ee
20 examples

Scheme 20. Enantioselective *N*-alkylation of indoles with *N*-benzyl-protected α,β-unsaturated γ-lactam.[31]

degree of enantioselectivity (70–99% ee). It was postulated that the efficient stereocontrol is due to the chiral NAI-phosphoric acid ion pair, combined with hydrogen-bonding interactions between the acid catalyst and the substrates.

As an alternative to dehydratation reactions, cyclic NAIs can be generated also from α,β-unsaturated γ-lactam by acid-promoted migration of their double bonds. Thus, Huang and coworkers have shown that in the presence of CPAs, 1*H*-pyrrol-2(5*H*)-ones first evolve into cyclic chiral *N*-acyliminium/phosphate ion pairs, whose reactions with indoles provide enantioenriched, 5-substituted pyrrolidinones via regioselective alkylation of the N-atom of indole (Scheme 20).[31] The nature of the protecting group on the N-atom of the pyrrolones proved to be crucial to obtain *N*-alkylation of indoles selectively. Indeed, *N*-alkylated indoles were obtained from *N*-benzyl-pyrrolones, whereas C3-alkylated indoles were obtained instead, when a phenyl or a Boc-protecting group was used. Several substituents were tolerated on the C2-, C3-, C4-, C5-, or C6-positions of indole, as well as cyclic units bridging the 2,3-positions. The desired pyrrolidinones were isolated in mainly good yields and enantioselectivities using either TRIP or the analogous thiophosphoric acid (80–95% ee). The mechanistic pathway, including the generation of the cyclic NAIs ion from the α,β-unsaturated γ-lactam, was supported by labeling experiments, and the formation of a defined chiral ion pair was supported by FTIR experiments and high-resolution mass spectroscopy (HRMS) analysis. It was also postulated that, concurrently, the conjugate base of the CPA activates the N-atom of indole via hydrogen bonding, at the N-alkylation step.

2. Nucleophilic additions of hydroxystyrenes and hydrazones

Beyond indoles, other carbon nucleophiles able to react with cyclic NAIs in enantioselective reactions include selected olefins and hydrazones.

The development of catalytic stereoselective metal-free alkenylation strategies remains an appealing and useful alternative to metal-mediated approaches especially when applications in medicinal chemistry are targeted. In this context, Shi and coworkers have elaborated a CPA-mediated enantioselective alkenylation method that involves polycyclic NAIs, generated from 13b-hydroxy-isoindolo-β-carbolines, and their vinylogous additions to o-hydroxystyrenes (Scheme 21).[32] The subsequent hydrogen elimination reactions provide styryl-substituted 13b-alkenylisoindolo-β-carbolines that are biologically relevant frameworks. The authors demonstrated that 10 mol% of BINOL-derived phosphoric acids are able to generate chiral N-acyliminium–phosphate ion pairs that direct the enantioselective addition step via hydrogen bonding interactions with the hydroxy group of o-hydroxystyrene. Several hydroxyisoindolo-β-carbolines and o-hydroxystyrenes are suitable for this transformation, leading to enantioenriched alkenylisoindolo-β-carbolines with an excellent control

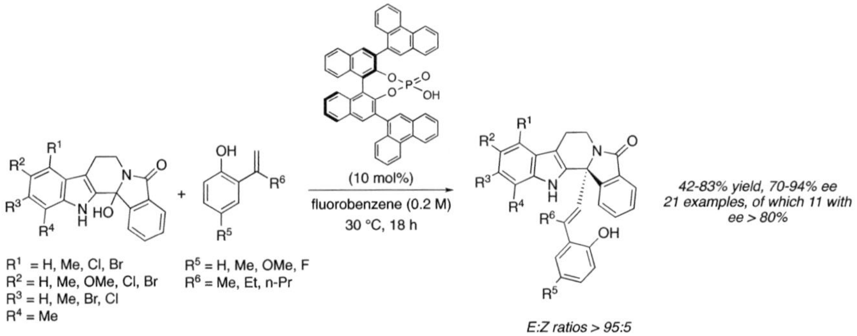

Scheme 21. Nucleophilic addition of o-hydroxystyrenes to 13b-hydroxyisoindolo-β-carbolines.[32]

of the E/Z ratios (E/Z > 95/5) and enantioselectivity (70–94% ee), in the presence of the 9-phenanthryl-substituted phosphoric acid. It was observed that C4-methyl-substituted hydroxyisoindolo-β-carboline give higher ee than C5-, C6-, or C7-methyl-substituted analogs, as well as that electron-donating substituents on the indole moiety tend to increase the enantioselectivity. Electron-rich or electron-neutral styrenes give better enantioselectivity than electron-poor derivatives. Also, moderate enantioselectivity is observed with hydroxystyrenes displaying ethyl or propyl groups at the vinylic position (R^6 = Et or n-Pr, ee = 74%), while high ees are obtained mainly for R^6 = Me.

Vicario, Uria, and coworkers have demonstrated that, under acidic conditions, N-acyldihydropyrroles can be precursors of quaternary NAIs, which will display then electrophilic reactivity toward suitable hydrazones. Indeed, in the presence of a BINOL-derived phosphoric acid, N-Boc-dihydropyrrole could be protonated and evolved to the corresponding chiral N-acyliminium as a phosphate ion pair. The NAI could be trapped by nucleophilic, donor-acceptor hydrazones to provide enantioenriched α-hydrazono-pyrrolidines in high yields and ee > 90% (Scheme 22(a)).[33] Overall, the reaction involves reversal of the electrophilic reactivity of the enamide induced by the acid catalyst, and takes advantage of the acyl-anion equivalent character of the hydrazone. Further screening demonstrated that dihydropyrroles with C(S)-NHAr functions on the nitrogen atom perform even better in terms of enantioselectivity (Scheme 22(b)). This is assigned to a beneficial effect of the NH function that can be engaged in H-bonding with the phosphate counterion. In these reactions, hydrazones with CO_2iPr, COMe, and CF_3 substituents give excellent >98% ee, while several aryl-substituted hydrazones give 82–97% ee (12 examples). This reaction has been performed also with 3-substituted dihydropyrroles resulting in the formation of α,β-disubstituted pyrrolidines with high ee (97–99% ee) but with moderate dr (3.2:1 to 2.5:1).

3. Nucleophilic additions of sulfur, phosphorus, and hydride nucleophiles

NAIs generated from cyclic N-acyl-aminals under acidic conditions can undergo enantioselective additions of noncarbon nucleophiles such as

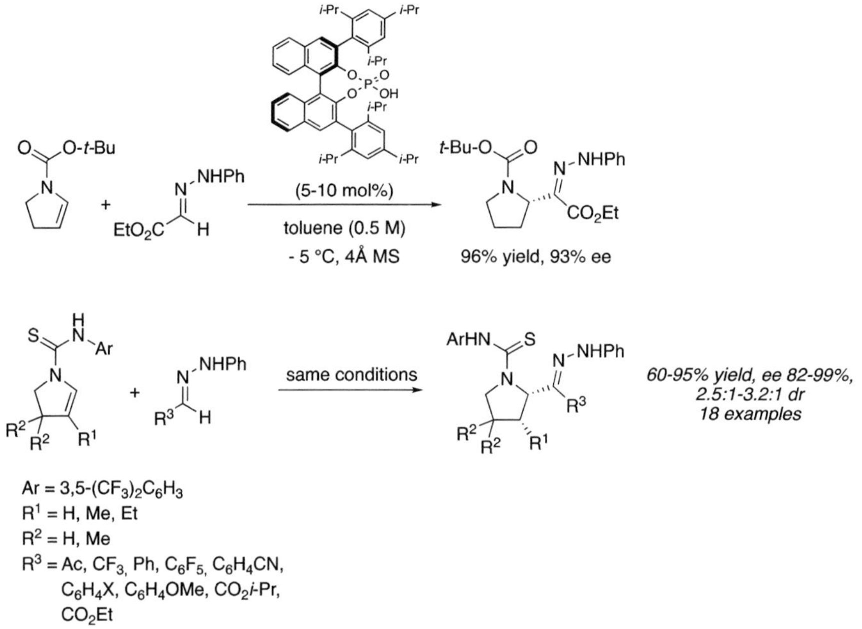

Scheme 22. Intermolecular enantioselective nucleophilic addition of hydrazones to N-acyldihydropyrroles.[33]

thiols, hydrides, and phosphorus nucleophiles. Three examples are given hereafter.

In 2016, Gredičak and coworkers have developed the first organocatalytic enantioselective reaction of thiols with 3-hydroxy-isoindolinones (Scheme 23).[34] This method addresses the lack of enantioselective strategies to prepare cyclic N(acyl),S-acetal scaffolds featuring a tetrasubstituted stereogenic carbon. These structural units are present notably in many bioactive molecules and natural products. By combining 3-hydroxy-isoindolinones with different thiols in the presence of the TRIP catalyst, the desired tetrasubstituted N(acyl),S-acetals were generally obtained in good yields (73–98%) and moderate to good enantioselectivities (up to 96% ee) via addition of the thiols to the intermediate NAI–phosphate pairs. The hydroxy-isoindolinone substrates feature aromatic or heteroaromatic substituents (R^1). High ees (>80%) were obtained with thiophenol and substituted thiophenols, except for some o-substituted species, but

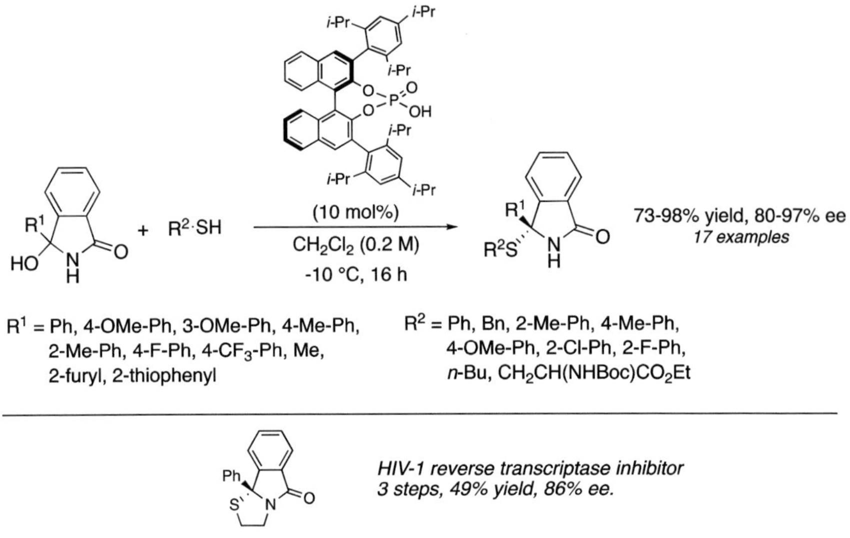

Scheme 23. Enantioselective synthesis of isoindolinone-derived N(acyl),S-acetals.[34]

also with linear alkyl thiols. The best enantioselectivities were obtained starting from p-fluorophenyl- or p-(trifluoromethyl)phenyl-3-hydroxy-3-isoindolinone and thiophenols (93–96% ee). It is worth noting that this method was also applied to the synthesis of a HIV-1 reverse transcriptases inhibitor that was obtained in 86% ee (Scheme 23).

Another example of reactions involving heteronucleophiles has been reported by Singh and coworkers who have envisioned to apply the organocatalytic strategy mentioned earlier to the enantioselective construction of isoindolinone-based α-amino-phosphonates (Scheme 24).[35] Thus, diphenyl phosphite was engaged in CPA-mediated phosphonylations of 3-substituted-3-hydroxyisoindolinones, by using the 9-anthryl-substituted BINOL-derived acid. The phosphorus nucleophile could be added enantioselectively to the chiral NAI intermediates to furnish the desired α-amino phosphonates in good yields (>80%). The enantioselectivity proved to be rather insensitive to the nature of the aryl group at the C-3 position of the isoindolinones, since ee > 85% were obtained in most cases (11 examples, up to 97% ee). The only exceptions are the p-CF$_3$- and p-F-phenyl substituted substrates that give slightly lower, 78%

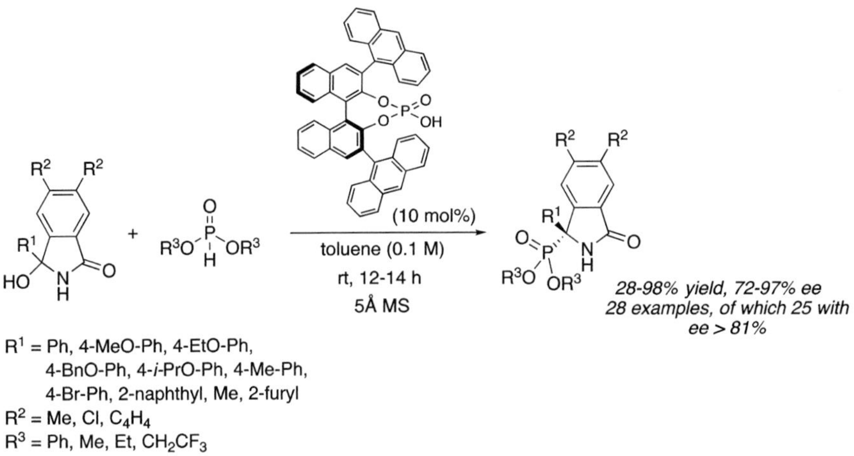

R¹ = Ph, 4-MeO-Ph, 4-EtO-Ph,
   4-BnO-Ph, 4-*i*-PrO-Ph, 4-Me-Ph,
   4-Br-Ph, 2-naphthyl, Me, 2-furyl
R² = Me, Cl, C₄H₄
R³ = Ph, Me, Et, CH₂CF₃

28-98% yield, 72-97% ee
28 examples, of which 25 with ee > 81%

Scheme 24. Enantioselective acid-catalyzed phosphonylation of isoindolinones.[35]

ee. Dialkyl phosphites could be used as well in these reactions: dimethyl- and diethylphosphite gave moderate yields only, while the more acidic bis(2,2,2-trifluoroethyl) phosphite gave the desired phosphonylation products with excellent yields and uniformly high ees (up to 97% ee).

Another type of CPA-mediated nucleophilic additions to cyclic *N*-acyliminium has been developed by Zhou and coworkers. These are reduction reactions carried out with organic hydrogen transfer reagents.[36] It was reported that 3-substitued 3-hydroisoindolinones could be reduced by an enantioselective hydrogenolysis reaction using catalytic amounts of 2,2′-diphenyl-(4-biphenanthrol) (VAPOL)-derived phosphoric acids (5 mol%) and the Hantzsch ester as the hydrogen source (Scheme 25). The optimized conditions (CH₂Cl₂, 35°C) led to the desired 3-substituted isoindolinones with moderate (61%) to very good (95%) enantioselectivity. The highest ee (93–95%) were obtained with 3-benzyl substituted isoindolinones. In some cases, the moderate yields are due to the formation of alkylidene isoindolinone by-products. Interestingly, it was demonstrated that these alkylidene isoindolinones can't be converted into the reduction products under these reaction conditions. Therefore, it is assumed that these reactions involve the direct hydrogenation of the iminium ion intermediates.

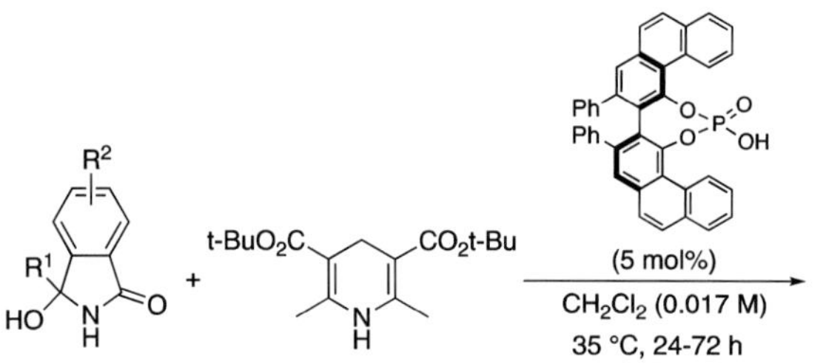

38-71% yield, 61-95% ee
15 examples, of which 11 with ee > 80%

R^1 = *n*-Bu, Me, Et, *n*-Pr, *i*-Pr, *n*-Hexyl, Cy, $PhCH_2CH_2$, Bn, 4-Me-Bn, 3-Me-Bn, 4-F-Bn, Ph
R^2 = H, 4-Me, 7-Me

Scheme 25. Hydrogenolysis of 3-hydroxyisoindolinones via enantioselective H-transfer reactions.[36]

B. Ion-paring Catalysis with Chiral Thioureas-based Catalysts

In straight continuation of their studies on intramolecular, chiral, thiourea-mediated enantioselective additions of indoles to cyclic NAIs disclosed in a previous section (see the Pictet–Spengler type reactions in Scheme 7), Jacobsen and coworkers reported in 2009 the intermolecular variant of the same process. The implementation of this strategy has required the use of a thiourea-Schiff base derivative as the chiral catalyst (Scheme 26).[37] The catalyst features an *anti*-1,2-diaminocyclohexane core and two additional stereogenic centers with defined relative configurations. After a very extensive optimization of the reaction conditions, the use of acetoxylactams as starting materials (better solubility with respect to hydoxylactams), TMSCl, a controlled amount of water (8 mol%) in TBME at –30°C, with 5 mol% of the chiral thiourea-Schiff base catalyst were found to be the most suitable conditions. The desired addition products were formed in high yields (up to 93%) and enantioselectivities (up to 95%) from electron-rich indoles. The authors suggested that the acetoxylactams react with HCl, generated *in situ* from TMSCl and water, to form the corresponding chlorolactams. The reaction proceeds then through an S_N1-type mechanism assisted by thiourea binding to the chloride anion.

In the case of electron-poor indoles, the $TMSCl/H_2O$ mixture was replaced by BCl_3 to obtain the corresponding indolyl-substituted adducts in useful yields (47–87%) and excellent ees (>90%).

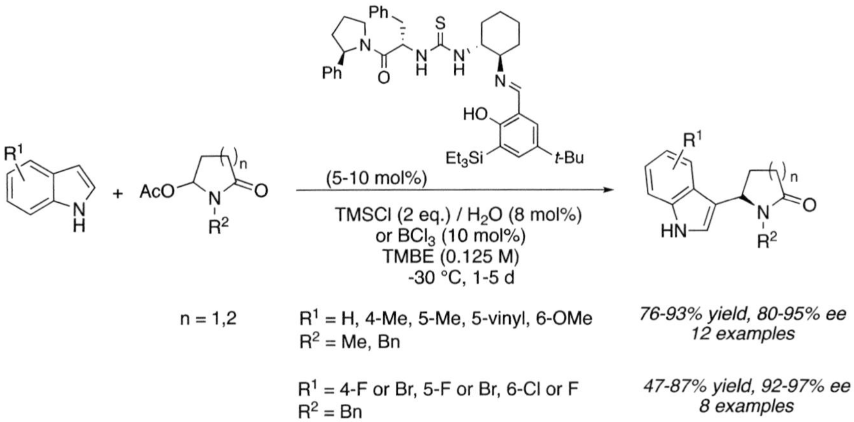

Scheme 26. Enantioselective additions of indoles to cyclic NAIs under thiourea catalysis.[37]

C. Enamine Catalysis

The intermolecular enantioselective Mannich reactions involving cyclic NAIs and chiral enamines generated *in situ* remain challenging, because of both the relative instability of the iminium intermediates, especially six-membered cyclic iminium ions, and the possible deactivation of the amine catalyst by acylation. In this context, Pineschi and coworkers reported in 2015 a one-pot strategy for the alkylation of quinoline *N,O*-acetals with aldehydes that includes the formation of the cyclic NAI from the quinoline and formation of the chiral enamine from the Hayashi–Jørgensen secondary amines. TsOH was employed as a cocatalyst. Acylation of the secondary amine catalyst was not competitive here, thanks to the synergistic effects of the Brønsted acid and the chiral amine in the activation of the aldehyde (Scheme 27).[38] Under these conditions, the Mannich reactions took place easily and, after reduction of the intermediate aldehyde with NaBH$_4$, the desired 2-alkylated *N*-carbamoyl dihydroquinolines were obtained as mixtures of *syn/anti* diasteromers (63/37–83/17), in good yields and ees up to 99% for the major *syn* isomer. Enantiocontrol was particularly successful with phenyl acetaldehyde (96–98% ee), while with acetaldehyde the ee dropped considerably (11–25% ee). Linear aliphatic aldehydes afforded the corresponding dihydroquinolines in good yield (52–88%) and high enantioselectivity (62–99%).

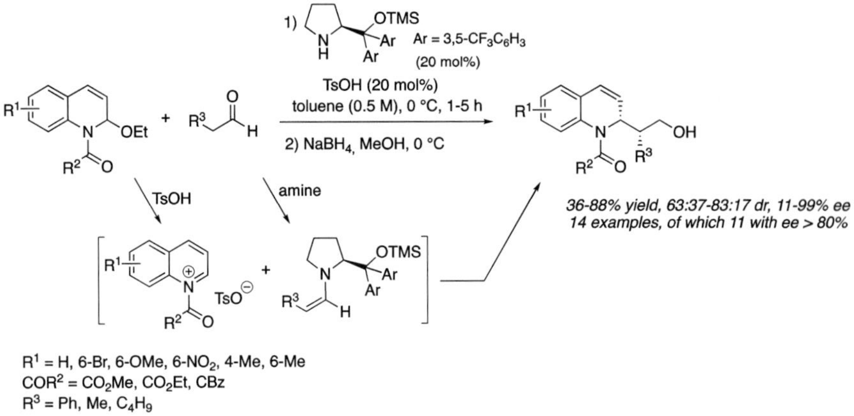

Scheme 27. Enantioselective alkylation of quinoline N,O-acetals with aldehydes under amine catalysis.[38]

The enantioselectivity was excellent for unsubstituted quinolines or quinolines substituted with both electron-withdrawing and electron-donating groups. Notably, 4-methyl-6-methyl and 6-bromo-substituted quinolines exhibited excellent enantioselectivity in their reactions with pentanal (99% ee).

The same year, Rueping and coworkers reported on analogous asymmetric additions of aldehydes to quinoline N,O-acetals promoted by a dual catalytic system involving indium triflate as a Lewis acid to activate the acetal, and a chiral imidazolidinone as a Lewis base catalyst. Enantioenriched dihydroquinolines were obtained in good yields and high enantioselectivity.[39]

The group of Pineschi also disclosed Mannich-type reactions involving nonaromatic NAIs obtained from γ-hydroxy-tetrahydropyridines (Scheme 28).[38,40] Taking advantage of the synergistic effects of Lewis acids $(Er(OTf)_3)$ and MacMillan organocatalyst, Pineschi and coworkers were able to synthesize the substituted tetrahydropyridine from phenylacetaldehyde, with an excellent control of the regioselectivity (98:2 α:γ substitution), diastereoselectivity (dr = 9:1) and enantioselectivity (ee = 93%). Furthermore, the resulting adduct could be employed as a precursor for the asymmetric synthesis of the blockbuster drug ritalin, a nervous system stimulant.

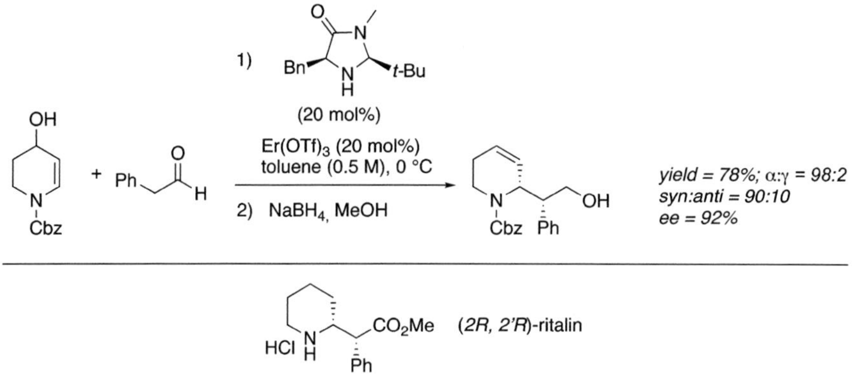

Scheme 28. Asymmetric addition of phenylacetaldehyde to a piperidine-based conjugated NAI.[40]

IV. Conclusion

Intra- and intermolecular enantioselective additions of nucleophiles to cyclic *N*-acyliminium is a powerful process to access highly enantioenriched heterocyclic compounds and structurally complex heteropolycyclic scaffolds. So far ion-pairing catalysis (asymmetric counterion directed catalysis) with CPAs proved to be a privileged approach, although chiral thioureas gave promising results also. In a different approach, chiral imidazolidinone and proline derivatives provided highly enantioselective processes via enamine intermediates. Key assets of these sustainable reactions are the easy availability of the starting materials and reaction partners, as well as the high efficiency demonstrated by well-known classes of organocatalysts. Another interesting feature is that the chemistry of *N*-acyl-iminium ions rests on simple protonation reactions that do not need sophisticated functional groups, but OH or olefin functions. These processes do not need metal derivatives either, although the reactivity of cyclic NAIs can be compatible with cascade reactions involving metal catalysis, thus giving sequential tandem reactions for the synthesis of more complex scaffolds.

The huge potential of the chemistry of NAIs for the synthesis of biorelevant and bioactive heterocyclic compounds has been largely demonstrated and further applications in total synthesis and medicinal chemistry can be easily anticipated. However, as shown in this short review, enantioselective variants are still limited. Thus, for instance, a number of intra- and intermolecular additions to olefins are known only as nonenantioselective reactions (see reviews in ref. 4 and other recent articles[41]). Therefore, organocatalytic enantioselective reactions are highly promising but remains largely untapped to date and deserve future focused studies.

V. Acknowledgments

Authors acknowledge the French Research Ministry, Aix-Marseille University and CNRS for financial support.

VI. References

1. Mannich, C.; Krösche, W. *Arch. Pharm.* **1912**, *250*, 647–667.
2. Pictet, A.; Spengler, T. *Ber. Dtsch. Chem. Ges.* **1911**, *44*, 2030–2036.
3. Belleau. B. *J. Am. Chem. Soc.* **1953**, *75*, 5765–5766.
4. (a) Wu, P.; Nielsen, T. E. *Chem. Rev.* **2017**, *117*, 7811–7856. (b) Yazici, A.; Pyne, S. G. *Synthesis.* **2009**, 339-368. (c) Yazici, A.; Pyne, S. G. *Synthesis.* **2009**, 513–541. (d) Maryanoff, B. E.; Zhang, H-C.; Cohen, J. H.; Turchi, I. J.; Maryanoff, C. A. *Chem. Rev.* **2004**, *104*, 1431–1628.
5. (a) Lee, Y. S.; Alam, Md. M.; Keri, R. S. *Chem. Asian J.* **2013**, *8*, 2906–2919. (b) Huang, P-Q. *Synlett.* **2006**, 1133–1149. (c) Royer, J.; Bonin, M.; Micouin, L. *Chem. Rev.* **2004**, *104*, 2311–2352.
6. (a) Brak, K.; Jacobsen, E. N. *Angew. Chem. Int. Ed.* **2013**, *52*, 534–561. (b) Mahlau, M.; List, B. *Angew. Chem. Int. Ed.* **2013**, *52*, 518–553. (c) Parmar, D.; Sugiono, E.; Raja, S.; Rueping, M. *Chem. Rev.* **2014**, *18*, 9047–9153.
7. Muratore, M. E.; Holloway, C. A.; Pilling, A. W.; Storer, R. I.; Trevitt, G.; Dixon, D. J. *J. Am. Chem. Soc.* **2009**, *131*, 10796–10797.
8. (a) Dai, J.; Dan, W.; Schneider, U.; Wang, J. *Eur. J. Med. Chem.* **2018**, *157*, 622–656. (b) Venault, P.; Chapouthier, G. *Sci. World. J.* **2007**, *7*, 204–223.
9. Holloway, C. A.; Muratore, M. E.; Storer, R. I.; Dixon, D. J. *Org. Lett.* **2010**, *12*, 4720–4723.
10. Cai, Q.; Liang, X-W.; Wang, S-G.; Zhang, J-W.; Zhang, X.; You, S-L. *Org. Lett.* **2012**, *14*, 5022–5025.
11. (a) Pilling, A. W.; Boehmer, J.; Dixon, D. J. *Angew. Chem. Int. Ed.* **2007**, *46*, 5428–5430. (b) Muratore, M. E.; Shi, L.; Pilling, A. W.; Storer, R. I.; Dixon, D. J. *Chem. Comm.* **2012**, *48*, 6351–6353.

12. Franzén, J.; Fisher, A. *Angew. Chem. Int. Ed.* **2009**, *48*, 787–791.
13. Aillaud, I.; Barber, D. M.; Thompson, A. L.; Dixon, D. J. *Org. Lett.* **2013**, *15*, 2946–2949.
14. (a) Gomez-SanJuan, A.; Sotomayor, N.; Lete, E.; *Tetrahedron Lett.* **2012**, *53*, 2157–2159. (b) Aranzamendi, E.; Sotomayor, N.; Lete, E. *ACS Omega.* **2017**, *2*, 2706–2718.
15. Allin, S. M.; Gaskell, S. N.; Towler, J. M. R.; Page, P. C. B.; Saha, B.; McKenzie, M. J.; Martin, W. P. *J. Org. Chem.* **2007**, *72*, 8972–8975.
16. Busschaert, N.; Caltagirone, C.; Van Rossom, W.; Gale, P. A. *Chem. Rev.* **2015**, *115*, 8038–8155.
17. Raheem, I. T.; Thiara, P. S.; Peterson, E. A.; Jacobsen, E. N. *J. Am. Chem. Soc.* **2007**, *129*, 13404–13405.
18. Raheem, I. T.; Thiara, P. S.; Jacobsen, E. N. *Org. Lett.* **2008**, *10*, 1577–1580.
19. Knowles, R. R.; Lin, S.; Jacobsen, E. N. *J. Am. Chem. Soc.* **2010**, *132*, 5030–5032.
20. Park, Y.; Schindler, C. S.; Jacobsen, E. N. *J. Am. Chem. Soc.* **2016**, *138*, 14848–14851.
21. Koley, D.; Krishna, Y.; Srinivas, K.; Khan, A. A.; Kant, R. *Angew. Chem. Int. Ed.* **2014**, *53*, 13196–13200.
22. Srinivas, K.; Singh, N; Das, D.; Koley, D. *Org. Lett.* **2017**, *19*, 274–277.
23. Rueping, M.; Nachtsheim, B. J. *Synlett*, **2010**, 119–122.
24. Courant, T.; Kumarn, S.; He, L.; Retailleau, P.; Masson, G. *Adv. Synth. Catal.* **2013**, *355*, 836–840.
25. Yu, X.; Lu, A.; Wang, Y.; Wu, G.; Song, H.; Zhou, Z.; Tang, C. *Eur. J. Org. Chem.* **2011**, 892–897.
26. Yu, X.; Lu, A.; Wang, Y.; Wu, G.; Song, H.; Zhou, Z.; Tang, C. *Eur. J. Org. Chem.* **2011**, 3060–3066.
27. Glavač, D.; Zheng, C.; Dokli, I.; You, S. L.; Gredicak, M. *J. Org. Chem.* **2017**, *82*, 8752–8760.
28. Zhang, L.; Wu, B.; Chen, Z.; Hu, J.; Zeng, X.; Zhong, G. *Chem. Commun.* **2018**, *54*, 9230–9233.
29. Aranzamendi, E.; Sotomayor, N.; Lete, E. *J. Org. Chem.* **2012**, *77*, 2986–2991.
30. Fang, F.; Hua, G.; Shi, F.; Li, P. *Org. Biomol. Chem.* **2015**, *13*, 4395–4398.
31. Xie, Y.; Zhao, Y.; Qian, B.; Yang, L.; Xia, C.; Huang, H. *Angew. Chem. Int. Ed.* **2011**, *50*, 5682–5686.
32. Zhang, H.-H.; Wang, Y.-M.; Xie, Y.-W.; Zhu, Z.-Q.; Shi, F.; Tu, S.-J. *J. Org. Chem.* **2014**, *79*, 7141–7151.
33. Zabaleta, N.; Uria, U.; Reyes, E.; Carrillo, L.; Vicario, J. L. *Chem. Commun.* **2018**, *54*, 8905–8908.
34. Sućc, J.; Dokli, I.; Gredičak, M. *Chem.Commun.* **2016**, *52*, 2071–2074.
35. Suneja, A.; Unhale, R. A.; Singh, V. K. *Org. Lett.* **2017**, *19*, 476–479.
36. Chen, M.-W.; Chen, Q. A.; Duan, Y.; Ye, Z.-S.; Zhou, Y.-G. *Chem. Commun.* **2012**, *48*, 1698–1700.

37. Peterson, E. A.; Jacobsen, E. N. *Angew. Chem. Int. Ed.* **2009**, *48*, 6328–6331.
38. Berti, F.; Malossi, F.; Marchetti, A.; Pineschi, M. *Chem. Comm.* **2015**, *51*, 13694–13697.
39. Volla, C. M. R.; Fava, E.; Atodiresei, I.; Rueping, M. *Chem. Comm.* **2015**, *51*, 15788–15791.
40. Berti, F.; Favero, L.; Pineschi, M. *Synthesis* **2016**, *48*, 2645–2652.
41. (a) Krishna, Y.; Shilpa, K.; Tanaka, F. *Org. Lett.* **2019**, *21*, 8444–8448. (b) Nash, A.; Qi, X.; Maity, P.; Owens, K.; Tambar, U. K. *Angew. Chem. Int. Ed.* **2018**, *57*, 6888–6891.

11 Enolate Surrogates and Unusual Nucleophiles in Stereoselective Iridium-catalyzed Allylic Substitutions

Pierre Bouillac,* Manuel Barday,* Thierry Constantieux* and Muriel Amatore*,†

*Aix Marseille University, CNRS, Centrale Marseille, Institut des Sciences Moléculaires de Marseille, UMR 7313, 52 Avenue Escadrille Normandie Niemen, 13013 Marseille, France
†muriel.amatore@univ-amu.fr

Table of Contents

List of Abbreviations	409
I. Introduction	411
II. Iridium-catalyzed Allylations of Silyl Enolates	415
A. Enantioselective α-Allylation of Silyl Enolates	416
1. α-allylation of silyl enol ethers as ketone surrogates	416
2. α-allylation of silyl ketene acetals as ester surrogates	417

         3. α-allylation of α,β-unsaturated silyl enol ethers   419
         4. α-allylation of silylated enols of vinylogous
            esters and amides   421
      B. Enantioselective γ-Allylation of Silyl Dienolates
         Derived from Dioxinones   422
      C. Conclusion   424
   III. Iridium-catalyzed Stereoselective and
        Stereodivergent Allylations of Other Enolate
        Surrogates   424
      A. Stereoselective α-Allylations   425
         1. α-allylation of enol ethers and enamines as
            aldehydes surrogates   425
         2. Enantioselective α-allylation of enamines,
            enamides, and other ketone surrogates   427
         3. Enantioselective α-allylation of orthoesters as
            ester surrogates   429
         4. Enantioselective α-allylation of ketene aminals
            as amide surrogates   430
      B. Stereodivergent α-Allylations Under Dual Catalysis   432
         1. Stereodivergent α-allylation of aldehydes
            under dual iridium and enamine catalysis   433
         2. Stereodivergent γ-allylation of α,β-unsaturated
            aldehydes under dual iridium and enamine
            catalysis   438
         3. Stereodivergent α-allylations under dual
            iridium and lewis base catalysis   438
      C. Conclusion   441
   IV. Iridium-catalyzed Enantioselective Allylation
       of Other Carbon Nucleophiles   443
      A. Enantioselective Allylation of Organometallic
         Reagents   443
         1. Allylation of organozinc reagents   443
         2. Allylation of boron, silicon, and silver-based
            reagents   447
      B. Stereoselective Allylation of Hydrazones and Imines   450
      C. Conclusion   452

V. Iridium-catalyzed Enantioselective Allylation
of Electron-rich Unsaturated Substrates and
Other Weak Nucleophiles 452
   A. Enantioselective Allylation of C=C Double Bonds 453
      1. Enantioselective allylation of olefins 453
      2. Stereoselective allylations of indoles and pyrroles 454
      3. Enantioselective *C*-allylation of anilines 463
      4. Enantioselective *C*-allylation of phenols 464
      5. Stereoselective allylations involving
cation-π cyclizations of polyenes 468
   B. Enantioselective Allylations of Benzylic
Nucleophiles and Allylic Aromatization of
Methylene-substituted Heterocycles 469
   C. Conclusion 471
VI. Conclusion 471
VII. Acknowledgments 472
VIII. References 472

List of Abbreviations

A	amine catalyst
aq.	aqueous
Ar	aryl
b/l	branched/linear
9-BBN	9-borabicyclo[3.3.1]nonane
BINAP	2,2′-*bis*(diphenylphosphino)-1,1′-binaphthyl
BINOL	1,1′-*bis*(2-naphthol)
Boc	*tert*-butoxycarbonyle
BTM	benzotetramisole
Bz	benzoyl
CBS	Corey-Bakshi-Shibata
cod	1,5-cyclooctadiene
DABCO	1,4-diazabicyclo[2.2.2]octane
dbcot	dibenzocyclooctatetraene
DBU	1,8-diazabicyclo[5.4.0]undec-7-ene
DCE	1,2-dichloroethane

DMAP	4-dimethylaminopyridine
DME	dimethoxyethane
DMF	*N*,*N*-dimethylformamide
dppf	1,1'-bis(diphenylphosphino)ferrocene
dr	diastereomeric ratio
EDG	electron-donating group
ee	enantiomeric excess
EWG	electron-withdrawing group
FG	functional group
gem-	geminal
HMDS	hexamethyldisilazane
L	ligand
LG	leaving group
M	metal
MS	molecular sieves
Ms	mesyl
NHC	*N*-heterocyclic carbene
NMM	*N*-methylmorpholine
Nu	nucleophile
PG	protecting group
Phth	phthalimide
Pin	pinacol
PMB	*para*-methoxybenzyl
PMHS	polymethylhydrosiloxane
quant.	quantitative
rr	regioisomeric ratio
rt	room temperature
SPINOL	1,1'-spirobiindane-7,7'-diol.
TBAB	tetrabutyl ammonium bromide
TBD	1,5,7-triazabicyclo[4.4.0]dec-5-ene
TBDMS	*tert*-butyldimethylsilyl
TFA	trifluoroacetic acid
THQphos	1,2,3,4-tetrahydroquinoline phosphoramidite
TMS	trimethylsilyl
Troc	trichloroethyl

I. Introduction

The enantioselective transition metal–catalyzed allylic substitution reactions, originally referred to as Tsuji–Trost reactions,[1] are of high importance for organic synthetic chemistry, as well as for its applications such as pharmaceutical chemistry, as they offer a powerful tool for the straightforward construction of C–C and C–X (X = O, N, S) bonds in a stereoselective manner. These reactions involve reactive allylic substrates bearing a leaving group (halides, esters, carbonates, or activated alcohols), which catalytically generate electrophilic π-allyl–metal complexes that undergo subsequently nucleophilic additions (Scheme 1).

The early work on stereoselective allylic substitution reactions has been mainly centered on palladium chemistry, with the development of numerous methodologies using readily available chiral ligands and a large panel of both electrophiles and nucleophiles.[2] Since then, other transition metals such as W,[3] Mo,[4] Ru,[5] Rh,[6] Ni,[7] and Cu[8] have revealed interesting reactivity in these reactions, with encouraging regio- and stereocontrol. However, among all transition metals, iridium complexes have received a particular attention as they offer high levels of regioselectivity in favor of branched products when using mono-substituted allylic substrates.[9] Thus, iridium-catalyzed allylic substitution reactions emerged as complementary methods to traditional palladium chemistry, mostly selective for linear compounds, allowing the formation of enantioenriched compounds from simple substrates (Scheme 2).

Since the preliminary work of Kashio and Takeuchi in 1997 probing the unusual efficiency and selectivity of iridium-based catalysts in allylic substitution reactions (racemic version),[10] both the design of chiral ligands and mechanistic studies have allowed the successful development of enantioselective and diastereoselective versions.[9c] Selected examples are given in Scheme 3 for carbon-based nucleophiles. Moreover, the use of prochiral nucleophiles enabled the stereoselective formation of vicinal

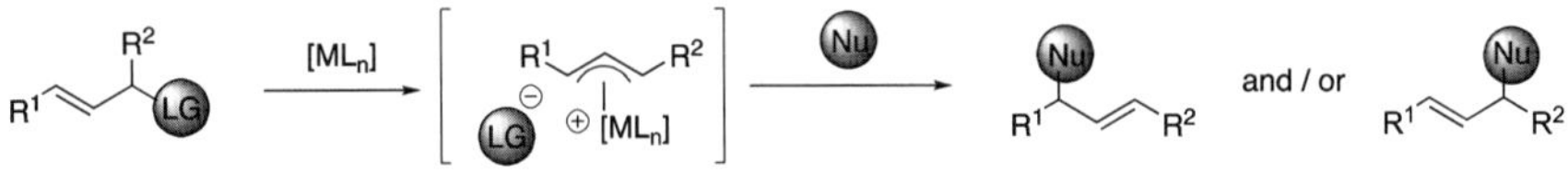

Scheme 1. Transition metal–catalyzed allylic substitution reactions.

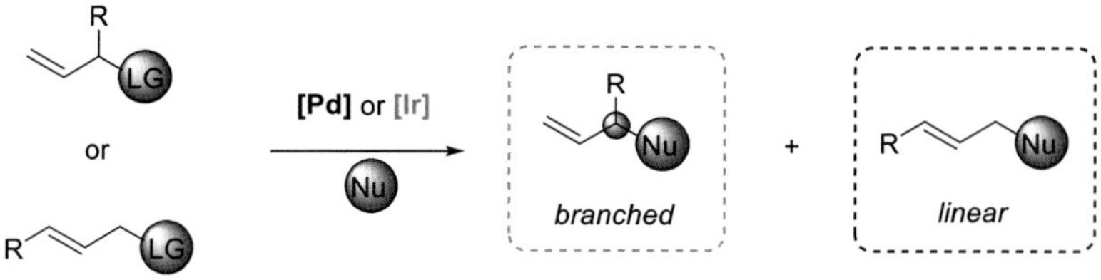

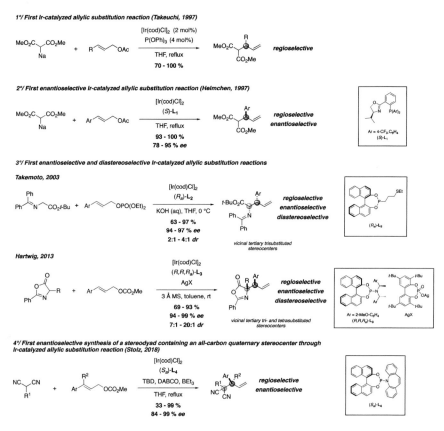

Scheme 2. Regioselectivity in palladium- and iridium-catalyzed allylic substitution reactions.

Scheme 3. Development of regio- and stereoselective iridium-catalyzed allylic substitution reactions (restricted to carbon-based nucleophiles) [See Ref. 10, 9c and references cited therein].

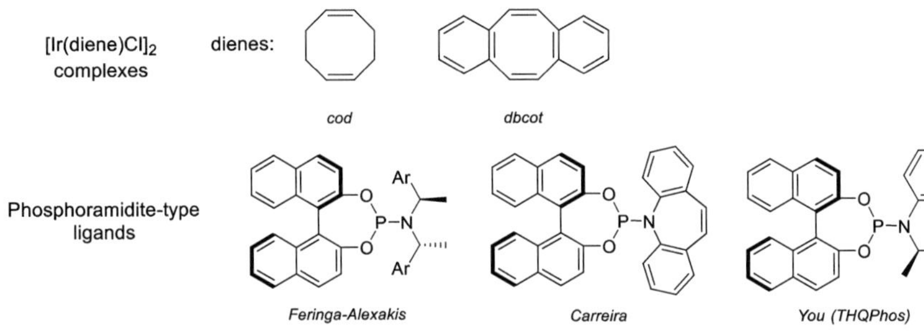

Scheme 4. Iridium complexes and phosphoramidite ligands for enantioselective allylic substitution reactions.

tertiary and/or quaternary stereogenic centers including all-carbon quaternary centers. The major breakthrough in this field is the impressive impact of chiral phosphoramidite ligands, which provide excellent levels of regio- and enantioselectivity.

Nowadays, iridium-catalyzed allylic substitution reactions are mainly carried out using dimeric [Ir(diene)Cl]$_2$ complexes as Ir(I) sources and phosphoramidite-type ligands featuring a 1,1'-bis(2-naphthol) (BINOL) framework (Scheme 4). Different combinations of dienes and ligands can be employed. Cycloocta-1,5-diene (cod) generally serves as standard spectator ligand, but dibenzocyclooctene (dbcot) can play the same exact role, producing highly efficient and more stable complexes that operate even under aerobic conditions. Feringa–Alexakis–type phosphoramidites are the most popular ligands but they reveal some limitations in terms of enantioselectivity with sterically hindered substrates.[11] Such limitations have been at the origin of the THQphos ligand, developed by You and coworkers, that displays a broader substrate scope.[12] Carreira's ligand is a good candidate as well for the allylation of sterically congested substrates (see above) but its principal application remains the direct activation of allylic alcohols as cationic π-allyl precursors.[13]

The *in situ* preparation of iridium catalysts based on Feringa–Alexakis–type and You's phosphoramidites ligands is realized under drastic dry and oxygen-free conditions. Basic conditions are required in order to promote a crucial cyclometalation step furnishing the corresponding iridacyles as active species, through insertion of the metal into C(sp^3)–H

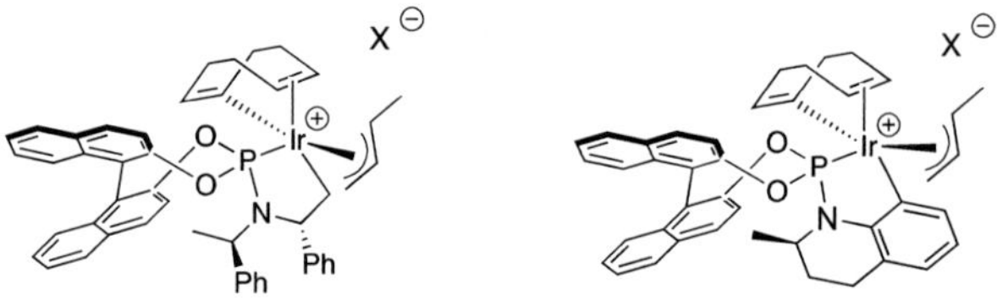

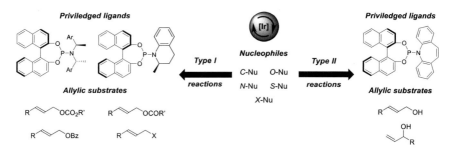

Scheme 5. (π-Allyl)Ir(III) complexes used as catalysts in enantioselective allylic substitution reactions.

Scheme 6. Type I and type II iridium-catalyzed allylic substitution reactions.

or C(sp^2)–H bonds of the ligands. In order to overcome the necessity of such basic conditions, the corresponding air stable, storable, and easy-to-handle cationic (π-allyl)Ir complexes can be synthesized easily and used directly as active catalytic species (Scheme 5).[14] On the other hand, Carreira's phosphoramidite ligand does not require basic conditions to be effective (olefin coordination takes place instead of C–H metalation) and even tolerates acidic conditions (Brønsted or Lewis acids).

These catalytic systems have enabled the expansion of the iridium-catalyzed stereoselective allylic substitution reactions to a large range of nucleophiles and pronucleophiles, stabilized or not (*C*-nucleophiles, *N*-nucleophiles, *O*-, *S*-nucleophiles and halide-nucleophiles). Two main approaches have been developed, depending on the nature of the allylic substrate (Scheme 6). Reactions of type I involve common allylic substrates such as carbonates, acetates, benzoyls, or even halides and are generally conducted using Feringa–Alexakis–type and You's phosphoramidite ligands. Reactions of type II refer to branched allylic alcohols as electrophilic π-allyl sources, and occur only in the presence of the

Carreira's phosphoramidite ligand that is insensitive to the acidic conditions required for the activation of the alcohols. In this case, the *in situ*-generated (π-allyl)Ir complexes exhibit strong electrophilicity and can react even with poor nucleophiles such as simple olefins, for example. Moreover, the use of simple and readily available allylic alcohols as substrates make these reactions greener in terms of atom economy as water is the only waste.

In the literature, many studies are devoted to the iridium-catalyzed regio- and stereoselective allylic substitutions with either heteroatomic nucleophiles or stabilized carbon-centered nucleophiles derived from 1,3-dicarbonyl compounds, such as malonates. On the contrary, the use of nonstabilized carbon-centered nucleophiles is less described, albeit such stereoselective reactions may offer a practical, highly enantioselective tool for target-oriented synthesis and total synthesis. Because of their intrinsic instability and the related nondesired side reactions, these nucleophiles are generally employed under their masked forms as "enolate surrogates." Moreover, organometallic reagents and formyl anion surrogates can act as nonstabilized carbon-centered nucleophiles in these reactions with excellent results. Finally, under specific conditions, weak nucleophiles such as olefins and aromatic compounds can also be subjected to iridium-catalyzed allylation.

The purpose of this chapter is to review the recent advances in the field of iridium-catalyzed stereoselective allylic substitution reactions with both enolate surrogates and other unusual nucleophiles. Most of the strategies involving enolate surrogates can be viewed as interesting alternatives to more conventional methods based on the direct use of enolates that lead to poor, if any, selectivity control. Although the given examples of allylations often refer to enantioselective reactions, a few examples are devoted to diastereoselective processes. Interesting reports on stereodivergent iridium-catalyzed allylations of enolate surrogates under dual catalysis will be discussed as well.

II. Iridium-catalyzed Allylations of Silyl Enolates

As mentioned before, most of the reported methodologies for the iridium-catalyzed enantioselective allylic substitutions with carbon-centered nucleophiles are described using malonate derivatives, β-dicarbonyls, and other equivalents. On the contrary, the use of nonstabilized carbon-centered nucleophiles, generated *in situ* under classical basic conditions, is less common due to their predisposition to undergo reactions with poor regio- and chemioselectivity, and also to favor multiple allylations. In order to overcome such issues, some methodologies have been designed that make use of nucleophiles surrogates, generally less basic than the corresponding alkali metal enolates, like silylated enolates.

A. Enantioselective α-Allylation of Silyl Enolates

Ketones, esters, and their α,β-unsaturated derivatives have been successfully submitted to iridium-catalyzed allylic substitution reactions thanks to the use of their silyl enol ether equivalents. In most cases, the challenge resides in a fine tuning of the conditions for the deprotection of the silylated group that will release the nucleophilic enolate in the medium. In type I reactions (Scheme 6), classical deprotection methods such as the use of fluoride sources or Lewis bases can be applied. In type II, either Brønsted or Lewis acids are employed for both deprotection of the silyl enol ether and activation of the allylic alcohol.

1. *α-allylation of silyl enol ethers as ketone surrogates*

Due to their nonstabilized character, enolates of simple ketones require a strong base to be generated. A convenient way to overcome this difficulty is to prepare the corresponding silyl enol ether, which readily delivers the desired enolate in a more controlled fashion, in milder conditions.

Hartwig and coworkers have reported the first iridium-catalyzed allylation of nonstabilized ketone enolates using the corresponding silyl enol ethers in 2005 (Scheme 7).[15] A combination of zinc and cesium fluoride proved to be the most efficient system to deprotect the silyl enol ethers. Their reactions with 3-substituted allylic carbonates provided the desired branched isomer in high regio- and enantioselectivities with good to excellent yields. No trace of the double allylation product is observed. The catalyst prepared *in situ* from [Ir(cod)Cl]$_2$ and the Feringa phosphoramidite afforded excellent enantiomeric excesses (ees) (91–96% ee).

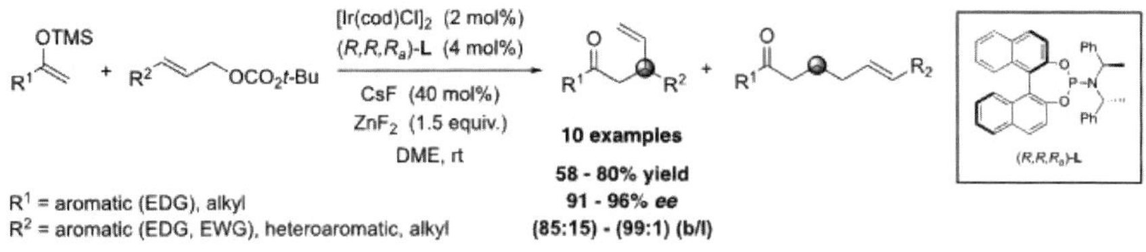

Enolate Surrogates and Unusual Nucleophiles in Allylic Substitutions 417

Scheme 7. First uses of silyl enol ethers as nucleophiles in iridium-catalyzed allylic substitutions.[15]

Silyl enol ethers can also be exploited in type II reactions, meaning using an allylic alcohol as π-allyl precursor. As allylic alcohols are less reactive than the corresponding carbonates, acidic conditions are required to activate the hydroxyl group and generate the corresponding π-allyl species. Thus, Yang and coworkers have envisioned using a Lewis acid as both a π-allyl promoter and a silyl deprotecting agent in the iridium-catalyzed α-allylation of trimethylsilyl (TMS) enol ethers with secondary allylic alcohols (Scheme 8).[16] Sc(OTf)$_3$ turned out to be the best promoter, giving the desired compounds in good yields. The catalyst prepared *in situ* from [Ir(cod)Cl]$_2$ and Carreira's phosphoramidite ligand furnished excellent enantio- and regioselectivities in most cases (13 examples, 91–99% ee). This method has been successfully applied to the stereodivergent total synthesis of calyxolanes A, B, and their respective enantiomers in a three-step sequence.

2. α-allylation of silyl ketene acetals as ester surrogates

Ester and lactone enolates represent also an important class of nonstabilized enolates that are used in Ir-promoted allylic substitutions as their silylated derivatives (silyl ketene acetals). Trimethylsilyloxyfuran, a lactone enolate surrogate, is an essential building block to access butenolides that are found notably in various natural products. Hartwig and coworkers described a rare C3 regioselective alkylation of trimethylsilyloxyfuran, by using allylic carbonates and iridium catalysts (Scheme 9).[17] A neutral preactivated iridium catalyst is employed, generated from C(sp^3)–H metalation of a Feringa–Alexakis–type phosphoramidite ligand and featuring ethylene as a supplementary ligand. Under these conditions, high regioselectivity is achieved as only traces of C5-alkylated product is observed. The C3-allylated product readily isomerizes to the corresponding

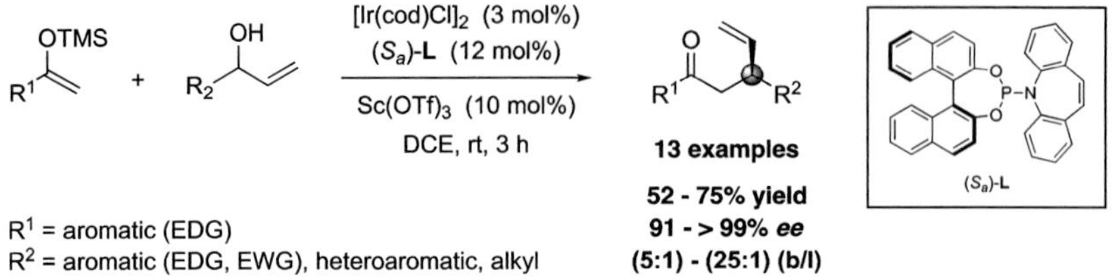

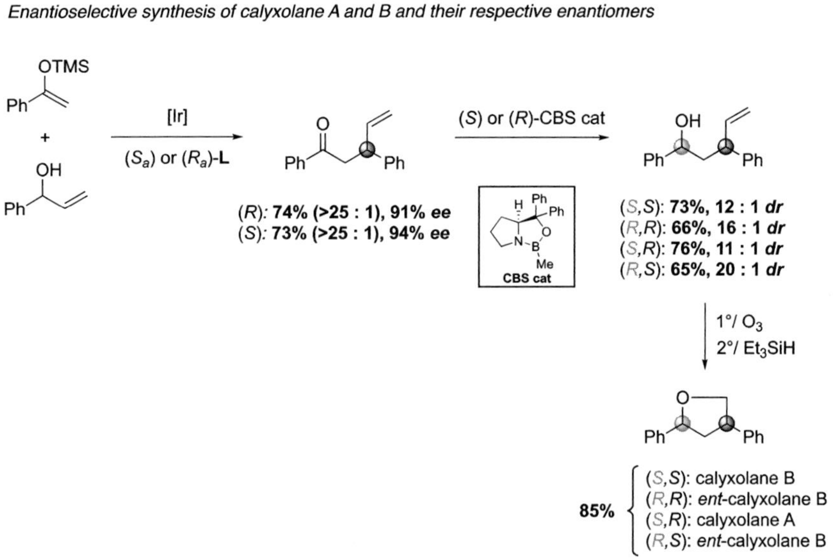

Scheme 8. Enantioselective α-allylations of silyl enol ethers with secondary alcohols and application to the total synthesis of calyxolanes A, B, and their enantiomers.[16]

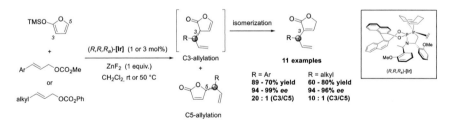

Scheme 9. Regio- and enantioselective allylation of trimethylsiloxyfurans as lactone enolate surrogates.[17]

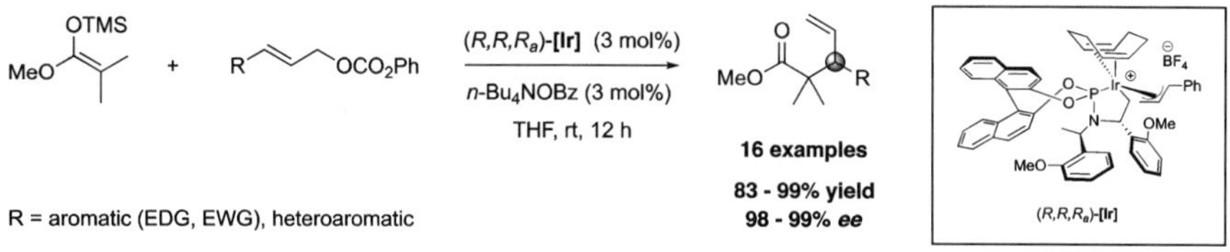

Scheme 10. Regio- and enantioselective allylation of a silyl ketene acetal as isobutyrate surrogate.[18]

conjugated lactone. Both electron-rich and electron-poor aromatic substituents on the aryl–allyl methyl carbonate are well tolerated, giving high yields (70–89%) and high enantioselectivity (ee up to 99%). Switching to phenyl carbonates, in combination with increased catalytic loadings and temperature, allows the alkylation of allyl units bearing aliphatic substituents, with similar yields and enantioselectivities. Finally 3-, 4- and 5-methyl-substituted trimethylsiloxyfurans are tolerated in this reaction, though with modifications of the reaction conditions. The fluoride source turned out to be critical as only ZnF_2 afforded the desired compounds in good yields. Control experiments suggested that the alkoxide released from the carbonate, following formation of the π-allyl species, frees the furan enolate by forming the corresponding TMS alcoholate.

Based on those observations, Hartwig and coworkers envisioned that the fluoride source could be replaced by a substoichiometric amount of tetrabutyl ammonium benzoate in order to release the desired enolate in the reaction medium (Scheme 10).[18] With this alternative methodology, the silyl ketene acetal derived from isobutyrate was successfully used as the ester surrogate in iridium-catalyzed allylic substitutions with various cinnamyl phenyl carbonates. The desired branched products featuring vicinal quaternary and tertiary centers were isolated in high yields and enantioselectivities. The reaction is easily transposed to other *gem*-dialkyl silyl ketene acetals derived from different aliphatic esters, with the same efficiency.

3. α-allylation of α,β-unsaturated silyl enol ethers

Often used as Michael acceptors, α,β-unsaturated ketones are rarely used in enantioselective allylation processes. The corresponding silyl enol ethers turned out to be efficient enolate surrogates to address this issue.

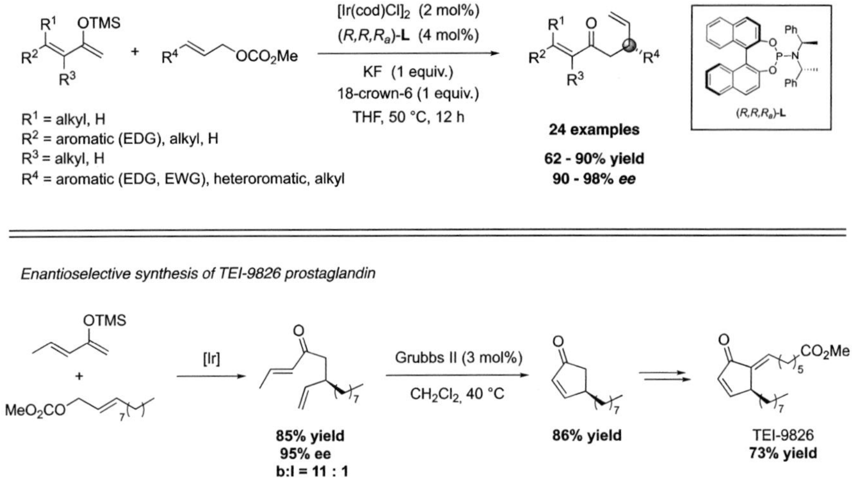

R¹ = alkyl, H
R² = aromatic (EDG), alkyl, H
R³ = alkyl, H
R⁴ = aromatic (EDG, EWG), heteroromatic, alkyl

24 examples
62 - 90% yield
90 - 98% ee

Enantioselective synthesis of TEI-9826 prostaglandin

85% yield
95% ee
b:l = 11 : 1

86% yield

TEI-9826
73% yield

Scheme 11. Enantioselective allylation of α,β-unsaturated silyl enol ethers and application to the total synthesis of TEI-9826 prostaglandin.[19]

Hartwig and coworkers have reported the first iridium-catalyzed α-allylation of α,β-unsaturated ketones by this method (Scheme 11).[19] Once again the initial deprotection of the TMS unit is critical. Previously described systems, such as mixtures of CsF and ZnF_2, are not successful. After fine tuning of the conditions, an equimolar combination of KF and 18-crown-6 ether afforded the desired allylated α,β-unsaturated ketones in excellent yields and enantioselectivities (24 examples, 90–98% ee). The products can serve as advanced intermediates for the synthesis of natural prostaglandins. With this in mind, the authors have reported the enantioselective synthesis of the prostaglandin derivative TEI-9826, a potent anti-cancer drug, by using their efficient methodology as key step. Intramolecular ring-closing metathesis and an additional three-step sequence led to the enantioenriched TEI-9826.

The method developed by Yang and coworkers (see Scheme 8) can also be applied to α,β-unsaturated silyl enol ethers in their reactions with allylic secondary alcohols (Scheme 12).[16] The allyl-substituted α,β-unsaturated ketones are obtained in modest to good yields, excellent enantioselectivity, and good regiocontrol toward the branched product in most cases.

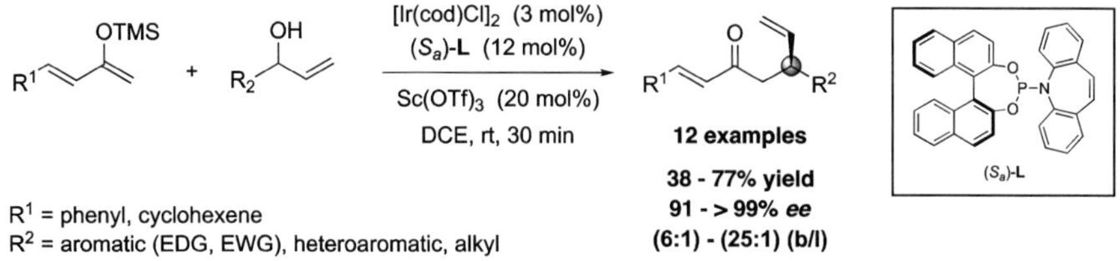

Scheme 12. Enantioselective allylation of α,β-unsaturated silyl enol ethers with allylic alcohols.[16]

4. α-allylation of silylated enols of vinylogous esters and amides

Danishefsky and Rawal dienes are enolates of vinylogous esters and amides, commonly used in Diels Alder [4+2] cycloadditions. Yet they are rarely employed as nucleophiles. The Hartwig group has reported the iridium-catalyzed allylic substitution of these dienes with allylic carbonates under different sets of conditions (Scheme 13).[20] Notably, the influence of additives such as KF, ammonium acetate, and alkoxide has been studied. A wide scope of aromatic and aliphatic allylic carbonates is tolerated in the optimized conditions. The stoichiometric combination of KF/18-crown-6 ether with methyl allyl carbonate as allyl precursor affords the desired vinylogous esters and amides in good to excellent yields and enantioselectivities (Scheme 13, conditions [a]). However, instead of behaving as enol-deprotecting agents (Lewis base role), those additives behave more as Brønsted bases and favor the C(sp$_3$-H) metalation of the ligand that generate the active catalytic species. To overcome this undesired role, new conditions have been developed using a substoichiometric amount of tetrabutyl ammonium acetate (Scheme 13, conditions [b]), which led to very similar yields and enantioselectivities. Finally, improved results were observed in the presence of a substoichiometric amount of KOMe/18-crown-6 ether (Scheme 13, conditions [c]). Finally, the use of the preformed metalacyclic ethylene–Ir complex minimized the impact of the additive on the *in situ* formation of the active catalyst (Scheme 13, conditions [d]). Under these conditions, no additives are required as the carbonate anion released during the reaction is able to deprotect the enolate. Finally, a nine-step sequence based on this enantioselective iridium-catalyzed α-allylation of Rawal's dienes has been

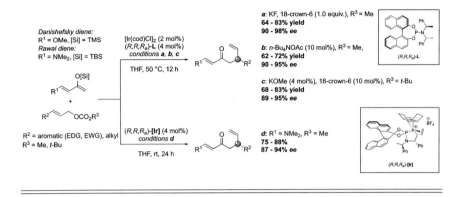

Scheme 13. Conditions for the enantioselective allylation of Danishefsky and Rawal dienes and application to the total synthesis of fesoterodine.[20]

designed by the authors to access the antimuscarinic drug fesoterodine. The key allylation step proceeded with 92% ee.

B. Enantioselective γ-Allylation of Silyl Dienolates Derived from Dioxinones

Albeit β-dicarbonyl compounds have been extensively studied in metal-catalyzed α-allylation, γ-allylation remains largely unexplored. Hartwig and coworkers have described such γ-allylation reactions by using the silyl enol ethers of dioxinones as β-ketoester surrogates (Scheme 14).[21] From a

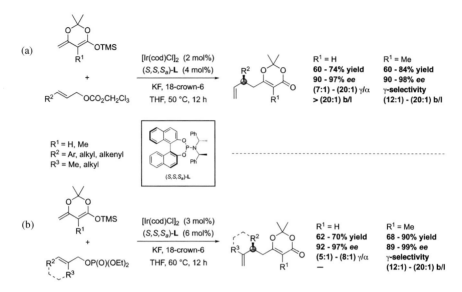

Scheme 14. (a) Regio- and enantioselective γ-allylation of silyl dienolates with disubstituted allylic carbonates (b) and trisubstituted allylic phosphates.[21,22]

synthetic point of view, the obtained allylated dioxinones offer versatile precursors to various motifs such as β-ketoesters, β-ketoamides, tetramic acid, or 1,3-oxazin-2,4-dione. The previously reported system KF/18-crown-6 in THF has proven to be the most efficient base in this case. The leaving group of the allylic substrate has a crucial impact on the reaction outcome. With simple carbonates such as methyl or t-butyl carbonates, a low γ/α regioselectivity is observed. However, the more reactive trichloro-ethyl (Troc) carbonates offer excellent reactivity: along with good yields, they provide high branched/linear (b/l) ratios, high γ-regioselectivities and high enantioselectivities (Scheme 14a). Due to their sterical hindrance, α-substituted silyl dienolates (R^1 = Me) offer complete γ-regioselectivity and b/l selectivities ranging from 12:1 to 20:1. Importantly, control experiments have shown that the γ/α-regioselectivity, as well as the b/l selectivity, is affected by both the phosphoramidite ligand and the allylic leaving group.

Moreover, the γ-allylation of silyl dienolates takes place also with trisubstituted allylic substrates (Scheme 14b).[22] Because of their steric hindrance, the formation of (π-allyl)Ir complexes from such substrates is thermodynamically disfavored. A simple way to address this issue is to employ a stronger allylic leaving group. With this in mind, the use of allylic diethyl phosphates was envisioned. These substrates, combined with slightly higher temperature (60°C) and amount of catalyst (3%), afforded the desired compounds in good yields and excellent ees (90–98% ee). While α-methyl-substituted dioxinone (R^1 = Me) show excellent γ-regioselectivity, a lower γ,α ratio is observed for the non-α-substituted dioxinones (R^1 = H). On the other hand, the steric hindrance of the allylic phosphate does not change significantly the regioselectivity. Overall, a wide variety of allylic substrates are tolerated and this process could be easily scaled up to a 2 mmol scale.

C. Conclusion

Because aliphatic ketones, esters, and α,β-unsaturated carbonyl compounds are known to afford highly unstable enolates, their use as nucleophiles is often associated with undesired side reactions. However, silyl enol ethers are convenient surrogates to these carbonyl enolates and offer a pleasant alternative in order to develop highly selective allylation reactions. Recent literature demonstrates indeed that they permit efficient chemo-, regio- and enantioselective iridium-catalyzed allylation reactions in mild conditions, with wide substrate scope.

III. Iridium-catalyzed Stereoselective and Stereodivergent Allylations of Other Enolate Surrogates

Since the use of silyl enol ethers as enolate surrogates in stereoselective allylation reactions under iridium catalysis has been well documented, additional methodologies involving less common enolate surrogates have emerged with success. Alkyl enol ethers, enamines, enamides, enecarbamates, and β-ketocarboxylates have been designed in order to suppress the main drawbacks associated to the use of highly reactive enolates generated from aldehydes or ketones. Similarly, esters and amides, which present an attenuated acidity, can be engaged in allylation reactions under their privileged masked forms of orthoacetates and ketene aminals, respectively. In this field, some original enantioselective and diastereoselective α-allylation reactions, giving excellent efficiency and selectivity, have been reported. Interestingly, the use of enamines as surrogates for aldehydes and ketones appears as a milestone for the development of highly valuable stereodivergent methodologies, for a straightforward access to all stereoisomers of a given compound under dual iridium and amine catalysis. Stereodivergent allylation reactions have been applied also to esters and α,β-unsaturated aldehydes by combining iridium and Lewis base catalysis.

A. Stereoselective α-Allylations

Most of the examples described in this section are related to the use of enol ethers, enamines, enamides, and enecarbamates as enolate surrogates for the development of stereoselective α-allylation reactions of highly reactive substrates such as aldehydes or ketones. Drawbacks from the use of highly reactive enolates, as seen in the previous section, can be overcome by employing such more stable masked forms. Orthoesters and ketene aminals are also considered as ester and amide enolate surrogates, respectively.

1. α-allylation of enol ethers and enamines as aldehydes surrogates

The enantioselective α-allylation reactions of aldehydes are poorly described, because the use of aldehyde enolates is associated to several issues. Indeed, rapid self-aldolization as well as O-alkylation side reactions cannot be suppressed, leading to low reaction efficiency.

Very recently, Carreira and coworkers have postulated that ethylene glycol mono-vinyl ether could be an efficient acetaldehyde anion surrogate in enantioselective reactions with allylic carbonates.[23] The authors assumed that the vinyl ether might act as a nucleophile toward the *in situ*-generated (π-allyl)Ir complex, furnishing a short-living oxonium intermediate prone to rapid cyclization to afford the corresponding acetal-protected γ-vinylaldehyde (Scheme 15).

These reactions have been realized in the presence of $ZnBr_2$ as a Lewis acid, with the Carreira's phosphoramidite ligand that is highly

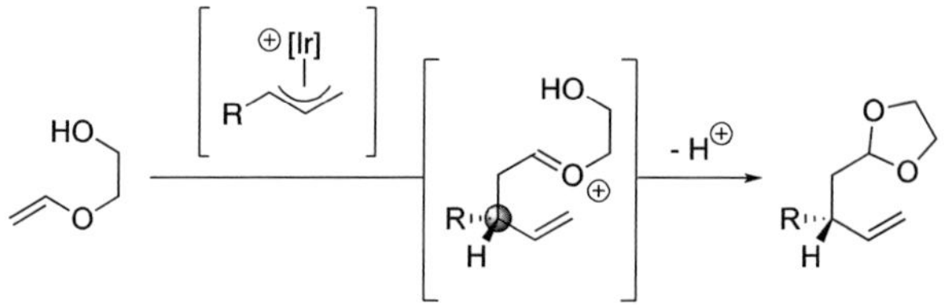

Scheme 15. Postulated mechanism for the iridium-catalyzed α-allylation of ethylene glycol mono-vinyl ether.[23]

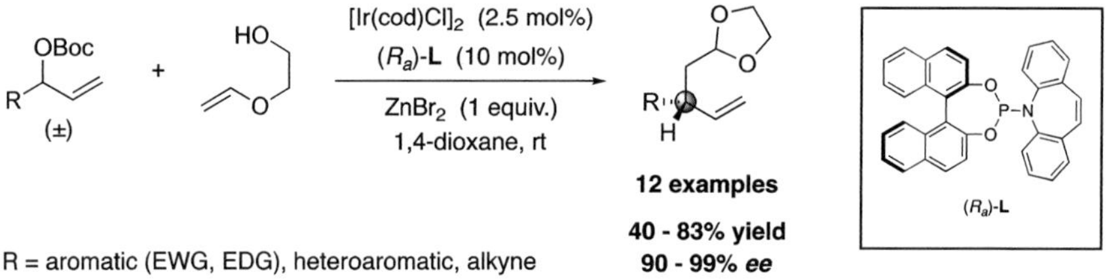

Scheme 16. Enantioselective α-allylations of ethylene glycol mono-vinyl ether.[23]

resistant under such reaction conditions. The substrate scope of this reaction is quite large, but mainly restricted to aromatic and heteroaromatic allylic carbonates, except for a 1,4-enyne substrate. Moderate to excellent yields and high enantioselectivities, up to 99% ee, were observed. Substrates bearing electrowithdrawing groups appeared as the best candidates (Scheme 16).

Enamines are also valuable enolate surrogates for reactive aldehydes. In their efforts to develop original synthetic strategies, Carreira and co-workers demonstrated that enamines generated *in situ* from chiral allylic amino alcohols and aldehydes can undergo intramolecular allylation reactions in the presence of iridium catalysts and acid additives. Reduction of the intermediate tetrahydropyridines affords enantioenriched 3,4-*trans*-disubstituted piperidines with high diastereocontrol (Scheme 17).[24] This approach represents a direct and efficient access to these valuable compounds, if compared to classical methodologies based on hydroaminations, dearomatizations, or lactam reductions, among others.

Optimization studies using [Ir(cod)Cl]$_2$ and the achiral Carreira's phosphoramidite ligand have revealed that 3,5-dichlorobenzoic acid is the best promoter for the formation of both the (π-allyl)Ir species and the

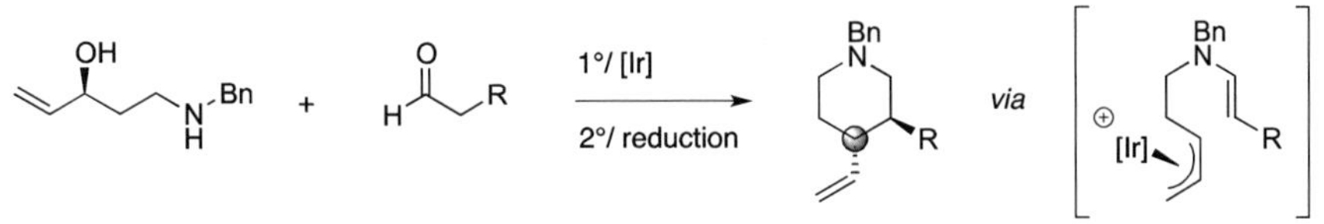

Scheme 17. Iridium-based strategy for the diastereoselective preparation of 3,4-*trans*-disubstituted piperidines.[24]

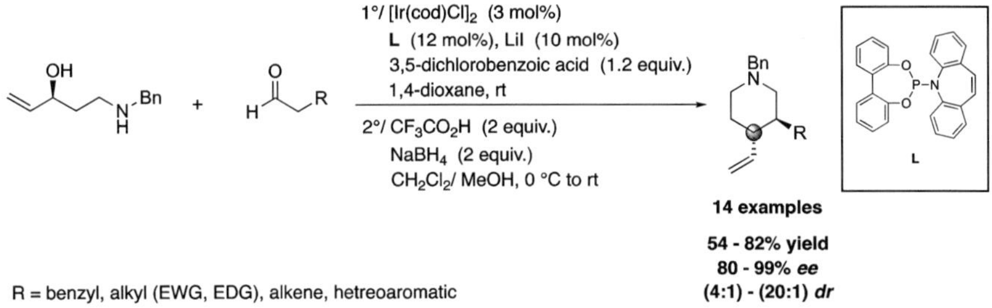

Scheme 18. Diastereoselective synthesis of 3,4-*trans*-disubstituted piperidines via intramolecular α-allylation of enamines.[24]

enamine intermediate. The addition of lithium iodide is necessary in order to improve the reaction selectivity. This one-pot iridium-catalyzed cyclization/reduction procedure is highly tolerant to functional groups, it offers thus a large substrate scope regarding the aldehyde partner, affording the corresponding piperidines as valuable building blocks, in good yields and stereoselectivities (Scheme 18). The same reaction could be applied to the straightforward preparation of bicyclic piperidines. Finally, further synthetic transformations of the obtained chiral piperidines give access to quinuclidine or decahydroisoquinoline frameworks for potential applications in natural product synthesis or drug discovery.

2. *Enantioselective α-allylation of enamines, enamides, and other ketone surrogates*

The iridium-catalyzed α-allylation of ketone-derived enamines was reported in 2007 by Hartwig and coworkers as an additional alternative to the direct allylation of nonstabilized enolates.[25] As shown in Scheme 19,

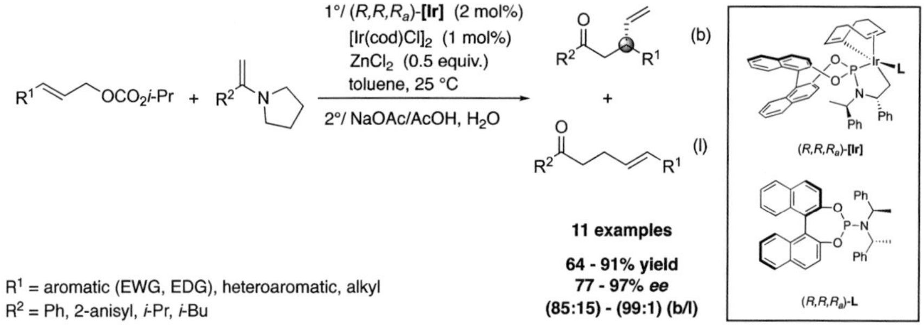

Scheme 19. Enantioselective allylation of ketone-derived enamines.[25]

a mixture of a preformed chiral iridium complex and [Ir(cod)Cl]$_2$ is employed in order to increase the reaction rate in the allylation of enamines with isopropyl allyl carbonates. The presence of ZnCl$_2$ as an additive is needed to trap the isopropanol released from the allylic carbonate. After a final enamine hydrolysis, the desired allylated ketones are recovered with yields up to 91% and excellent regio- and enantioselectivities (Scheme 19). Aromatic and aliphatic pyrrolidine enamines can be used with no real change in reaction efficiency, only the electronic and steric properties of the allylic carbonate derivative have an influence on the reaction rate.

Due to their higher stability, enamides and enecarbamates represent valuable ketone enolate surrogates as well. This novel type of nucleophiles has been involved in the enantioselective iridium-catalyzed alkylation of branched allylic alcohols to furnish the corresponding homoallylic ketones.[26] The use of a Lewis acid promoter (Sc(OTf)$_3$ for enamides and Zn(OTf)$_2$ for enecarbamates) in combination with [Ir(cod)Cl]$_2$ and the Carreira's phosphoramidite ligand allowed the synthesis of a large range of homoallylic ketones with moderate to high yields and enantioselectivities up to 99% (Scheme 20). Aromatic enamides and enecarbamates are the best candidates for these reactions and various secondary allylic alcohols bearing an aryl substituent can be employed with good efficiency.

An elegant strategy based on the *in situ* generation of ketone enolates from γ-substituted allyl β-ketocarboxylates has been developed by You and coworkers. The reaction takes place in the presence of catalytic amounts of [Ir(cod)Cl]$_2$, Feringa–Alexakis–type phosphoramidite

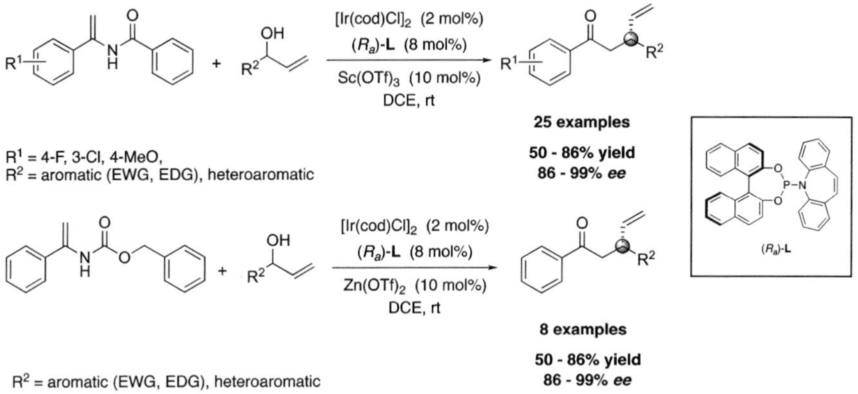

Scheme 20. Enantioselective α-allylation of enamides and enecarbamates.[26]

and 1,8-diazabicyclo[5.4.0]undec-7-ene (DBU) as additive. It would proceed through decarboxylation of the starting material leading to the corresponding (β-oxo-alkyl) (π-allyl)Ir intermediate (Scheme 21).[27]

The substrate scope of these regio- and enantioselective allylic alkylation reactions is large but most examples relate to ketocarboxylates derived from γ-aryl substituted allylic alcohols. These substrates give b/l ratios >98:2 and ees in the range 91–95%. γ-Alkyl allyl alcohols (R^2 = Me or n-C_5H_{11}) are also suitable substrates for this reaction, albeit with slightly lower regio- and enantioselectivities (Scheme 22). When starting from an equimolar mixture of two different allyl ketocarboxylates, four distinct allylated ketones were recovered (branched isomers). This crossover experiment allowed to exclude that the products are formed from an intramolecular rearrangement of the (β-oxo-alkyl)(π-allyl)Ir intermediate.

3. Enantioselective α-allylation of orthoesters as ester surrogates

In parallel to their work on the use of ethylene glycol mono-vinyl ether as aldehyde enolate surrogate, Carreira and collaborators have also reported that trimethyl orthoacetate can be employed as a performant acetate enolate surrogate for the enantioselective alkylation of allylic carbonates under the same conditions (Scheme 23).[23]

The substrate scope of the two reactions is quite similar and satisfactory results could be observed in terms of yield and enantioselectivity with

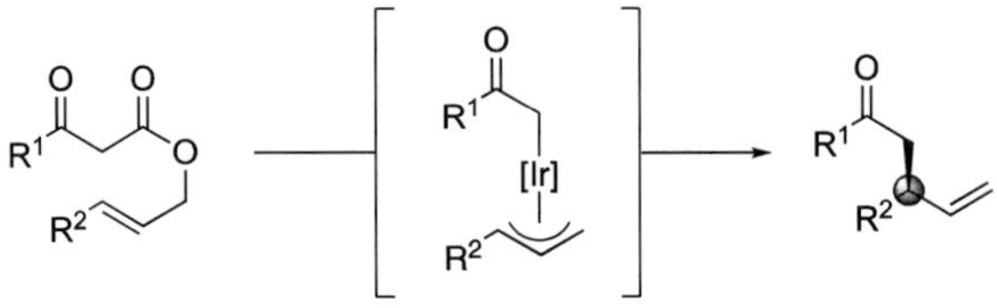
[Ir]

Scheme 21. General mechanism for the decarboxylative allylation of γ-substituted allyl β-ketocarboxylates.[27]

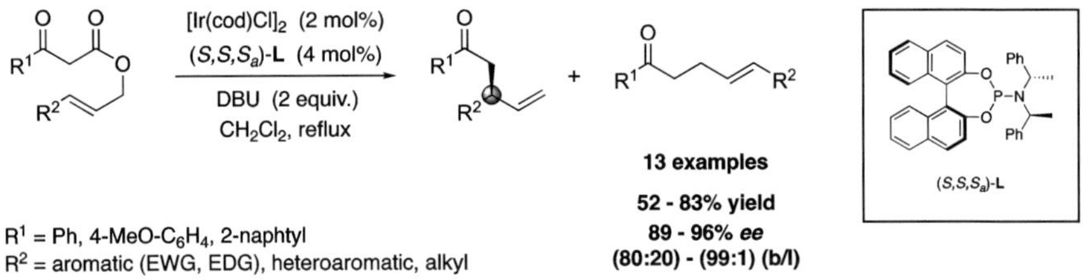
[Ir(cod)Cl]2 (2 mol%)
(S,S,Sa)-L (4 mol%)
DBU (2 equiv.)
CH2Cl2, reflux
13 examples
52 - 83% yield
89 - 96% ee
(80:20) - (99:1) (b/l)
(S,S,Sa)-L
R1 = Ph, 4-MeO-C6H4, 2-naphtyl
R2 = aromatic (EWG, EDG), heteroaromatic, alkyl

Scheme 22. Enantioselective decarboxylative allylation of γ-substituted allyl β-ketocarboxylates.[27]

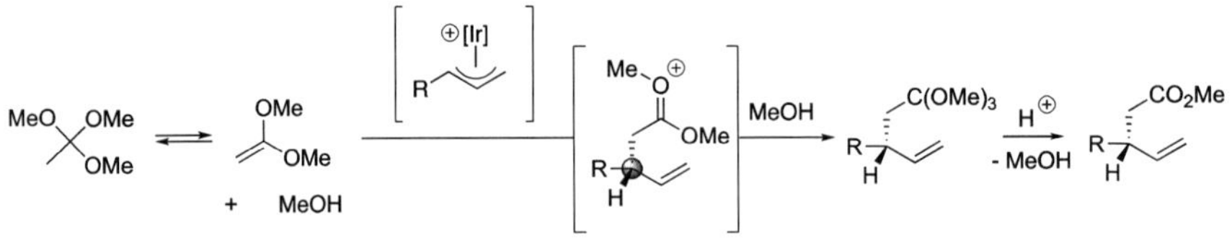
MeOH
- MeOH

Scheme 23. General mechanism for the α-allylation of trimethyl orthoacetate.[23]

allylic carbonates bearing aromatic groups with both electron-withdrawing and donating substituents, heteroaromatic groups, and even a boron-substituted aromatic groups for additional synthetic transformations. This original methodology was highlighted with a formal enantioselective synthesis of (+)-conicol (Scheme 24).

4. *Enantioselective α-allylation of ketene aminals as amide surrogates*

Because of the low C–H acidity of amide derivatives, their use as nucleophiles in iridium-catalyzed enantioselective allylation reactions is rare. An

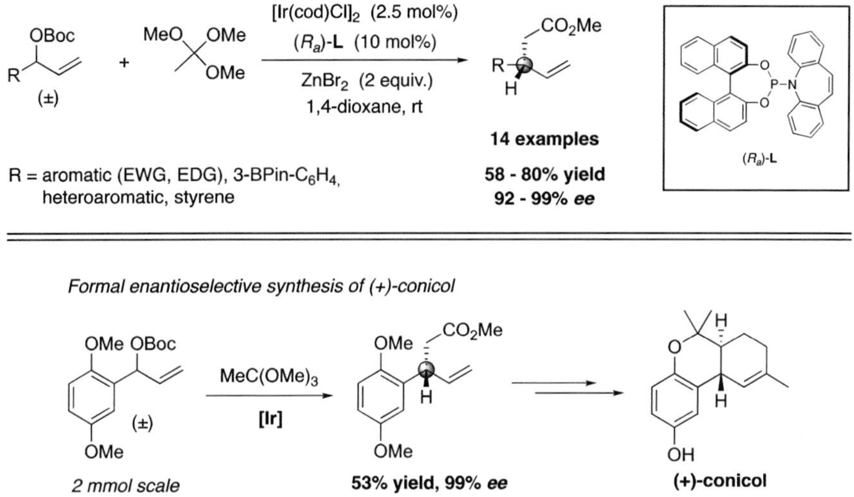

Enolate Surrogates and Unusual Nucleophiles in Allylic Substitutions   431

Scheme 24. Enantioselective α-allylation of trimethyl orthoacetate and application to the formal enantioselective synthesis of (+)-conicol.[23]

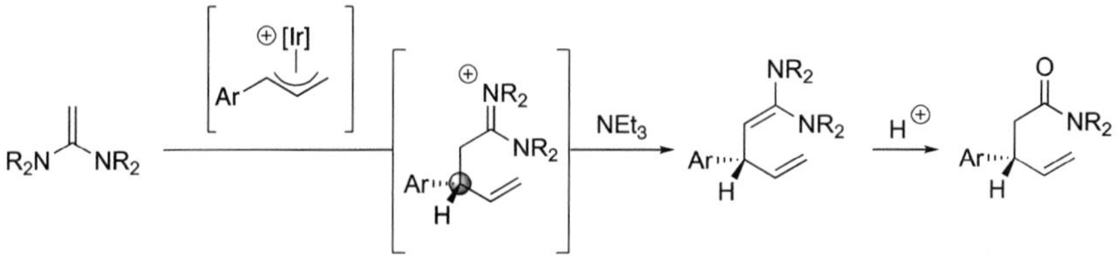

Scheme 25. General mechanism for the iridium-catalyzed α-allylation of ketene aminals.[28]

alternative has been proposed by Carreira and coworkers and allows the simple preparation of enantioenriched β-substituted γ,δ-unsaturated amides starting from the corresponding morpholine ketene aminals as amide enolate surrogates (Schemes 25 and 26).[28]

The scope of the reaction is limited to the use of branched allylic carbonates featuring an aromatic or heteroaromatic group as the substituent. Because morpholine ketene aminals are prompted to decomposition under the acidic conditions (Lewis and Brønsted acids) generally used for the activation of the carbonate moiety, triethylamine has been selected as additive of choice. In this work, two strategies have been reported for the

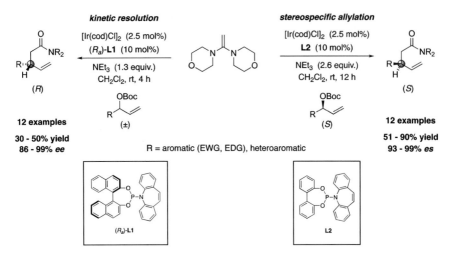

Scheme 26. Enantioselective α-allylation of a morpholine ketene aminal through kinetic resolution and stereospecific allylation reactions.[28]

preparation of both enantiomers of the β-substituted γ,δ-unsaturated amides from racemic allylic carbonates. On one hand, kinetic resolution of the racemic allylic carbonates has been carried out using a chiral ligand **L1**. On the other hand, the unreacted optically pure carbonate has been subjected to a stereospecific allylation reaction with an achiral ligand **L2** (Scheme 26). Furthermore, the obtained amides could be engaged as valuable building blocks for the preparation of ketones, acylsilanes, cyclopentenones, or amines.

B. Stereodivergent α-Allylations Under Dual Catalysis

Nowadays, a major challenge in stereoselective synthesis is to develop straightforward methodologies to access all stereoisomers of a defined molecule bearing multiple stereogenic centers, starting from a single combination of achiral or racemic substrates. Many stereodivergent strategies have been reported and most of them involve the use of complementary chiral catalysts to control the stereogenic centers of the final products. Two different approaches have been applied.

Considering the stereocontrolled construction of two stereogenic centers, the first approach relies on two distinct chiral catalysts that

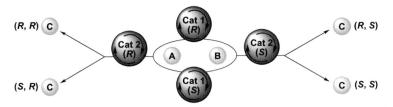

Scheme 27. Stereodivergent dual catalysis principle.

operate separately and form the two stereocenters in a sequential manner. Such strategy implies that a first enantioselective reaction occurs, with the resulting product becoming the substrate for a second diastereoselective step. The second approach, named dual catalysis, is totally different as the two stereogenic centers are formed simultaneously. Both catalysts are committed in one single transition state in which the match/mismatch effects of the chiral catalysts pair will play a key role in stereochemical control. The dual catalysis approach allows the development of fully stereodivergent pathways to the desired stereoisomers, by using defined combinations of chiral catalysts (Scheme 27).

Most of the methods reported in this section are relying on dual metal-organic catalysis combining formal iridium and amine enantioselective catalysis in order to perform the stereodivergent allylation reaction between two highly face-selective catalytic species, the enantioenriched (π-allyl)Ir and the *in situ*-generated enamine partners. Only a few examples describe the merging of iridium and Lewis base or *N*-heterocyclic carbene catalysis.

1. *Stereodivergent α-allylation of aldehydes under dual iridium and enamine catalysis*

Stereodivergent α-allylation of aldehydes has been mainly developed by Carreira and coworkers since their early works in 2013.[29] The reactions between aldehydes and branched allylic alcohols (reactions of type II) were carried out in the presence of Carreira's phosphoramidite ligand and either cinchona alkaloid-derived primary amines or chiral secondary amines, such as the Jørgensen diarylprolinol silyl ether.[30] The main strategy is based on a stereocontrolled activation of both aldehyde and allylic

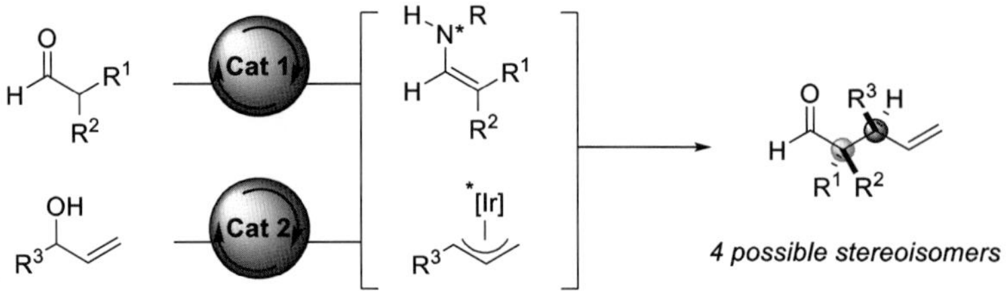

Scheme 28. Strategy for the stereodivergent α-allylation of aldehydes with chiral amine (Cat 1) and iridium (Cat 2) catalysts.[29]

alcohol, furnishing diastereomeric transition states. The straightforward preparation of the four stereoisomers of the final product is achieved by changing the configurations of the two catalysts (Scheme 28).

In their initial report,[29] the authors have first confirmed that both the chiral iridium catalyst and the amine are necessary to observe high relative (diastereomeric ratio) and absolute strereocontrols in the α-allylation of branched aldehydes ($R^1 \neq R^2 \neq H$) with phenylvinyl carbinol and trichloroacetic acid as promoter (Scheme 29). The challenging formation of vicinal quaternary/tertiary stereocenters is achieved with excellent results. More experiments have revealed that such reaction can be fully stereodivergent when using four designed combinations of the enantiomers of the Carreira's phosphoramidite ligand with a cinchona alkaloid-derived amine or its pseudoenantiomer. The substrate scope is quite large and includes allylic alcohols substituted with various aromatic groups, as well as aldehydes bearing functional groups. The stereodivergence of these reactions has been ascertained by preparing the different stereoisomers of representative products featuring contrasting electronic properties. By employing a diallyl alcohol substrate, the reaction yields the corresponding diene with complete regiocontrol (allylation occurring at the more hindered position) and high enantio- and diastereocontrol for the generated stereodyad. According to this result, the authors could postulate an outer sphere transition state in which the two reactive partners present their diastereofaces in an opposite manner diminishing the matched/mismatched effects.

This reaction of α-allylation under dual catalysis has been rapidly applied to the more challenging linear aldehydes by switching from the cinchona alkaloid-derived primary amine to the Jørgensen diarylprolinol silyl ether catalyst.[31] Reaction conditions are quite similar albeit the

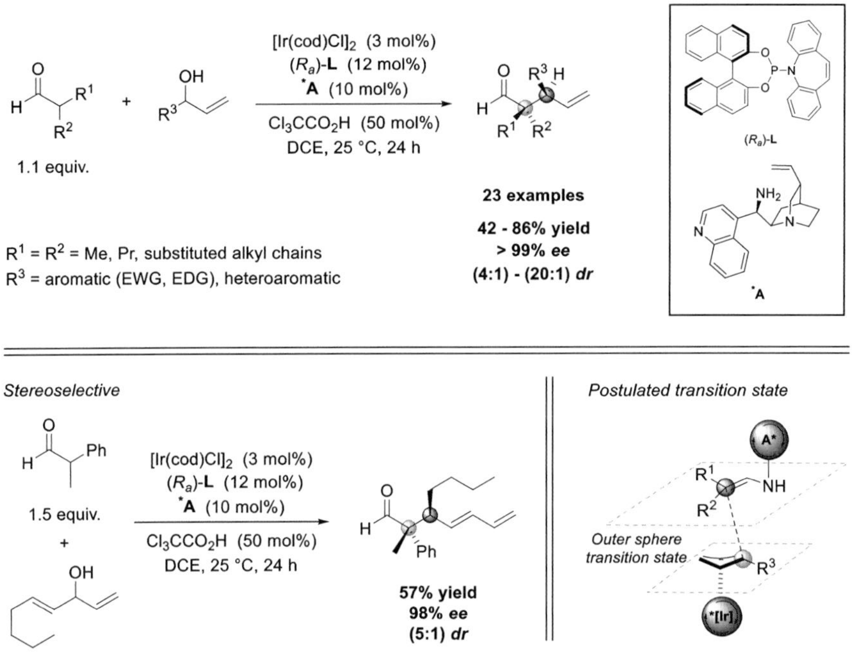

Scheme 29. Enantio- and diastereodivergent α-allylation of branched aldehydes under dual catalysis.[29]

crucial acid promoter has been changed to dimethylhydrogen phosphate. Once again, various branched allylic alcohols bearing an aromatic substituent are employed with a set of aliphatic linear aldehydes to afford the corresponding adducts with satisfactory yields and high enantio- and diastereoselectivities (ee > 99%). The stereodivergence of the catalytic system has been verified by a simple selection of catalyst pairs. Finally, the methodology has been efficiently applied to the stereoselective synthesis of (-)-paroxetine (Scheme 30).

Under similar reaction conditions, this stereodivergent catalytic system has been applied with success to the α-allylation of protected α-amino and α-hydroxyacetaldehydes derivatives to furnish valuable building blocks for further transformations.[32] Moreover, this stereodivergent methodology allows the straightforward access to all four stereoisomers of Δ^9-tetrahydrocannabinols, including the two natural diastereomers, starting from identical substrates and using the same reaction procedures in

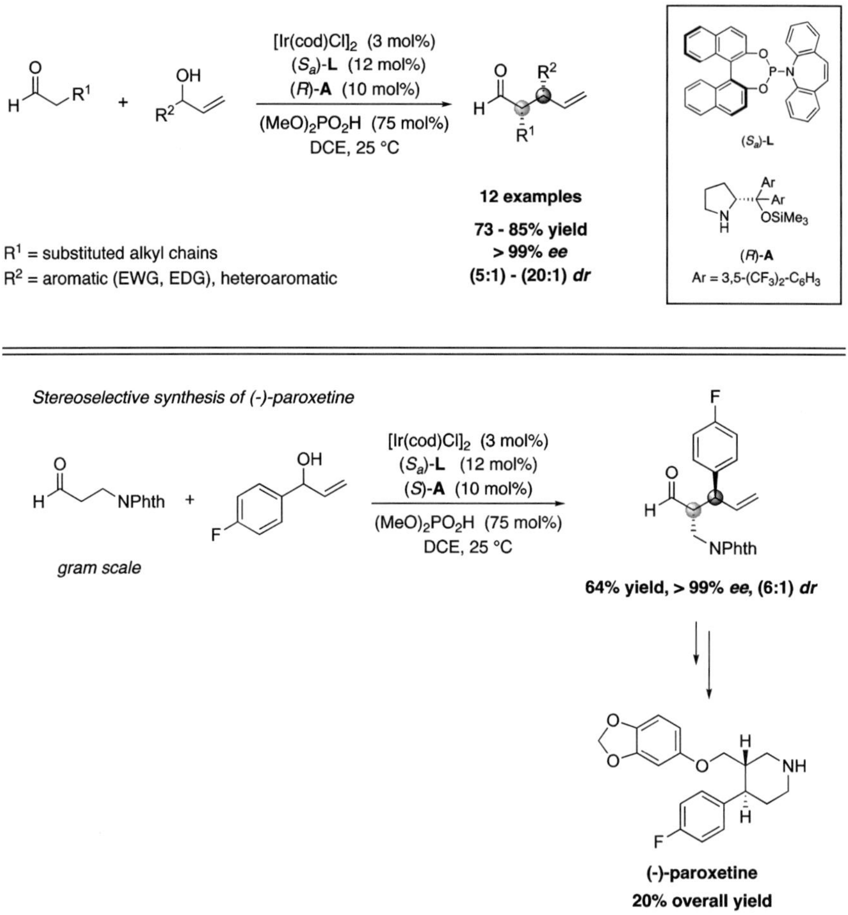

Scheme 30. Sterodivergent α-allylation of linear aldehydes under dual catalysis and application to the stereoselective synthesis of (-)-paroxetine.[31]

each case.[33] Once again, this stereodivergent reaction is controlled by a precise combination of Carreira's phosphoramidite ligands, Jørgensen diarylprolinol silyl ether organocatalysts and catalytic amounts of $ZnBr_2$ salts as efficient promoters (Scheme 31).

Recently, Carreira and coworkers have transposed such methodology to the stereoselective α-allylation of various unstable or volatile nucleophiles like α-chloroaldehydes, α-bromoaldehydes, glutaraldehyde, and acetaldehyde.[34] These aldehydes are highly reactive and undergo rapid self-condensation, polyacetalization, and polymerization. Their use in

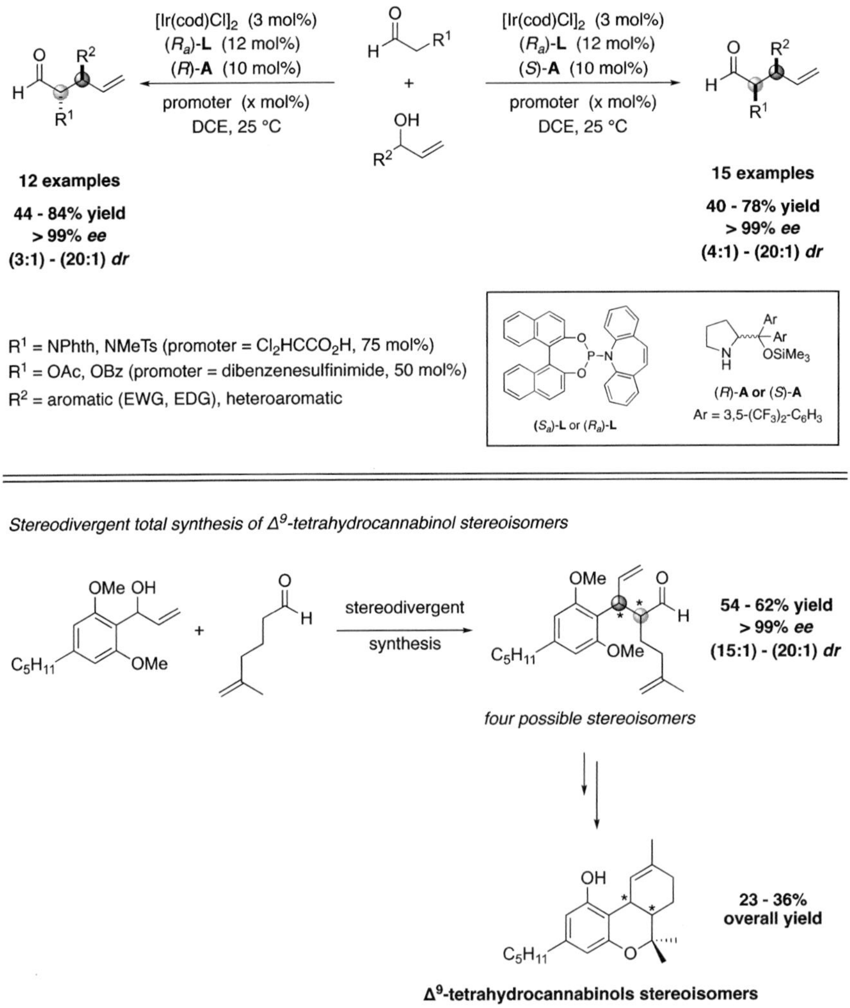

Scheme 31. Sterodivergent α-allylation of protected α-amino and α-hydroxyacetaldehydes derivatives and stereodivergent total synthesis of Δ⁹-tetrahydrocannabinol stereoisomers under dual catalysis.[32,33]

aqueous solutions is an elegant way to take advantage of their nucleophilic properties in carbon–carbon bond forming reactions, while avoiding side reactions. Although no stereodivergence studies have been reported, the authors have demonstrated that the dual catalytic system based on Carreira's phosphoramidite ligand and Jørgensen diarylprolinol silyl ether organocatalyst is efficient in these α-allylation reactions using branched allylic alcohols. Satisfactory yields and enantioselectivities have been obtained. In the special case of α-chloroaldehydes, α-bromoaldehydes, and glutaraldehyde, an efficient diastereocontrol is observed as well. Besides the possible diastereocontrol offered by the prolinol organocatalyst, the authors have shown that these species stabilize the reactive nucleophiles through enamine formation.

2. Stereodivergent γ-allylation of α,β-unsaturated aldehydes under dual iridium and enamine catalysis

Given the ability of dual catalytic systems combining iridium and amino catalysis to promote stereodivergent α-allylation reactions of aldehydes with branched allylic alcohols, as described by Carreira and coworkers, the Jørgensen group has considered applying such catalytic systems to the stereodivergent γ-allylation of α,β-unsaturated aldehydes.[35] With these aldehydes, several challenges have to be examined like the α/γ-selectivity (*via* the transient dienamine intermediate), the regioselectivity (branched vs linear products), the *E*/*Z*-selectivity (due to the transient dienamine intermediate as well), and the stereoselectivity. With this in mind, the authors have developed an optimal dual catalytic system based on Carreira's phosphoramidite ligand and a Jørgensen diarylprolinol silyl ether organocatalyst that allows the stereoselective γ-allylation of cyclic α,β-unsaturated aldehydes. Using branched allylic alcohols with aromatic substituents excellent results in terms of yields, regio-, enantio-, and diastereoselectivities have been attained (Scheme 32). The diastereodivergency of the proposed reaction has been ascertained by using catalysts with opposite configurations. To conclude their study, the authors have reported a complementary general access to linear allyl-substituted products by replacing the iridium catalyst by a palladium catalyst. Satisfactory yields, regio- and enantioselectivities were obtained.

3. Stereodivergent α-allylations under dual iridium and lewis base catalysis

Both chiral tertiary amines and *N*-heterocyclic carbenes represent efficient organocatalysts, in combination with iridium catalysts, for the development

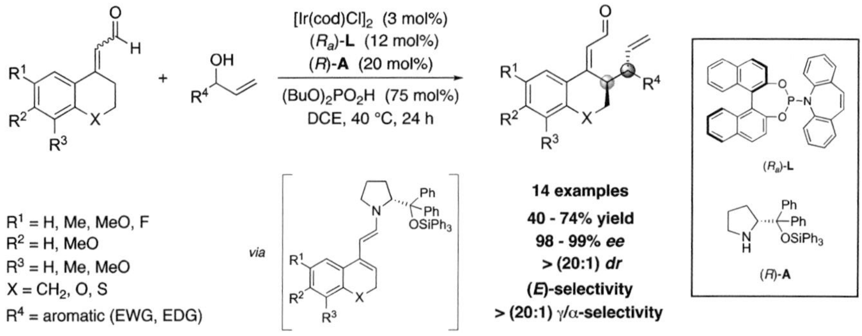

Scheme 32. Stereodivergent γ-allylation of cyclic α,β-unsaturated aldehydes under dual catalysis.[35]

of stereodivergent allylic substitution reactions involving carbonyl derivatives. Both strategies rely on the *in situ* formation of a nucleophilic enolate surrogate. Indeed, NHC organocatalysts are known to perform straightforward α-functionalization of carbonyl compounds, with high stereocontrol, *via* the so-called Breslow intermediates. In the case of tertiary amines, the α-functionalization of carbonyl compounds is also possible via ammonium enolates, as long as the catalyst can be regenerated through acyl transfer to an appropriate nucleophile. Such enolate surrogates can react with electrophilic (π-allyl)Ir complexes to furnish the desired α-allylated products in a stereodivergent fashion, if employing the right combination of chiral catalysts.

It is well documented that chiral tertiary amines are able to form configurationally stable ammonium enolates by reacting with acyl derivatives. Moreover, the resulting enolates undergo electrophilic additions with high facial differentiation. Thus, one can imagine to catalytically generate an enolate starting from an activated ester and the appropriate Lewis base, and to carry out then allylation of this intermediate by using the Hartwig iridium catalytic system, since this catalyst is not sensitive to basic conditions. According to the proposed strategy, the ammonium enolate should act as a stronger nucleophile than the alkoxide or phenoxide released from the ester. It should react with the electrophilic (π-allyl)Ir complex in a stereocontrolled fashion. Finally, the alkoxide or phenoxide would end up the catalytic cycle by regenerating the Lewis base in the reaction medium. By choosing the correct combinations of the two catalysts, all possible stereoisomers can be obtained selectively (Scheme 33).[36]

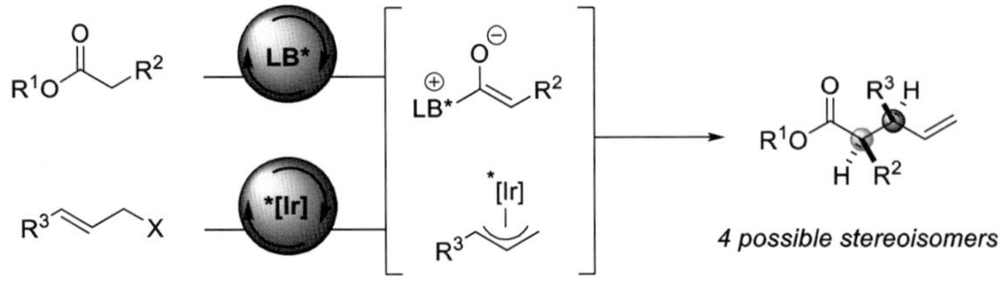

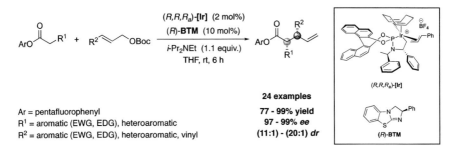

Scheme 33. Strategy for the stereodivergent α-allylation of esters under iridium and Lewis base catalysis.[36]

Scheme 34. Stereodivergent α-allylation of esters under dual iridium–Lewis base catalysis.[36]

With this idea in mind, Hartwig and his collaborators have designed a dual catalytic system based on a preformed cationic (π-allyl)Ir complex and chiral benzotetramisole (BTM) as the Lewis base. Under these conditions, efficient stereodivergent reactions have been observed, giving excellent results in terms of yields and stereoselectivities. The scope of the reaction is wide, but limited to arylacetic acid pentafluorophenyl esters and aromatic allylic carbonates (Scheme 34).

α,β-Unsaturated azolium enolates are generated from the condensation of an α,β-unsaturated aldehyde with an NHC organocatalyst. These intermediates can react with electrophilic allylic species obtained from a cyclic vinyl carbonate under iridium catalysis, to easily generate α,β-disubstituted-γ-butyrolactones as valuable frameworks. The sequence involves α-allylation of the NHC-enolate with a defined (π-allyl)-alcoholate iridium complex and subsequent lactonization. It results in a formal and highly stereocontrolled (3+2) cycloaddition process, in which two new vicinal stereocenters are formed. Because the absolute and relative

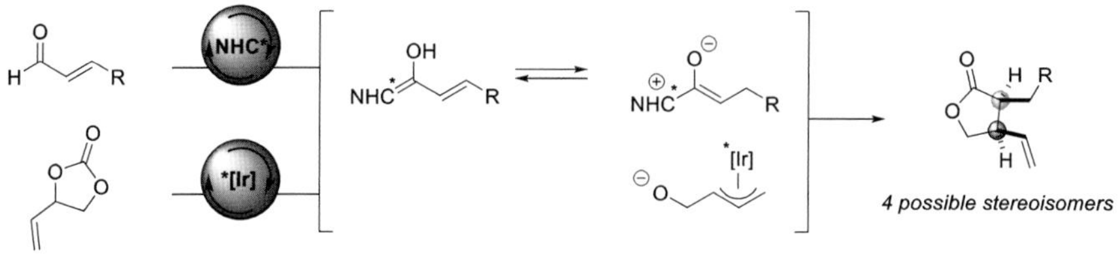

Scheme 35. Strategy for the stereodivergent synthesis of α,β-disubstituted γ-butyrolactones under iridium and NHC catalysis.[37]

configurations of these stereocenters can be controlled by simply using two distinct chiral catalysts, a fully stereodivergent synthesis of the desired α,β-disubstituted-γ-butyrolactones becomes possible (Scheme 35).[37]

This strategy has been developed very recently by Glorius group and relies on the use of a triazolium NHC precatalyst with a classical iridium-based catalytic system. The expected reactivity is only observed when using a specific SPINOL-derived phosphoramidite (SPINOL = 1,1'-spirobiindane-7,7'-diol). The reaction has been investigated with various α,β-unsaturated aldehydes bearing either aromatic or aliphatic substituents and vinylethylene carbonate. It demonstrated high functional group tolerance, since it gave satisfactory yields and excellent enantio- and diastereoselectivities in many instances. Some issues, that is, longer reaction times and partial conversion, are observed with the use of electron-rich α,β-unsaturated aldehydes, which could be solved by switching to the corresponding α-chloro-aldehydes, more reactive toward NHC activation. The full or partial stereodivergence of the process has been verified in several cases, albeit the NHC precatalyst had to be optimized to give good results. Thus, this original dual catalytic system promotes the stereoselective and straightforward formation of substituted δ-lactams featuring two vicinal stereocenters. Finally, its synthetic utility has been highlighted with the total synthesis of the natural lignin (-)-hinokinin (Scheme 36).

C. Conclusion

The use of enol ethers, enamines, enamides, esters, orthoformates, and aminals as enolate surrogates has been explored in order to expand the field of iridium-catalyzed stereoselective allylation reactions. Such

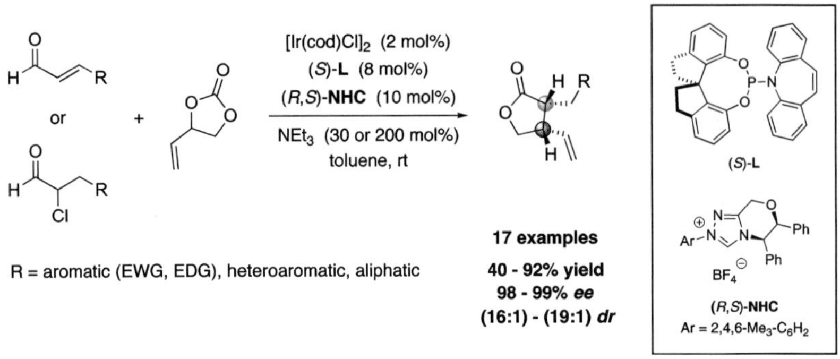

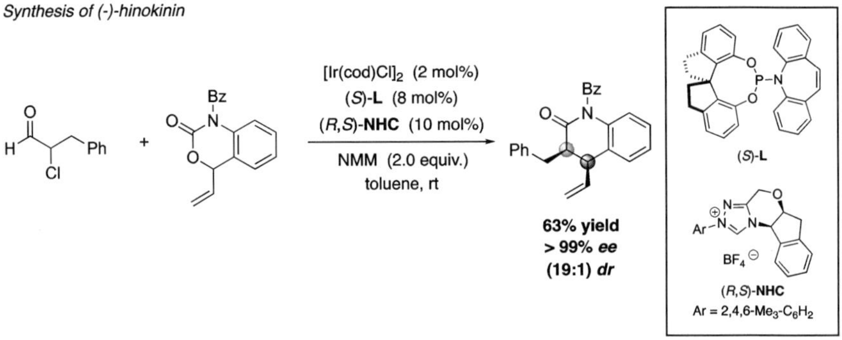

Scheme 36. Stereodivergent α-allylation of α,β-unsaturated aldehydes and related compounds under dual iridium/NHC catalysis, and application to the stereoselective synthesis of (-)-hinokinin.[37]

strategies allowed to enlarge the scope of these reactions, as highly or poorly reactive carbonyl compounds can be employed under a privileged masked form, giving access to original frameworks without undesired side reactions. The required enolate surrogates can be readily obtained from commercially available starting materials or from stable and easy to handle precursors.

Enolate surrogates of these series can be also generated *in situ*, in catalytic amounts, by means of either amines or NHCs, and participate then in the iridium-promoted process. Thus, the general tolerance of iridium catalysts enables chiral amines, Lewis bases, or *N*-heterocyclic carbenes to operate simultaneously, leading to the development of dual catalysis. Especially, stereodivergent processes could be implemented in this way. These powerful methodologies offer a straightforward access to all stereoisomers of a desired compound, opening the way to the stereo-controlled synthesis of valuable products for industrial, pharmaceutical, and medicinal applications.

IV. Iridium-catalyzed Enantioselective Allylation of Other Carbon Nucleophiles

Although the iridium-catalyzed allylation of stabilized nucleophiles has seen a gain in interest, other nucleophiles such as organometallic species remain understudied. The following section will be dedicated to the use of hard organometallic nucleophiles as well as to nucleophilic C=N derivatives.

A. Enantioselective Allylation of Organometallic Reagents

1. *Allylation of organozinc reagents*

In 2007, Alexakis and coworkers have reported the first examples of iridium-catalyzed enantioselective allylations using organozinc partners (Scheme 37).[38] The organometallic species are prepared *in situ* from aryl Grignard reagents and zinc bromide. The catalyst generated from [Ir(cod)Cl]$_2$ and a chiral phosphoramidite led to high enantioselectivities (56–99% ee), albeit with poor regioselectivities, leading to mixtures of branched and linear adducts. Nevertheless, the authors have used their methodology to access a (–)-sertraline precursor,[39,40] which could be transformed into the biologically active target in four more steps, in good yields.

The same group has expanded then this organozinc chemistry and shown that the active zinc species can be generated easily from readily available organohalides *via* the Knochel halogen-metal exchange with magnesium or lithium, followed by transmetalation with zinc bromide (Scheme 38).[41,42] The allylation reactions with allylic carbonates exhibited excellent conversion rates, as well as high ees, but the lack of regioselectivity remained a drawback. Moreover, the authors have demonstrated that π-allyl species can be generated also from halides and acetates as

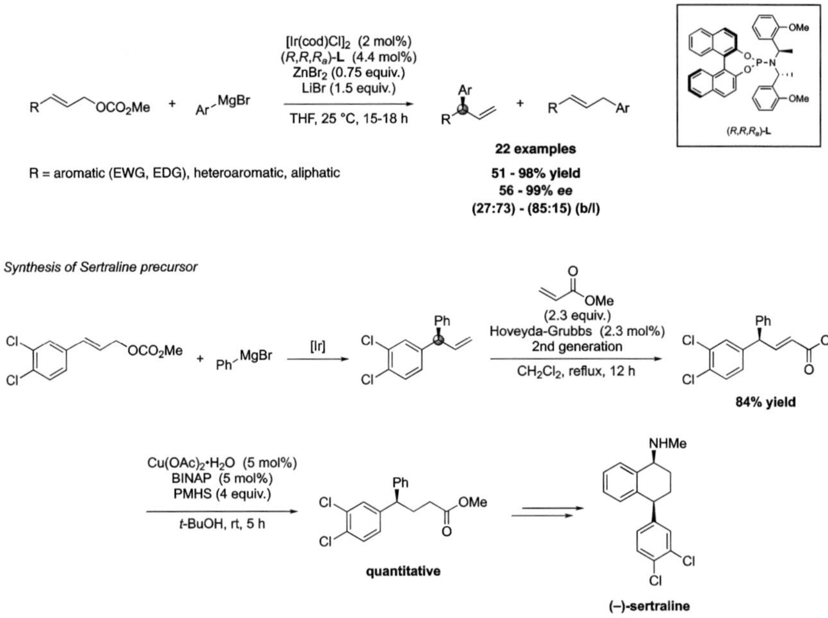

Scheme 37. Enantioselective allylations of organozinc derivatives and application to the synthesis of (−)-sertraline.[38]

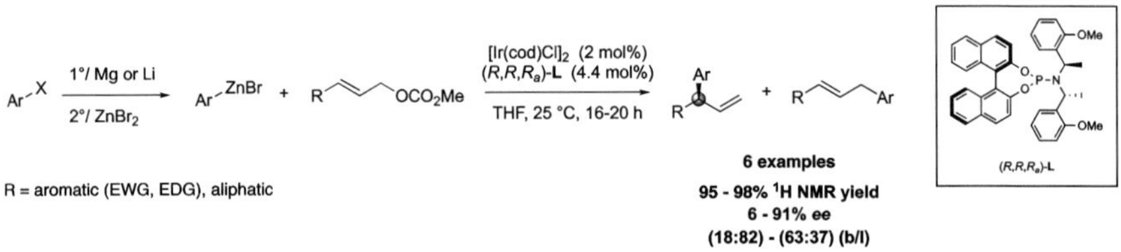

Scheme 38. Enantioselective allylations of organozinc species generated via halogen-metal exchange.[41,42]

leaving groups, and the corresponding reactions proceed with good yields and enantioselectivities.

Carreira and his group have managed to perform the same allylations of organozinc species with total regiocontrol, starting from the branched Boc-protected allylic alcohols (Scheme 39).[43] The final products are obtained in good to excellent yields as single regioisomers. High enantioselectivities could be accessed using the Carreira's phosphoramidite

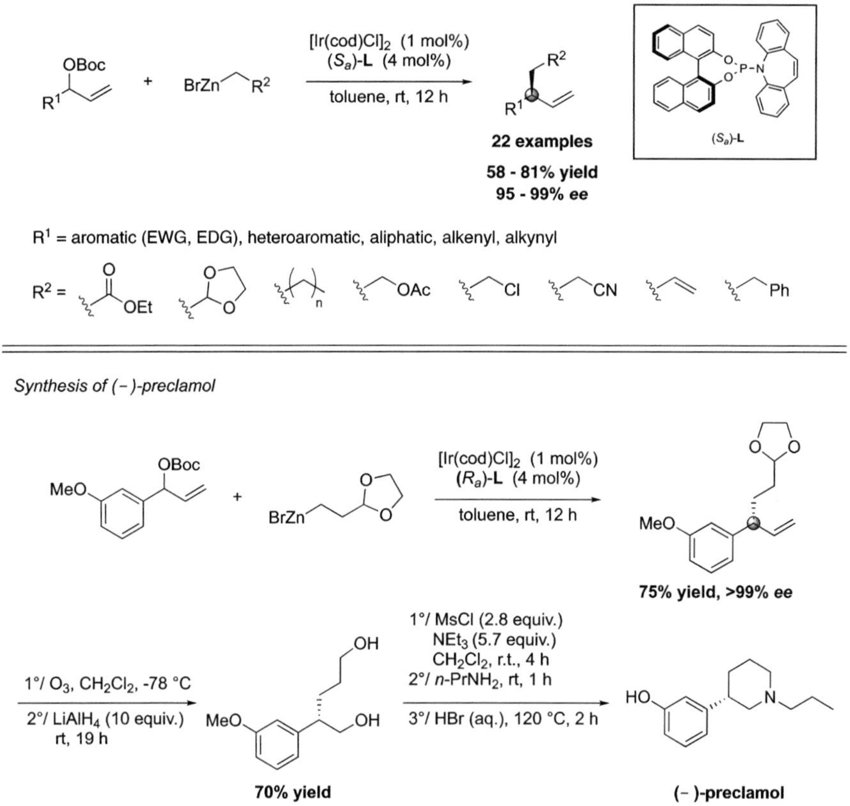

Scheme 39. Regiocontrolled enantioselective allylations of organozinc species.[43]

ligand. Furthermore, a wide range of functional groups are tolerated in this transformation. The allylation reaction has been used as the starting point for the six-step synthesis of (−)-preclamol, a dopamine receptor agonist.[44]

The group of Carreira has extended then its method to the enantioselective α-alkylation of allenes featuring a suitable leaving group. The iridium-catalyzed alkylation of allenes had been first reported by Takeuchi and coworkers in 2004, whereby the authors could access achiral functionalized allenes using malonates as nucleophiles.[45] Thus, Carreira and coworkers have implemented these iridium-catalyzed alkylations using organozinc species as nucleophiles (Scheme 40).[46] Interestingly, complete regiocontrol is observed toward the branched compounds over the possible

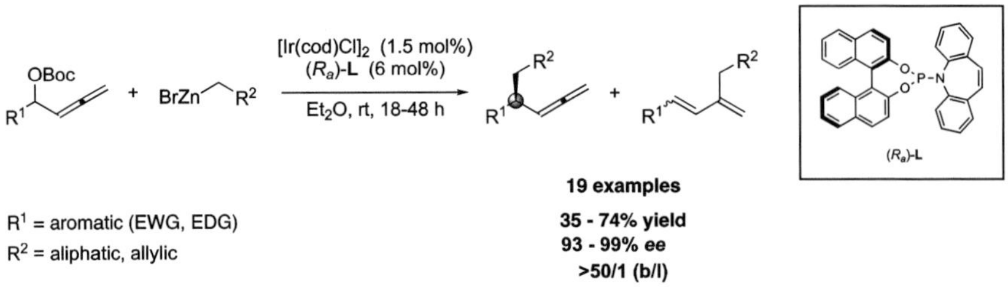

Scheme 40. Enantioselective alkylation of allenes using organozinc species.[46]

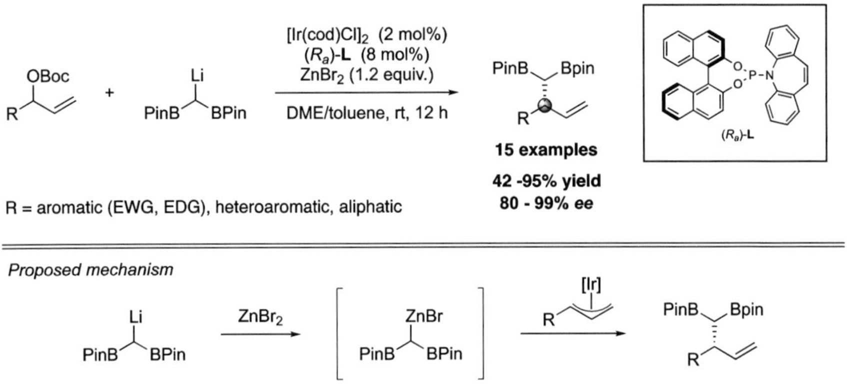

Scheme 41. Enantioselective allylation of *gem*-diboryl-methyllithium.[48]

linear dienes, which are generally favored in the palladium-catalyzed alkylations of allenes with hard nucleophiles.[47] Under these conditions, the enantiomerically enriched allenes (93–99% ee) could be obtained in moderate to good yields. These reactions demonstrated a wide scope and high functional group tolerance.

Recently, Cho and coworkers have exploited the relative stability of lithiated *gem*-diboronates by performing their iridium-catalyzed allylation with branched Boc-protected allylic alcohols, in the presence of zinc bromide (Scheme 41).[48] The reaction is believed to proceed *via* transmetallation to generate the active organozinc species prior to the addition onto the π-allyl intermediate. The corresponding enantioenriched *gem*-diborylalkanes are obtained in good yields and excellent enantioselectivities. This methodology provides a synthetic handle for further functionalizations.

2. Allylation of boron, silicon, and silver-based reagents

Examples of iridium-catalyzed allylation reactions using organometallic coupling partners other than zinc derivatives are rather scarce, and only a handful of reports have been disclosed until now.

While working on the reactivity of allylic alcohols, Carreira and his coworkers have developed the iridium-catalyzed vinylation of unprotected racemic allyl alcohols with potassium vinyltrifluoroborates (Scheme 42).[49] Extremely harsh acidic conditions (hydrofluoric acid) are necessary to achieve good conversions. Nevertheless, the corresponding 1,4-dienes are isolated as sole products in good yields with high ees (85–99% ee) when the Carreira's phosphoramidite is employed as the chiral ligand. The synthesis of two bioactive molecules, (-)-nyasol and (-)-hinokiresinol, could be achieved using this methodology as the first step of the synthetic route.

Shortly after, the same group has expanded its methodology to alkynyltrifluoroborates to access enantioenriched 1,4-enynes *via* the iridium-catalyzed alkynylation of unprotected allylic alcohols (Scheme 43).[50] In this case, the use of hydrofluoric acid is no longer required and can be replaced by less acidic conditions. A wide scope of allylic alcohols as well

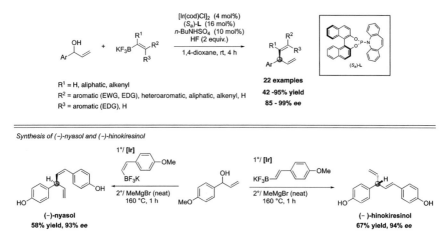

Scheme 42. Enantioselective allylation of potassium vinyltrifluoroborates and application to the enantioselective synthesis of (-)-hinokiresinol.[49]

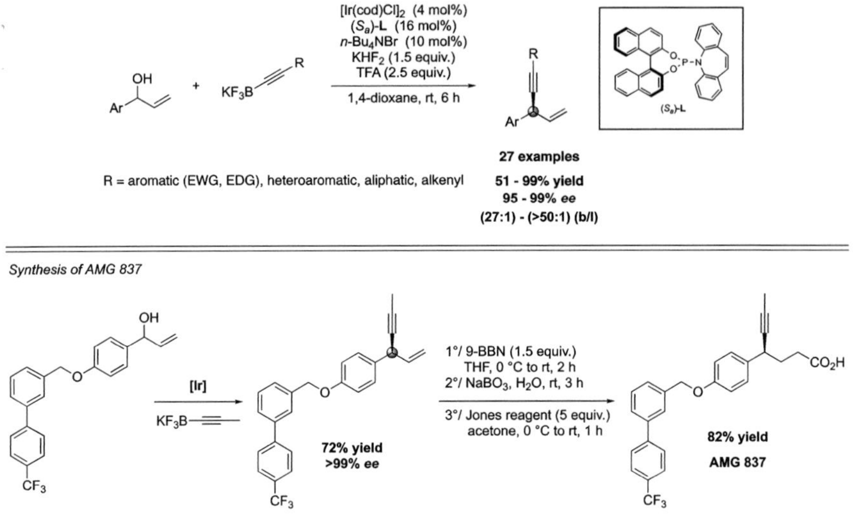

Scheme 43. Enantioselective iridium-catalyzed alkynylation of allyl-alcohols using organotrifluoroborates and application to the enantioselective synthesis of AMG 837.[50]

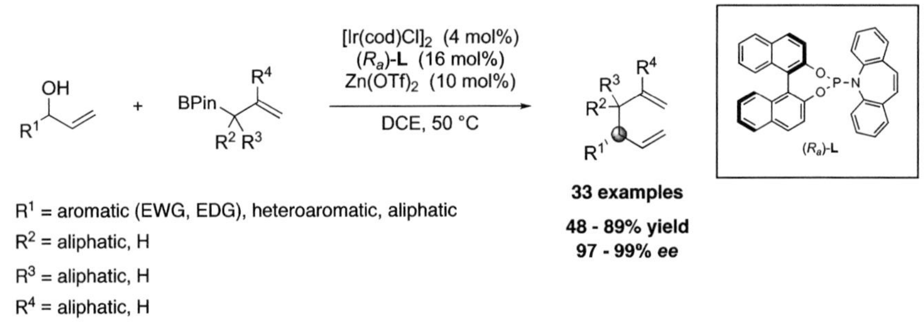

Scheme 44. Enantioselective allyl–allyl cross-couplings involving allyl–boronic esters.[52]

as several alkynes are tolerated to give the corresponding 1,4-enynes in good to excellent yields, high enantioselectivities, and regioselectivities. Furthermore, the authors have applied this alkynylation process to the synthesis of AMG 837, a GPR40 agonist that is active against type 2 diabetes.[51]

Finally, Yang and his group have developed the iridium-catalyzed allyl–allyl cross-coupling of unprotected allyl-alcohols with allyl–boronic esters (Scheme 44).[52] A wide series of 1,5-dienes has been synthesized in

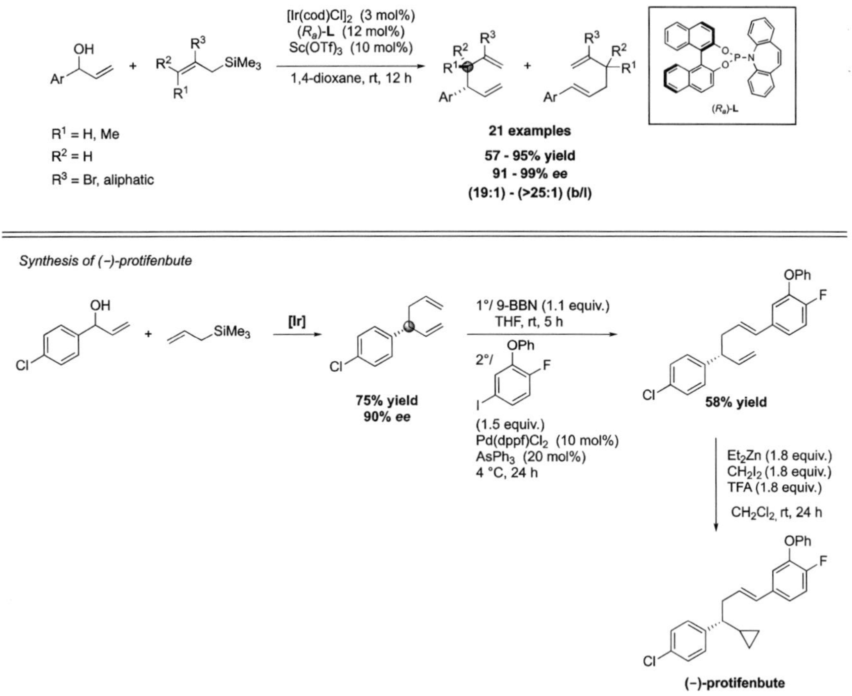

Scheme 45. Enantioselective allylation of allyl-alcohols using allylsilanes and enantioselective synthesis of (-)-protifenbute.[53]

average to high yields with good enantioselectivities (97–99% ee) using the Carreira's phosphoramidite as the chiral ligand.

Carreira's group has shown that allylsilanes are good coupling partners as well for the synthesis of enantioenriched 1,5-dienes from allylic alcohols (Scheme 45).[53] Activation of the allylic alcohol with a Lewis acid such as Sc(OTf)$_3$ is necessary to access high yields. Moreover, under these conditions, the formation of the undesired regioisomer, the linear 1,5-diene, could be prevented. To showcase the utility of the methodology, the authors have carried out the enantioselective synthesis of (-)-protifenbute in four steps with 41% overall yield.

Finally, in 2016, Niu and coworkers have disclosed a new methodology for the allylation of organometallics that involves the generation of a silver species from *bis*[(pinacolato)boryl]methane (Scheme 46).[54] Both branched and linear Boc-protected allylic alcohols are tolerated in the process, though the ligand has to be changed, depending on the substrates,

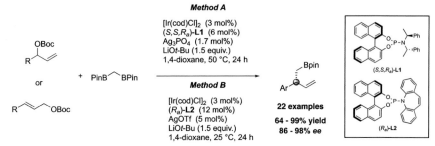

Scheme 46. Enantioselective allylation of bis[(pinacolato)boryl]methane via silver intermediates.[54]

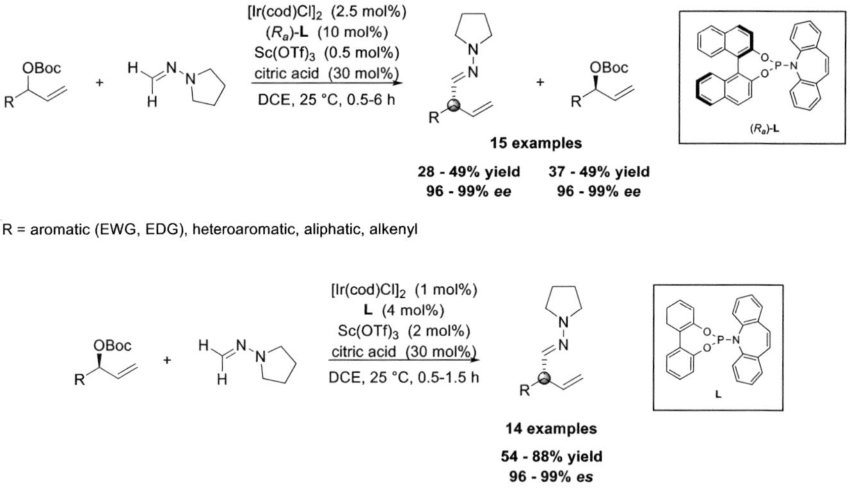

Scheme 47. Enantioselective allylation of hydrazones via kinetic resolution.[55]

in order to achieve good conversions and selectivities. The Feringa–Alexakis phosphoramidite **L1** and the Carreira phosphoramidite **L2** have been used in methods A and B, respectively.

B. Stereoselective Allylation of Hydrazones and Imines

Carreira and his team have disclosed the first and only report on iridium-catalyzed allylation reactions involving hydrazones as formyl anion surrogates (Scheme 47).[55] In the presence of the chiral phosphoramide–iridium catalyst, kinetic resolution of branched Boc-protected allylic alcohols afforded the desired allylic hydrazones in good yields and enantioselectivities via stereospecific substitution reactions. Addition of both citric acid and scandium triflate is necessary in order to achieve suitable conversions. Moreover, the enantioenriched starting material is recovered and further reacted with the hydrazone in a stereospecific fashion to yield the opposite enantiomer of the allylic hydrazone, with excellent ee. This strategy allows an easy access to enantioenriched aldehydes and other derivatives from their hydrazone precursors.

Similarly to hydrazones, imines have raised some interest in iridium-catalyzed allylation reactions. The umpolung of *N*-fluorenylimines *via* a 2-azaallyl anion is employed in the enantioselective synthesis of homoallyl-amines by aminoalkylation of allylic carbonates. The final amines are isolated as either the corresponding Boc-protected derivatives or the ammonium salts (Scheme 48).[56,57] The reaction is believed to go through the formation of the 2-azaallyl anion **I**, upon deprotonation of the starting imine. After alkylation at the most hindered position of the allylic substrate (intermediate **III**), a 2-aza-Cope rearrangement occurs to give the intermediate **IV**, which leads to the desired amine after workup.

Shortly after, in 2017, Han and coworkers have published a similar umpolung strategy for the diastereoselective and enantioselective allylation of imino esters catalyzed by a mixture of the iridium dimer, an

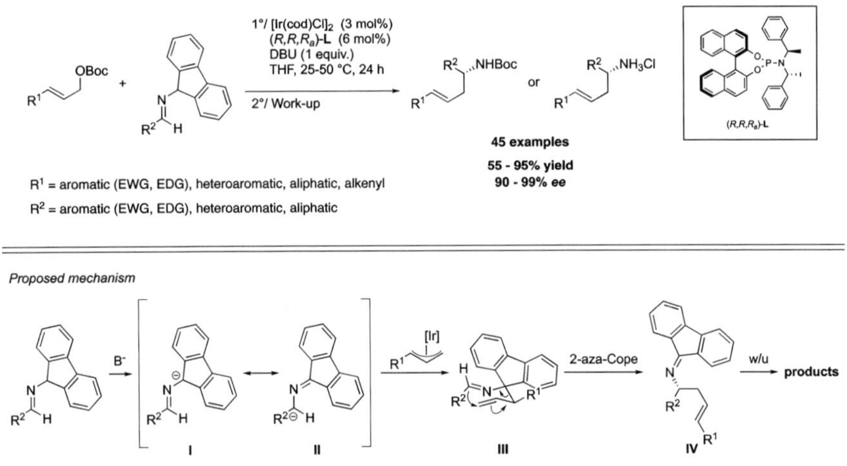

Scheme 48. Enantioselective allylation of imines *via* an aza-Cope rearrangement.[56]

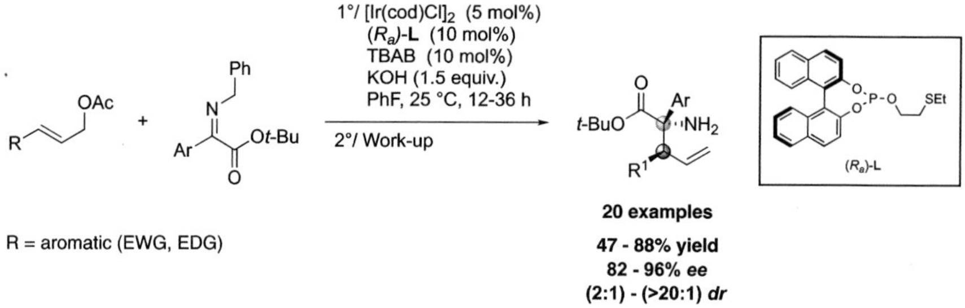

Scheme 49. Diastereo- and enantioselective allylation of imino esters.[58]

ammonium salt as a phase transfer catalyst, and the chiral Takemoto's phosphite **L** (Scheme 49).[58,59] The corresponding quaternary amino acids, displaying two contiguous stereogenic centers, are obtained through this process in good to excellent yields, with an excellent control over both the diastereoselectivity and the enantioselectivity.

C. Conclusion

Although not extensively used in iridium-catalyzed allylation reactions, organometallic reagents, hydrazones, and imines have proven their usefulness. Interesting frameworks can be accessed, allowing the synthesis of several biologically active molecules. In most cases, a total control of the stereoselectivity as well as of the regioselectivity is achieved. Moreover, even hard nucleophiles such as organometallic species give good conversion rates and yields, showing that even those compounds prone to undergo side reactions can be reacted in a controlled manner. The work presented earlier paves the way for the development of novel reagents, beyond the classical carbon nucleophiles used in iridium-catalyzed allylation.

V. Iridium-catalyzed Enantioselective Allylation of Electron-rich Unsaturated Substrates and Other Weak Nucleophiles

The abundance of electron-rich C=C double bonds, in both olefins and (hetero)aromatic compounds, makes them potential nucleophiles of choice in iridium-catalyzed allylation reactions. However, due to their reduced nucleophilicity as compared to other species, the development of efficient methods remains limited. This section will cover in a general way the use of C=C double bonds as coupling partners for the stereoselective iridium-catalyzed allylation reactions.

A. Enantioselective Allylation of C=C Double Bonds

The allylation of C=C double bonds under iridium catalysis remains challenging due to the reduced nucleophilicity of electron-rich olefins and aromatic compounds, as compared to other nucleophiles presented previously.

1. *Enantioselective allylation of olefins*

In 2009, during their studies of the iridium-catalyzed allylations of amines, You and coworkers have stumbled upon the unprecedented formation of a skipped (Z, E)-diene, instead of the desired aminated product, when reacting an aniline derivative bearing a styrene unit (Scheme 50).[60] Good yields are observed for this reaction, only linear products are formed, but a mixture of regioisomeric dienes is obtained in some cases. These unexpected results demonstrating the ability of olefins to act as nucleophiles in iridium-catalyzed allylation reactions have then opened a new area of research in the field, notably for the development of enantioselective methods.

With their knowledge about iridium-catalyzed allylation reactions, You and collaborators have shown that the allyl–alkene cross-coupling

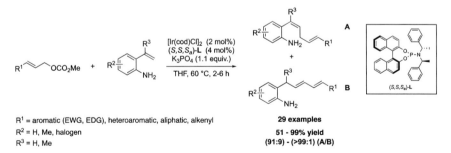

Scheme 50. Iridium-catalyzed coupling of styrene derivatives with allyl carbonates.[60]

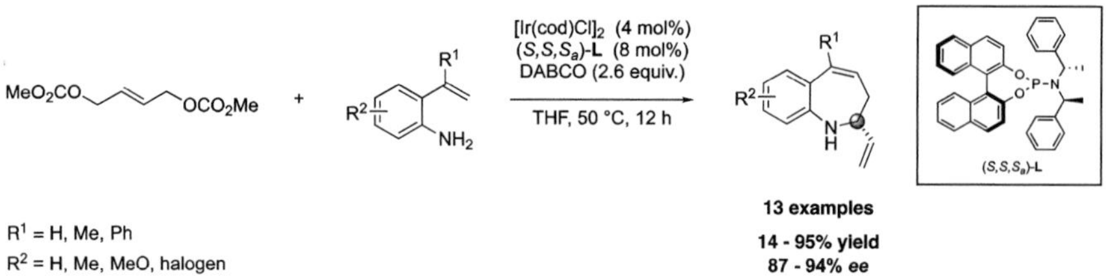

Scheme 51. Enantioselective allylic olefination/amination sequence.[61]

mentioned earlier can be combined to the well-known iridium-catalyzed amination of allylic carbonates to access medium-sized nitrogen heterocycles. The key starting material is an allylic biscarbonate that undergoes two successive allylic substitution reactions when reacting with o-vinylanilines. Mechanistic studies proved that the olefination step occurs prior to the intramolecular amination step. In this way, benzoazepines were obtained in excellent yields and high enantioselectivities (Scheme 51).[61,62]

Carreira and his group have reported the enantioselective access to 1,5-dienes starting from unprotected allylic alcohols and disubstituted olefines, with a catalytic mixture made of the iridium dimer, the chiral phosphoramidite ligand, and a sulfonimide cocatalyst (Scheme 52).[63] The desired product is obtained in good to excellent yields and high enantioselectivities. In some instances, the reaction has proven to be regioselective, though another regioisomer of the diene can be isolated in other cases. Unfortunately, the cross-coupling is limited to *gem*-disubstituted olefins. Regardless, the methodology has been applied to the synthesis of JNJ-40418677, a candidate for the treatment for the Alzheimer's desease.[64]

2. Stereoselective allylations of indoles and pyrroles

Indoles and pyrroles are the most prolific heterocycles used in iridium-catalyzed allylation reactions. However, due to the competition between three nucleophilic positions on the heteroaromatic core (nitrogen atom, C2 position, and C3 position), the development of a chemo-, regio- and enantioselective coupling remains challenging.

The first C3 allylation of indole with iridium as the catalyst has been reported by You and his group in 2008. The intermolecular coupling of allyl carbonates with indoles proceeds well and affords the C3-functionalized

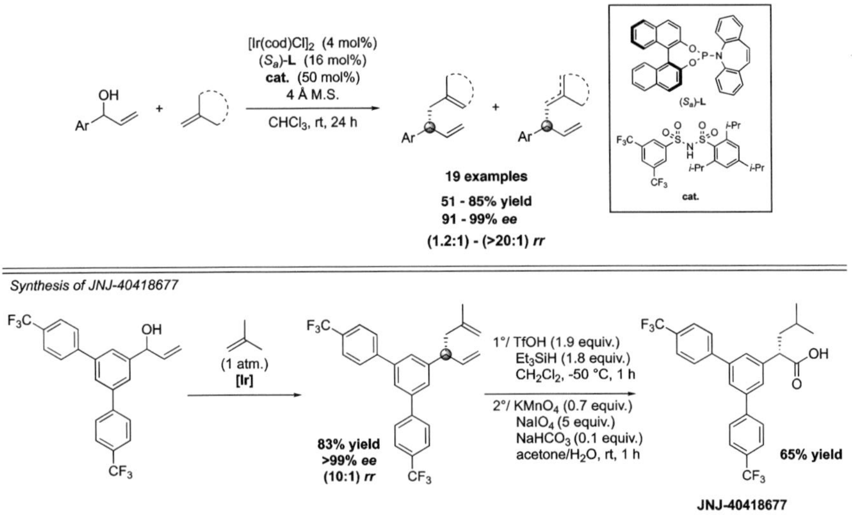

Scheme 52. Enantioselective allylation of *gem*-bisalkyl olefins and application to the synthesis of JNJ-40418677.[63]

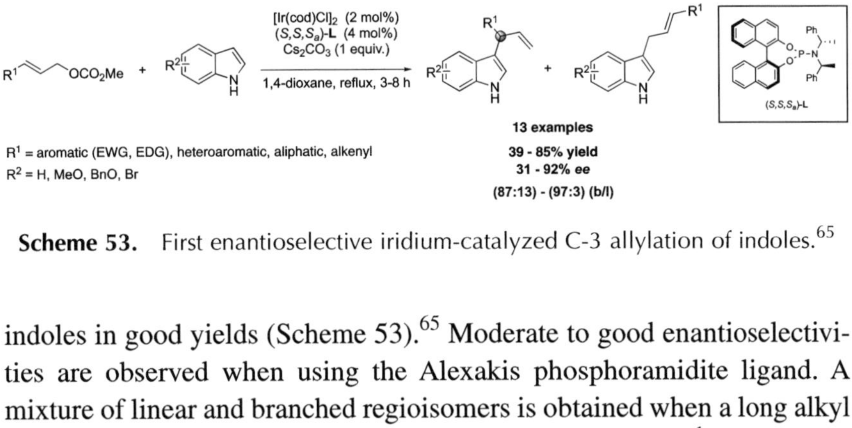

Scheme 53. First enantioselective iridium-catalyzed C-3 allylation of indoles.[65]

indoles in good yields (Scheme 53).[65] Moderate to good enantioselectivities are observed when using the Alexakis phosphoramidite ligand. A mixture of linear and branched regioisomers is obtained when a long alkyl chain is selected as substituent on the allylic carbonate (R^1 group). The scope of the reaction has been further expanded by the same group shortly after.[66–68]

A series of spirocycles has been synthesized *via* intramolecular allylation reactions involving N-allyl–tryptamine derivatives as substrates (Scheme 54).[69] This interesting framework was isolated in average to excellent yields, with high enantioselectivities (88–97% ee), by using the

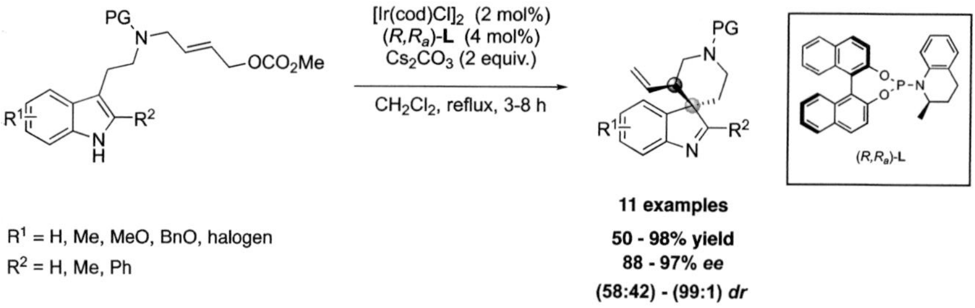

Scheme 54. Enantio- and diastereoselective intramolecular allylation of indoles.[69]

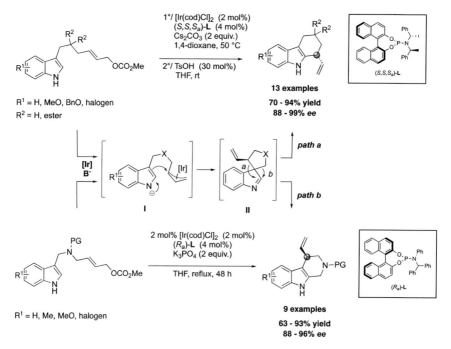

Scheme 55. Complementary intramolecular C2 and C3 allylation of indoles.[71]

chiral You's phosphoramidite as the ligand. Moreover, excellent diastereoselectivities were observed for R^2=H, though a significant decrease of the diastereomeric ratio occurred with C2-substituted indoles.

In 2012 and 2013, You and his team have developed two complementary methodologies in order to access C2 and C3 allyl-functionalized indoles from substrates displaying the same heteroaromatic backbones and C3-tethered allyl units (Scheme 55).[70,71] The mechanism would proceed *via* the formation of the spirocyclic intermediate **II** upon intramolecular nucleophilic addition onto the π-allyl complex **I**. The acid-promoted migration of the allylic fragment, through **pathway a**, would lead to the C2-allylated tetrahydrocarbazole.[72] On the other hand, a nitrogen functional group embedded within the tether would promote a spontaneous migration according to **pathway b** to yield the C3-allylated tetrahydrocarbazole derivative.

With nitrogen containing tethers, the highly reactive spirointermediate is not isolated and spontaneously undergoes alkyl migration. The authors have shown later that spirocyclic derivatives can be obtained in good yields, enantio- and diastereoselectivities, after the iridium-catalyzed allylation, if an aromatic substituent lies in C2 position (Scheme 56).[73] Purification led, however, to degradation of the product, and a reduction step was required in order to circumvent this drawback. Furthermore, the addition of an aromatic substituent on the linker allowed the authors to access a tryptamine derivative in good yield and high enantioselectivity (93% ee) *via* the spiranic intermediate, through a retro-Mannich step followed by hydrolysis of the iminium intermediate.

A similar reactivity has been developed with pyrroles by You and his group. They have reported the dearomatization of C2 functionalized pyrroles *via* intramolecular iridium-promoted allylations, giving access to six-membered spirocyclic pyrroles (Scheme 57).[74] Excellent

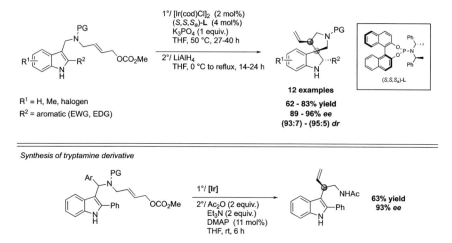

Scheme 56. Synthesis of spiroindolines and an enantioenriched tryptamine derivative.[73]

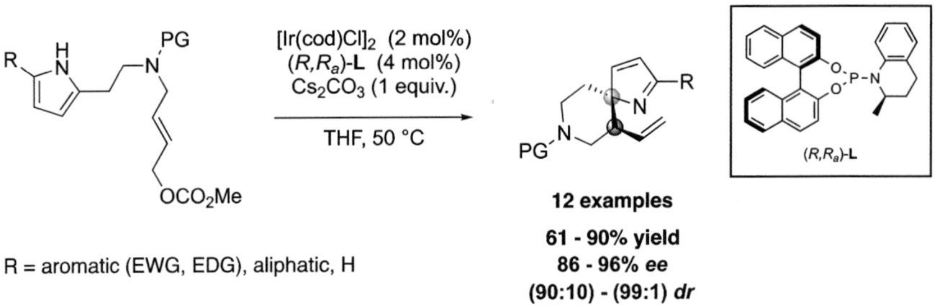

Scheme 57. Synthesis of six-membered spiropyrroles *via* intramolecular allylations.[74]

stereoselectivities and yields can be achieved under the reaction conditions given in the scheme.

The same methodology has been applied with success to the synthesis of five-membered spiropyrroles (Scheme 58).[75] Enantioenriched pyrrole derivatives are isolated in good yields as well as diastereoselectivities. When these spiropyrroles were subjected to the previously optimized migratory conditions, in acidic medium, bicyclic pyrrole derivatives were obtained in good yields. Importantly, the enantiomeric purity is kept intact during this process.

Further studies on the iridium-catalyzed allylation of indoles have led You's group to highlight the intramolecular C3 allylation of specific indole compounds bearing allylic moieties tethered to their C5 position (Scheme 59).[76] Overall, the C3 allylation occurs in good yields and leads to the synthesis of the corresponding polycyclic indole derivatives in high enantioselectivities (91–97% ee). In general, the functionalization in position 2 has no influence on the reaction outcome. However, C3-substituted indoles undergo allylation at the C6 position and lead to the corresponding tricyclic indole products in moderate to good yields. Interestingly, a significant decrease of the enantioselectivity is observed when R^2 is an aromatic substituent.

Shortly after, You's group has reported the enantioselective synthesis of the indole-annulated medium-sized rings shown in Scheme 60.[77] According to the proposed mechanism, C3 allylation of the starting tetrahydrocarbazole takes place first giving intermediate **I**. A *retro*-Mannich/hydrolysis sequence, similar to that mentioned above in Scheme 56, leads then to the

Enolate Surrogates and Unusual Nucleophiles in Allylic Substitutions 459

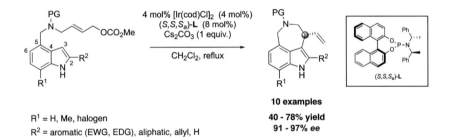

Scheme 58. Synthesis of five-membered spiropyrroles and acid-promoted allyl migration giving bicyclic pyrrole derivatives.[75]

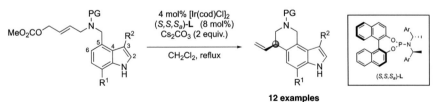

Scheme 59. Intramolecular allylations of C5-functionalized indoles.[76]

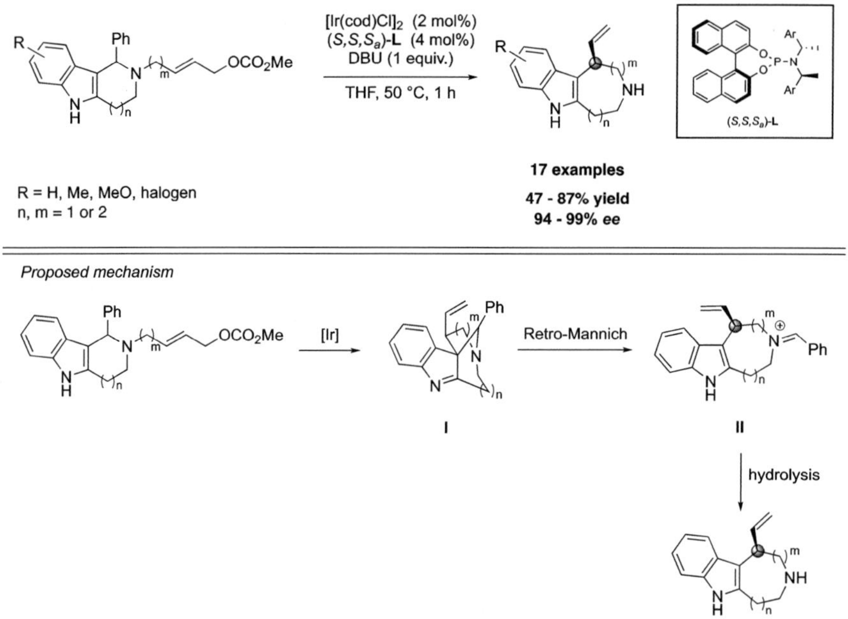

Scheme 60. Homologation of tetrahydrocarbazoles through an allylation/retro-Mannich/hydrolysis sequence.[77]

formation of seven- and eight-membered cyclic indole derivatives in moderate to good yields, with excellent enantioselectivities. Substitution of the tetrahydrocarbazole substrate with a phenyl group is required to promote the *retro*-Mannich step and to stabilize the iminium intermediate **II**. Without this functionalization, the same substrate affords quinuclidine derivatives in moderate to good yields with excellent enantioselectivities and diastereoselectivities.[78] Moreover, the authors have shown that the direct C3-allylation of indoles with allyl units tethered to their C2 position is not a viable method to access the corresponding eight-membered annulated frameworks, demonstrating the value of the allylation/retro-Mannich/hydrolysis sequence starting from tetrahydrocarbazoles.[79]

You and coworkers have applied the same homologation strategy to suitably substituted bicyclic pyrroles (Scheme 61).[80] Both seven- and eight-membered bicyclic pyrroles ($n = 1, 2$) are obtained in moderate to good yields with high enantioselectivities.

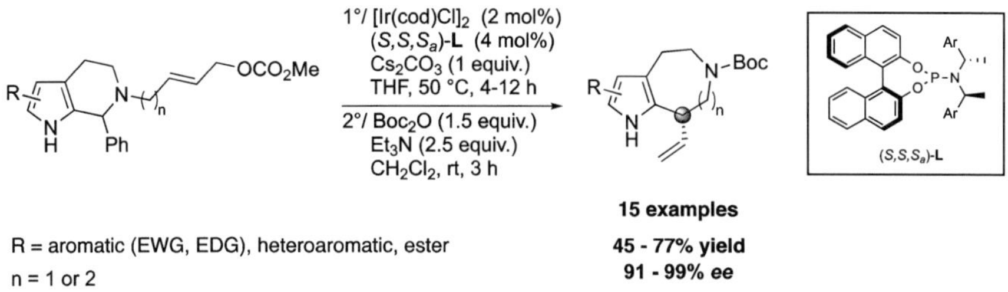

Scheme 61. Homologation of bicyclic pyrroles through an allylation/retro-Mannich/hydrolysis sequence.[80]

In 2016, You and coworkers have reported the enantio- and diastereoselective iridium-catalyzed synthesis of tetrahydrocarbazoles bearing two stereogenic centers, by starting from C3 functionalized indoles and by following the same strategy as in Scheme 55 (Scheme 62).[81] The use of a strong acid (HCl) is necessary in order to reach good conversions in the course of this one-pot iridium-catalyzed allylation/carbon–carbon migration process. Diastereomeric ratios up to 6:1 have been observed. The same approach has been applied to the desymmetrization of a substrate displaying a *gem-bis*indole unit tethered to an allyl carbonate unit (Scheme 62).[82] Under the given conditions, the allylation step yields spirocyclic indolines with high enantioselectivity. Then, under acidic conditions, an enantiospecific alkyl migration takes place. Thus, seven-membered indole derivatives can be synthesized in good yields, high enantiospecifity, and diastereoselectivity over the two steps.

The value of the iridium-catalyzed allylation of indoles has been highlighted by Yang and his team with the enantioselective synthesis of two biologically relevant compounds, aspidophylline A and alstoscholarisine A (Scheme 63).[83,84] The former has shown some ability to reverse resistance in drug-resistant KB cells, while the latter is found to promote the development of adult neuronal stem cells.[85,86] These compounds have been obtained from suitably functionalized indoles that display allylic alcohol units. In both cases, the iridium-catalyzed allylation reactions, in the presence of Lewis acid additives, represent a key step and furnishes the desired polycyclic products in good yields and excellent enantioselectivities.

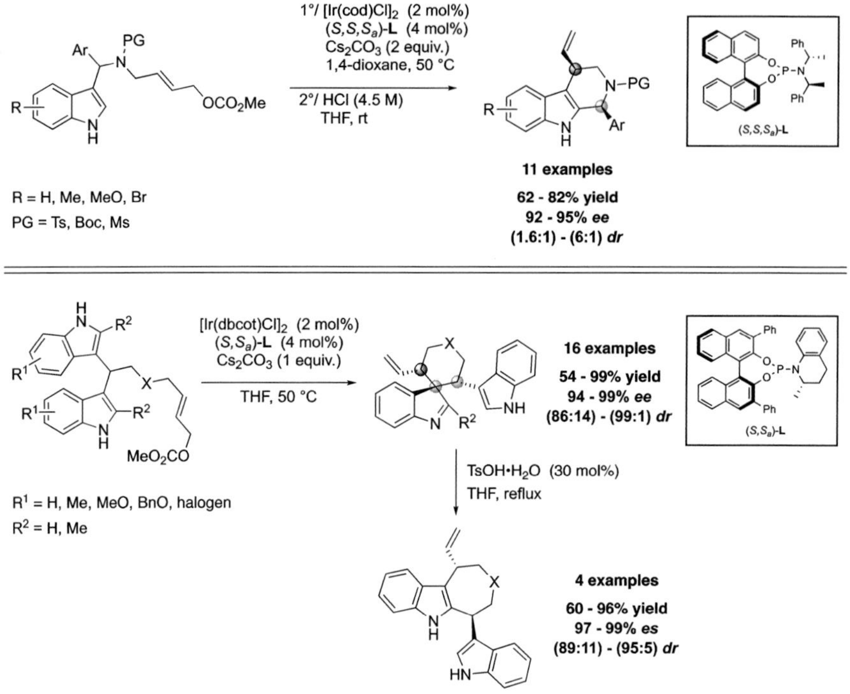

Scheme 62. Enantio- and diastereoselective synthesis of carbazole derivatives and application to the desymmetrization of *gem-bis*indole allyl carbonates.[81,82]

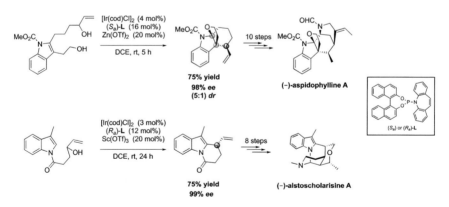

Scheme 63. Enantioselective synthesis of (−)-aspidophylline A and (−)-alstoscholarisine A.[83,84]

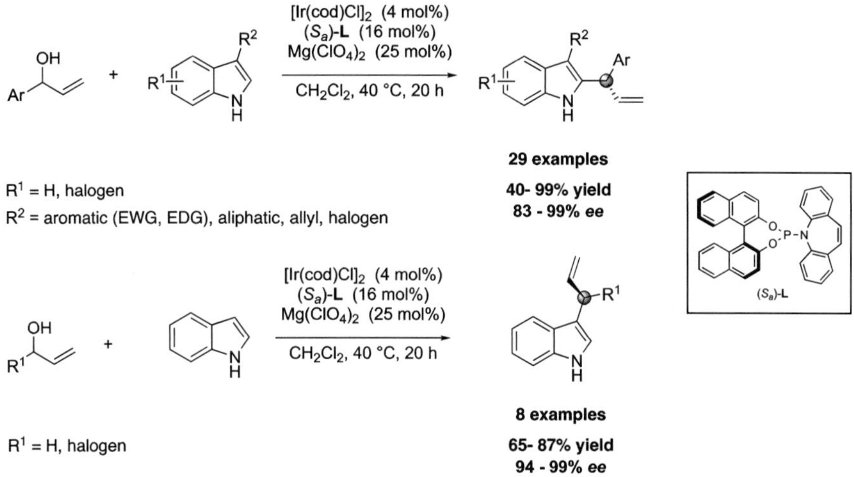

Scheme 64. Enantioselective intermolecular C2 and C3 allylation of indoles.[87]

Very recently, You and his group have reported the intermolecular direct C2 allylation of indoles with aryl-substituted allylic alcohols (Scheme 64).[87] The C3 position of the indole has to be masked in order to prevent the previously described C3 electrophilic allylation reaction. A cheap Lewis acid such as magnesium perchlorate can be used for the activation of the starting allylic alcohol. Depending on the substitution patterns of the substrates, the desired C2 substituted indoles are isolated in moderate to quantitative yields and high enantioselectivities. However, when the naked indole (R^2=H) is used under those conditions, unsurprisingly, C3-functionalized products are obtained in good yields.

3. Enantioselective C-allylation of anilines

Although extremely important in nature, anilines have been understudied as potential carbon nucleophiles in the iridium-catalyzed allylation reactions. Only two examples have been reported up to now, both in 2017.

Fu and his group have been the first to use *N,N*-dialkylaniline derivatives and allylic alcohols as substrates to obtain, under iridium catalysis, the corresponding *para*-substituted anilines in good yields and high enantioselectivities (Scheme 65).[88] A catalytic amount of a Lewis acid is required in order to activate the allylic alcohol partner.

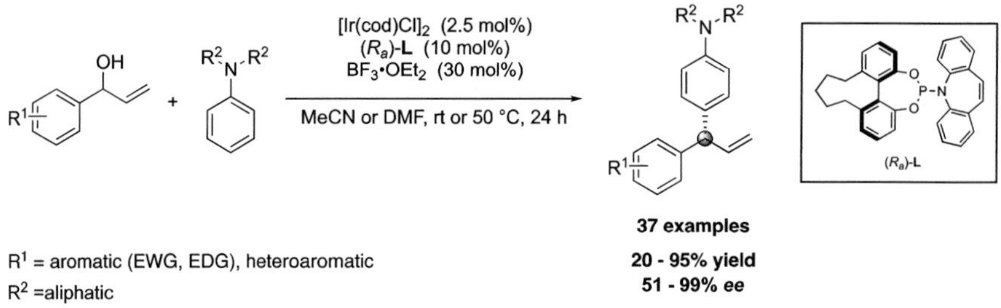

Scheme 65. Enantioselective iridium-catalyzed *para*-allylation of N,N-dialkylanilines.[88]

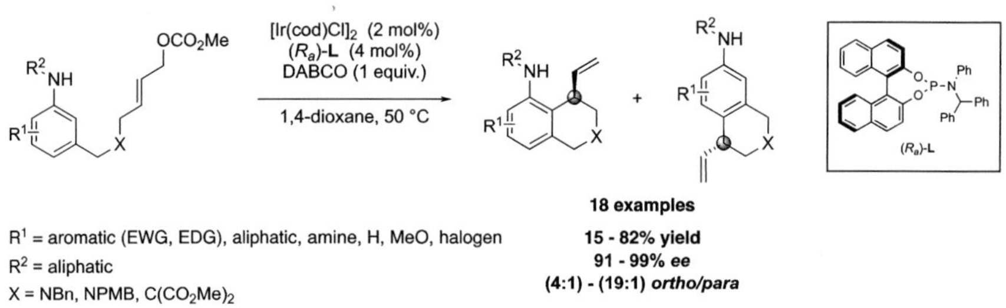

Scheme 66. Enantioselective intramolecular *ortho*-allylation of anilines.[89]

Shortly after, You and coworkers have used anilines with allyl functions tethered to the *meta*-position, to access tetrahydroisoquinolines through an intramolecular iridium-catalyzed allylation reaction (Scheme 66).[89] The authors have optimized the reaction in order to obtain almost exclusively the *ortho*-allylation products (up to 19:1 *ortho/para* ratios) in good yields and high enantioselectivities. It is noteworthy that protection of the aniline NH function is not required for the reaction to proceed well, and a free nitrogen atom in C5 position is tolerated in the final product.

4. *Enantioselective C-allylation of phenols*

Phenols as well have raised some interest in the scientific community as nucleophilic coupling partners for the development of iridium-catalyzed C-allylation reactions.

In 2011, You and coworkers have reported the first example of iridium-catalyzed intramolecular dearomatization of phenols *via* allylation

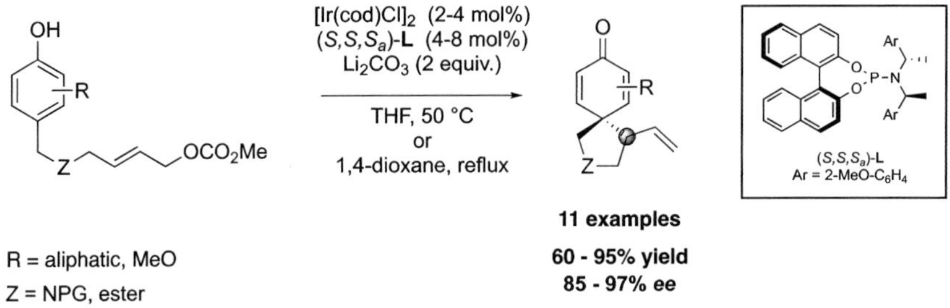

Scheme 67. Enantioselective dearomatization of phenols via intramolecular allylations.[90]

reactions, leading to spirocyclohexadienones (Scheme 67).[90] The starting phenols display allyl carbonate units tethered to their *para* positions. The bicyclic compounds are obtained in good to excellent yields, with high enantioselectivities. It is noteworthy that high yields and enantio- and diastereoselectivities can be also observed when starting from unsymmetrical phenol derivatives.

The intramolecular dearomatization of phenols has been further explored by the same group in 2012. The intramolecular allylation of phenols bearing the allylic moiety in *meta* position leads to a mixture of enantioenriched regioisomers, which can be isolated in high yields and enantioselectivities (Scheme 68).[91] The *ortho*-allylated product remains the major regioisomer in most cases.

With β-naphthols as starting materials for the iridium-catalyzed allylation, You and his team have accessed spironaphthalenone derivatives in good yields (Scheme 69).[92] A more advanced ligand derived from THQ-Phos is required to obtain high enantioselectivities. The ligand displays additional phenyl substituents in the 3,3′-positions of the binaphthyl unit.

Starting from an *ortho*-substituted α-naphthol, a mixture of *C*- and *O*-substituted isomers is obtained. Although the yield is significantly decreased, both enantio- and diastereoselectivities remain good. Finally, the *para*-substituted α-naphthol affords the desired spiroproduct in excellent yields and selectivities.

Carreira and his team have shown that protected phenols are also suitable substrates for iridium-catalyzed intramolecular allylation reactions

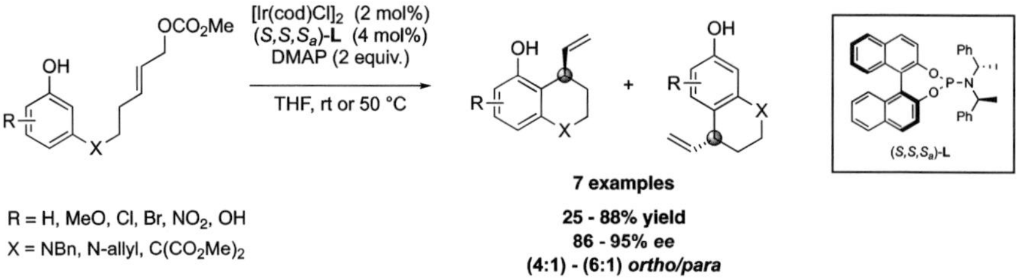

Scheme 68. Enantioselective synthesis of bicyclic phenols *via* intramolecular allylations.[91]

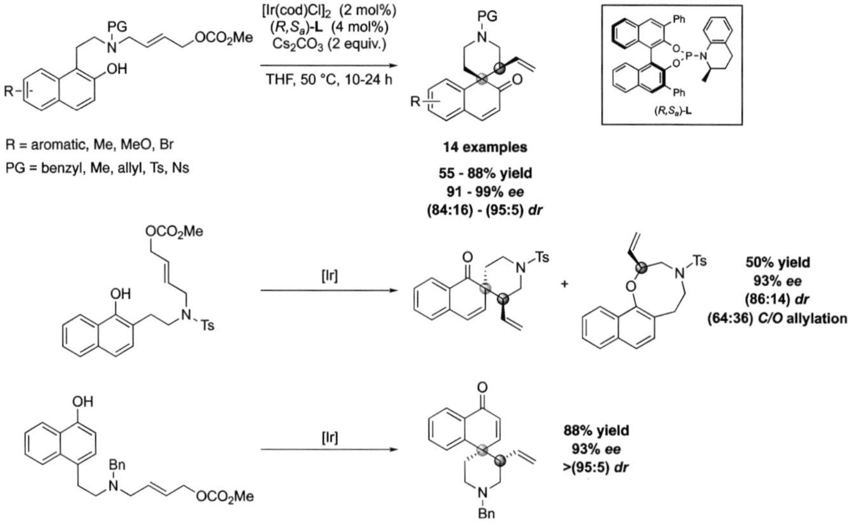

Scheme 69. Enantioselective synthesis of spironaphthalenone derivatives *via* intramolecular *C*-allylation.[92]

(Scheme 70).[93] They have obtained five- and six-membered benzofused cyclic compounds from suitably functionalized aryl ethers (R^1 = alkyl), *via* a *C*-allylation/aromatization sequence. Excellent ees are observed in this case. Indoles, pyrroles, and benzofuranes can also be employed as substrates in this process. It is worth mentioning that the enantioselective synthesis of erythrococcamide A and B has been performed using this methodology.

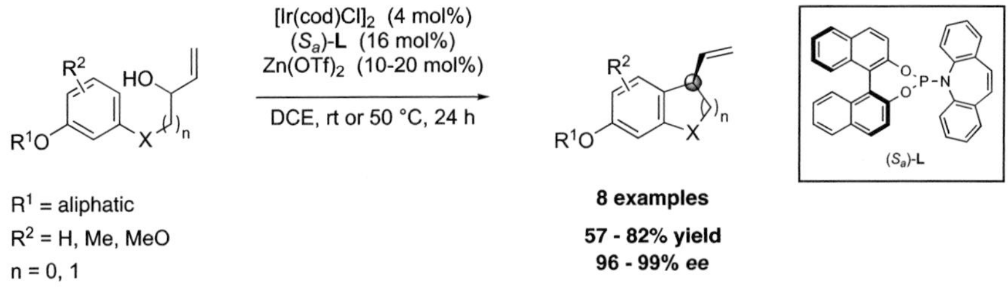

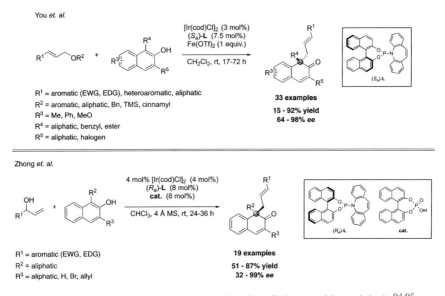

Scheme 70. Enantioselective intramolecular allylation of protected phenols.[93]

Scheme 71. Enantioselective intermolecular allylation of β-naphthols.[94,95]

Finally, Zhong and You have independently reported the intermolecular catalytic dearomatization of β-naphthols (Scheme 71).[94,95] In both cases, unprotected allylic alcohols are used for the transformation with either a stoichiometric amount of a Lewis acid or a catalytic amount of a Brønsted acid. Interestingly, You and coworkers have used for the first time TMS-protected allylic alcohols and allylic ethers as coupling partners for the reaction, though with decreased yields. Moreover, in both reports, naphthols displaying unsubstituted benzofused rings ($R^3 = H$) still furnish the α-allylation products that display all carbon quaternary centers, without traces of other regioisomers.

5. Stereoselective allylations involving cation-π cyclizations of polyenes

Iridium-catalyzed allylation reactions involving a cation-π cyclization of polyene frameworks has been less documented and only two reports have been published by the Carreira's team only.

The first example has been disclosed in 2012, in which the authors are able to cyclize polyenes displaying branched allyl alcohol functions, to tricyclic products (Scheme 72).[96] The iridium catalyst, combined with a Lewis acid catalyst, initiates a cascade reaction that involves allylation of the first olefinic unit, followed by an electrophilic addition of the cationic intermediate to the C=C bond of an heteroaromatic moiety. Several aromatic and heteroaromatic rings are tolerated in this transformation and provide the corresponding tricyclic compounds with high ees when using the Carreira's phosphoramidite as the chiral ligand. The methodology has been expanded also to a triple cyclization process, leading to the corresponding tetracyclic product in moderate yield but high enantioselectivity.

The usefulness of the methodology has been highlighted by the enantioselective synthesis of (+)-asperolide C. The appropriate unprotected allylic alcohol, which displays a 1,5-dienic framework, is subjected to the iridium-catalyzed allylation process to give the desired bicyclic adduct in good yield and high diastereo- and enantioselectivities (Scheme 73).[97] In

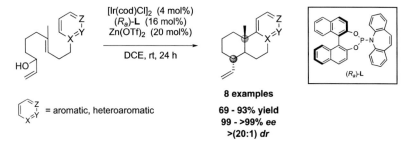

Scheme 72. Stereoselective polyene cyclization involving an olefin allylation step.[96]

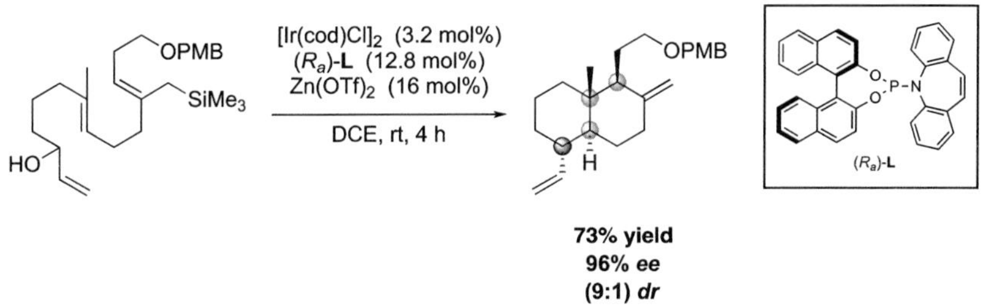

Scheme 73. Polyene cyclization for the stereoselective synthesis of (+)-asperolide C.[97]

this process, four stereogenic centers are generated with excellent stereocontrol in a single step. The biologically active target is then obtained in 10 more steps from the decalin framework.

B. Enantioselective Allylations of Benzylic Nucleophiles and Allylic Aromatization of Methylene-substituted Heterocycles

After the pioneering work from Trost and Walsh groups on palladium-catalyzed allylations of benzylic positions of methylpyridine and toluene derivatives, the corresponding iridium-promoted version has been developed by You and coworkers.[98,99] Using a preformed, stable (π-allyl)Ir complex derived from the Feringa–Alexakis phosphoramidite, the authors were able to isolate enantioenriched α-allylated methyl pyridines in moderate to excellent yields (Scheme 74).[100] The preformed (π-allyl)Ir complex is necessary to achieve the highest ees, though it gives similar yields as the *in situ*-generated active species. The transformation has been used as the first step of the enantioselective synthesis of the biologically active (−)-lycopladine A.

The Mukherjee group has reported the iridium-catalyzed enantioselective allylation of methylated coumarins and quinolones with Boc-protected allylic alcohols (Scheme 75).[101–102] Overall, good yields and enantioselectivities are obtained.

Finally, You and coworkers have shown that methylene-substituted (hetero)cycles can be used as equivalents of benzylic nucleophiles in the iridium-catalyzed alkylations of allylic alcohols (Scheme 76).[103] A wide variety of methylene-substituted substrates are tolerated, such as oxazole,

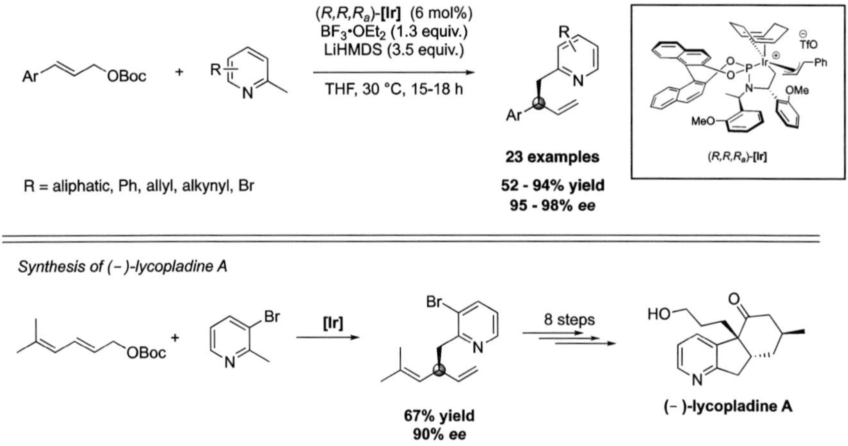

Scheme 74. Enantioselective α-allylation of methyl pyridines and application to the synthesis of (-)-lycopladine A.[100]

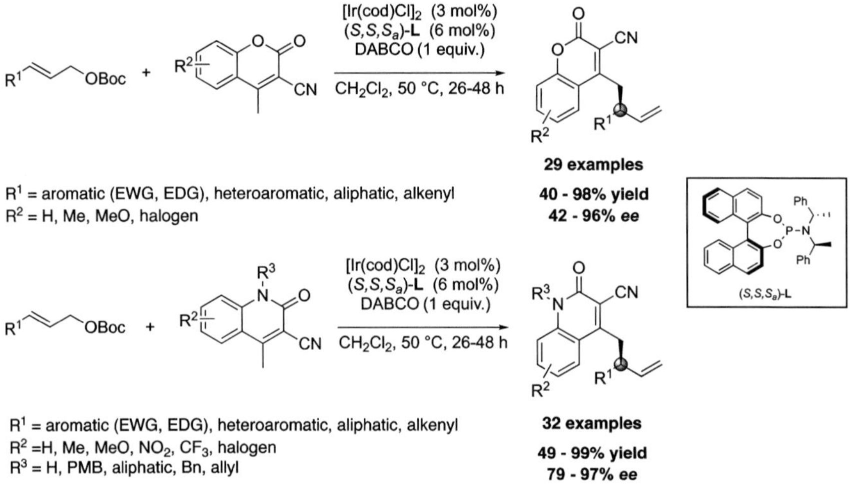

Scheme 75. Enantioselective allylation of methylated coumarins and quinolones.[101,102]

thiazole, furan, benzothiophene, and indole derivatives. Allylation of their exocyclic methylene units induces concomitant aromatization of these heterocyclic substrates. In the presence of $Fe(OTf)_2$ as an additive and the Carreira phosphoramidite as the chiral ligand, these reactions provided the

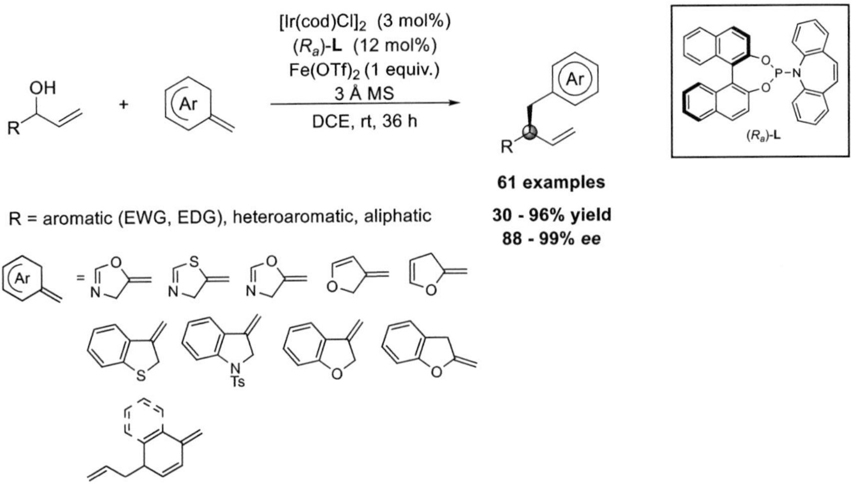

Scheme 76. Enantioselective allylation of methylene-substituted (hetero)cycles.[103]

desired aromatic products in low to excellent yields. The ee is high in all cases. This strategy has been applied to the synthesis of an amiodarone analog.

C. Conclusion

This section has shown that less nucleophilic species can still be used as efficient partners for the development of stereoselective iridium-catalyzed allylation reactions. A huge scope of nucleophiles, from simple alkenes to heteroaromatic compounds, is tolerated, giving access to a remarkable variety of frameworks. Such aspect has been demonstrated by concise syntheses of many interesting and biologically active targets. Both stereoselectivities and yields are high overall. The ongoing research on nucleophilic C=C double bonds will bring new and powerful tools for the synthesis of more challenging and complex chiral heterocycles.

VI. Conclusion

Despite being great functional groups to create new C–C bonds, enolates derived from simple carbonyl compounds usually require harsh conditions to reveal their nucleophilic reactivity in iridium-catalyzed allylation reactions. To circumvent this issue, many surrogates have been developed. Organometallic compounds, alkenes, and other miscellaneous substrates have been used as nucleophiles in those processes. Enantioselective variants of these reactions have been implemented successfully, thanks to the use of BINOL-derived and SPINOL-derived phosphoramidites as chiral ligands. The design and tuning of the phosphoramidite scaffolds have enabled great regio- and stereoselectivities, making the use of iridium-catalyzed allylation reactions particularly interesting in the stereocontrolled synthesis of valuable building blocks for total synthesis. Introduction of dual catalysis has created a breakthrough in the stereodivergence field. In the case of the addition of a prochiral nucleophile on a prochiral π-allyl precursor, two stereogenic centers are independently and simultaneously created by a combination of iridium- and organocatalysts, providing one discrete stereoisomer among the four possible. However, each method has its scope and limitations and a general procedure has yet to be found.

VII. Acknowledgments

This work was supported by the CNRS, MESRI, Fondation de la Maison de la Chimie and Aix-Marseille University which we gratefully acknowledge.

VIII. References

1. (a) Tsuji, J.; Takahashi, H.; Morikawa, M. *Tetrahedron Lett.* **1965**, *6*, 4387–4388. (b) Trost, B. M.; Fullerton, T. J. *J. Am. Chem. Soc.* **1973**, *95*, 292–294.
2. (a) Trost, B. M.; van Vankren, D. L. *Chem. Rev.* **1996**, *96*, 395–422. (b) Trost, B. M. *Chem. Pharm. Bull.* **2002**, *50*, 1–14. (c) Trost, B. M.; Crawley, M. L. *Chem. Rev.* **2003**, *103*, 2921–2944. (d) Milhau, L.; Guiry, P. J. *Top. Organomet. Chem.* **2011**, *38*, 95–153.
3. Lloyd-Jones, G. C.; Pfaltz, A. *Angew. Chem. Int. Ed. Engl.* **1995**, *34*, 462–464.
4. (a) Belda, O.; Moberg, C. *Acc. Chem. Res.* **2004**, *37*, 159–167. (b) Trost, B. M. *Org. Process Res. Dev.* **2012**, *16*, 185–194. (c) Moberg, C. *Org. React.* **2014**, *84*, 1–74. (d) Trost, B. M.; Hachiya, I. *J. Am. Chem. Soc.* **1998**, *120*, 1104–1105. (e) Malkov, A. V.; Gouriou, L.; Lloyd-Jones, G. C.; Starý, I.; Langer, V.; Spoor, P.; Vinader, V.; Kočovský, P. *Chem. Eur. J.* **2006**, *12*, 6910–6929. (f) Trost, B. M.; Zhang, Y. *J. Am. Chem. Soc.* **2007**, *129*, 14548–14549. (g) Trost, B. M.; Zhang, Y. *Chem. Eur. J.* **2010**, *16*, 296–303. (h) Trost, B. M.; Zhang, Y. *Chem. Eur. J.* **2011**, *17*, 2916–2922. (i) Ozkal, E.; Pericàs, M. A. *Adv. Synth. Catal.* **2014**, *356*, 711–717.

5. (a) Matsushima, Y.; Onitsuka, K.; Kondo, T.; Mitsudo, T.; Takahashi, S. *J. Am. Chem. Soc.* **2001**, *123*, 10405–10406. (b) Onitsuka, K.; Matsushima, Y.; Takahashi, S. *Organometallics* **2005**, *24*, 6472–6474. (c) Onitsuka, K.; Okuda, H.; Sasai, H. *Angew. Chem. Int. Ed.* **2008**, *47*, 1454–1457. (d) Trost, B. M.; Rao, M.; Dieskau, A. P. *J. Am. Chem. Soc.* **2013**, *135*, 18697–18704. (e) Kawatsura, M.; Uchida, K.; Terasaki, S.; Tsuji, H.; Minakawa, M.; Itoh, T. *Org. Lett.* **2014**, *16*, 1470–1473. (f) Kanbayashi, N.; Hosoda, K.; Kato, M.; Takii, K.; Okamura, T.; Onitsuka, K. *Chem. Commun.* **2015**, *51*, 10895–10898.
6. (a) Leahy, D. K.; Evans, P. A. Rhodium(I)-Catalyzed Allylic Substitution Reactions and their Applications to Target Directed Synthesis. In *Modern Rhodium-Catalyzed Organic Reactions*; Evans, P. A. Ed. John Wiley & Sons, Inc.: New York, **2005**, pp. 191–214. (b) Evans, P. A.; Nelson, J. D. *J. Am. Chem. Soc.* **1998**, *120*, 5581–5582. (c) Hayashi, T.; Okada, A.; Suzuka, T.; Kawatsura, M. *Org. Lett.* **2003**, *5*, 1713–1715. (d) Kazmaier, U.; Stolz, D. *Angew. Chem. Int. Ed.* **2006**, *45*, 3072–3075. (e) Sidera, M.; Fletcher, S. P. *Nat. Chem.* **2015**, *7*, 935–939. (f) Li, C.; Breit, B. *Chem. -Eur. J.* **2016**, *22*, 14655–14663. (g) Parveen, S.; Li, C.; Hassan, A.; Breit, B. *Org. Lett.* **2017**, *19*, 2326–2329.
7. (a) Didiuk, M. T.; Morken, J. P.; Hoveyda, A. H. *J. Am. Chem. Soc.* **1995**, *117*, 7273–7274. (b) Chung, K.-G.; Miyake, Y.; Uemura, S. *J. Chem. Soc. Perkin Trans. 1.* **2000**, 15–18. (c) Kita, Y.; Kavthe, R. D.; Oda, H.; Mashima, K. *Angew. Chem. Int. Ed.* **2016**, *55*, 1098–1101.
8. (a) Langlois, J. B.; Alexakis, A. *Top. Organomet. Chem.* **2011**, *38*, 235–268. (b) Malda, H.; van Zijl, A. W.; Arnold, L. A.; Feringa, B. L. *Org. Lett.* **2001**, *3*, 1169–1171. (c) Van Veldhuizen, J. J.; Campbell, J. E.; Giudici, R. E.; Hoveyda, A. H. *J. Am. Chem. Soc.* **2005**, *127*, 6877–6882. (d) Yoshikai, N.; Zhang, S.-L.; Nakamura, E. *J. Am. Chem. Soc.* **2008**, *130*, 12862–12863. (e) Selim, K. B.; Matsumoto, Y.; Yamada, K.; Tomioka, K. *Angew. Chem. Int. Ed.* **2009**, *48*, 8733–8735. (f) Langlois, J.-B.; Alexakis, A. *Adv. Synth. Catal.* **2010**, *352*, 447–457. (g) Shi, Y.; Jung, B.; Torker, S.; Hoveyda, A. H. *J. Am. Chem. Soc.* **2015**, *137*, 8948–8964. (h) You, H.; Rideau, E.; Sidera, M.; Fletcher, S. P. *Nature* **2015**, *517*, 351–355. (i) Rideau, E.; You, H.; Sidera, M.; Claridge, T. D.W.; Fletcher, S. P. *J. Am. Chem. Soc.* **2017**, *139*, 5614–5624.
9. Selected reviews: (a) Scho, S. E.; Cheng, Q.; Tu, H.-F.; Zheng, C.; Qu, J.-P.; Helmchen, G.; You, S. L. *Chem. Rev.* **2019**, *119*, 1855–1969. (b) Rössler, S. L.; Petrone, D. A.; Carreira, E. M. *Acc. Chem. Res.* **2019**, *52*, 2657–2672. (c) Schockley, S. E.; Hethcox, J. C.; Stolz, B. M. *Synlett.* **2018**, *29*, 2481–2492. (d) Yuan, C.; Liu, B. *Org. Chem. Front.* **2018**, *5*, 106–131. (e) Qu, J.; Helmchen, G. *Acc. Chem. Res.* **2017**, *50*, 2539–2555. (f) Hethcox, J. C.; Shockley, S. E.; Stoltz, B. M. *ACS Catal.* **2016**, *6*, 6207–6213. (g) Madrahimov, S. T.; Li, Q.; Sharma, A.; Hartwig, J. F. *J. Am. Chem. Soc.* **2015**, *137*, 14968–14981. (h) Helmchen, G. *Mol. Catal.* **2014**, 239–253. (i) Tosatti, P.; Nelson, A.; Marsden, S. P. *Org. Biomol. Chem.* **2012**, *10*, 3147–3163. (j) Sundararaju, B.; Achard, M.; Bruneau, C. *Chem. Soc. Rev.* **2012**, *41*, 4467–4483. (k) Liu, W.-B.; Xia, J.-B.; You, S.-L. Iridium-catalyzed Asymmetric Allylic

Substitutions. In *Topics in Organometallic Chemistry*; Andersson, P. G. Ed. Springer-Verlag: Berlin, Germany, **2011**, Vol. 38, pp. 155–208. (l) Hartwig, J.-F.; Pouy, M. J. Iridium-Catalyzed Allylic Substitution. In *Topics in Organometallic Chemistry*; Andersson, P. G. Ed. Springer-Verlag: Berlin, Germany, **2011**, Vol. 34, pp. 169–208. (m) Hartwig, J. F.; Stanley, L. M. *Acc. Chem. Res.* **2010**, *43*, 1461–1475. (n) Helmchen, G . Iridium-Catalyzed Asymmetric Allylic Substitutions. In Iridium Complexes in Organic Synthesis; Oro, L. A.; Claver, C. Eds. *Wiley-VCH Verlag GmbH & Co. KGaA, Weinheim,* **2009**, pp. 211–250. (o) Helmchen, G.; Dahnz, A.; Dübon, P.; Schelwies, M.; Weihofen, R. *Chem. Commun.* **2007**, 675–691. (p) Takeuchi, R.; Kezuka, S. *Synthesis* **2006**, 3349–3366. (q) Miyabe, H.; Takemoto, Y. *Synlett* **2005**, 1641–1655.

10. Takeuchi, R.; Kashio, M. *Angew. Chem. Int. Ed. Engl.* **1997**, *36*, 263–265.
11. (a) Teichert, J.-F.; Feringa, B. L.; Phosphoramidites: Privileged Ligands in Asymmetric Catalysis. *Angew. Chem. Int. Ed. Engl.* **2010**, *49*, 2486–2528. (b) Tissot-Croset, K.; Polet, D.; Alexakis, A. *Angew. Chem. Int. Ed. Engl.* **2004**, *43*, 2426–2428. (c) de Vries, A. H. M.; Meetsma, A.; Feringa, B. L. *Angew. Chem. Int. Ed. Engl.* **1996**, *35*, 2374–2376.
12. Zhang, X.; You, S.-L. *Chimia* **2018**, *72*, 589–594.
13. Defieber, C.; Ariger, M. A.; Moriel, P.; Carreira, E. M. *Angew. Chem. Int. Ed.* **2007**, *46*, 3139–3143.
14. (a) Spiess, S.; Raskatov, J. A.; Gnamm, C.; Brödner, K.; Helmchen, G. *Chem. -Eur. J.* **2009**, *15*, 11087–11090. (b) Raskatov, J. A.; Spiess, S.; Gnamm, C.; Brödner, K.; Rominger, F.; Helmchen, G. *Chem. -Eur. J.* **2010**, *16*, 6601–6615. (c) Raskatov, J. A.; Jäkel, M.; Straub, B. F.; Rominger, F.; Helmchen, G. *Chem. -Eur. J.* **2012**, *18*, 14314–14328. (d) Madrahimov, S. T.; Marković, D.; Hartwig, J. F. *J. Am. Chem. Soc.* **2009**, *131*, 7228–7229. (e) Liu, W.-B.; Zheng, C.; Zhuo, C.-X.; Dai, L.-X.; You, S.-L. *J. Am. Chem. Soc.* **2012**, *134*, 4812–4821.
15. Graening, T.; Hartwig, J. F. *J. Am. Chem. Soc.* **2005**, *127*, 17192–17193.
16. Liang, X.; Wei, K.; Yang, Y.-R. *Chem. Commun.* **2015**, *51*, 17471–17474.
17. Chen, W.; Hartwig, J. F. *J. Am. Chem. Soc.* **2012**, *134*, 15249–15252.
18. Jiang, X.; Hartwig, J. F. *Angew. Chem. Int. Ed.* **2017**, *56*, 8887–8891.
19. Chen, M.; Hartwig, J. F. *Angew. Chem. Int. Ed.* **2014**, *53*, 8691–8695.
20. Chen, M.; Hartwig, J. F. *J. Am. Chem. Soc.* **2015**, *137*, 13972–13979.
21. Chen, M.; Hartwig, J. F. *Angew. Chem. Int. Ed.* **2014**, *53*, 12172–12176.
22. Chen, M.; Hartwig, J. F. *Angew. Chem. Int. Ed.* **2016**, *55*, 11651–11655.
23. Sempere, Y.; Carreira, E. M. *Angew. Chem. Int. Ed.* **2018**, *57*, 7654–7658.
24. Sandmeier, T.; Krautwald, S.; Carreira, E. M. *Angew. Chem. Int. Ed.* **2017**, *56*, 11515–11519.
25. Weix, D. J.; Hartwig, J. F. *J. Am. Chem. Soc.* **2007**, *129*, 7720–7721.
26. Yue, B.-B.; Deng, Y.; Zheng, Y.; Wei, K.; Yang, Y.-R. *Org. Lett.* **2019**, *21*, 2449–2452.
27. He, H.; Zheng, X.-J.; Li, Y.; Dai, L.-X.; You, S.-L. *Org. Lett.* **2007**, *9*, 4339–4341.

28. Sempere, Y.; Alfke, J. L.; Rössler, S. L.; Carreira, E. M. *Angew. Chem. Int. Ed.* **2019**, *58*, 9537–9541.
29. Krautwald, S.; Sarlah, D.; Schafroth, M. A.; Carreira, E. M. *Science* **2013**, *340*, 1065–1068.
30. (a) Hayashi, Y.; Gotoh, H.; Hayashi, T.; Shoji, M. *Angew. Chem. Int. Ed.* **2005**, *44*, 4212–4215. (b) Franzén, J.; Marigo, M.; Fielenbach, D.; Wabnitz, T. C.; Kjærsgaard, A.; Jørgensen, K. A. *J. Am. Chem. Soc.* **2005**, *127*, 18296–18304.
31. Krautwald, S.; Schafroth, M. A.; Sarlah, D.; Carreira, E. M. *J. Am. Chem. Soc.* **2014**, *136*, 3020–3023.
32. Sandmeier, T.; Krautwald, S.; Zipfel, H. F.; Carreira, E. M. *Angew. Chem. Int. Ed.* **2015**, *54*, 14363–14367.
33. Schafroth, M. A.; Zuccarello, G.; Krautwald, S.; Sarlah, D.; Carreira, E. M. *Angew. Chem. Int. Ed.* **2014**, *53*, 13898–13901.
34. (a) Sandmeier, T.; Goetzke, F. W.; Krautwald, S.; Carreira, E. M. *J. Am. Chem. Soc.* **2019**, *141*, 12212–12218. (b) Sandmeier, T.; Carreira, E. M. *Org. Lett.* **2020**, *22*, 1135–1138.
35. Nœsborg, L.; Halskov, K. S.; Tur, F.; Mønsted, S. M. N.; Jørgensen, K. A. *Angew. Chem. Int. Ed.* **2015**, *54*, 10193–10197.
36. Jiang, X.; Beiger, J. J.; Hartwig, J. F. *J. Am. Chem. Soc.* **2017**, *139*, 87–90.
37. Singha, S.; Serrano, E.; Mondal, S.; Daniliuc, C. G.; Glorius, F. *Nat. Catal.* **2020**, *3*, 48–54.
38. Alexakis, A.; El Hajjaj, S.; Polet, D.; Rathgeb, X. *Org. Lett.* **2007**, *9*, 3393–3395.
39. Corey, E. J.; Grant, T. J. *Tetrahedron Lett.* **1994**, *35*, 5373–5376.
40. Vukics, K.; Fodor, T.; Fischer, J.; Fellegvari, I.; Lévai, S. *Org. Process. Res. Dev.* **2002**, *6*, 82–85.
41. Polet, D.; Rathgeb, X.; Falciola, C. A.; Langlois, J.-B.; El Hajjaji, S.; Alexakis, A. *Chem. Eur. J.* **2009**, *15*, 1205–1216.
42. Boudier, A.; Bromm, L. A.; Lotz, M.; Knochel, P. *Angew. Chem. Int. Ed.* **2000**, *39*, 4414–4435.
43. Hamilton, J. Y.; Sarlah, D.; Carreira, E. M. *Angew. Chem. Int. Ed.* **2015**, *54*, 7644–7647.
44. Hjorth, S.; Carlsson, A.; Wikström, H.; Lindberg, P.; Sanchez, D.; Hacksell, U.; Arvidsson, L.-E.; Svensson, U.; Nilsson, J. L. G. *Life Sci.* **1981**, *28*, 1225–1238.
45. Kezuka, S.; Kanemoto, K.; Takeuchi, R.Centre *Tetrahedron Lett.* **2004**, *45*, 6403–6406.
46. Petrone, D. A.; Isomura, M.; Franzoni, I.; Rossler, S. L.; Carreira, E. *J. Am. Chem. Soc.* **2018**, *140*, 4697–4704.
47. (a) Kleijn, H.; Westmijze, H.; Vermeer, P. *Recl. Trav. Chim. Pays-Bas* **1983**, *102*, 378–380. (b) Djahanbuni, D.; Cazes, B.; Gore, J. *Tetrahedron Lett.* **1984**, *25*, 203–206. (c) Nokami, J.; Maihara, A.; Tsuji, J. *Tetrahedron Lett.* **1990**, *31*, 5629–5630.
48. Lee, Y.; Park, J.; Cho, S. H. *Angew. Chem. Int. Ed.* **2018**, *57*, 12930–1293.
49. Hamilton, J. Y.; Sarlah, D.; Carreira, E. M. *J. Am. Chem. Soc.* **2013**, *135*, 994–997.

50. Hamilton, J. Y.; Sarlah, D.; Carreira, E. M. *Angew. Chem. Int. Ed.* **2013**, *52*, 7532–7535.
51. Akerman, M.; Houze, J.; Lin, D. C. H.; Liu, J.; Luo, J.; Medina, J. C.; Qiu, W.; Reagan, J. D.; Sharma, R.; Shuttleworth, S. J.; Sun, Y.; Zhang, J.; Zhu, L. *Int. Appl. PCT, WO 2005086661,* **2005**.
52. Zheng, Y; Yue, B.-B.; Wei, K.; Yang, Y.-R. *Org. Lett.* **2018**, *20*, 8035–8038.
53. Hamilton, J. Y.; Hauser, N.; Sarlah, D.; Carreira, E. M. *Angew. Chem. Int. Ed.* **2014**, *53*, 10759–10762.
54. Zhan, M.; Li, R.-Z.; Mou, Z.-D.; Cao, C.-G.; Liu, J.; Chen, Y.-W.; Niu, D. *ACS Catal.* **2016**, *6*, 3381–3386.
55. Breitler, S.; Carreira, E. M. *J. Am. Chem. Soc.* **2015**, *137*, 5296–5299.
56. Liu, J.; Cao, C.-G.; Sun, H.-B.; Zhang, X.; Niu, D. *J. Am. Chem. Soc.* **2016**, *138*, 13103–13106.
57. Cram, D. J.; Guthrie, R. D. *J. Am. Chem. Soc.* **1966**, *88*, 5760–5764.
58. Su, Y.-L.; Li, Y.-H.; Chen, Y.-G.; Han, Z.-Y. *Chem. Commun.* **2017**, *53*, 1985–1988.
59. Kanayama, T.; Yoshida, K.; Miyabe, H.; Takemoto, Y. *Angew. Chem. Int. Ed.* **2003**, *42*, 2054–2056.
60. He, Hu; Liu, W.-B.; Dai, L.-X.; You, S.-L. *J. Am. Chem. Soc.* **2009**, *131*, 8346–8347.
61. He, Hu; Liu, W.-B.; Dai, L.-X.; You, S.-L. *Angew. Chem. Int. Ed.* **2010**, *49*, 1496–1499.
62. He, Hu; Liu, W.-B.; Dai, L.-X.; Helmchen, G.; You, S.-L. *J. Am. Chem. Soc.* **2011**, *133*, 19006–19014.
63. Hamilton, J. Y.; Sarlah, D.; Carreira, E. M. *J. Am. Chem. Soc.* **2014**, *136*, 3006–3009.
64. Van Broeck, B.; Chen, J.-M.; Tréton, G.; Desmidt, M.; Hopf, C.; Ramsden, N.; Karran, E.; Mercken, M.; Rowley, A. *Br. J. Pharmacol.* **2011**, *163*, 375–389.
65. He, Hu; Liu, W.-B.; Dai, L.-X.; You, S.-L. *Org. Lett.* **2008**, *10*, 1815–1818.
66. He, Hu; Liu, W.-B.; Dai, L.-X.; You, S.-L. *Synthesis.* **2009**, *12*, 2076–2082.
67. Zhang, X.; Han, L.; You, S.-L. *Chem. Sci.* **2014**, *5*, 1059–1063.
68. Zhang, X; Liu, W.-B.; Tu, H.-F.; You, S.-L. *Chem. Sci.* **2015**, *6*, 4525–4529.
69. Wu, Q.-F.; He, H.; Liu, W.-B.; You, S.-L. *J. Am. Chem. Soc.* **2010**, *132*, 11418–11419.
70. Wu, Q.-F.; Zheng, C.; You, S.-L. *Angew. Chem. Int. Ed.* **2012**, *51*, 1680–1683.
71. Zhuo, C.-X.; Wu, Q.-F.; Zhao, Q.; Xu, Q.-L.; You, S.-L. *J. Am. Chem. Soc.* **2013**, *135*, 8169–8172.
72. Zheng, C.; Wu, Q.-F.; You, S.L. *J. Org. Chem.* **2013**, *78*, 4357–4365.
73. Zhuo, C.-X.; Zhou, Y.; Cheng, Q.; Huang, L.; You, S.-L. *Angew. Chem. Int. Ed.* **2015**, *54*, 14146–14149.
74. Zhuo, C.-X.; Liu, W.-B.; Wu, Q.-F.; You, S.-L. *Chem. Sci.* **2012**, *3*, 205–208.
75. Zhuo, C.-X.; Cheng, Q.; Liu, W.-B.; Zhao, Q.; You, S.-L. *Angew. Chem. Int. Ed.* **2015**, *54*, 8475–8479.
76. Xu, Q.-L.; Dai, L.-X.; You, S.-L. *Chem. Sci.* **2013**, *4*, 97–102.
77. Huang, L.; Dai, L.-X.; You, S.-L. *J. Am. Chem. Soc.* **2016**, *138*, 5793–5796.

78. Huang, L.; Cai, Y.; Zhang, H.-J.; Zheng, C.; Dai, L.-X.; You, S.-L. *CCS Chem.* **2019**, *1*, 106–116.
79. Xu, Q.-L.; Zhuo, C.-X.; Dai, L.-X.; You, S.-L. *Org. Let.* **2013**, *15*, 5909–5911.
80. Huang, L.; Cai, Y.; Zheng, C.; Dai, L.-X.; You, S.-L. *Angew. Chem. Int. Ed.* **2017**, *56*, 10545–10548.
81. Wu, Q.-F.; Zheng, C.; Zhuo, C.-X.; You, S.-L. *Chem. Sci.* **2016**, *7*, 4453–4459.
82. Wang, Y; Zheng, C.; You, S.-L. *Angew. Chem. Int. Ed.* **2017**, *56*, 15093–15097.
83. Jiang, S.-Z.; Zeng, X-Y.; Liang, X.; Lei, T.; Wei, K.; Yang, Y.-R. *Angew. Chem. Int. Ed.* **2016**, *55*, 4044–4048.
84. Liang, X.; Jiang, S.-Z.; Wei, K.; Yang, Y.-R. *J. Am. Chem. Soc.* **2016**, *138*, 2560–2562.
85. Subramaniam, G.; Hiraku, O.; Hayashi, M.; Koyano, T.; Komiyama, K.; Kam, T.-S. *J. Nat. Prod.* **2007**, *70*, 1783–1789.
86. Yang, X.-W.; Yang, C.-P.; Jiang, L.-P.; Qin, X.-J.; Liu, Y.-P.; Shen, Q.-S.; Chen, Y.-B.; Luo, X.-D. *Org. Lett.* **2014**, *16*, 5808–5811.
87. Rossi-Ashton, J. A.; Clarke, A. K.; Donald, J. R.; Zheng, C.; Taylor, R. J. K.; Unsworth, W. P.; You, S.-L. *Angew. Chem. Int. Ed.* **2020**, *59*, 2–9.
88. Tian, H; Zhang, P.; Peng, F.; Yang, H.; Fu, H. *Org. Lett.* **2017**, *19*, 3775–3778.
89. Zhao, Z.-L.; Gu, Q.; Wu, X.-Y.; You, S.-L. *Chem. Asian J.* **2017**, *12*, 2680–2683.
90. Wu, Q.-F.; Liu, W.-B.; Zhuo, C.-X.; Rong, Z.-Q.; Ye, K.-Y.; You, S.-L. *Angew. Chem. Int. Ed.* **2011**, *50*, 4455–4458.
91. Xu, Q.-L.; Dai, L.-X.; You, S.-L. *Org. Lett.* **2012**, *14*, 2579–2581.
92. Cheng, Q.; Wang, Y.; You, S.-L. *Angew. Chem. Int. Ed.* **2016**, *55*, 3496–3499.
93. Schafroth, M. A.; Rummelt, S. M.; Sarlah, D.; Carreira, E. M. *Org. Lett.* **2017**, *19*, 3235–3238.
94. Tu, H.-F.; Zheng, C.; Xu, R. Q.; Liu, X.-J.; You, S-L. *Angew. Chem. Int. Ed.* **2017**, *56*, 3237–3241.
95. Shen, D.; Chen, Q.; Yan, P.; Zeng, X.; Zhong, G. *Angew. Chem. Int. Ed.* **2017**, *56*, 3242–3246.
96. Schafroth, M. A.; Sarlah, D.; Krautwald, S.; Carreira, E. M. *J. Am. Chem. Soc.* **2012**, *134*, 20276–20278.
97. Jeker, O. F.; Kravina, A. G.; Carreira, E. M. *Angew. Chem. Int. Ed.* **2013**, *52*, 12166–12169.
98. Trost, B. M.; Thaisrivongs, D. A. *J. Am. Chem. Soc.* **2008**, *130*, 14092–14093.
99. Mao, J.; Zhang, J.; Jiang, H.; Bellomo, A.; Zhang, M.; Gao, Z.; Dreher, S. D.; Walsh, P. J. *Angew. Chem. Int. Ed.* **2016**, *55*, 2526–2530.
100. Liu, X.-J.; You, S.-L. *Angew. Chem. Int. Ed.* **2017**, *56*, 4002–4005.
101. Sarkar, R.; Mitra, S.; Mukherjee, S. *Chem. Sci.* **2018**, *9*, 5767–5772.
102. Sarkar, R.; Mukherjee, S. *Org. Lett.* **2019**, *21*, 5315–5320.
103. Liu, X.-J.; Zheng, C.; Yang, Y.-H.; Jin, S.; You, S.-L. *Angew. Chem. Int. Ed.* **2019**, *58*, 10493–10499.

© 2022 World Scientific Publishing Company
https://doi.org/10.1142/9789811248436_bmatter

Index

η^6-arene ruthenium complexes, 289, 315, 320
additive manufacturing technologies (3D printing), 199–245
additive manufacturing, 199
allosterism in non-natural systems, 62
allylation of organozinc, 408, 443
allylation of silyl enolates, 407, 416
allylation reactions, 74, 77–78, 118 (Pd), 411–471 (Ir)
allylic alcohols, 413–415, 417, 428, 429, 433–435, 438, 444, 446, 447, 449, 451, 454, 463, 467, 469, 470
allylic aminations, 16–17, 45–46
allylic carbonates, 416, 417, 421, 423, 425, 426, 429, 430–432, 440, 443, 451, 454
artificial metalloenzymes, 315–316
asymmetric counterion-directed catalysis, 95–120, 354–355, (organometallic catalysis), 371 374–379, 387–400 (organocatalysis)
azide, 260, 261, 264

bifunctional catalysts, 281, 282, 285, 286, 287, 310, 319, 348, 354, 356
bifunctional hydrogenation catalysts, 284–357
biocatalysis, 235–236, 241–243, 263–266
boron, 408, 430, 447

catalyst libraries, 62, 75, 90
catalytic Mitsunobu reactions, 186–191
catalytic wittig reaction, 161–185
catalytically active solid objects, 220
C–H activation, 270
C–H amination, 256–278
C–H insertion, 267
chiral amines, 143
chiral boranes, 126, 129, 131, 132, 133, 140
chiral borates, 102
chiral cofactors, 91
chiral diarylprolinol silyl ethers, 433, 434, 436, 437
chiral inducers, 91
chiral lewis acids, 123, 128, 134, 141, 142, 152

chiral lewis base, 142, 143, 145, 432–438
chiral phosphines, 142, 167, 172, 174, 175, 179, 181, 192
chiral phosphoramidites, 472
chiral phosphoric acids/phosphates/ phosphoramidates, 76–78, 100–120, 354–355, 373–379, 387–400
chiral regulating agents, 60, 70
chiral thiourea, 380, 382, 383, 400
chiral triazoliums, 441
chiral-at-metal complexes, 261, 316–318
chirogenesis, 62, 90
Chromium, 69–70
Cobalt, 262–264, 270–271
continuous stirred tank reactors, 232
continuous-flow synthesis, 223
cooperative metal-ligand catalysis, 294
copper based lewis acid catalysis, 26
Copper, 5–16, 18–23, 82–84, 109–110, 114, 210–214, 241
covalent anchoring, 13
cu and fe promoted
cyclic N-acyliminium ions, 371–404
cycloadditions, 23–25, 39, 40, 85, 96, 115–119, 212–220, 440
cycloisomerization, 95, 107, 111, 114, 116
cyclopropanations, 12–13, 15, 20–21

desymmetrization, 79, 107–108, 172–176, 184–185, 461–462
diamination, 95, 109, 110
diastereoselective, 411, 415, 424, 427, 433, 451, 456, 461, 462
diazaphosphenium cations, 141

diel-alder reactions, 5–11, 14–16, 18–20
dioxazolone, 266–270, 272, 278
DNA-based biohybrid catalysts, 3
DNA-based catalysts, 1–27
double stereodifferenciation, 118, 119
dual catalysis, 415, 433–437, 439, 441, 472
Dual organo-organometallic catalysis, 432–441

enamides, 408, 424, 425, 427, 428, 429, 441
enamine catalysis, 372, 384, 387, 401
enol ethers, 407, 408, 416–422, 424, 425, 441
enolate surrogates, 411–472
enzymatic catalysis, 263

flow chemistry, 223
friedel-crafts alkylations, 9–11, 14, 15, 20, 23
frustrated Lewis pairs (FLP), 124–155
fused deposition modeling, 200

gold, 101–109, 112, 115, 116, 221–225
G-quadruplex binders, 17
grove binders, 1, 5, 8

5-hydroxypyrrolidinone, 379, 386
(hetero)aromatizations, 452
helical assemblies, 81, 85
heteroatom substituted secondary phosphine oxides (HASPO), 32, 36
heterocyclic compounds, 371, 378, 403, 409, 454, 469, 457

photo-induced [2+2] cyclizations, 25
polyenes, 409, 468

radical, 262–264, 266
regioselective, 417, 454
regulation agent, 60, 61, 62
Rhodium, 39, 41–43, 49–53, 66–69, 72–76, 81–85, 101–103, 116–119, 270–276, 282, 284, 287, 289–291, 293, 299, 307–312, 315, 319, 332–345, 356
ruthenium, 119, 120, 260, 261, 268, 273, 283–307, 310, 311, 312, 315, 318–332, 344, 346, 356

secondary phosphine oxides (SPO), 31, 33–54, 343–344
selective laser sintering, 200
self-assembled ligands, 40
sergeants-and-soldiers effect, 84
Shvo-type catalysts 305–307, 319–320
shvo-type ruthenium catalysts, 306
silanes, 163, 167, 174, 192

silicon reagents, 447
silver, 117–118, 224, 275–277, 449–450
solid supported DNAs, 2, 17, 23
staudinger reaction, 176
stereodivergent and stereoselective reactions, 415
stereolithography, 200
substrate-ligand ion pairs, 342
supramolecular anchoring, 4, 13
supramolecular regulation, 61–90

tandem transformations, 114, 163, 183
three-dimensional-printed devices, 197–199
transfer hydrogenation, 41, 283–318 (ketones), 306, 315 (C=N bonds), 316 (olefins)
TRIP, 97, 100, 101, 104–107, 110, 111, 113–120

vinylogous mannich reactions, 132

hydrazones, 408, 450, 451, 452
hydroalkoxylations/
 hydrocarboxylations, 104–108
hydroamination, 102, 104–108
hydroboration, 126–130, 132, 133, 141–143
hydroformylations, 66–69, 75
hydrogen bonding interaction, 268
hydrogenation, 38–43, 46, 50–53, 72, 82, 100, 102, 103, 118, 120, 125, 126, 129–147, 149–155, 221, 229, 234, 319–338, 340, 342–345 (olefins) 346–350 (ketones), 351–356 (C=N bonds)
hydrogenations, organocatalytic, 126–139, 144–155, 399
hydrophobic pocket, 268, 269, 274
hydrosilylation, 82–85, 126–127, 129–156 (ketones and imines)
hydrovinylation, 96, 119, 120
hypervalent iodine, 259, 271, 272, 275, 278

imines, 334, 451, 452
iminium ions, 371
in situ phosphine oxide reduction, 168
indoles, 409, 454–456, 458–461, 463
intercalators, 1, 5, 7
ion-pair based catalysts, 76–80, 98, 102–103, 342–344, 349–353, 400–401
ion-paired ligands, 77, 78, 80, 91
iridium, 16–17, 38, 79, 112, 113, 267–270, 282, 287, 290, 293, 312–317, 319, 341, 344–356, 407–471
iron, 13, 20, 264–266

mannich reactions, 131–132, 136–137, 218, 384, 401–403
michael addition reactions, 20
mitsunobu reaction, 162, 163, 186, 187, 189

N-Acyliminium Ions, 369, 371
nickel catalysts, 42
Nickel, 42, 49, 209
nitrene, 259, 264, 266, 269, 270, 272, 274
nitrogen-containing heterocycles, 177, 179, 185
non-covalent interactions, 356
noyori-type catalysts, 282, 287, 320
Noyori-type Ru-catalysts, 287–304, 320–331

Organocatalysis, 124–155 (frustrated Lewis pairs catalysis), 162–190 (phosphine catalysis), 374–379, 387–399 (phosphoric acid catalysis), 380–383, 400 (thiourea catalysts), 384–386, 401–403 (enamine catalysis), 85–90, 241 (miscellaneous)
organocatalysis, 60, 85, 161, 209
overman rearrangement, 95, 110, 111

P(III)/P(V) redox cycling, 162–191
palladium, 16, 37–40, 45–46, 64–65, 76–78, 100, 110–111, 212–216, 221, 236, 412
P-chirogenic ligands, 33–36, 38, 39, 40, 43, 46–48, 53
phosphine-catalyzed reactions, P(III)/P(V) redox catalysis, 163
photocatalysis, 23–25